Lecture Notes in Business Information Processing 218

Series Editors

Wil van der Aalst
Eindhoven Technical University, Eindhoven, The Netherlands
John Mylopoulos
University of Trento, Povo, Italy
Michael Rosemann
Queensland University of Technology, Brisbane, QLD, Australia
Michael J. Shaw
University of Illinois, Urbana-Champaign, IL, USA
Clemens Szyperski
Microsoft Research, Redmond, WA, USA

More information about this series at http://www.springer.com/series/7911

Bogumił Kamiński · Gregory E. Kersten
Tomasz Szapiro (Eds.)

Outlooks and Insights on Group Decision and Negotiation

15th International Conference, GDN 2015
Warsaw, Poland, June 22–26, 2015
Proceedings

 Springer

Editors
Bogumił Kamiński
Warsaw School of Economics
Warsaw
Poland

Tomasz Szapiro
Warsaw School of Economics
Warsaw
Poland

Gregory E. Kersten
Concordia University
Montreal, QC
Canada

ISSN 1865-1348   ISSN 1865-1356   (electronic)
Lecture Notes in Business Information Processing
ISBN 978-3-319-19514-8   ISBN 978-3-319-19515-5   (eBook)
DOI 10.1007/978-3-319-19515-5

Library of Congress Control Number: 2015940410

Springer Cham Heidelberg New York Dordrecht London

Printed on acid-free paper

Springer International Publishing AG Switzerland is part of Springer Science+Business Media
(www.springer.com)

Preface

Group decision and negotiation (GDN) refers to the academic and professional discipline that focuses on gaining an understanding of collective decision-making processes. It is involved with the formulation of rules, models, and procedures to improve these processes. The range of GDN research reflects the breath of the strategic and tactical, social–psychological and economic, individual and group, conflict and cooperation, and software-supported and software-conducted processes. The field encompasses theory building and testing, laboratory and online experiments, as well as observations in the field. Therefore, GDN researchers are involved in the theoretical, experimental, and applied studies as well as in the development, testing, and implementation of support systems, decision aids, and software agents. They aim at helping decision makers, advisors, facilitators, and third parties to deal with difficult problems, make better decisions, and/or delegate certain decisions to software.

GDN meetings bring together researchers and practitioners from the fields of humanities, social sciences, economics, law, management, engineering, and computer science. These diverse areas reflect the breath of GDN research. The meetings' participants discuss and compare different paradigms, methods of inquiry, and objectives that they employ in their research. What is common to all participants is their interest in the difficult decision problems that involve conflicts and/or cooperation and the challenges that people face when they attempt to find satisficing agreements and to reach consensuses.

Researchers from the Americas, Asia, Europe, Africa, and Oceania participate in GDN meetings. They have a stimulating variety of backgrounds and represent a wide range of disciplines. While many of us come from different traditions, we all share a common passion: research into complex decision making and negotiation involving multiple stakeholders, different perspectives, issues, and emotions, requiring decision and negotiation support for both process and content.

The Group Decision and Negotiation (GDN) conference series started in Glasgow, Scotland, UK, in 2000 and was hosted by Colin Eden. At that time, Mel Shakun—the founding member of the section and its chairperson from 1995 until 2014—assumed that the next conference would take place only after several years. There was so much interest, however, that the second meeting took place just one year later. It was organized by Alain Checroun and held in La Rochelle in 2001. Mohammed Quaddus organized the next meeting in Perth (2002). Then, from Western Australia we moved to Istanbul (2003) and the following year to Banff (2004); these latter two meetings were held as a meeting-within-a-meeting at larger INFORMS-affiliated conferences.

The memorable GDN meetings that took place in Vienna and Karlsruhe were hosted by Rudolf Vetschera (2005) and Christof Weinhardt (2006), respectively. The 2007 GDN meeting was organized by Gregory Kersten at Mont Tremblant in Quebec, Canada. João Climaco and João Paulo Costa hosted GDN 2008 in Coimbra. Then, Gwendolyn Kolfschoten organized GDN 2010 in Delft.

Amer Obeidi did a lot of work on the organization of GDN 2011 in Amman, Jordan. Unfortunately, this meeting did not take place because of the events in neighboring countries at that time. The next year, Adiel Teixeira de Almeida organized GDN 2012 in Recife, Pernambuco, Brazil. GDN 2013 was hosted by Bilyana Martinovski in Stockholm and it was followed by the GDN 2014 meeting in Toulouse, which was hosted by Pascale Zarate.

GDN 2015 was the 15th meeting organized by the INFORMS section on Group Decision and Negotiation. The conference was hosted by Tomasz Szapiro at the Warsaw School of Economics in Warsaw. During this meeting we revived the Young Researcher Award that was first given at the 2007 meeting. The award was given to a student researcher who authored and presented the best paper at the conference. In addition to this award, young researchers also participated in the Doctoral Consortium. Ofir Turel and Rudolf Vetschera served as the consortium's chairs and hereby we acknowledge their contribution.

At the 2014 GDN meeting two volumes of proceedings were introduced; one volume published by Springer in the LNBIP series [1] and the second volume published by Toulouse University [2]. The GDN 2015 proceedings are also in two volumes: the present volume and the accompanying volume [3].

In both volumes we have introduced thematic streams of sessions. Researchers who participated in the organization of the streams wrote introductions to each stream. These introductions are included in the separate section "Introductions" (pp. XIII–XLVI). They briefly discuss the streams' contributions published in both volumes thus making them better integrated. We hope that this will give the readers a more comprehensive overview of all contributions.

The contributions in this volume and in the proceedings [3] reflect the richness of GDN scholarship. Using a variety of research approaches including real organizational settings and laboratory situations, they focus on the development, application, and evaluation of concepts, theories, methods, and techniques.

The contemporary political landscape abounds in situations of multidimensional conflicts which mix military, economic, and social dimensions. Troops and tanks, economic measures and sanctions, as well as massive violent protests may become destructive means of conflict resolution. Wisdom armed with values, knowledge, and methods will assist politicians in the creation of new instruments for effective group decisions and negotiations. These widely shared expectations challenge researchers and simultaneously direct their efforts in creation and dissemination of ethically driven, knowledge-based applicable findings. The multicultural and interdisciplinary GDN community presents their results on progress in this area.

"Collaboration leads to growth, which engenders accomplishment." [2, p. VIII]. The GDN 2015 conference and its proceedings were made possible through the collaboration of many researchers, students, and support staff. Their dedication and support were exceptional. We are grateful to all of them; to those who made contributions, presented papers, prepared the proceedings, maintained the conference website, and undertook many other necessary tasks. Their contributions, including help in the organization of the streams and the sessions as well as the accompanying events, were key to the success of this meeting. We thank the reviewers for their work. It is thanks to their in-depth reviews that we are able to maintain the high academic standard of the GDN meetings. The stream organizers' and reviewers' work is greatly appreciated, particularly because often they were given very little time. Their reviews provided the authors with much-needed feedback. Thank you:

Fran Ackerman, Yasir Aljefri, Adiel Almeida, Marek Antosiewicz, Reyhan Aydogan, Deepinder Bajwa, Martin Bichler, Réal Carbonneau, Wojciech Cellary, João Clímaco, Grazia Concilio, Ana Paula Costa, Suzana Daher, Luis Dias, Colin Eden, Verena Dorner, Liping Fang, Mario Fedrizzi, Michael Filzmoser, Florian Hawlitschek, Shawei He, Keith Hipel, Masahide Horita, Michał Jakubczyk, Marc Kilgour, Mark Klein, Grzegorz Koloch, Beata Koń, Sabine Koszegi, Kevin Li, Jan Machowski, Yasser Matbouli, Paul Meerts, Danielle Morais, José Maria Moreno-Jiménez, Hannu Nurmi, Amer Obeidi, Pierpaolo Pontrandolfo, Ewa Roszkowska, Anne Rutkowski, Mareike Schoop, Roman Słowiński, Rangaraja Sundraraj, Przemysław Szufel, David Tegarden, Timm Teubner, Ernest Thiessen, Sathyanarayanan Venkatraman, Rudolf Vetschera, Doug Vogel, Tomasz Wachowicz, Christof Weinhardt, Dariusz Witkowski, Paweł Wojtkiewicz, Shi Kui Wu, Yinping Yang, Bo Yu, Yufei Yuan, Pascale Zaraté, Mateusz Zawisza, John Zeleznikow, and Daniel Zeng.

The quality of the presentations is associated with the excellence of the papers. It is also affected by the venue and the overall organization of the meeting and its associated events. The local Organizing Committee was responsible for these aspects of the meeting and they did everything to make the meeting pleasant and memorable. Thank you:

Przemyslaw Szufel, Marek Antosiewicz, Michał Jakubczyk, Grzegorz Koloch, Beata Koń, Tomasz Kuszewski, Jan Machowski, Paweł Wojtkiewicz, and Karolina Zakrzewska-Szlichtyng.

Finally, we thank Michel J. Shaw, an editor of the LNBIP series, who helped us to get the GDN proceedings into the series, and Ralf Gerstner, an executive editor at Springer, who guided us through preparation and submission of the proceedings.

We hope that you find the contents of this book as well as the contents of the accompanying volume [3] useful and interesting. The authors' effort in clarifying complex problems and proposing innovative solutions should help you to cope with numerous challenges that are faced by researchers of group decision and negotiations. We also hope that the meeting and the contributions foster collaboration among the meeting's attendees as well as joint projects with researchers who were not able to come to Warsaw and participate in GDN 2015.

April 2015

Bogumił Kamiński
Gregory E. Kersten
Melvin F. Shakun
Tomasz Szapiro

References

1. Zaraté, P., Kersten, G.E., Hernández, J.E. (eds.) GDN 2014. LNBIP, vol. 180. Springer, Heidelberg (2014)
2. Zaraté, P., Camilleri, G., Kamissoko, D., Amblard, F., (eds.) Group Decision and Negotiation 2014: Proceedings of the Joint International Conference of the INFORMS GDN Section and the EURO Working Group on DSS, 2014, Tolouse University, Tolouse. 363 p. (2014)
3. Kamiński, B., Kersten, G.E., Szapiro, T. (eds.) GDN 2015. LNBIP, vol. 218. Springer, Heidelberg (2015)

Organization

General Chairs

Melvin F. Shakun — New York University, USA
Tomasz Szapiro — Warsaw School of Economics, Poland

Program Chairs

Bogumił Kamiński — Warsaw School of Economics, Poland
Gregory (Grzegorz) E. Kersten — Concordia University, Canada

Organizing Chair

Przemyslaw Szufel — Warsaw School of Economics, Poland

Doctoral Consortium Chairs

Ofir Turel — California State University, USA
Rudolf Vetschera — University of Vienna, Austria

Program Committee

Fran Ackerman — Curtin University, Australia
Adiel Almeida — Federal University of Pernambuco, Brazil
Reyhan Aydogan — Delft University of Technology, The Netherlands
Deepinder Bajwa — Western Washington University, USA
Martin Bichler — TUM, Germany
Tung Bui — University of Hawaii, USA
Christer Carlsson — Abo Akademi University, Finland
Wojciech Cellary — Poznań University of Economics, Poland
João Climaco — University of Coimbra, Portugal
Grazia Concilio — Politecnico di Milano, Italy
Suzana F.D. Daher — Federal University of Pernambuco, Brazil
Luis Dias — University of Coimbra, Portugal
Colin Eden — University of Strathclyde, UK
Liping Fang — Ryerson University, Canada
Mario Fedrizzi — University of Trento, Italy
Raimo Pertti Hamalainen — Aalto University, Finland
Keith Hipel — University of Waterloo, Canada
Masahide Horita — University of Tokyo, Japan

Takayuki Ito	Nagoya Institute of Technology, Japan
Michal Jakubczyk	Warsaw School of Economics, Poland
Catholjin Jonker	Delft University of Technology, The Netherlands
Marc Kilgour	Wilfrid Laurier University, Canada
Mark Klein	MIT, USA
Sabine Koeszegi	Vienna University of Technology, Austria
Hsiangchu Lai	NTNU, Taiwan
Michael Lewis	University of Pittsburgh, USA
Kevin Li	University of Windsor, Canada
Ivan Marsa-Maestre	Universidad de Alcala, Spain
Paul Meerts	Clingendael Institute, The Netherlands
Ugo Merlone	University of Turin, Italy
Daniel Mittleman	DePaul University, USA
Danielle Morais	Federal University of Pernambuco, Brazil
José Maria Moreno-Jiménez	Zaragoza University, Spain
Hannu Nurmi	University of Turku, Finland
Amer Obeidi	Cybernetics Consultants, Canada
Pierpaolo Pontrandolfo	Politecnico di Bari, Italy
Ewa Roszkowska	University of Białystok, Poland
Anne-Françoise Rutkowski	Tilburg University, The Netherlands
Mareike Schoop	Hohenheim University, Germany
Wei Shang	Chinese Academy of Sciences, China
Gheorghe Cosmin Silagh	Babeş-Bolyai University, Romania
Rangaraja Sundraraj	IIT Madras, India
Katia Sycara	Carnegie Mellon University, USA
Przemyslaw Szufel	Warsaw School of Economics, Poland
David P Tegarden	Virginia Tech, USA
Ernest M. Thiessen	SmartSettle, Canada
Ofir Turel	California State University, USA
Rustam Vahidov	Concordia University, Canada
Rudolf Vetschera	University of Vienna, Austria
Gert-Jan de Vrede	University of Nebraska-Omaha, USA
Tomasz Wachowicz	University Economics in Katowice, Poland
Christof Weichardt	KIT, Germany
Haiyan Xu	Nanjing University of Aeronautics and Astronautics, China
Yinping Yang	AStar, Singapore
Yufei Yuan	McMaster University, Canada
Pascale Zarate	University of Toulouse 1, France
John Zeleznikow	Victoria University, Australia

Doctoral Consortium Committee

Keith W. Hipel	University of Waterloo, Canada
Masahide Horita	Graduate School of Frontier Sciences, Japan
Pawel Kalczynski	California State University, USA

Gregory (Grzegorz)   Concordia University, Canada
  E. Kersten
José Maria Moreno-Jiménez   Zaragoza University, Spain
Hannu Nurmi   University of Turku, Finland
Tomasz Szapiro   Warsaw School of Economics, Poland

Organizing Committee

Marek Antosiewicz   Warsaw School of Economics, Poland
Michał Jakubczyk   Warsaw School of Economics, Poland
Grzegorz Koloch   Warsaw School of Economics, Poland
Beata Koń   Warsaw School of Economics, Poland
Tomasz Kuszewski   Warsaw School of Economics, Poland
Jan Machowski   Warsaw School of Economics, Poland
Paweł Wojtkiewicz   Warsaw School of Economics, Poland
Karolina   Warsaw School of Economics, Poland
  Zakrzewska-Szlichtyng

GDN | 2015

GROUP DECISION AND NEGOTIATION

Warsaw School of Economics

June 22–26, 2015
Warsaw, POLAND

www.sgh.waw.pl/gdn2015/

Introductions

The Conference Streams and the Proceeding Sections

The papers submitted to GDN 2015 were organized somewhat differently than in past years. There were nine streams at the conference with each stream constituting one section of the Springer 218 LNBIP Proceedings as well as one section of the accompanying Proceedings published by the Warsaw School of Economics.

The multidisciplinary aspect of research on group decision and negotiation processes poses challenges for organizers. These include, but are not limited to: extending invitations to renowned researchers to deliver invited lectures; approaching colleagues to review submissions; and maintaining an overview of the process.

Our colleagues who generously agreed to be the Streams Organizers succeeded in attracting many renowned scholars to the conference. They facilitated the assessment of submissions and reviewed many papers. They also wrote introductions for each stream providing unique insights into the current directions and findings in group decision and negotiations. All of this work was done under time pressure as the deadlines for preparing the proceedings were very tight.

Each of the two volumes of the GDN 2015 proceedings has nine sections. Correspondingly, you will find here nine introductions. We wish to express our gratitude to the Stream Organizers as well as the authors of the introductions. Our thanks go to:

Fran Ackermann and Colin Eden;
Tomasz Wachowicz;
Adiel T. de Almeida, Ewa Roszkowska, and Tomasz Wachowicz;
João Climaco;
Hannu Nurmi;
Mareike Schoop, Sabine Koeszegi, and Rudolf Vetschera;
Keith W. Hipel, D. Marc Kilgour, Liping Fang, and Amer Obeidi;
R.P. Sundarraj; and
Verena Dorner, Timm Teubner, and Christof Weinhardt.

Bogumił Kamiński and Gregory E. Kersten
Program Chairs

Group Problem Structuring and Negotiation

Fran Ackermann[1(⊠)] and Colin Eden[2]

[1] Curtin University, Perth, Australia
[2] Strathclyde Business School, Strathclyde, UK
fran.ackermann@curtin.edu.au

1 Overview

Welcome to the stream focusing on group problem structuring and negotiation. We are delighted to have received so many interesting papers reflecting the vibrancy and relevance of the area. All of the papers focus, to some extent, on the behaviours within small groups: small group problem solving and decision making, managing conflict and multiple perspectives, and developing competences.

There are important emerging themes showing the research effort in this area takes a number of different but also related directions:

– Taking a 'human' approach to the topic rather than focusing only on an analytical approach to negotiation. Consequently many of the papers discuss work with groups recognising the need to attend to the socio-political aspects as well as supporting decision making. This is evident in papers where we see research being carried out which (a) explores and supports the negotiation between multiple collaborators who may also be competitors; (b) aims to support the management of conflict, (c) recognises the importance of considering procedural justice authentically, and (d) seeks to enhance the negotiation abilities of staff within organizations [1–5].
– Exploring new angles relating to problem structuring through (a) use of group support systems adopting causal modelling and facilitation to ensure procedural justice is fully supported and views can be widely contributed, (b) the interaction between consultant and client where the use of productive dialogue can aid the development of an effective relationship and affect the trajectory and outcomes of the workshops and (c) unpacking complexity associated with the practice of problem structuring [4–5, 7–8].
– Focussing on application with papers discussing work in the area of disaster management planning involving community groups, in strategy making in relation to the use of artefacts to support effective sense-making, in supporting etc., in health care planning of an aging population where group support systems are used to ensure a more effective use of data, in encouraging organizations to view negotiation as a corporate competence, with UK clinical strategy making groups helping to improve outcomes, and in social housing in relation to the assessment of which housing projects to fund to meet the technical and social conditions [1, 8, 9–10].

Notably the papers reflect work being done in different locations: UK, Italy, Sweden, Australia, China; and within different types of organization from public sector (health and housing) to private sector (conflict management, competence development).

We hope that you will find the themes of interest and consequently join us at the conference.

References

1. Ackermann, F., Eden, C., Alexander, J.: Collaboration through negotiation: experiences and lessons from the field. In: [12]
2. Burns, T.R., Corte, U., Machado Des Johansson, N.: Toward a universal theory of the human group: sociological systems framework applied to the comparative analysis of groups and organizations. In: [12]
3. Carreras, A., Franco, L.A., Papadopoulos, T.: Managing the relationship between Clients and Consultants. In: [12]
4. Kaur, P., Carreras, A.: Managing the relationship between Clients and Consultant. In: [12]
5. Tavella, E.: Negotiating meaning through artefacts: a micro-level analysis of strategy discourse. In: [12]
6. Franco, L.A., Greiffenhagen, C.: Unpacking the complexity of group problem structuring. In: [12]
7. Lami, I., Abastante, F., Ingaramo, L., Lombardi, P.: Social housing allocation: a problem structuring analysis. In: [12]
8. Chosokabe, M. Tsuguchi, Y., Sakakibara, H. Nakayama, T. Mine, S., Kamiya, D., Yamanaka, R., and Miyaguni, T: Effects of small group discussion: case study of community disaster risk management in Japany. In: Kamiński, B., Kersten, G.E., Szapiro, T. (eds.) GDN 2015. LNBIP, vol. 218, pp. 3–12. Springer, Heidelberg (2015)
9. Eden, C., Ackermann, F.: Two-party conflict resolution in 55 minutes! In: [12]
10. Vogel, D.: Group support for healthcare data utilization. In: [12]
11. White, L. Yearworth, M., Burger, K.: Understanding PSM interventions through sensemaking and the mangle of practice lens. In: Kamiński, B., Kersten, G.E., Szapiro, T. (eds.) GDN 2015. LNBIP, vol. 218, pp. 13–27. Springer, Heidelberg (2015)
12. Kamiński, B., Kersten, G.E., Szapiro, T. (eds.) GDN 2015. LNBIP, vol. 218. Springer, Heidelberg (2015)

Negotiation and Group Processes Support

Tomasz Wachowicz[1]([⊠]) and Gregory E. Kersten[2]

[1] Department of Operations Research, University of Economics
in Katowice, Katowice, Poland
[2] InterNeg Research Centre, J. Molson School of Business
Concordia University, Montreal, PQ, Canada
tomasz.wachowicz@ue.katowice.pl

1 Overview

Negotiation and group processes are complex interactions among the parties conducted in an effort to arrive at a decision that they accept and are willing to implement. These interactions are based on communication, both verbal and nonverbal, that aims at educating the participants about ones needs, preferences, and limitations. The communication is framed by the negotiation strategies and tactics, including promises, assurances, and threats. In addition, third parties and other stakeholders as well as a broader context and external events, are likely to affect the discourse between the parties.

Negotiation and group processes are decision making processes that require that the participants assess the alternatives and evaluate offers made by other participants. Often, the participants need to make concession while searching for potential improvements of the negotiation results. These processes exhibit both socio-psychological, economic, and decision-analytic aspects making them difficult to organize and manage.

Both behavioural and formal approaches to negotiations and group decisions resulted in numerous studies. Behavioural approaches use methodologies and test models formulated in anthropology, psychology, sociology, communication, and organization science. They often focus on such aspects of negotiation and group processes as the context, the stages of the process of conflict solving, the relationship between parties, the parties' reputation, behaviour, strategies and tactics. They examine the influences of the participants' personal, demographic, cultural, or professional traits on their actions and the outcomes. Some studies aim to build theories and to formulate procedures for effective management of conflicts and for the construction of checklists that are to help the parties organize their tasks.

Formal approaches have been developed within the fields of economics, management science, decisions science, game theory, and econometrics. They assume that the participants are rational or at least logical decision-makers with a value-seeking perspective on the process. They rely on formal models of the processes and develop methods for aiding and supporting decision-making in both individual and collaborative settings. These models can be used to facilitate the analysis of the problem and the participants prior to their interactions. They also can be used during the process, in order to analyze one's own and the counterparts' decisions and to provide alternative courses of action.

Many formal methods rely on the game-theoretical concepts and formulate normative recommendations for negotiators who are efficient and rational. They suggest solutions that allow the negotiators to achieve optimal outcomes. They also provide the tools for formal statistical analysis of the experimental results.

Studies that aim at presenting a more comprehensive view include both approaches. This includes research which aims at experimental or in the field verification of formal procedures for conflict management and resolution as well as formal models embedded in software. The approaches rely on computer science, management information systems and software engineering to provide development tools and platforms for the design and construction of group support systems, negotiation and e-negotiation systems, online mediators, decision aids, and negotiation software agents.

The papers mentioned below are included in this proceedings (Section 3) and in the accompanying proceedings published by the Warsaw School of Economics [1]. Their authors study the problems of how the negotiation and group process and the results can be influenced by:

- negotiation strategies and tactics which the parties employ;
- negotiators' personal and demographic characteristics and external factors; and
- the facilitation procedures, models and frameworks applied to support the negotiators' activities.

2 Negotiation Strategies and Tactics

The difficulty in defining an effective negotiation strategy that best fits the negotiation problem and context and results in most profitable outcomes is one of the most important tasks in the pre-negotiation preparation. Such a strategy determines not only the general behaviour and tactics used by the parties, but also specific moves such as opening offers, response rules, concession paths etc. In [2] the effects of using the door-in-the-face tactic is studied. The authors prove that it leads to feelings of mistreatment by the opponents, who, however, may use such an approach in future negotiation to make larger demands and achieve better outcomes. The relationship between purchasing managers' negotiation styles and tactics is examined in [3]. The authors confirm that the long-term orientation of purchasing negotiators had an impact on their applied negotiation tactics.

The following two papers analyze the effect of frames and anchors in the negotiation process. In [4] the use of language to frame the negotiation as integrative or distributive while holding the offers and payoffs constant was studied. The second paper is focused on analyzing the importance and effects of first concessions made by parties [5], being the anchors in the negotiation process. It appears that the party who submitted the first concession achieved a better individual outcome and, furthermore, that the first concession influenced the opponent's concession behavior in terms of the reward theory.

3 Personal Characteristics and External Factors

The second group of papers is focused on analyzing various factors that may influence the negotiation and group process or the participants' behavior and outcome. The influence of demographic factors, process measures, and individual and joint outcomes on the desire of the participants to negotiate again with their counterparts is studied in [6]. The interesting finding is, that post-negotiation perceptions of honesty and individual outcome had differential effects on the desire to negotiate again, depending on whether or not an agreement was reached.

The main personal traits are also studied in [7], but from the viewpoint of hindering the facilitation of cooperative negotiations in familial disputes.

The participants' creativity and their cognitive limitations, such as a need for closure, are identified in [8]; their impact on negotiation outcome is studied by using a Dynamical Negotiation Networks model.

An important issue of negotiation data collection and its relevance in negotiation research is studied in [9]. The transcribed video recordings of negotiations are compared with the negotiators' statements included in post-surveys in order to determine the negotiators' recall of their performance and to find how well they remember their negotiation.

4 Frameworks, Models, and Procedures

From the viewpoint of effective management of the negotiation and group processes as well as for the purpose of their support it is highly important to develop models, procedures and frameworks. The process formal representations can be implemented in software and provide prescriptive or normative recommendations.

Modelling may be done at the choice and decision-level as well as at the meta-choice level. The issue of a procedural meta-choice problem is discussed in [10]. Such problem may appear if a group of decision-makers cannot agree on a decision rule. The authors propose a relation-valued procedural choice rule and discuss the advantages and limitations of such a rule.

The complexities faced by the intergovernmental organizations (IGO) during post-conflict reconstruction are studied in [11]. This paper discusses the added-value of social responsibility in the context of a "comprehensive approach," to better grasp the organizational design of the latter.

The processes that involve intergovernmental organizations are also discussed in [12]. The paper addresses the effects of the International Criminal Court (ICC) interventions on negotiated peace processes. The paper offers an analytical framework which aims at the identification and assessment of the effects of the ICC on conflict and peace processes.

A conceptual model on the role and impact of cultural intelligence on conflict and its management and on negotiation behaviors in culturally diverse environments is presented in [13]. A general process model focuses on the goal-oriented balancing process. It describes the necessity for negotiators to continuously balance the opposing forces in order to reach the goal. It is an interactive model that tries to incorporate all the important dimensions that exist in negotiation processes.

References

1. Kamiński, B., Kersten, G.E., Szapiro, T. (eds.) GDN 2015. LNBIP, vol. 218. Springer, Heidelberg (2015)
2. Wong, R.: The hidden costs of the door-in-the-face tactic in negotiations. In: Kamiński, B., Kersten, G.E., Szapiro, T. (eds.) GDN 2015. LNBIP, vol. 218, pp. 61–74. Springer, Heidelberg (2015)
3. Herbst, U., Hotait, A., Preuss, M.: What is really behind all this? The relationship between negotiation styles and negotiation tactics. In: [1]
4. Seferagic, H., Griessmair, M.: Framing in Negotiation. In: [1]
5. Herbst, U., Kemmerling, B., Voeth, M.: First come, first served? – The impact of the first concession on negotiation outcome. In: [1]
6. Fleck, D., Volkema, R., Pereira, S., Levy, B., Vaccari, L.: Back to the future: an examination of factors affecting desire to negotiate again. In: [1]
7. Araszkiewicz, M., Łopatkiewicz, A., Zienkiewicz, A.: Personal traits that hinder cooperative negotiations Rrgarding familial disputes and the usage of modern informational technology. In: [1]
8. Jochemczyk, L., Pietrzak, J.: Dynamical negotiation networks. In: [1]
9. Herbst, U., Knöpfle, T.A., Borchardt, M.T.: Do it by surveying - rethinking methods in negotiation research. In: [1]
10. Suzuki, T., Horita, M.: How to order the alternatives, rules, and the rules to choose rules: when the endogenous procedural choice regresses. In: Kamiński, B., Kersten, G.E., Szapiro, T. (eds.) GDN 2015. LNBIP, vol. 218, pp. 47–59. Springer, Heidelberg (2015)
11. Gans, B., Rutkowski, A.-F.: Social consciousness in post-conflict reconstruction. In: Kamiński, B., Kersten, G.E., Szapiro, T. (eds.) GDN 2015. LNBIP, vol. 218, pp. 31–45. Springer, Heidelberg (2015)
12. Kersten, M.: Negotiating peace, conflict and justice - an analytical framework. In: [1]
13. Åge, L.-J.: Goal oriented balancing - a general model of negotiation processes. In: [1]

Preference Analysis and Decision Support

Adiel T. de Almeida[1], Ewa Roszkowska[2], and Tomasz Wachowicz[3]

[1] Management Engineering Department, Federal University of Pernambuco
Recife, Pernambuco, Brazil
[2] Faculty of Economy and Management, University of Bialystok
Bialystok, Poland
[3] Department of Operations Research, University of Economics in Katowice
Katowice, Poland
almeidaatd@gmail.com, erosz@o2.pl,
tomasz.wachowicz@ue.katowice.pl

1 Overview

The outcomes the parties achieve in group decision and negotiation and the efficiency of the results they obtain are of major importance for economics and management science. To measure the quality of the negotiation agreements or the decisions made by the groups, the preferences of all parties involved in the bargaining process need to be elicited first. This requires that the negotiation problem is represented formally and the negotiation template is designed and evaluated. Based on the evaluation a scoring system can be built and used to evaluate the negotiation offers and alternatives for the agreement. Such a system allows to support the parties to analyze the negotiation progress, measure the scale of concessions made by the parties; and to visualize the negotiation history and the negotiation dance. It also allows to conduct the proactive mediation and/or arbitration and to search for a fair solution for all the parties involved.

Various formal methods, techniques and models may be used to support decision-makers to define their goals, elicit preferences and construct scoring systems. These methods are derived from the fields of multiple criteria decision-making (MCDM) and game theory. These methods, however, need to be modified and adopted to fit the decision context that is characteristic to negotiations and group decision-making, e.g. deciding under the pressure of time, and/or when the negotiation space is imprecisely defined, reservation and aspiration levels are changeable and many decision makers are involved. Moreover, the negotiators' cognitive and perceptional capabilities as well as their formal knowledge and skills for using different mechanisms and tools (for negotiation support) need to be taken into account in redesigning the existing and designing new methods and algorithms for preference elicitation and decision support in negotiation and group decision making. This often requires that new the software solutions such as the negotiation and group decision support systems be built.

The main contribution of this section is bringing together the perspectives of researchers and practitioners (in the field of group decision and negotiation analysis) on recent developments and findings in the areas of preference analysis and decision support. We have contributions on both theoretical and empirical aspects of designing and using formal models and techniques for preference analysis and decision support in negotiation and group decision making.

The papers included in this section and in the accompanying volume [1] are divided into the following four groups:

1. Methodological issues of preference analysis.
2. Application of MCDM methods in negotiation and group decision support.
3. Applications in real-world negotiations and group decision making problems.
4. Group decisions based on partial information or imprecise and vague preference.

2 Methodological Issues of Preference Analysis

The quality of results obtained in group decision processes depends on the foundations of preference analysis, so methodological issues play an important role in the area. The understanding and the use of group decision analysis model is of particular relevance. The concepts and intuitive logic for the group decision model is approached in [2], including some practical aspects of applying it. One of the issues is related to preference strength, which is considered in [3]. Surrogate weights are associated with the fact that decision-makers often possess more information regarding the relative strengths of the criteria to be incorporated in the preference analysis process.

Group preference management in social choice and in the recommended systems is considered in [4], which presents a comparative study of preference management. There are also two papers which discuss the problem of the effective usage of SAW in order to construct a negotiation offer scoring system. The issue of inaccuracy in defining preferences by the electronic negotiation system users is studied in [5]. The authors consider the elicitation of the negotiators' preferences with a simple additive weighting method. The linkages between the scale of inaccuracy and the negotiation profiles are verified [6]. The methodological differences between two alternative methods are discussed in the last paper in this grouping [7]. The authors compare MARS and GRIP from the perspective of the holistic evaluation of the negotiation template.

3 MCDM Methods

This group of articles deals with MCDM methods and their application to the negotiation and group decision contexts. An MCDM model is used to compare subjective and objective evaluation in [8], including an application to analyze the graduate's leaning ability. A well know MCDM method, ELECTRE III is considered for a group decision-making in [9], in which inference of pseudo criteria parameters are worked out. The dominance-based rough set approach is considered in [10] for an MCDM group decision model for supporting operations in intelligent electrical power grids. Using an additive weighting method is considered in [11] as a part of an algorithm for evaluation of the stakeholders in the sustainability reporting process. Finally, the issue of universal judgments in human groups and communities concerning procedural fairness and just outcomes is discussed in [12], aiming to legitimize group decisions and outcomes and to generate group equilibria.

4 Empirical Applications

Papers in this section deal with applications of formal decision support tools to facilitate real-world negotiation and group decision making problems.

A procedure for finding compromises among the watershed communities is proposed in [13]. The algorithm described by the authors applies ELECTRE II for supporting individual choices and then aggregates them through a weighted voting system based on classification by quartile. The idea of a new model for subcontractor selection applying different support algorithms for high and low costs of hiring contracts is presented in [14].

5 Partial Information and Imprecise Preference

There are situation in which preferences cannot be precisely defined. The papers in this section deal with such situations in the group-decision making context. An approach hybridizing the notion of veto and adjusting function incorporated into the additive model, and trapezoidal fuzzy numbers to solve group decision making problems is proposed in [15]. A different approach, one that stems from linguistic fuzzy rough sets, is presented in [16]. The model is enriched by introducing the linguistic hedges with the inclusive interpretation.

The notion of hesitant fuzzy sets is applied to TOPSIS algorithm [17]. The authors recommend that the algorithm be used to determine the weights of criteria in group decision-making problems. Fuzzy environment is also considered in [18], where classic PROMETHEE is adopted for the problem of selecting a facility location.

Acknowledgements. The organization of the Preference Analysis and Decision Support Conference stream, Section 3 in this volume, and Section 3 in [1] was supported by the grant from Polish National Science Centre (DEC-2011/03/B/HS4/03857).

References

1. Kamiński, B., Kersten, G.E., Szapiro, T. (eds.) GDN 2015. LNBIP, vol. 218. Springer, Heidelberg (2015)
2. Keeney, R.L.: Understanding and using the group decision analysis model. In: Kamiński, B., Kersten, G.E., Szapiro, T. (eds.) GDN 2015. LNBIP, vol. 218, pp. 77–86. Springer, Heidelberg (2015)
3. Danielson, M., Ekenberg, L.: Using surrogate weights for handling preference strength in multi-criteria decision. In: Kamiński, B., Kersten, G.E., Szapiro, T. (eds.) GDN 2015. LNBIP, vol. 218, pp. 107–118. Springer, Heidelberg (2015)

4. Naamani-Dery, L.: Group preference management: elicitation and aggregation in social choice and in recommender systems. In: [1]

5. Roszkowska, E., Wachowicz, T.: Inaccuracy in defining preferences by the electronic negotiation system users. In: Kamiński, B., Kersten, G.E., Szapiro, T. (eds.) GDN 2015. LNBIP, vol. 218, pp. 125–134. Springer, Heidelberg (2015)

6. Kersten, G.E., Roszkowska, E., Wachowicz, T.: Do the negotiators' profiles influence iccuracies of their scoring systems? In: [1]

7. Roszkowska, E., Wachowicz, T.: Holistic evaluation of the negotiation template – comparing MARS and GRIP approaches. In: [1]

8. Chen, Y., Li, Y., Sun, W., Xu, H.: A multiple criteria model for comparison of subjective-objective evaluations and its application. In: Kamiński, B., Kersten, G.E., Szapiro, T. (eds.) GDN 2015. LNBIP, vol. 218, pp. 89–99. Springer, Heidelberg (2015)

9. Alvarez, P.A., Morais, D., Leyva, J.C., Teixeira De Almeida, A.: Inferring pseudo criteria parameters in a procedure for group decision-making. In: [1]

10. Daher, S.: A multicriteria group decision model for supporting operations in intelligent electrical power grids. In: [1]

11. Bellantuono, N., Pontrandolfo, P., Scozzi, B.: Stakeholders' engagement in sustainability reporting. In: [1]

12. Burns, T., Machado, N., Roszkowska, E.: Distributive justice, legitimizing collective choice procedures and the production of normative equilibria in social groups: towards a theory of social order. In: Kamiński, B., Kersten, G.E., Szapiro, T. (eds.) GDN 2015. LNBIP, vol. 218, pp. 87–98. Springer, Heidelberg (2015)

13. Urtiga, M., Morais, D.: Group approach to support decision making in watershed committees. In: [1]

14. Palha, R., Teixeira De Almeida, A., Costa Morais, D.: Group decision model for subcontractors selection in construction industry. In: [1]

15. Sabio, P., Jiménez-Martín, A., Mateos, A.: Veto values within MAUT for group decision making on the basis of dominance measuring methods with fuzzy weights. In: Kamiński, B., Kersten, G.E., Szapiro, T. (eds.) GDN 2015. LNBIP, vol. 218, pp. 119–130. Springer, Heidelberg (2015)

16. Wang, W., Lu, Q.-A., Yang, L.: Multiple attribute group decision making under hesitant fuzzy environment. In: Kamiński, B., Kersten, G.E., Szapiro, T. (eds.) GDN 2015. LNBIP, vol. 218, pp. 171–182. Springer, Heidelberg (2015)

17. Tavakkoli-Moghaddam, R., Gitinavard, H., Meysam Mousavi, S., Siadat, A.: A new interval-valued hesitant fuzzy TOPSIS method to determine weights of criteria in group decision-making problems. In: Kamiński, B., Kersten, G.E., Szapiro, T. (eds.) GDN 2015. LNBIP, vol. 218, pp. 157–169. Springer, Heidelberg (2015)

18. Tavakkoli-Moghaddam, R., Sotoudeh-Anvari, A., Siadat, A.: A new multi-criteria group decision making approach for facility location selection using PROMETHEE under a fuzzy environment. In: Kamiński, B., Kersten, G.E., Szapiro, T. (eds.) GDN 2015. LNBIP, vol. 218, pp. 145–156. Springer, Heidelberg (2015)

Formal Models

João Climaco

Institute for Systems Engineering and Computers at Coimbra, INESC-Coimbra
University of Coimbra, Coimbra, Portugal
jclimaco@fe.uc.pt

1 Introduction

Decisions made by collectives constitute a major issue in our civilization. Nowadays, global governance is crucial, mainly because of economic, environmental and social challenges, such as the food shortages, the increasing inequalities worldwide, the environmental and climate changes, and security problems. Broadly speaking, the following quote characterizes the present situation: "if we accept the point that we are living in the time of changing civilization eras, and conceptual change is one of the main ingredients of the civilization change, up to the formation of a new episteme, then the need of new concepts and approaches, even new hermeneutical horizons also within group decisions and negotiation theory is evident" [1].

In these circumstances it is crucial to re-invent the global governance which implies a parallel revolution in the framework of collective decision procedures at the local and global levels. Of course, the new communication technologies, and in particular the Internet, open bright horizons enabling the interactive combination of human intervention aided by computerized decision aids. However, it must be emphasized that the analysis and support of group decisions as well as of negotiation processes are complex, multi-disciplinary tasks involving psychological, sociological, cognitive and political issues. Therefore, the real improvement of group decision and negotiation on a global scale is a major challenge in the XXI century.

Mathematically based models have been developed in the framework of operations research, systems science, game theory, etc., and they are an essential part of many group decision and negotiation support systems.

2 Framework

Kilgour and Eden in the introduction to the Handbook on GDN [2] note that:

> "The use of formal procedures for reaching a collective decision-making can be 'improved' by a systematic approach or by a kind of group support. Group decision and negotiation is the academic and professional field that aims to understand, develop, and implement these ideas in order to improve collective decision processes."

In order to foresee the potential and the limitations of the use of any mathematical model in this framework it seems that some ideas must be first articulated, namely:

– The range of the field is very broad and studied from very diversified perspectives, including not only a wide type of situations involving collaboration/conflicting, tactics/strategies, cognitive/emotional, social/cultural issues, but also the cross-fertilization of a large number of disciplinary areas, such as theory of the organizations, political science, sociology, psychology, telecommunications/internet, systems science, operations research, information systems, decision support systems, etc.

- Generally speaking, the developed approaches range from the theoretical analysis of the specific types of problems to the process oriented prescriptive support tools, and also the descriptive approaches. The intended help does not consist of showing the various actors involved in the course to follow, but rather of constructing a set of coherent recommendations that contribute to the clarification of the process. Thus, the models' goals and values do not run the risk of being replaced by any calculated rationale.

- Davey and Olson [3] observe that: "Decision making groups can range from cooperative, with very similar goals and outlooks, to antagonistic, with diametrically opposed objectives. Even in cooperative groups, conflict can arise during the decision process". In order to clarify the meaning of the co-existence of *collaboration* and *conflict* in group decision and in negotiation it is recommended that contrasting characteristics of these concepts be considered [4].

In group decisions we deal mostly with common sets of alternatives and objectives, while in negotiations proposals are sequentially presented by parties, which involves making concessions. This peculiar interdependence among actors, "rather than conflict, distinguishes negotiation from other forms of decision making" [5]. Furthermore, sharing information is characteristic of group decisions, contributing to the reduction of uncertainty and ambiguity; in negotiations information, values and beliefs of the parties are hidden. In group decisions leaving a group is not usual, and, inversely, the group cohesion is promoted. Finally, negotiation involves competition, while group decisions are mostly based in deliberative processes.

3 Mathematically Based Models

3.1 Models and Reality – Are Simplifications Acceptable?

A very old drawback concerning real world applications of operations research is the mistrust in mathematical models, particularly prevalent in group decisions and negotiations where the complexity and the range of scientific, cultural, social and behavioral issues are even more relevant. Rapoport [6] observed that:

> "The mathematical model is a set of assumptions. We know that every assumption is false. Nevertheless we make them, for our purpose at this point is not to make true assertions about human behavior but to investigate consequences of assumptions, as in any simulation or experimental game."

Kersten [7] noted that Rapaport presented a conundrum that is particularly troublesome when the models and systems are used by the end-users, i.e., the decision-makers rather than by the analysts and OR specialists. He commented that:

"The above quote, while controversial, suggests that formal models and support systems in which they are embedded may suffer from false assumptions or from assumptions that either seem unreasonable or are difficult to accept".

He proposed that to overcome this blocking situation an outreach strategy could be used [7]. This new strategy should pursue new hermeneutical horizons [1]. Rather than the continuation of the traditional path a new paradigm is proposed. In the outreach strategy assumptions and simplifications of mathematical models are still necessary, but they should be validated by the actors of the group decision and negotiation process; building systems integrating several complementary approaches is advisable; and those systems cannot forget social and behavioral issues, etc.

In what follows the papers included in this track are discussed.

3.2 Multi-criteria Analysis

In recent years, multi-criteria models integrated in group decision and negotiation systems have undergone major development, and, in our opinion, in many cases, the most adequate are the models rooted in constructivism. The use of multi-criteria models allows us to avoid one of the problems that has followed us over time, the aggregation of the preferences of decision agents in a single criterion, which reduces everything to just one measure. Some multi-criteria approaches propose the combination of algorithmic protocols and the experience and intuition of the actors intervening in the process of preference aggregation. However, if only formalized procedures are used to aggregate preferences of criteria and decision actors, these can be interactive, and oftentimes they should not be compensatory. Furthermore, it must be remarked that aggregation always implies loss of information, therefore it means that it needs to be done carefully and the resulting simplification needs to be assessed.

Different categories of models have been used in the past, i.e. multi-attribute models, including value functions and outranking approaches; and mathematical programming models, highlighting goal programming approaches. As the lack of adequate information is particularly relevant when we integrate multi-criteria models in group decision and negotiation aiding systems, we would like to register/draw the readers' attention to the use of models using incomplete/imprecise information. See, for example, an additive model based system dedicated to using incomplete information regarding the scaling constants and integrated in a GDSS – VIP Analysis [4], and a GDSS – IRIS integrating an aggregation/disaggregation approach for the ELECTRE TRI method [8].

This track includes two papers that present different multi-criteria models.

- A cooperative group multi-attribute analysis of routing models for a telecommunication network is discussed in [9]. The proposed method is grounded in GDSS – VIP Analysis which allows for incomplete information regarding scaling constants. This method is used to support a group of experts in the evaluation of alternative options of decentralized routing models.
- An interactive evolutionary multiple objective optimization model for group decision problems is proposed in [10]. The user interacts with the model via ordinal regression in order to identify the set of Pareto-optimal alternatives. The authors

propose an interactive meta-heuristic approach dedicated to a multiple objective optimization problem where preference information is provided by several decision makers and incorporated into the evolutionary search. The interaction is based on ordinal regression building value functions. In our opinion the added value of this paper is the careful experimentation with several variants of the interactive procedure exploiting conjointly the preference information provided by the decision makers.

3.3 Game Theory

Game theory is dedicated to the choice of optimal behavior of two or more rational players interacting strategically. Costs and benefits of each option for one player depend on the choices of the other players. It is clearly the most rigorous approach to dealing with conflicts. In this context it must be emphasized that this type of mathematically based models are the root of many group decision and negotiation theoretical and methodological approaches–in many cases, the analysis of the stability of outcomes is one of its key issues.

Many researchers have exploited a great number of cooperative and non-cooperative game models– some are considered in the following four papers of the GDN 2015 track. On the one hand game theory is a very important and productive field, but on the other hand it has been misused in many situations. We decided that before summarizing the papers of the track integrating Game Models we will pay attention to its limitations/ weaknesses. The following two quotes depict the problem very astutely:

1. "Unfortunately, game models must usually abstract one or a few specific features from a real world situation, drastically simplifying the rest, in order to avoid problems of complexity and tractability. In most cases, realistic game models are impossible to analyze." [8]
2. "The weaknesses of game-theoretic approaches include the treatment of the process and its impact on the game itself, and strict rationality assumptions which, for numerous reasons, rarely hold (e.g., imperfect information, parties' cognitive limitations, and deception)... Thus, while game-theoretic methods have a significant role to play in the prior or posterior analysis of the group decision or negotiation problems, their usefulness as a support tool during the process is limited". [6]

We believe that the above lines give an accurate picture of the problem. The four papers which rely on game-theoretical models are briefly discussed below.

– A fiscal-monetary non-cooperative game can be studied with the use of a dynamic macroeconomic model [11]. The fiscal and monetary authorities' strategic moves and the Nash equilibrium are analyzed. The simulation of the results enables to conclude that, as in many other situations, that in general the Nash equilibrium is not Pareto optimal. In these circumstances, looking for a Pareto optimal negotiation outcome is necessary. The paper, in general is interesting but of a particular interest are the computer simulations for various states of the economy and the discussion. As the Nash equilibrium is not Pareto optimal, the proposal to promote negotiations based on a bargaining problem which is analyzed using multi-criteria optimization tools is also interesting.

– A stochastic dynamic cooperative game which represents interaction among decision agents who control a dynamic system is discussed in [13]. The agents represent economic and financial entities such as real-estate market and regional economic, and social networks. The authors study the dependence among the characteristics of the trajectory of the aggregate outcomes, the behavior of the decision agents (namely the interaction among decision agent preferences) and the importance of the localization of the decision agents in respect to specific local centers. The usefulness of the proposed game is discussed.

Acknowledgments. This work was financially supported by the EU Community Support Framework III Program and national funds (Portuguese Foundation for Science and Technology) under PEst- OE/EEI/UI0308/2014.

References

1. Wierzbicki, A.: Group decisions and negotiations in the knowledge civilization era. In: Kilgour, M. and Eden, C. (eds.) Handbook of Group Decisions and Negotiation, pp. 11–24. Springer, Heidelberg (2010)
2. Kilgour, M., Eden, C. (eds.): Handbook of Group Decisions and Negotiation, vol. 4. Springer Science & Business Media (2010)
3. Davey, A., Olson, D.: Multiple criteria decision making models in group decision support. Group Decis. Negot. **7**, 55–77 (1998)
4. Dias, L., Clímaco, J.: Dealing with imprecise information in group multicriteria decisions: a methodology and a GDSS architecture. Eur. J. Oper. Res. **160**, 291–307 (2005)
5. Kersten, G., Cray, D.: Perspectives on representation and analysis of negotiation: towards cognitive support systems. Group Decis. Negot. **5**, 433–467 (1996)
6. Rapoport, A.: Strategy and Conscience. Harper & Row, New York (1964)
7. Kersten, G. Support for group decisions and negotiations – an overview. In: Clímaco, J. (ed.) Multi-criteria Analysis, pp. 332–346. Springer, Heidelberg (1997)
8. Damart, S, Dias, L., and Mousseau, V.: Supporting groups in sorting decisions: methodology and use of a multi-criteria aggregation/disaggregation. Decis. Support Syst. **43**, 1464–1475 (2007)
9. Clímaco, J., Craveirinha, L., Martins, L.: Cooperative group multi-attribute analysis of routing models for telecommunication networks. In: [14]
10. Kadziński, M., Tomczyk, M.: Using ordinal regression for interactive evolutionary multiple objective optimization with multiple decision makers. In: Kamiński, B., Kersten, G.E., Szapiro, T. (eds.) GDN 2015. LNBIP, vol. 218, pp. 185–198. Springer, Heidelberg (2015)
11. Kilgour, M.: Game models of negotiation and arbitration. In: Encyclopedia of Life Support Systems, EOLSS (2013)
12. Kruś, L., Woroniecka-Leciejewicz, I.: Fiscal-monetary game analyzed with use of a dynamic macroeconomic model. In: Kamiński, B., Kersten, G.E., Szapiro, T. (eds.) GDN 2015. LNBIP, vol. 218, pp. 199–208. Springer, Heidelberg (2015)
13. Kosiorowski, D., Zawadzki, Z.: Locality, robustness and interactions in simple cooperative dynamic game. In: [14]
14. Kamiński, B., Kersten, G.E., Szapiro, T. (eds.) GDN 2015. LNBIP, vol. 218. Springer, Heidelberg (2015)

Voting and Collective Decision-Making

Hannu Nurmi

University of Turku
Turku, Finland
hnurmi@utu.fi

1 Overview

Voting is an important way to make group decisions. It has been used in a wide variety of contexts ranging from highly regulated and formalized – e.g. political elections – to informal ad hoc settings used in deciding leisure activities in small groups. Often the elections are considered as essential elements of democratic governance.

The specific procedures of voting, however, vary greatly, not only between countries, but also within countries. For instance, one system regulates in the election of the head of state, while another one is followed in electing the members of legislature. Or, one system is used in electing the leaders of religious communities, while another is resorted to in electing the presidents of universities. This variety of procedures has given rise to a rich literature on the desiderata associated with procedures. For example, which precise properties of procedures pertain to democratic group decision making or to collective rationality?

The theory of voting and collective decision making is based on the social choice theory. Its best-known results tend to be of negative nature; they demonstrate incompatibilities among various desirable choice-theoretic properties. Some of the incompatibilities are surprising, counterintuitive or paradoxical. While these results are unquestionably important, it is important to study their relevance in real world collective decision making. The context in which the procedures are being used as well as the plausibility of their underlying assumptions are important determinants of the relevance. Which goals are the procedures intended to serve? To what extent are these reconcilable with the goals of the participants? Are the expected outcomes of procedures likely to be welfare increasing or divisive? These are some of the issues discussed in this stream of presentations.

The papers included in this section as well as the papers included in the accompanying volume [1] can be thematically divided into the following three groups:

1 The direct vs. indirect (representative) aggregation of opinions;
2. Alignments, power and bargaining; and
3. The choice of rule.

2 The Direct vs. Indirect Aggregation of Opinions

Many results in the social choice theory pertain to aggregation of opinions. One of them, the referendum paradox, is the phenomenon whereby the outcome of collective decision making involving just two alternatives (yes-no) crucially depends on the order in which the aggregation takes place. The possibility of this paradox opens new vistas for strategic behavior among participants [2]. Strategic behavior is often viewed as intentional strive for individually beneficial outcomes, but it can also be related to the more permanent personality traits of voters. It is, therefore, worthwhile to study the expected consequences of the prevalence of specific personality traits among voting population [3].

3 Alignments, Power, and Bargaining

Both voting and bargaining are the mechanisms that aim at working out universally acceptable outcomes when the interests of the participants differ. The setting where there are only two participants with different opinions regarding two options already captures the some essential differences and similarities of the two mechanisms [4]. Various procedures have obvious implications for power distribution among participants with varying resources. These have been extensively studied in dichotomous settings. However, with three or more alternatives considered simultaneously, the measurement of a priori voting power becomes more complicated [5].

4 The Choice of Rule

Historically and analytically the choice of the procedure differs from the application of the chosen procedure in determining policy or the composition of the representative body. Are there any general principles one could resort to in designing a voting rule to be applied in business decisions or in informal settings [6]? The existing – relatively rich – literature focuses on dichotomous choices (rule x *versus* rule y) and often assumes voter preferences regarding the outcomes that result from the application of rules. It is, however, also possible to address the problem via the criteria that various procedures satisfy or fail to satisfy [7]. This renders the rule choice an instance of a general MCDM problem and may seem a plausible way of augmenting the current recommendation systems [8].

References

1. Kamiński, B., Kersten, G.E., Szapiro, T. (eds.) GDN 2015. LNBIP, vol. 218. Springer, Heidelberg (2015)
2. Dindar, H., Laffond, G., Lainé, J.: Vote swapping in representative democracy. In: Kamiński, B., Kersten, G.E., Szapiro, T. (eds.) GDN 2015. LNBIP, vol. 218, pp. 227–239. Springer, Heidelberg (2015)

3. Sosnowska, H., Przybyszewski, K.: Do some characteristics of personality influence decision-making in approval voting? In: [1]
4. Bánnikova, M.: Gathering support from rivals: the two rivals case. In: [1]
5. Mercik, J., Ramsey, D.: A formal a priori power analysis of the security council of the United Nations. In: [1]
6. Teixeira de Almeida, A., Nurmi, H.: A framework for aiding the choice of a voting procedure for a business decision problem. In: Kamiński, B., Kersten, G.E., Szapiro, T. (eds.) GDN 2015. LNBIP, vol. 218, pp. 211–225. Springer, Heidelberg (2015)
7. Nurmi, H.: The choice of voting rules based on preferences over criteria. In: Kamiński, B., Kersten, G.E., Szapiro, T. (eds.) GDN 2015. LNBIP, vol. 218, pp. 241–252. Springer, Heidelberg (2015)
8. Naamani-Dery, L., Teixeira de Almeida, A., Nurmi, H.: Choosing a voting procedure for a Leisure Group activity. In: [1]

Conflict Resolution in Energy and Environmental Management

Keith W. Hipel[1,2], D. Marc Kilgour[1,3], Liping Fang[1,4], and Amer Obeidi[1,5]

[1] Department of Systems Design Engineering, University of Waterloo, Waterloo, Canada
[2] Centre for International Governance Innovation (CIGI), Waterloo, Canada
[3] Department of Mathematics, Wilfred Laurier University, Waterloo, Canada
[4] Department of Mechanical and Industrial Engineering, Ryerson University,
Toronto, Canada
[5] Department of Management Sciences, University of Waterloo, Waterloo, Canada
{kwhipel,amer.obeidi}@uwaterloo.ca, mkilgour@wlu.ca,
lfang@ryerson.ca

1 Overview

The recent development of new technologies that can help analysts understand strategic conflicts and provide strategic support to negotiators has been a great benefit for many decision makers. New theoretical issues are being explored, and at the same time new software systems are making modeling easier and analytical results clearer. Environmental management, including energy projects, is a natural area of application of technologies for the analysis of strategic conflict, and has motivated both theoretical and practical advances. This stream collects contributions highlighting new advances in the Graph Model for Conflict Resolution (GMCR) and other methodologies that have been influenced by issues arising in environmental management, energy development and other types of disputes. More specifically, one major thrust contained in the paper is the presentation of techniques for modelling preferences within the GMCR using probability theory, fuzzy sets and grey numbers. A second set of papers deal with basic structures of conflict which can be expressed within a GMCR structure: hierarchical conflicts, misunderstanding in disputes and multi-level options. Finally, one paper is concerned with fairly allocating water among competing users employing decentralized optimization.

2 Group Decision and Negotiation

As described by authors such as Kilgour and Eden [1] and Hipel [2, 3], a rich range of formal techniques and methodologies are available for modelling controversies arising in group decision and negotiation. Of particular interest here is the Graph Model for Conflict Resolution for investigating real world disputes occurring in energy development, water resources, environmental management, international trade, industrial development and many other areas [4, 5]. This methodology can be applied in practice to actual conflicts by using the decision support systems such as GMCR II [6, 7] or GMCR+ [8].

A GMCR model consists of three key pieces of information: decision makers (DMs), the options or courses of action under the control of each DM and the relative preferences among the feasible states or scenarios in the conflict. Because preferences are relatively difficult to obtain in practice and there may be high uncertainty contained in them a number of mathematical approaches have been proposed for capturing their uncertainty including unknown preference information [9], fuzzy sets [10], probability theory [11] and grey numbers [12]. Another approach for dealing with situations in which a DM may greatly prefer one situation (ex. peace deal is reached) over another (war breaks out) is called strength or level of preference [13, 14]. As explained in the next section papers contained within this stream contain a range of advances in modelling uncertain preferences within the GMCR paradigm. These contributions can be operationalized by including them within expanded versions of DSSs for GMCR or developing new systems. The matrix formulation of a conflict [15] can reduce computational time within a DSS for implementing GMCR.

Some basic structures could also be embedded within GMCR to further enhance its applicability. For instance, in some cases, a hierarchical structure of conflicts may be present and hence one may wish to reflect this within GMCR [16]. In situations in which misunderstanding or misperceptions are present, one may wish to take into account what is called a hypergame framework [17, 18]. For some disputes one may want to allow for levels in an option such as having a high, medium or low level of water supply available. These types of advancements are addressed in this stream of papers.

The fair allocation of resources constitutes an important problem in many fields such as fairly distributing bandwidth among broadcasting stations in the communications industry and equitably allocating water among competing users in a river basin. Based upon concepts from hydrology, economics and cooperative game theory within an overall large scale optimization problem, Wang et al. [19] developed a comprehensive model for fairly allocating water among users with application to the South Saskatchewan River Basin in the Canadian Province of Alberta and the Aral Sea.

3 Contributions Contained in This Stream of Papers

The first 19 references contained in the bibliography are additional references used in this introduction. The last ten references refer the papers contained in this stream. References 20 to 22 are full papers while references 23 to 29 are extended abstracts. The paper by Kornis et al. [20] nicely demonstrates how GMCR can be applied to study the dispute over fluctuating water levels in the Great Lakes using the DSS GMCR II [6, 7]. Next, Hou et al. [21] model how three-levels of preferences can be obtained using what is called option prioritization.

For the papers in which extended abstracts are provided, the first set of papers is mainly concerned with preference uncertainty. In particular, after defining a solution concept called symmetric sequential stability, Rego and Vieira [22] extend it for employment with uncertain, probabilistic and fuzzy preferences. The same authors [23] then furnish matrix methods for calculating stability when preferences can be probabilistic.

The next three papers deal with different structures that can be handled within GMCR. Specifically, within a hierarchical graph model, He et al. [24] present option prioritization methods for determining preferences in a higher level conflict from lower ones. This is followed by Aljefri et al. [25] show how misperception of options by DMs can be formally handled within GMCR. Matbouli et al. [26] then explain how options can be split into levels within a graph model structure. Finally, Xiao et al. [27] develop a modified penalty based decentralized optimization method for employment in fairly allocating water among competing stakeholders.

References

1. Kilgour, D.M., Eden, C. (eds.) Handbook of Group Decision and Negotiation. Springer, Dordrecht (2010)
2. Hipel, K.W. (ed.) Conflict Resolution, vol. 1. Eolss Publishers, Oxford (2009)
3. Hipel, K.W. (ed.) Conflict Resolution, vol. 2. Eolss Publishers, Oxford (2009)
4. Kilgour, D.M., Hipel, K.W., Fang, L.: The graph model for conflicts. Automatica, 23(1), 41–55 (1987)
5. Fang, L., Hipel, K.W., Kilgour, D.M.: Interactive Decision Making: The Graph Model for Conflict Resolution. Wiley, New York (1993)
6. Fang, L., Hipel, K.W., Kilgour, D.M., Peng, X.: A decision support system for interactive decision making, part 1: model formulation. IEEE Trans. Syst. Man Cybern. Part C: Appl. Rev. 33(1), 42–55 (2003)
7. Fang, L., Hipel, K.W., Kilgour, D.M., Peng, X.: A decision support system for interactive decision making, part 2: analysis and output interpretation. IEEE Trans. Syst. Man Cybern. Part C: Appl. Rev. 33(1), 56–66 (2003)
8. Kinsara, R.A., Petersons, O., Hipel, K.W., Kilgour, D.M.: Advanced decision support system for the graph model for conflict resolution. J. Decis. Syst. accepted for publication on March 12, 2015
9. Li, K.W., Hipel, K.W., Kilgour, D.M., Fang, L.: Preference uncertainty in the graph model for conflict resolution. IEEE Trans. Syst. Man Cybern. Part A: Syst. Hum. 34(4), 507–520 (2004)
10. Bashar, M.A., Kilgour, D.M., Hipel, K.W.: Fuzzy preferences in the graph model for conflict resolution. IEEE Trans. Fuzzy Syst. 20(4), 760–770 (2012)
11. Rego, L.C., Santos, A.M.: Probabilistic preferences in the graph model for conflict resolution. IEEE Trans. Syst. Man Cybern. Syst. (2015). doi:10.1109/TSMC.2014.2379626
12. Kuang, H., Bashar, M.A., Hipel, K.W., Kilgour, D.M.: Grey-based preference in a graph model for conflict resolution with multiple decision makers. IEEE Trans. Syst. Man Cybern. Syst. (2015). doi:10.1109/TSMC.2014.2387096
13. Hamouda, L., Kilgour, D.M., Hipel, K.W.: Strength of preference in graph models for multiple decision-maker conflicts. Appl. Math. Comput. 179(1), 314–327 (2006)
14. Xu, H., Hipel, K.W., Kilgour, D.M.: Multiple levels of preference in interactive strategic decisions. Discrete Appl. Math. 157(15), 3300–3313 (2009)
15. Xu, H., Hipel, K.W., Kilgour, D.M.: Matrix representation of solution concepts in multiple decision maker graph models. IEEE Trans. Syst. Man Cybern. Part A: Syst. Hum. 39(1), 96–108 (2009)

16. He, S., Kilgour, D.M., Hipel, K.W., Bashar, M.A.: A basic hierarchical graph model for conflict resolution with application to water diversion conflicts in China. INFOR: Inf. Syst. Oper. Res. **51**(3), 103–119 (2013)

17. Bennett, P.G.: Toward a theory of hypergames. Omega **5**(6), 749–751 (1977)

18. Wang, M., Hipel, K.W., Fraser, N.M.: Modelling misperceptions in games. Behav. Sci. **33** (3), 207–223 (1988)

19. Wang, L., Fang, L., Hipel, K.W.: Basin-wide cooperative water resources allocation. Eur. J. Oper. Res. **190**(3), 798–817 (2008)

20. Karnis, M., Bristow, M., Fang, L.: Controversy over the international upper great lakes study recommendations: pathways towards cooperation. In: Kamiński, B., Kersten, G.E., Szapiro, T. (eds.) GDN 2015. LNBIP, vol. 218, pp. 255–267. Springer, Heidelberg (2015)

21. Hou, Y., Jiang, Y., Xu, H.: Option prioritization for three-level preference in the graph model for conflict resolution. In: Kamiński, B., Kersten, G.E., Szapiro, T. (eds.) GDN 2015. LNBIP, vol. 218, pp. 269–280. Springer, Heidelberg (2015)

22. Rego, L.C., Vieira, G.I.A.: Symmetric sequential stability in the graph model for conflict resolution. In: [28]

23. Rego, L.C., Vieira, G.I.A.: Matric representation of solution concepts in the graph model for conflict resolution with probabilistic preferences. In: [28]

24. He, S., Hipel, K.W., Kilgour, D.M.: Option prioritization methods in the general hierarchical graph model. In: [28]

25. Aljefri, Y.M., Bashar, M.A., Hipel, K.W., Fang, L.: Generating hypergame states within the paradigm of the graph model for conflict resolution. In: [28]

26. Matbouli, Y.T., Hipel, K.W., Kilgour, D.M.: Multi-level options in the graph model for conflict resolution. In: [28]

27. Xiao, Y., Hipel, K.W., Fang, L.: A decentralized optimization method for water resources allocation. In: [28]

28. Kamiński, B., Kersten, G.E., Szapiro, T. (eds.) GDN 2015. LNBIP, vol. 218. Springer, Heidelberg (2015)

Negotiation Support Systems and Studies

Mareike Schoop[1], Sabine Koeszegi[2], and Rudolf Vetschera[3]

[1] University of Hohenheim, Stuttgart, Germany
[2] Vienna University of Technology, Vienna, Austria
[3] University of Vienna, Vienna, Austria
m.schoop@uni-hohenheim.de, Sabine.Koeszegi@tuwien.ac.at,
Rudolf.Vetschera@univie.ac.at

1 Introduction

Business and personal interactions increasingly take place online. Such interactions vary from personal communications (e.g. using email, twitter, skype) to formal organisational processes such as procurement, sale, or marketing. What they have in common is that they all deal with communication of different sorts. Concentrating on the business context, communication gets more structured. Mutual understanding becomes a prime goal in order to enable effective business interactions.

Electronic negotiations are an archetype of organizational communication processes that involve decision making and conflict management at the same time. Whilst negotiators in organisational e-negotiation processes might use general communication systems such as email or skype, there are also systems that are more specifically targeted at e-negotiations. They can be support tools as part of business systems or dedicated electronic support systems (NSSs). NSSs support communication, decision making, document management, and/or conflict resolution in business contexts. Over the past decades, we have seen sophisticated NSSs that provide holistic support of all of the above negotiation elements. They have been tested in various experiments and have been shown to improve both process and outcome.

The papers of this section as well as of the accompanying volume [1] show the work of researchers, developers, and practitioners who design and develop NSSs, study their use in the laboratories and in the field, or incorporate NSS components into negotiation, mediation and facilitation. In particular, the papers deal with: (1) communication and language aspects, (2) behavioural aspects, (3) system and media aspects, and (4) new applications of NSSs.

2 Communication Negotiation Support Systems and Studies

Communication is the core functionality of negotiations present in any context, by any stakeholder, using any medium or system. Thus, communication support must be a core functionality of an NSS.

The keynote by Schoop [2] addresses this need for dedicated communication support. With two communication theories as a firm basis, different aspects of communication support in electronic negotiations such as semantic and pragmatic message elements and validity claims as meta-communication are introduced. The theoretical constructs have been implemented in the negotiation support system Negoisst which is also discussed in the paper. Schoop shows the role communication support plays for electronic negotiations.

The paper by Schoop et al. [3] provides the basis for communication support in NSSs by analysing the role of ontologies in electronic negotiations. Since the overall goal of communication support is mutual understanding, ontologies can provide the means to achieve understanding on the syntactic as well as the semantic level. Combined with pragmatic support in an NSS, this would enable a complete support of negotiation communication on all semiotic levels.

The paper by Kersten [4] analyses how negotiations can contribute to the acquisition of English as a second language in a university course. The students of the course were provided with academic negotiation publications and used an NSS to try out the concepts in practice. Communication practice is of prime importance when learning a new language. Together with joint problem solving, these negotiation components helped the students in their learning tasks.

3 Behavioural Aspects in Negotiation Support Systems and Studies

Negotiations involve at least two stakeholders in interaction processes. These negotiators make decisions and concessions, show emotions, and behave in different ways during the negotiation process based on their cultural context.

The paper by Vetschera [5] addresses the interdependence of behaviour of negotiators, in particular the sequence of offers they are making. It extends the Actor-Partner Interdepence Model, which was specifically developed for the analysis of data resulting from dyadic interactions, to the specific situation of negotiations. Results from applying this model to two data sets identify some robust patterns, but also indicate that interaction processes are strongly dependent on the negotiation task.

The paper by Etezadi and Kersten [6] studies multi-bilateral negotiations, in which one buyer simultaneously negotiates with multiple sellers and analyses how the negotiation tactics of the buyer influence behaviour of the sellers. The authors estimate a simultaneous equations model using data of 229 experimental negotiations. Their findings confirm the asymmetric role of reciprocity, in that competitive tactics are reciprocated, but sellers try to exploit cooperative buyers.

The paper by Sundarraj and Morais [7] raises the question how culture determines a particular behavioural issue, namely time preference. They envision testing time preference in a cross-cultural experiment with students from Brazil and India.

The paper by Gettinger and Köszegi [8] deals with aspects of affective complexity in electronic negotiations. Its management is fundamental for negotiators to reach mutual understanding in communication and a positive relationship. They propose to support electronic negotiations with communication tools that facilitate the contextualization of communication by providing emoticons.

4 Medium and System Aspects of Negotiation Support Systems

Negotiation Support Systems exploit the potential of information and communication technology to enable or to improve electronic negotiation processes and lead to better outcomes.

The paper by Moura and Costa [9] introduces an NSS called NegPlace that considers personality traits of negotiators for the support of electronic negotiation processes. The ultimate aim is to improve the negotiation by considering individual styles of the participants.

The paper by Sugimoto et al. [10] discusses a study of decision making in crisis management. Japanese and British students. The authors compare face-to-face scenarios to online scenarios in this context and analyse the differences. They show the need for dedicated ICT support.

5 New Applications of Negotiation Support Systems

The paper by Lenz et al. [11] introduces the field of requirements analysis to electronic negotiations and vice versa. Electronic requirements negotiations involve multiple stakeholders and are multi-attribute negotiations by nature. Surprisingly, the majority of previous work on such negotiations stems from the requirements engineering community. The authors discuss the particulars of requirements negotiations and show that NSSs can provide the means for support.

References

1. Kamiński, B., Kersten, G.E., Szapiro, T. (eds.) GDN 2015. LNBIP, vol. 218. Springer, Heidelberg (2015)
2. Schoop, M: The role of communication support for electronic negotiations. In: Kamiński, B., Kersten, G.E., Szapiro, T. (eds.) GDN 2015. LNBIP, vol. 218, pp. 283–287. Springer, Heidelberg (2015)
3. Schoop, M., Bumiller, S., Fernandes, M.: Ontologies in electronic negotiations. In: [1]
4. Kersten, M.: Negotiations and second language acquisition. In: [1]
5. Vetschera, R.: Applying the APIM model to concession patterns in electronic negotiations. In: [1]
6. Etezadi, J., Kersten, G.E.: The effect of buyers' negotiation approach on sellers' attitude and behavior. In: [1]
7. Sundarraj, R., Morais, D.: Analysis of Cross-cultural Behavioural Time-preference, and its Impact on E-negotiation. In: [1]
8. Gettinger, J., Koeszegi, S.T.: More than words: the effect of emoticons in electronic negotiations. In: Kamiński, B., Kersten, G.E., Szapiro, T. (eds.) GDN 2015. LNBIP, vol. 218, pp. 289–305. Springer, Heidelberg (2015)

9. Moura, J., Costa, A.P.: NegPlace platform a web negotiation support system that incorporates negotiator's styles and personality. In: [1]

10. Sugimoto, Y., Papamichail, N., Greenhill, A.G.: Face-to-face versus computer-mediated collaborative decision making process in extreme events: a comparative study. In: [1]

11. Lenz, A., Schoop, M., Herzwurm, G.: Requirements analysis as a negotiation process. In: [1]

Online Collaboration and Competition

R.P. Sundarraj

Indian Institute of Technology Madras, Chennai, India
rpsundarraj@iitm.ac.in

1 Overview

Large corporations have been leveraging the Internet for melding functional silos and for creating a ubiquitous set of inter-organizational business partners who collaborate with one another on various planning and operational activities. With the entry of social media, the scope for online collaboration and competition has expanded even further, in terms of both the geographic spread, as well as the real-time nature of the decisions that need to be taken. All of these necessitate the incorporation of newer methodologies.

One theoretical base for online technologies is rooted in the area of psychology. For example, what constructs are crucial in determining the onset and dynamics of an online interaction? Further on, how can such constructs be modeled mathematically and what techniques can be used to take real-time decisions on the basis of such models? Novel collaboration tools can be applied in a variety of contexts. The relevant question then is how can generic models be adapted to a particular known application area? Are there newer forms of applications that have arisen as a result, and if so, what models are most adaptable? Finally, there is the issue of social and customer acceptance of online systems. That is, what factors lead to the acceptance and adoption of various online technologies?

The aforementioned discussion leads us to thematically divide the papers in Section 8 of this volume and in [1] into the following three groups:

1. Online Technology Constructs and Models
2. Applications online technologies
3. Acceptance of online technologies.

2 Models and Constructs

One common approach to developing online tools is agent-based systems. Agents are software programs that act autonomously on behalf of a user and interact with other users or agents. Robertson and Franco [2] consider the question of how knowledge can be transferred through the use of inter-group interaction, and employ an agent-based approach for this purpose. Multi-agent coordination is also the underlying mechanism in an online system for managing consumer-collectives of renewable energy. Algorithms for the demand-management of this collective, and salient computational results with the algorithms are given in [3]. Online systems today provide a unique way for organizations to elicit the participation of the public (e.g., through crowdsourcing). Antecedents constructs for which crowdsourcing becomes useful for the organization are proposed in [4].

3 Applications

At the organizational level, collaborative technologies can be used for decision-making by multiple stakeholders. A multicriteria decision-making model for assessing cloud-computing investment decisions is given in [5]. Online collaborative technologies allow for the possibility of engaging the mass population at large. Thus, these multicriteria methods can also be employed at the end-customer level as well, especially with the prevalence of online shopping. One such application of an AHP-based approach to determine the factors that influence customer shopping can be found in [6].

4 Acceptance

Collaborative technologies offer new marketing possibilities, for example, that of attracting customer through location-based electronic coupons. A customer's intention to re-purchase electronic coupons is affected by the quality of service of both the coupon-distributor and that of the store [7]. In addition to marketing, online tools are also influencing human courtships [8]: a survey of online-dating-site users found that looks and temporary physical encounters are not important for both men and women, although men are more in hurry to find a mate.

A number of research studies have brought out how trust and its sub-constructs are important in determining the acceptance of online collaborative technologies. In the case of longitudinal use of one such group decision-support-system, one trust sub-factor, namely risk-perception, decreased with the usage of a system [9]. Trust can also have an influence on global teams that engage in virtual collaboration [10].

Finally, the question is how does the acceptance of collaborative technology change across time and region? Using the US and Australia as examples, it is shown that even though technology access varies across these geographies, the perceived impact is more affected by the length of use rather than by the end-user's regional origin [11].

References

1. Kamiński, B., Kersten, G.E., Szapiro, T. (eds.) GDN 2015. LNBIP, vol. 218. Springer, Heidelberg (2015)
2. Poblet, M., Rowe, M., Thomson, J.: Creating Value through Crowdsourcing: the antecedent conditions. In: Kamiński, B., Kersten, G.E., Szapiro, T. (eds.) GDN 2015. LNBIP, vol. 218, pp. 345–355. Springer, Heidelberg (2015)
3. Zheng, R., Xu, Y., Chakraborty, N., Lewis, M., Sycara, K.: Demand management with energy generation and storage in collectives. In: Kamiński, B., Kersten, G.E., Szapiro, T. (eds.) GDN 2015. LNBIP, vol. 218, pp. 369–381. Springer, Heidelberg (2015)
4. Robertson, D., France, L.: An agent-based model of knowledge transfer. In: [1]

5. Luz, N., Poblet, M., Silva, N., Novais, P.: Defining human-machine micro-task workflows for constitution making. In: Kamiński, B., Kersten, G.E., Szapiro, T. (eds.) GDN 2015. LNBIP, vol. 218, pp. 333–344. Springer, Heidelberg (2015)

6. Venkataraman, S., Sundarraj, R.P.: On integrating an IS success model and multicriteria preference analysis into a system for cloud-computing investment decisions. In: Kamiński, B., Kersten, G.E., Szapiro, T. (eds.) GDN 2015. LNBIP, vol. 218, pp. 357–368. Springer, Heidelberg (2015)

7. Lai, H., Hsu, S.: Intention to repurchase group coupon service: the intertwined effect of service quality of vendors and service providers. In: Kamiński, B., Kersten, G.E., Szapiro, T. (eds.) GDN 2015. LNBIP, vol. 218, pp. 321–332. Springer, Heidelberg (2015)

8. Orger, S., Aleti, T., Zelezikow, J.: Shopping for love: do men and women do it differently online? In: [1]

9. Cheng, A., Fu, S., and Peng, Y.: Longitudinal case study on risk factor in trust development of facilitated collaboration. In: Kamiński, B., Kersten, G.E., Szapiro, T. (eds.) GDN 2015. LNBIP, vol. 218, pp. 309–320. Springer, Heidelberg (2015)

10. Cheng, X., Huang, J., de Vreede, G., Fu, S.: Exploring trust factors in global hybrid virtual collaboration: a case study of a Chinese multi-national firm. In: [1]

11. Bajwa, D., Lewis, F., Pervan, G.: Comparing collaborative information technologies assimilation and impacts across Australian and US organizations: an exploratory study. In: [1]

Market Mechanisms and Their Users

Verena Dorner, Timm Teubner, and Christof Weinhardt

Karlsruhe Institute of Technology (KIT), Karlsruhe, Germany
verena.dorner@kit.edu, timm.teubner@kit.edu,
chrstof.weinhardt@kit.edu

1 Overview

Online marketplaces represent ideal playing fields for market engineers and researchers in decision-making on mechanisms and aspects of information presentation. Rules and mechanisms for markets vary widely and continue to evolve, especially with the increasing amount of computer power which permits developing and implementing complex, memory-intensive mechanisms like combinatorial auctions. The other major factor in exchange transactions are the market participants and their interaction with such systems and the mechanisms. How humans react to specific set of market rules and principles determines the success or failure of this market. This introduces behavioral aspects to market design considerations.

Explaining and predicting user reactions to market design as well as designing new and more efficient forms of market mechanisms are therefore two of the most urgent tasks for researchers from many fields—including information systems, operations research, economics, and social and political sciences. Phenomena like auction fever, overbidding, information overload, social competition, and other forms of social preferences and behavioral biases illustrate this notion. The papers included in this section as well as the papers included in the accompanying volume [1] address such aspects and they can be divided into the following three groups:

1. negotiations and auctions;
2. peer-to-peer markets; and
3. emotions in markets.

2 Negotiations and Auctions

Complex market mechanisms, especially in business-to-business auctions, have attracted a good deal of attention in research and practice in the last decades. Auctions and formal negotiation structures represent a possible means of reaching agreements between the involved parties, where usually auctions follow well defined rules and negotiations often do not. In this volume, a series of suggestions are made for applying or combining negotiations and auctions to relevant problems.

A multi-attribute view on the auction market place eBay is considered in [2]. The authors use multiple (parallel) auctions on the same product type with different feature properties, taking advantage of large auction numbers, the price variation therein, and different feature characteristics that fit the bidders' specific preferences.

Another type of auctions is the first-price sealed-bid auction. A special case, in which the responsibility for bid and payment is split is discussed in [3]. Such a division of responsibility may be due to the principal-agent relationship that arises between the bidder and the payer. In two-unit two-bidder scenarios, the effect of overall allowances for one and two units on bidding equilibria are compared. Overall allowances for two units result in an equilibrium with no single-unit bids; only if single-unit allowances exceed two-unit allowances do the agents place single-unit bids.

Two contributions consider supply chain environments. A two-stage multi-echelon supply chain model is introduced in [4]. A numerical example is used to compare the different order and production lot determination approaches that involves central planning (optimization) as well as decentralized negotiations.

Inefficiencies in reverse procurement auctions in just-in-time production environments due to non-linear contract curves with interrelated product or service attributes are discussed in [5]. In order to improve auction efficiency, in terms of outcomes for both sides, post-auction multi-bilateral negotiations are suggested.

3 Peer-to-Peer Markets

Another evolving form of online markets are peer-to-peer platforms. Today's e-commerce landscape experiences the development of a broad variety of such markets. Whereas the last decade was mainly characterized by B2C e-commerce, we now see an increasing number of C2C platforms: private persons share goods and services in large scale peer-to-peer networks. Ebay, for instance, may be regarded as one of the early pioneers in provisioning and managing such a C2C market platform. The spectrum of sharing activities nowadays shifts from mere resale of spare goods to other forms (e.g., co-usage and renting. The proponents of these markets often claim that they offer a more social and sustainable alternative to traditional forms of consumption.

Knowledge of the factors that are used to determine pricing help us to better understand its functioning. The factors that determine prices on the apartment sharing platform Airbnb can be determined with the standard regression analysis [6]. The prices are set by the individual providers, usually private persons, and thus reflect a wide range of influences. The model explains app. 35% of the listing prices' variation, which includes size and location of the place as well as city-specific aspects like population and the general rent price level. The approach allows to obtain such insights into city structures as listing density and spatial price variations.

Another key issue in sharing economy is trust. Using the case of Airbnb the formation of trust is discussed in [7]. The authors propose a model that captures trust-relevant factors such as the hosts' ratings, activity, and trustworthiness as conveyed by their profile pictures and links those factors to booking intention, i.e., the economic manifestation of trust.

4 Emotions in Markets

It is common consensus by now that emotions play a large part in decision-making. Determining the exact circumstances under which certain emotions arise and how they shape the behavior of market participants, however, is still a relatively novel research field today. Inexpensive and small sensors, which are now commonly available, can be used to help consumers improve their decision making, e.g. in a purchase situation. In turn, market providers may use this information to improve their mechanisms, e.g., to choose less (or more) stress- or excitement-inducing auction formats.

In the context of C2C market platforms, a research model of the relationships between cues to trust, trust, emotions and purchase intention in order to increase understanding of C2C market stability is proposed [8]. Cues to trust are further differentiated as heuristic cues—e.g. interest similarity between consumers—or independent cues which have no connection to the consumer's actual decision or purchase situation (e.g. shared birthdays). One major part in the proposed research is better understanding the role of emotions in cue processing and trust formation.

Further developments in this area include adaptive systems with biofeedback applications which adapt to individual and situational consumer needs [8]. Such systems pose new challenges for businesses in terms of data analysis and market engineering: Structuring, processing and interpreting consumers' behavioral data enhanced with biodata is a complex task. Designing stable markets and systems based on such data requires highly skilled analysts and (adaptive) analytics systems.

A framework for integrating NeuroIS methods into business analytics to improve corporate processing and analyze large volumes of consumer and market data is proposed in [9]. The authors suggest adaptive analytics systems, e.g., systems based on biofeedback, to help business analysts improve their decision-making skills.

References

1. Kamiński, B., Kersten, G.E., Szapiro, T. (eds.) GDN 2015. LNBIP, vol. 218. Springer, Heidelberg (2015)
2. Carbonneau, R., Rustam V.: Back-end bidding for front-end negotiation: a model. In: [1]
3. Paulsen, P., Bichler, M.: First-price package auctions in a principal-agent environment. In: [1]
4. Filzmoser, M.: Lot-rolling – supply chain negotiation in a two-stage multi-echelon system. In: Kamiński, B., Kersten, G.E., Szapiro, T. (eds.) GDN 2015. LNBIP, vol. 218, pp. 395–401. Springer, Heidelberg (2015)
5. Kersten, G.E.: Procurement auctions: improving efficient winning bids through multi-bilateral negotiations. In: Kamiński, B., Kersten, G.E., Szapiro, T. (eds.) GDN 2015. LNBIP, vol. 218, pp. 403–416. Springer, Heidelberg (2015)
6. Kriegisch, N.: Airbnb: how the characteristics of listings drive the listings' prices. In: [1]
7. Hawlitschek, F., Lippert, F.: Whom to trust? assessing the role of profile pictures on sharing economy platforms. In: [1]

8. Lux, E., Hawlitschek, F., Hariharan, A., Adam, M.: Happy birthday! emotions and cues to trust on consumer-to-consumer market platforms. In: [1]
9. Hariharan, A., Kunze von Bischhoffshausen, J., Adam, M.: Leveraging the potential of NeuroIS for business analytics. In: [1]

Contents

Formal Models

Voting and Collective Decision-Making

Conflict Resolution in Energy and Environmental Management

Negotiation Support Systems and Studies

Online Collaboration and Competition

Market Mechanisms and Their Users

L   Contents

Group Problem Structuring and Negotiation

Warsaw School of Economics

Effects of Small Group Discussion: Case Study of Community Disaster Risk Management in Japan

Madoka Chosokabe[1](✉), Yukino Tsuguchi[2], Hiroyuki Sakakibara[3],
Takanobu Nakayama[4], Shota Mine[5], Daisuke Kamiya[5],
Ryo Yamanaka[6], and Toshiaki Miyaguni[6]

[1] Wakayama University, Wakayama, Japan
madoka.chosokabe@gmail.com
[2] Kyushu Regional Development Bureau, Kyushu, Japan
[3] Yamaguchi University, Yamaguchi, Japan
[4] Ibaraki Prefectural Government, Mito, Japan
[5] University of the Ryukyus, Nishihara, Japan
[6] Chuo Kensetsu Consultant Co., Ltd., Urasoe, Japan

Abstract. Community governance needs a small group discussion among community people to identify their concerns, to share them each other and to generate better alternatives for solving problems. A planner should manage the discussion to achieve these objectives. This study analyzed the small group discussion in the community disaster risk management by using text mining. Correspondence analysis was applied to the text data of the discussion. Analytical results revealed the characteristics and effects of small group discussion.

Keywords: Community governance · WS discussion · Disaster risk management · Correspondence analysis · Text mining

1 Introduction

Community governance is an appropriate process of decision-making used to solve problems in a community. Somerville [9], referring to Clarke and Stewart [4, 5], describes community governance as giving the public the right to participate in, and wherever possible, determine issues affecting a community through direct control and through such institutions as neighborhood forums or community councils. Bowles and Gintis [1] state that communities play a role in good governance because they address certain problems that cannot be handled either by individuals acting alone or by markets and governments. Members of community must participate in solving the problems in such areas as environment, transportation, and disaster risk. Community governance, therefore can achieve sustainable community development.

Small group discussion such as workshop (WS) discussions have an important role in community governance. First, such discussions can identify the community's concerns and needs, on the one hand, and the information requirements, on the other hand. WS discussions, therefore, help to determine the problems that exist in the community. Second, these discussions facilitate the possibility for community members to share

B. Kamiński et al. (Eds.): GDN 2015, LNBIP 218, pp. 3–12, 2015.
DOI: 10.1007/978-3-319-19515-5_1

their concerns, needs and information. Lastly, small group discussions may generate a set of alternative solutions for the community. Prior to decision making, such discussion are very important in identifying the problems, and generating a set of alternative solutions for the community.

Generation of better alternative solutions depends on the management of these types of discussions. Planners should facilitate the discussion to reveal the concerns, needs and information shared in the community. Information from not only community people but also planners is necessary to generate better alternatives for community. Planners are expected to provide people in their communities with critical and professional information which they do not know, but which is important for solving the problems that the community faces.

Understanding the characteristics of the small group discussion will aid the planner in managing these types of discussions. It is important to analyze what the members of the community discuss and understand their shared concerns and the information that is presented. This type of analysis help to develop plans for the management of future small group discussions.

This study focuses on the communicative process among members of a community and the process between members of a community and planners. The purpose of this study is to clarify the similarities and the differences of concerns between these two groups. An analysis of the content of small group discussion and the effects of small group discussions in community governance is done. Small group discussions were held for disaster risk management in the Japanese community and the process of discussion was observed and analyzed.

2 Community Governance and Disaster Risk Management

2.1 Community Disaster Risk Management of Kunigami Village in Okinawa, Japan

Community governance is extremely important in disaster risk management in Japan. Two types of disaster prevention plans have been implemented in Japan. The first is the "Basic Disaster Management Plan" which is a comprehensive and long-term plan at the national level. The second is the "Local Disaster Management Plan (LDMP)" which is developed by each prefectural and municipal disaster management council, subject to local circumstances and based on the Basic Disaster Management [2].

After the Great East Japan Earthquake in 2011, the Japanese government devised a new plan named the "Community Disaster Management Plan (CDMP)" in 2013. This plan supports the residents and the employers to develop a disaster management plan for their community. The aim of CDMP is to promote the enhancement of disaster reduction activities at the community level. This plan is expected to make members of the community help themselves and each other should a disaster occur.

The basic principles of CDMP are:

1. CDMP should be made by members of the community using a bottom-up approach;
2. CDMP should depend on the social, environmental and geographical characteristics of a community;

3. CDMP should promote activities depending on the disaster situation; and
4. CDMP should improve community preparedness continuously.

These principles imply the function of community governance in disaster risk management.

The Kunigami village is located in the north of Okinawa Island in Japan. There are 20 districts within Kunigami village. The municipal government of Kunigami started to review LDMP in 2013 before CDMP were devised. The municipal government held workshop discussions at Yona district in 2013.

2.2   Basic Model of Community Governance

Chosokabe et al. [3] proposed a basic model of community governance as shown in Fig. 1. The set of alternatives is constrained by the social context (I). A community chooses an alternative from the set of alternatives (II) and subsequently implements it (III). The result of the implementation, in turn, affects the social context (IV).

In disaster risk management, members of the community need to identify their strengths and weaknesses on disasters at Phase I. They should understand the geographical, meteorological, historical and social characteristics of their community, such as the critical location, evacuation route and people needing assistance should an earthquake and/or tsunami happen. The planners are expected to give information about the disaster and its location, as well as safety recommendations. The small group discussions help them to learn and share the information with the community.

The contents of discussion should be revealed (in a way that both the members of the community and the planner can understand it) in Phase I. The observable contents help them to understand what information is and is not available. Analysis of the text data of the discussion gives a better understanding of the process of community governance.

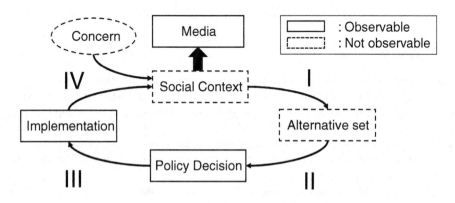

Fig. 1. The basic model of community governance [3].

3   Methodology

The purpose of the analysis is to clarify what members of the community (participants) and the planners tend to express in these small group discussions. The analysis is applied to explain the similarities and the differences between what each participant mentioned and what planners mentioned in the discussion.

Correspondence analysis is applied to the text data of WS discussions. Correspondence analysis is a multivariate analysis technique for exploring cross tabular data by converting such tables into graphical displays, called 'maps', and related numerical statistics [6]. It is one of the easiest to use methods for classification of text data [7]. This study focused on what a speaker mentioned and what groups of speakers mentioned. The method is expected to reveal characteristics of the words used by each speaker and each group of speakers.

Figure 2 shows the process of analysis, which includes four phases. The phases are: (i) making a transcript from voice data; (ii) applying morphological analysis to the transcript and extracting nouns, adjectives and verbs from it; (iii) calculating the frequency of words used for each speaker or each group of speakers and making cross tabulation table; (iv) applying correspondence analysis to the table. Correspondence analysis is applied to the cross tabulation table. The row header represents each word, and column header represents the name of the speaker or group. The table shows how many times the speaker or the group mentioned the word in the discussion.

Fig. 2. The four steps of the analysis process.

4   Results

4.1   Data Collection

WS discussions were held at three times for residents at Yona district in 2013. 14 residents, 9 males and 5 females, participated in the discussion. Participants were sorted into three groups for discussion. Group A and B were male groups and group C was a female group. 7 planners, 4 university staff members, 2 engineering consultants and a Kunigami village official, facilitated each group discussion.

This study analyzed data from 6 dialogues of each group and each WS discussion. Table 1 shows the details of the data.

This study also focused on what each speaker remarked at the discussion and analyzed the remarks of the members of group A; resident a, b, c and facilitator (Fa) as shown in Table 2.

Table 1. The number of utterances of the groups at the 1st and the 2nd discussions. (1A refers to data from group A at the 1st WS discussion, 1B to data from group B at the 2nd, etc.)

	1A	1B	1C	2A	2B	2C
Residents	5	4	4	4	4	4
Facilitators	2	2	3	2	2	2
Utterances	900	765	844	1327	1841	1659

Table 2. The number of utterances of each group member (a, b and c) and the facilitator (Fa) at the 1st and the 2nd discussions

Speaker	1a	1b	1c	1Fa	2a	2b	2c	2Fa
Utterances	82	170	76	162	325	267	177	253

4.2   Speaker-Based Analysis

Figure 3 shows the result of correspondence analysis for each speaker and each WS discussion. The analysis applied to the table whose row headers represent words and column headers represent speakers at the 1st and the 2nd WS discussions. In Fig. 3, circular markers represent the words and "x" marks represent the speakers. For example, "1Fa" represents the facilitator at the 1st discussion and "2a" represents the speaker "a" at the 2nd discussion. In this analysis, the remarks of two facilitators are integrated. The top 73 words in frequency at the 1st and the 2nd discussions were used for the analysis.

Vertical axis divided the "x" marks into the 1st discussion (Left) and 2nd discussion (Right). Horizontal axis tended to divide them into the residents (Upper) and the facilitator (Bottom). There is the distance between the "x" marks of the speakers at the 1st discussion, especially between speaker "a" and facilitator. The results show that the words remarked by each speaker were different. The examples of the words each speaker especially tended to remark at the 1st discussion were given as follows.

- Speaker a: "*Adan*," "rice field" and "river." "*Adan*" is a species of Pandanus that grows in the local area. He tended to express the information of the plants in the local area.
- Speakers b and c: "different," "place," "sand" and "clogged." These words suggest the contents "the sand-clogged location in the local area".
- Facilitator: "evacuation," "remark," "tsunami," "government," "earthquake" and "drill." These words suggest "disaster prevention" when earthquake and tsunami occur.

On the other hand, the distance between "x" marks became to be close at the 2nd discussion. The examples of the words each speaker especially tended to remark were given as follows.

- Speakers a, b and c: "concrete," "walk," "big," "family name," "cooperative store," "healthy" and "vacant house." These words suggested the detailed information about the community such as the place and peoples' names or conditions.
- Facilitator: "high," "leave," "association" and "community center."

The result of speaker-based analysis suggested the following two points.

1. The residents informed each other about risky locations and about weak people at the time of a disaster ("typhoon"). On the other hand, the facilitators (planners) informed residents about the different types of disasters (ex. "earthquake," "tsunami") and the management of the community for the disaster planning ("evacuation," "drill").
2. At the 2nd discussion, the distance between the residents and the facilitator was reduced. The transition showed that the residents and facilitators used more similar words than at the 1st discussion.

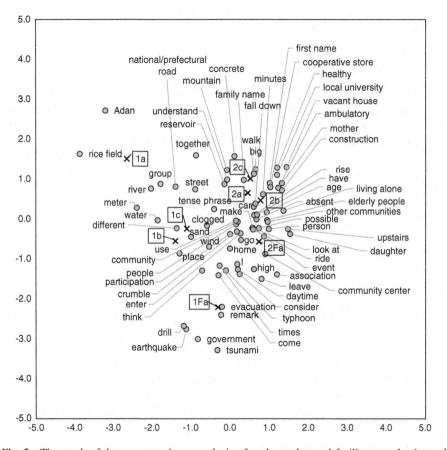

Fig. 3. The result of the correspondence analysis of each speaker and facilitator at the 1st and 2nd discussion.

4.3 Group-Based Analysis

The results of the correspondence analysis for each group and each WS discussion are shown in Fig. 4. The analysis was applied to the table whose row headers represent words and column headers represent groups at the 1st and the 2nd WS discussions.

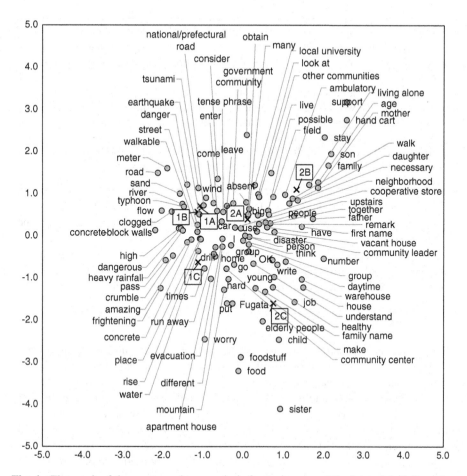

Fig. 4. The result of the correspondence analysis for each group at the 1st and 2nd discussion.

Circular markers represent the words and "x" marks represent the groups. The top 105 words in frequency at the 1st and the 2nd discussions were used for the analysis.

The vertical axis divided the "x" marks into the 1st (left) and the 2nd (right) discussion. The horizontal axis divided them into groups A and B (male groups) (top) and the group C (female group) (bottom). Examples of words each group tended to use are given below.

These words suggest that Group C focused on food preparedness during a time of disaster and on vulnerable people such as children and elderly people.

The contents discussed by groups A and B (male group) at the 1st discussion were similar to each other as seen in the close distance of their points (shown in Fig. 4). On the other hand, the point of the group C, female group, was far from the points of groups A and B. The results of this group-based analysis suggest the following:

- Groups A and B at the 1st discussion: "meter," "road," "sand," "danger," "earthquake," "tsunami," "river," "typhoon" and "national/prefectural road." These words suggest the types of disasters and possible damage in disaster situations.
- Group B at the 2nd discussion: "support," "hand cart," "stay," "son" and "family." These words suggest that Group B focused on the people who need support to evacuate at the time of a disaster, such as elderly people who were living alone and the physically disabled.
- Group C at the 1st and the 2nd discussion: "worry," "food," "foodstuff," "sister," "child" and "elderly people." "Sister" means "elderly female" in their community.

1. The male and female groups spoke from different viewpoints at the first discussion. For example, the female group (Group C) focused on food-related issues at the time of disaster.
2. Each group discussed from a different viewpoint at the 2nd discussion. The groups discussed detailed information such as, associations (group A), individual information of weak people (group B) and food-related issues (group C).

5 Discussion

This study applied correspondence analysis to the text data of small group discussion on the disaster risk management. Speaker-based analysis showed the different viewpoints on a disaster between the residents and the planners. The residents especially mentioned the risky place and the weak people at the time of typhoon disaster in their community. In other words they showed the detailed information about their community, which they only knew but the planners did not know. In contrast, the facilitators (planners) mentioned the risk management for an earthquake disaster and a tsunami disaster. Facilitators reminded the residents that they needed to prepare against these disasters. For example, the residents knew the risky places at the flood disaster but they didn't know how to reduce such risks until the next disaster occurred.

Speaker-based analysis also revealed the process of mitigating the perception gaps between the residents and planners. The distance of the "x" marks between the residents and the facilitator became to be close at the 2nd discussion (Fig. 3). The transition suggested that they became to mention more similar topics than they mentioned at the 1st discussion. In other words, they had shared their information and concerns each other through the discussion.

Group-based analysis clearly revealed the different viewpoints between male and female group. The male groups (Group A and B) mentioned the types of disasters and the damage in disaster situations. In contrast, the female group (Group C) especially mentioned the foodstuff at the time of disaster. The analysis also revealed the different viewpoints of each group at the 2nd discussion. The plotted words suggested that each group discussed the details about the disaster such as association (group A), individual information of weak people (group B) and foodstuff (group C) in Fig. 4.

Fishkin [8] points out that the quality of the deliberative process can be explained in terms of five conditions; (a) information, (b) substantive balance, (c) diversity,

(d) conscientiousness and (e) equal consideration. The five conditions are helpful and important to evaluate the effect of small group discussion. The analyses can help the evaluation of the conditions (a), (b) and (c).

The position and distance of plotted word by correspondence analysis revealed the similarities and differences of information at the discussion. The plotted words suggest many different kinds of information mentioned by various participants such as a male and female resident and a planner. The planner can evaluate a diversity of information and can also control the balance of information from the planner's viewpoints based on the plotted words. However, the analyses cannot evaluate the conditions (d) and (e). The other scales are needed to reveal conscientiousness and equal consideration of a speaker at a discussion.

6   Conclusion

This study analyzed the contents of small group discussions on disaster risk management in a Japanese local community. The correspondence analysis laid out the words remarked by each resident and facilitator (as seen in Figs. 3 and 4). The results revealed (i) the different and various information shared by the residents and facilitator (planner), (ii) the process of sharing information between residents and facilitator and (iii) the different viewpoints between male and female residents.

The plotted words characterized the contents of the discussion by each speaker or group. The different and various words were observed during remarks of each speaker and each group. The residents informed each other from their own viewpoints that had not been shared before. It was also observed that the viewpoints were different between male and female residents. The observations showed the effects of small group discussions.

The changes of distances between the residents and facilitator, as shown in Fig. 3, suggest the process of sharing information and concerns through discussion. In disaster risk management, the information from the planner is important because the residents have little information about the disaster risk. In contrast, the planner does not have enough local information, such as, risky locations and frail people who may need extra attention at the time of a disaster. The results revealed the information by analyzing the text data from these discussions.

The results suggest the small group discussions help the communicative process between residents and planner. Our future work will focus on examining other patterns of information sharing among speakers, testing the effectiveness of our method from the viewpoints of speakers and eventually using our method for facilitating discussions.

References

1. Bowles, S., Gintis, H.: Social capital and community governance. Econ. J. **112**(483), 419–436 (2002)
2. Cabinet Office, Government of Japan: Disaster Management in Japan. http://www.bousai.go. jp/1info/pdf/saigaipanf_e.pdf (2011)

3. Chosokabe, M., Takeyoshi, H., Sakakibara, H.: Study on temporal change of social context: in the case of bicycle riding issue in Japan. In: GDN Proceedings, LNBIP, vol. 180, pp. 315–322 (2014)
4. Clarke, M., Stewart, J.: The local authority and the new community governance. Reg. Stud. **28** (2), 201–219 (1994)
5. Clarke, M., Stewart, J.: Handling the wicked issues: a challenge for government. Discussion paper, Birmingham, School of Public Policy, University of Birmingham (1997)
6. Greenacre, M., Blasius, J.: Correspondence Analysis in the Social Sciences. Academic Press, London (1994)
7. Ishikawa, S., Maeda, T., Yamazaki, M. (eds.): Statistics guide for language study. Kuroshio, Tokyo (2010). (in Japanese)
8. Fishkin, J.S.: When the People Speak: Deliberative Democracy and Public Consultation. Oxford University Press, Oxford (2009)
9. Somerville, P.: Community governance and democracy. Policy Polit. **33**(1), 117–144 (2005)

Understanding PSM Interventions Through Sense-Making and the Mangle of Practice Lens

Leroy White[1(✉)], Mike Yearworth[2], and Katharina Burger[2]

[1] Warwick Business School, University of Warwick, Coventry, UK
Leroy.white@wbs.ac.uk
[2] Faculty of Engineering, University of Bristol, Bristol, UK
{Mike.yearworth,Katharina.burger}@bris.ac.uk

Abstract. In this paper we seek to understand how individuals, as part of a group facilitated modelling setting, commit themselves to a set of actions, as a basis of sense-making, sense-giving and coordinated actions. For this we introduce Pickering's Mangle of Practice to understand the practice of a group facilitated modelling setting. Using video data from a group modelling building exercise, we analyze how individual actors framed their circumstances in communication with one another and how through facilitated model building this affected their subsequent interpretation and decisions as the process unfolds. We show how, through the models as objects enhanced the interaction between verbal communication, expressed and felt emotion and material cues led to collective behavior within the group. With our study we extend prior research and elaborate on the role of objects and materiality as part of group decision making.

Keywords: Group decision making · Problem structuring methods · Sense-making · The mangle · Collective behavior

1 Introduction

Understanding the impact of PSM interventions is problematic because they are complex settings involving the interaction of actors and modelling devices [1]. Specifically, since a great deal of OR interventions are one-off and temporary, it becomes necessary to devise techniques to ensure an appropriate evaluation of the efficacy of the approaches [2, 3].

In this paper, we introduce and explore the use the use of sense-making and sense giving as one means to study interventions as complex interactions of people, models and context. We take as our example a case study on the participatory planning of Smart City experiments for energy efficient city district redevelopment. We apply PSMs in the case, but we are faced with the question: Is there a means for understanding how individuals working together perform effectively as an ensemble? As with all PSM methods the process of the intervention is conducted in a group, where the process is consultative and iterative. Behaviorally, the process provides a succession of models delivering different perspectives, which contribute to a deepening understanding of the problem as new insight emerges. Also, the process uses the sense

© Springer International Publishing Switzerland 2015
B. Kamiński et al. (Eds.): GDN 2015, LNBIP 218, pp. 13–27, 2015.
DOI: 10.1007/978-3-319-19515-5_2

of unease among the problem owners about the present representation of the problem as a signal that further modelling may be needed. However, the idea that PSMs and their models can mediate behavior within groups is not a new idea. This has been acknowledged to some degree, for example by [4] who suggested that models in soft OR represents a facilitative device. Also, Franco and Montibeller [5] discuss models as boundary objects. However, these are examples of a loose coupling of models and the actual situation and therefore it is difficult to infer any theory of behavior through representation. Recently, Ackermann and Eden [6] suggested that the PSM is a process of collective sharing, understanding, and negotiation. They explored this via principles suggested by Fisher and Ury [7], and show how PSMs enable cognitive and social negotiation. We build on this research and suggest a sense-making approach may be appropriate.

Addressing how sense-making and structuring shape the outcome of a PSM intervention is important because we will be better able to understand the processes of PSMs. In doing so, our aim is to understand why attempts to enact PSM interventions often fail to bring about desired results or rarely lead to the substantive claims that are intended, but in many cases lead to unintended outcomes. We draw on the Mangle of Practice (henceforth called the Mangle) proposed by Ormerod [8] and based on Pickering [9, 10]) to conceptualize the interactions within a PSM intervention, and thus we explore the multi-dimensional nature of PSM interventions.

Thus, our paper makes the following contributions. First, we build on recent interest in behavior and OR, as a means for understanding how individuals working together perform effectively as an ensemble through the mediating role of the model. Second, we contribute to the literature on methods to evaluate the process of OR, by employing the concepts from the Mangle [9, 10]) to explain the complex outcomes of (collective) OR processes, namely extended learning. Finally, many scholars of OR will agree that the nature of the link between OR processes and outcomes has yet to be definitively proven. In our paper we test the idea that OR interventions creates the conditions for collective behavior evidenced through a sense-making approach [6, 11].

2 Theoretical Considerations

Over the last 20 years or so, OR scholars have devoted significant attention to understanding processes that shape interventions [4]. In this context, PSM interventions may be conceptualized as creating small scale test beds for understanding how better collective decision-making may arise. An increasing interest in gaining access to understanding the dynamics of PSM interventions in situ has led to important insights regarding theory, behavior and outcomes pertaining to (particularly soft) OR processes [2, 3, 12, 13]. However, significant methodological and epistemological challenges remain in the study of OR interventions [3, 14].

Relatedly, scholars have examined how the importance of a theory driven approach helps to understand how patterns of interactions shape the PSM process [3, 15]. This line of research acknowledges that actors through sense-making shape the interactions which in turn shape the structuring of the problem and so on, but there remain two important gaps. First, existing studies have focused on actors' individual characteristics

[16] and, in doing so, have not examined how the group setting shape sense-making and structuring. Second, studies have neglected the multidimensional and interactive nature of PSMs' contexts, typically examining only single dimensions such as group membership. Hence the richness and importance of the interactions remains largely empirically unexplored and under-theorized. To help address these gaps we draw on sense-making concepts and employ Pickering's notion of the Mangle to understand the sense-making structuring cycle in PSM interventions.

2.1 Sense-Making

The concept of sense-making and sense-giving derive from Weick's work [17], with numerous scholars highlighting they are key elements of the problem structuring process (e.g. [3, 6, 15]). Sense-making is the primary site where meanings materialize to inform, and constrain identity and action [17]. It is described as a process which is 'retrospective, social, on-going, and driven by plausibility' [17]. In contrast, sense-giving, a sense-making variant (Weick et al., 2005), involves attempts to influence the sense-making and meaning construction of others towards a preferred re-definition of reality [18, 19]. As such, we suggest that sense-making and sense-giving are concepts central to PSM interventions, particularly in the face of increased complexity, ambiguity and uncertainty that contemporary organizations face, where the need to create and maintain coherent understandings that sustain relationships and enable collective action is especially important and challenging [17].

Weick [17] argues that sense-making and sense-giving are important in any context where there is a need to create and maintain coherent understandings that sustain relationships and enable collective action. Similarly, Maitlis and Lawrence [19] suggest that sense-making and sense-giving are triggered in a broad range of contexts, particularly in environments characterized by uncertainty and complexity, and where issues are deemed to be significant to stakeholders. Therefore, the need for sense-making, and hence the potential for sense-giving, is often heightened under conditions of equivocality [17]. In line with this, we examine the process shaping of a PSM intervention drawing on a case where models were used to increase the group's sense-making. We highlight that the literature is cognitive in orientation, in that on the one hand, through sense-giving efforts, formal group members can work to enact shared meaning to other group members as a basis for organized action [20]. And on the other hand, recipients of sense-making activities do not merely accept new ideas, rather they interpret them through their existing cognitive frames [21].

Sense-making is not to studying OR interventions [6]. In their earlier work, Eden and Ackermann provided guidance on how to explore the relationship between users' sense-making and the negotiation by organizational members drawing on cognitive mapping [22]. Cognitive mapping was a useful approach for eliciting and clarifying users' sensemaking in negotiation [23]). Recently, within a communication frame, Franco [15] perceived structuring as a process of sense-making framework in a cyclical pattern indicating a loop where, as the problem is being structured, participating individuals engage in the sense-making of the problem, and as the change in their understanding is achieved, individuals engage in further structuring. The result of the cyclical pattern of structuring and sense-making is an agreed accommodation of an understanding of the problem. We regard this as sense-making in the first order. While important, it does not entirely lead us to critically and analytically understand the processes at play in how accommodation is reached. Thus we introduce the Mangle as a socio-cultural learning theory to understand sense-making in the second order.

2.2   Pickering's The Mangle

Ormerod has brought to the attention of the OR community Pickering's theoretical work on the 'Mangle of Practice' [8–10] Following Ormerod's interpretation, Pickering's contribution was to move beyond the representational idiom of understanding 'science-as-knowledge' towards an explanation of scientific practice and culture using a 'performative idiom'. Developed from Latour's Actor Network Theory (ANT) [24] this new idiom does not disregard science-as-knowledge but enhances this perspective with an understanding of science as "*the field of powers, capacities, and performances situated in machinic captures of material agencies*" [10]. Whilst Pickering was primarily concerned with moving the sociology of scientific knowledge (SSK) to a new theoretical foundation, Ormerod's insight was to see that the Mangle was equally applicable to understanding OR interventions. What then is the Mangle, this performative idiom, and how does it apply to improving our understanding of OR interventions?

The Mangle describes the constant interplay between material and human agency. Pickering's break with pure ANT was the recognition that human agency and material agency are not equivalent things, that human agency is imbued with purpose, whereas material agency is not so; "*Human intentionality…appears to have no counterpart in the material realm*" [10]. This break in symmetry opens up an interpretation of scientific endeavor as the constant drive, the human purpose of science, to wrest knowledge from material whose intrinsic agency can be viewed as resistance to our attempts, the "*dance of agency*" [10]. The Mangle is not about the knowledge that we gain, although this is the ultimate purpose of science, but the *narrative* of the struggle to arrive at that knowledge. The Mangle is a wonderfully descriptive term for this.

Ormerod picks up on this narrative element and reminds us that the practice of 'normal' scientific publishing, including OR, discourages practitioners from writing about the trials and labors to obtain results. Our outputs are generally sanitized accounts of methodology and results. Ormerod's thesis is that the OR practitioner can learn more by reading about the details of how other practitioners obtained their results. His conclusion is a plea for "*more informative case studies of 'technical' projects*". Thus as OR practitioners we should be just as concerned with the analyses of the *process of OR* as in the results of the actual interventions. Perhaps more so since we use OR in the realm of 'wicked problems' [25] or 'messy ones' [26] such as in the case we analyze.

We thus suggest that in order to understand the effectiveness of PSM interventions in the realm of tackling 'wicked problems' it is necessary to gain a deeper understanding of the dynamics of the sense-making processes in groups [15, 17]. More specifically, a critical examination of "sense-making [as] mental activity which involves the interpretation and understanding of [problem structuring as an articulating framework], and the actions that seem to be suggested by it, mean for an individual in relation to the world in which he/she acts" [15]. We propose sense-making to study the micro processes that form the evolving shape of the collective PSM intervention, which following Pickering we regard as a *narrative* of the struggle to arrive at that knowledge and to sustain the affective aspects of an intervention.

The Mangle provides a theoretical lens for understanding the sense-making in the co-construction of object-oriented agency, which PSMs are an example [27]. The Mangle also presents us with the opportunity to re-establish the sociological underpinnings of PSMs. As an illustration, Checkland originally developed SSM without reference to any explicit theory [28]. He did however review SSM against Churchman's enquiring systems and Vickers' appreciative systems theories leading to a formulation of SSM as an *"...enquiring system whose mode of operation provides a formal means of initiating and consciously reflecting upon the social process of 'appreciation'"*. From this Checkland concluded that SSM is not a version of functionalism but rather a *"phenomenological investigation into the meanings which actors in a situation attribute to the reality they perceive"* and thus to the *"philosophical/ sociological tradition of interpretive social science"* [27]. We do not need to delve further into interpretivism (as did Checkland) but seek instead links between Checkland's view of SSM (and hence of perhaps all PSMs) as phenomenological investigations into meanings and Pickering's performative idiom. This would simplify things enormously. We believe the link exists; Pickering makes specific reference to the need for *"phenomenal accounts"* as necessary *"conceptualizations of the aspects of the material world"* that are supported by experimentation [10].

We now move on to empirically examine through the Mangle how sense-making and sense-giving underpin PSM interventions. Prior to this, we set out details of our case context and research design.

3 The Case Study: The STEEP Project

Our case study is based on the STEEP (Systems Thinking for Energy Efficient Planning) project, which is an EU FP7 project that seeks to identify innovative policy experiments for city district energy planning in Bristol, South West England and the partnering cities.[1]

The context for the case is the growing concern for securing sustainable, reliable, and affordable energy systems in EU-countries that seek to address climate change risks by meeting EU 2020 carbon reduction targets. Opportunities for change towards lower carbon energy systems arise from the convergence of ubiquitous IT systems with decentralised energy technologies that create new complementarities of small-scale, local technologies with traditional networked infrastructures. Cities are perceived to be ideal test beds due to their limited scale, their diversity hence opportunities for learning about the complexity of socio-cultural practice change that accompanies technology transitions [29].

---

[1] Further details can be found at http://smartsteep.eu.

The City of Bristol, in South-West England, is one of the partnering cities in the project, and a representative of the local council manages the project as part of the council's smart city programme portfolio. In its Smart City Programme, the City Council formulates that aim that Bristol aims to be in the top 20 European cities by 2020 and has made a clear commitment to create a world-class and inclusive green-digital economy. The aim is to "...*use smart technologies to meet our ambitious target to reduce CO_2 emissions by 40 % by 2020 from a 2005 baseline, as well as our social and economic objectives*" [39].

The project partners in Bristol comprised the local university, an engineering consultancy, a third sector organisation with expertise in energy modelling, and the local council. The project proposal document states that the project's specific objectives are

- To enable all participants cities and partners to learn from the successful and unsuccessful experiences of other cities and experts
- To integrate all stakeholders in smart city plan definition: public administrations, policy makers, technology providers, financial organisations, enterprises and citizens
- To better understand the complex energy, resources, social and economic flows and their relationships
- To have a clear picture on the number, effectiveness, cost and interdependence of the possible smart city interventions and projects
- To disseminate or application plan to other similar cities at the European scale

The Bristol Temple Quarter Enterprise Zone (TQEZ) was chosen at the city district for the STEEP approach in Bristol. The TQEZ is a designated regeneration area that aims to attract businesses through reduced business (tax) rates, encourages development through a relaxed planning application processes, and enables regeneration through enabling infrastructure such as investment in transport and heating systems.

3.1 PSM Workshop Description

Group model building workshops were held as part of the STEEP project to facilitate the exploration of aspects relevant to systemic energy planning for the District Modelling of Bristol's TQEZ.

STEEP employed a form of Soft Systems Methodology [30] and Hierarchical Process Modelling (HPM) [31, 32] as a Problem Structuring Method following the generic constitutive definition of Yearworth and White [14]. Modelling a transformation as a system using HPM requires a top-level process to be identified that acts as a descriptor, or the purpose of the system [31–33]. A diagrammatic view of the problem structuring method adapted from [31, 32] and originally adapted from [34] is shown in Fig. 1, demonstrating how HPM, which has been enhanced with Issue Based Information System (IBIS) capabilities (referred to in this methodology as Evidential Discourse in ENgineering (EDEN) [35], are integrated in an SSM learning cycle. Detailed information about the methodology in practice can be found at http://smartsteep.eu.

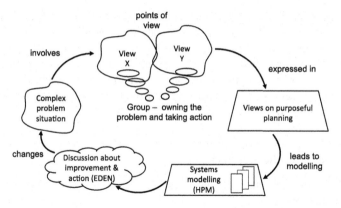

Fig. 1. Diagrammatic view of the problem structuring method. Adapted from [31, 32] and originally adapted from [34]

The workshop was attended by representatives of technology manufacturing companies, infrastructure operators, third-sector organisations with an interest in energy and low carbon development in Bristol, consultancies (multi-disciplinary engineering and architects), local authority employees and University academics. The invitations to participate were sent to a variety of organisations who were known to the project partners as having an interest in redevelopment projects in Bristol. The methodology allowed participants to take a wide view of 'energy', including building types and usage profiles, infrastructure systems and technology, movement/transport mix, thereby considering social practices – the changing nature of work in networking hubs, increasing awareness of sustainable energy behaviours and environmentally, and health-friendly travel choices. The workshop began with a representative from the city council setting the context, followed by an explanation of the methodology by a University academic.

4 Method of Analysis

For the ethnographic fieldwork within our empirical case, observational data was collected through unobtrusive video recordings of the activity during the workshop. Our analysis of the video data focused on key incidents [36]. According to Emerson [36] *"A key incident attracts a particular field researcher's immediate interest, even if what occurred was mundane and ordinary to participants. This 'interest' is not a full-blown, clearly articulated theoretical claim, but a more intuitive, theoretically sensitive conviction that something intriguing has just taken place."* Incidents of interest were defined by (i) two or more participants explaining terminology, models, or other 'tools for thinking' to each other, thereby transferring knowledge and creating a common semantic space, (ii) two or more participants disagreeing about the representation of a problem in the model, engaging in an argumentation process (issues-options-arguments) and (iii) two or more participants scaffolding collective agency through the co-construction (reciprocal contributions) of activities targeted at the resolution of a problem. For brevity we focus on one key incident from the study. The incidents illustrate the theoretical lenses presented in the previous sections, i.e. (i) sense-making and structuring loops [15], (ii) the co-construction of object-oriented agency through a mangle of argumentation processes, and (iii) commitment to coordinated action as an indication of collective behaviour. The key incident presented in this paper is taken from the first STEEP workshop, which took place in March 2014.

4.1   Presentation of the Key Incident

The incident we focus on illustrates the resolution of conflict between two participants, seeking each other out in its resolution and is illustrated in Fig. 2. The role of the model and the method (requiring co-constructed ideas to be brought onto post it notes and paper) is relevant in mediating the conversation [15].

Well I can think of a way of carbon offsetting [...] the offsetting actually on the homes of the people who are working in the area.. you know...
...THAT I will accept..

do you see what I'm saying?
K: *Yes, that was what I was trying to get to*

And, and, and... to come back to that, more K,.., I argue for having the people,... the Green DNA in the building. The DNA has got to be thinking about whole cell. (...) And we were talking about the city region as well[...] At the very least the offsetting has got to be in the city region – but ideally it'd have something to do with the people working in the buildings.

Fig. 2.   Key Incident - Socio-material sense-making and structuring

In our key incident, the first participant strongly disagrees with a suggestion about practice in the enterprise zone, made by another. The second participant, holding the post-it note and thereby expressing his intention to represent (and influence through his wording) the developed ideas, then creates an ultimatum (*we are going to have to do some [offsetting]*) for the first participant, requiring a compromise. This may be seen as an incident demonstrating the structuring of an agreement through the use of an abstract model that extends the problem over a different spatial-temporal realm, invoking materiality outside the workshop: given a shared target (zero carbon), and the perceived collective inability to achieve it in practice in the TQEZ, the model of (carbon offsetting in the community/the region) is employed to produce an acceptable approach for all participants in the group. The extended argument that took place can be represented in an iterative sense-making-structuring process.

4.2   Interpretation of the Incident Through a Combination of Sense-Making and The Mangle

The incident shows a process of identifying a contradiction between the zero carbon vision and the politico-economic purpose of attracting businesses and investment by property developers through the instrument of enterprise zones. In a process of sense-making and structuring (mangling) the participants then find a mutually acceptable way of maintaining the contradicting objectives of zero carbon development and economic friendliness of the investment proposition, by defining a form of 'carbon offsetting in the community' as the shared 'mental model' or tool for thinking. Similar sense-making and structuring processes took place in the different groups regarding different contradictions present in the *problematique* in context. As a result of two STEEP workshops, stakeholders committed themselves to the following top three actions: Understanding property developer business models, Funding models that address local objectives, and Mapping stakeholders [37].

The incident demonstrates how tensions introduced through the existence of a power base (current policy) influence the structuring of the *problematique* (a low carbon zone) and shape problem structure, as well as feasible and desirable commitments to collective behaviour (developing funding models that address local objectives). The material reality of policy thus influences the sense-making process in the workshop through introduction of boundaries, e.g. by suggesting something is unavoidable (we'll have to), and by establishing criteria for a desirable solution (resilient and flexible) in the structuring process. This is further explored in a more detailed application of the mangle to the incident in the following section.

Armed with the Mangle as an analytical framework it is tempting to take it at face value and view the process of OR enacted in the workshop as our target of analysis. The *dance of agency* certainly seems to capture the busy activity in the workshop as participants are modelling. However, we need to be careful here. Pickering's development of the Mangle was in response to a need to re-theorise SSK and originates in an *asymmetric* interpretation of ANT. The dance is in fact the interplay of human *intentional* agency and material agency, which is devoid of intention. This direct interaction between human and material agency is certainly the stuff of the experimental scientist, but not the OR practitioner. Therefore, it would be inappropriate to apply the Mangle as an analytical lens on the interplay of human-to-human agency in the workshop as this would deviate from its theoretical intentions.

The application of the Mangle in this case requires more effort. Certainly we can think in terms of the performative idiom by focussing on the *actions* within the workshop, literally examining the *doing*, the process of OR, and ignoring for now the epistemic gains that resulted. However, how is the interaction with material agency to be brought into the workshop? The answer seems to be that the material world, the eventual implementation of interventions leading to energy efficiencies in the target development zone, are represented in the workshop in the form of a *proxy*. The stakeholders are bringing into the workshop their own expertise and knowledge about the difficulties, challenges and actual physical limitations of achieving the agreed transformation. So each participant represents symbolically, and in terms of actual knowledge, an aspect of the material world, which must be voiced to other workshop participants in order to not mislead the workshop about the feasibility, or otherwise of processes that are being discussed in the modelling as possible answers to the question "*how is this to be achieved?*" Therefore, based on the Mangle the atomic unit of analysis would thus be exchanges of this form

Participant 1 "we should try to do this"
Participant 2 either "yes, this is feasible" or "no, this would not work"

Here, Participant 1 is expressing a possible performative act imbued with human agency and representing a creative, exploratory attempt to shape the world. Participant 2 counters with expertise of the material world and represents it to the group, knowing that the putative action is either possible or not.

The incident presented above illustrates the proxy representation of material agency into the conversation. The participant **P** asserts

> "*If you're going to say that it's a zero carbon development, then you're only going to take on developers who are committed to zero carbon development...*"

and participant **K** responds with

> "*(shaking his head) It's just not possible.. I mean, I love the idea and I think it'd be amazing if you could, but it's just not possible. Understanding planning and unless you want the HCA to go against its own government... Network Rail,... power station...*"

P is taking a position aligned with the original goal of the transformation "*Achieving a zero carbon development of the TQEZ*" by asserting an intention to shape the world that requires developers to be similarly aligned. Participant **K** states the material agency limitation concretely by a very clear assertion that this is not possible. Participant **P** receives this push-back from material agency and repeats it to show that the message has been heard. However, the dance is not over yet. **P** now asserts intentionality again, by saying that the obstacle must be removed.

> "*...So, so, that's interesting, that's good... I want to hear somebody say 'we can't get there from here', and then I would say 'well then we will have to remove the obstacle'*"

From here, the direct physical achievement of zero carbon is now pushed to one side and another aspect of material agency enters the discussion, that of carbon offsetting.

We can thus see that the *dance of agency* that Pickering describes is actually taking place, with the participants behaving like *avatars* of material agency as well as asserting their own intentional agency. On each occasion where material agency needs to assert itself to the group in response to human intentionality, it was enacted in proxy by one of the participants. Of course, we have been liberal in our interpretation of material agency. This is not just the *physical* properties of the world, the known physical laws that govern such things as (say) the possible energy efficiency of a given technical solution. There is also the material agency of such things as available capital, interest rates, regulations, and so forth and in this case the notion of zero carbon and carbon offsetting. They all need to be represented in proxy form at appropriate points in the discussion. The net effect is to shape the overall direction of exploration by the group as it runs up against the material limits of the world it is trying to *shape*. Note that we keep returning to the notion of shaping as an intentional act. This particular project is firmly in the realm of *engineering*, the physical limits in the broad sense discussed here limit the scope of intentional engineering action focussed on the technologies associated with the carbon emissions of the TQEZ development. However, engineering is a necessary but still not sufficient element needed to enact transformation. The project is also undoubtedly political; the scope for taking action is more likely to be constrained by the full gamut of PESTEL realities than by technical feasibility. Whilst human agency was focussed on achieving the desired transformation "*Achieving a zero carbon development of the TQEZ*", the Mangle can thus explain why the group emerged from the first workshop with the modified transformational goal of "*Achieving an operational low carbon development of the TQEZ*". The atomic unit of analysis carried out above illustrated just one of the individual incidents during the workshop where participants' representations of the material limitations exerted themselves in the direction which modelling was taken and thus, eventually, in the change in goal.

5 Discussion

The objectives of our paper were to explore how sense-making and structuring shape an intervention and to understand the processes of PSMs. Ormerod [8] has directed us towards the Mangle as a potential useful analytical device to better understand the process of OR. Ormerod's interpretation of the Mangle led him to propose the need for more informative case studies with an "*emphasis on the interaction through time of material, human and conceptual components of a research programme*". To aid the OR researcher, Ormerod has proposed a set of desirable characteristics of case studies that would help draw out and reveal these interactions. Indeed, the STEEP project has already taken Ormerod's suggestion on board in the design and implementation of its evaluation and have collected a number of such narratives for the benefit of other researchers [37].

However, our contribution is to go beyond this narrative over time and have demonstrated the value of the Mangle for the micro-level analysis of participant interaction in-group model building workshops. This has been achieved by returning to Pickering's original theoretical conception of the Mangle as an asymmetric interpretation of ANT and his proposal for a shift to the ontological performative idiom. In doing so we have recognized that material agency enters the discussion in the workshop in proxy form. Participants represent material agency as avatars and literally push-back on other participants when they recognize that assertions of human intent are not possible based on their expert knowledge of limitations. Pickering's image of the

Mangle as a dance of material and human agency is actually observable, as we have demonstrated in our analysis of the incident from the case.

Reflecting on our micro-level analysis using the Mangle we can see that the notion of the performative idiom provides insight into workshop participants' commitment to action. The 'dance' in fact reveals viable action as the process of participants searching for pathways forward in a complex landscape *bounded* by material agency. Pickering stresses "the temporal *emergence* of plans and goals and their transformability in encounters with material agency" [10]; the avatars of material agency are literally saying "no, not that way" and the path forward emerges as a consequence. This adds further support for our claim that action is an emergent property and is not a result of the facilitator leading or suggesting the way forward.

Consistent with more recent studies (e.g. [14, 38]) our research demonstrates the possibilities of a theory informed view of the micro-processes of PSM practice. We argue therefore that it is important to look beyond group membership to understand how sense-making and structuring shape an intervention and, in doing so, it is possible to describes the conditions that lead to intended and unintended effects. This position also means that it is possible to study the collective intent of an intervention [6], opening up the possibility to study PSMs as collective phenomena. PSMs have been applied widely to scaffold the resolution of multi-voiced, multi-perspectival problem situations. However, their emphasis on consensus seeking dialogue as a requisite principle for change and the associated lack of a critical examination of representations that mediate discussions, results in concerns around their adequacy for contested social innovation processes. Specifically, the possibility of accommodating foundational conflict and the constant struggle for dispositional power amongst the participants involved in integrative negotiation is not explicitly considered. We introduce the Mangle as a specific means to address the multi-dimensional aspect of PSM interventions.

Having used the mangle and sensemaking lenses to gain insight into the dynamics in a micro-episode within a soft OR intervention, it seems relevant to position the episode in the multi-layered context of socio-technical transitions: as part of the workshop, the intervention project (STEEP), the smart city Bristol programme, and the national policy of devolution. In the UK context a traditionally very powerful central government, is beginning to devolve some control to local authorities, thereby increasing their agency to develop locally sustainable solutions for transitions.

The City of Bristol thus pursues its own Smart City Bristol Programme with the aim to be "*in the top 20 European cities [that use smart technologies to help deliver a cleaner environment, a higher quality of life and a vibrant economy] by 2020 [having] made a clear commitment to create a world-class and inclusive green-digital economy*" [39]. Furthermore, in Bristol, voters have expressed their desire to be involved in and shape a 'Bristolian' future by choosing a directly elected Mayor in Bristol.

Thus, considering the sense-making and mangle episode from the perspective of local stakeholders striving for greater self-determination of transition policies, it becomes possible to interpret the collective learning facilitated through the problem structuring workshop in the context of socio-technical transition theory (e.g. [29]). The need to further develop collective steering competence [40] is explicitly stated as the following quote by a representative of an engineering consultancy (also a STEEP project partner) at a University collaboration facilitation meeting exemplifies:

"There is no panacea for the challenges Bristol as a city is facing. A range of multi-agency, multi-faceted programmes is needed to tackle these" (March 2014).

To facilitate the emergence of goal-oriented collective steering competence, the dynamic ability to interact on the basis of collectively developed desirable futures and adaptive collective behaviour through self-evaluation processes, need to be scaffolded (Kemp et al., 2007). The following quote illustrates how sensemaking and mangling in the workshop were effective in identifying shared areas to focus collective effort for change:

"If we hadn't have gone through the process of modelling, if we hadn't have got stakeholders into the room, we wouldn't have discovered that [...] there was lack or lack at the moment of a coherent vision around carbon reduction. So the modelling process itself flagged up what these barriers were, what these issues were, as it is supposed to do as a model. It doesn't assume that there is consensus but it highlights where the issues or the gaps might be in that consensus". (City Council Project Representative at STEEP Consortium meeting, September 2014).

In order for goal-directed collective agency to arise in sustainability transitions, such as for the energy efficient planning of the enterprise zone that is the aim of the STEEP project, processes of *"goal-oriented modulation: between planning and incrementalism"* [40] take place. Hence, as a next step to inform practice change through the STEEP workshop learning processes, goal-oriented collective behaviour is foreseen.

Considering this process in the context of the smart city programme, further evidence for the emergence of this collective steering competence is seen through the increasing number of formalised cross-organisational projects for transitions in Bristol between stakeholders from the local authority, the University, businesses and NGOs. Several real-world' laboratories for socio-technical sensemaking and mangling are being set up with new technologies in people's homes in Bristol (e.g. SPHERE (e-health)) and scaled-up IT test-beds (Bristol is Open). Zooming back in to the micro-episode presented in this paper we thus suggest that it offers a in-depth view of the collective sensemaking and mangling processes that – embedded in programmes of projects with related problem structuring interventions in Bristol – facilitate the development of collective steering competence. Over time, policy influence and integration may thus result from locally developed shared notions, ideas and instruments for transitions that emerged from micro-episodes of the problem structuring in workshops.

As such, collective sensemaking and the mangle may effectively challenge unsustainable routines and practices, especially when the resulting adaptive collective behaviour is maintained in wider programmes of goal-directed problem structuring practice, such as in the context of the collaborative projects cited above, which include developmental monitoring and evaluation processes in the pursuit of a locally sustainable approach to transition management.

Finally, there are various implications and benefits to OR practice that we can see from our work. Our first contribution is that our approach enables a more dynamic interpretation of the problem structuring setting. Rather than taking a static view we can see that the problem context naturally enters dynamically in group model building in a proxy form as participants represent material agency as avatars. Our second contribution has been to show the need for awareness by the facilitator of the possibility of the transformation being modelled changing as individual participants each represent a partial embodiment of material agency, the group thus collectively recognising after some time what limits are placed on their collective intention by material agency. We thus conclude with the suggestion that in order to understand the effectiveness of such complex PSM processes, it is necessary to gain a deeper understanding of the dynamics of negotiations in co-learning processes in groups.

Acknowledgments. This work was supported in part by the EU FP7-ENERGY-SMARTCI-TIES-2012 (314277) project STEEP (Systems Thinking for Comprehensive City Efficient Energy Planning).

References

1. Mingers, J., Rosenhead, J.: Problem structuring methods in action. Eur. J. Oper. Res. **152**, 530–554 (2004)
2. Eden, C.: On evaluating the performance of "wide-band" GDSS's. Eur. J. Oper. Res. **81**, 302–311 (1995)
3. White, L.: Evaluating problem-structuring methods: developing an approach to show the value and effectiveness of PSMs. J. Oper. Res. Soc. **57**, 842–855 (2006)
4. Ackermann, F.: Problem structuring methods "in the Dock": arguing the case for Soft OR. Eur. J. Oper. Res. **219**, 652–658 (2012)
5. Montibeller, G., Franco, A.: Decision and risk analysis for the evaluation of strategic options. In: O'Brien, F.A., Dyson, R.G. (eds.) Supporting Strategy: Frameworks, Methods and Models, pp. 251–284. Wiley, West Sussex (2007)
6. Ackermann, F., Eden, C.: Negotiation in strategy making teams: group support systems and the process of cognitive change. Group Decis. Negot. **20**, 293–314 (2011)
7. Fisher, R., Ury, W.: Getting to yes. Hutchinson, London (1982)
8. Ormerod, R.: The mangle of OR practice: towards more informative case studies of 'technical' projects. J. Oper. Res. Soc. **65**, 1245–1260 (2014)
9. Pickering, A.: The mangle of practice - agency and emergence in the sociology of science. Am. J. Sociology **99**, 559–589 (1993)
10. Pickering, A.: The Mangle of Practice: Time, Agency, and Science. University of Chicago Press, Chicago (1995)
11. Weick, K.E., Roberts, K.H.: Collective mind in organizations: heedful interrelating on flight decks. Adm. Sci. Q. **38**, 357–381 (1993)
12. Connell, N.A.D.: Evaluating Soft OR: some reflections on an apparently "Unsuccessful" implementation using a soft systems methodology (SSM) based approach. J. Oper. Res. Soc. **52**, 150–160 (2001)
13. White, L.: Understanding problem structuring methods interventions. Eur. J. Oper. Res. **199**, 823–833 (2009)
14. Yearworth, M., White, L.: The non-codified use of problem structuring methods and the need for a generic constitutive definition. Eur. J. Oper. Res. **237**, 932–945 (2014)
15. Franco, L.A.: Forms of conversation and problem structuring methods: a conceptual development. J. Oper. Res. Soc. **57**, 813–821 (2006)
16. Franco, L.A., Meadows, M.: Exploring new directions for research in problem structuring methods: on the role of cognitive style. J. Oper. Res. Soc. **58**, 1621–1629 (2007)
17. Weick, K.E., Sutcliffe, K.M., Obstfeld, D.: Organizing and the process of sense-making. Organ. Sci. **16**, 409–421 (2005)

18. Gioia, D.A., Chittipeddi, K.: Sense-making and sense-giving in strategic change initiation. Strategic Manage. J. **12**, 433–448 (1991)
19. Maitlis, S., Lawrence, T.B.: Triggers and enablers of sense-giving in organizations. Acad. Manage. J. **50**, 57–84 (2007)
20. Smircich, L., Morgan, G.: Leadership: the management of meaning. J. Appl. Behav. Sci. **18**, 257–273 (1982)
21. Fiske, S.T., Taylor, S.E.: Social Cognition, 2nd edn., Xviii, 717 pp. Mcgraw-Hill Book Company, New York (1991)
22. Eden, C., Ackermann, F.: Cognitive mapping expert views for policy analysis in the public sector. Eur. J. Oper. Res. **152**, 615–630 (2004)
23. Eden, C., Ackermann, F.: Group decision and negotiation in strategy making. Group Decis. Negot. **10**, 119–140 (2001)
24. Latour, B.: Science in Action: How to Follow Scientists and Engineers Through Society. Harvard University Press, Cambridge (1987)
25. Rittel, H.W., Webber, M.M.: Dilemmas in a general theory of planning. Policy Sci. **4**, 155–169 (1973)
26. Ackoff, R.L.: The art and science of mess management. Interfaces **11**, 20–26 (1981)
27. White, L.: Behavioural Issues in PSMs [WWW Document]. IFORS Conf. Present (2014). http://www.ifors2014.org/files2/program-ifors2014.pdf
28. Checkland, P.: Systems Thinking, Systems Practice: Includes a 30-year Retrospective. Wiley, Chichester (1999)
29. Geels, F.W.: Ontologies, socio-technical transitions (to sustainability), and the multi-level perspective. Res. Policy **39**, 495–510 (2010)
30. Checkland, P., Scholes, J.: Soft systems methodology: a 30-year retrospective (1999)
31. Davis, J., MacDonald, A., White, L.: Problem-structuring methods and project management: an example of stakeholder involvement using Hierarchical Process Modelling methodology. J. Oper. Res. Soc. **61**, 893–904 (2010)
32. Yearworth, M., Schien, D., Burger, K.: D2.1 R1 Energy Master Plan Process Modelling STEEP PROJECT (314277) - Systems Thinking for Comprehensive City Efficient Energy Planning, p. 70. University of Bristol (2014)
33. Yearworth, M., Schien, D., White, L., Burger, K.: Sustainable urban energy planning: a development of problem structuring methodology (2015) (in review)
34. Hindle, G.A.: Case Article—Teaching soft systems methodology and a blueprint for a module. INFORMS Trans. Educ. **12**, 31–42 (2011)
35. Marashi, E., Davis, J.P.: An argumentation-based method for managing complex issues in design of infrastructure systems. Reliab. Eng. Sys. Safe. **91**, 1535–1545 (2006)
36. Emerson, R.: Working with 'key incidents'. In: Seale, C., Gobo, G., Gubrium, J.F., Silverman, D. (eds.) Qualitative Research Practice, pp. 427–442. Sage, London (2004)
37. Yearworth, M.: D2.5 Evaluation STEEP PROJECT (314277) - Systems Thinking for Comprehensive City Efficient Energy Planning, p. 56. University of Bristol (2014)
38. Franco, A.: Rethinking Soft OR interventions: models as boundary objects. Eur. J. Oper. Res. **231**, 720–733 (2013)
39. Bristol City Council: Bristol Smart City Programme [WWW Document] (2012). http://www.greendigitalcharter.eu/wp-content/uploads/2012/05/Smart-City-Bristol-Programme-April-2012-Briefing-Note.doc. Accessed on 27 November 2014
40. Kemp, R., Loorbach, D., Rotmans, J.: Transition management as a model for managing processes of co-evolution towards sustainable development. Int. J. Sust. Dev. World. **14**, 78–91 (2007)

Negotiation and Group Processes

Warsaw School of Economics

Social Consciousness
in Post-conflict Reconstruction

Ben Gans[(⊠)] and Anne-Françoise Rutkowski

Tilburg School of Economics and Management, Tilburg University,
Warandelaan 2, 5037 AB Tilburg, The Netherlands
{B.Gans,A.Rutkowski}@uvt.nl

Abstract. This paper sheds light on the complexities intergovernmental organizations are facing during post-conflict reconstruction. The article discusses the
added-value of Social Responsibility in the context of the Comprehensive
Approach, involving collaboration amongst defense, diplomacy and development. To better understand the role of public-private partnerships in enabling
Corporate Social Responsibility activities we conducted a single case study. The
aim is to better grasp the organizational design of the Comprehensive Approach
as well as to comprehend the type of relations during the decision-making
process. The results of the content analysis of 8 semi-structured interviews with
senior diplomats, military commanders, and civilian entrepreneurs support the
discussion. Particular attention is paid to the existing variety in norms relevant to
the involvement of the private sector, social consciousness, and the potential
role of public-private partnerships in enabling stabilization as well as reconstruction in post-conflict zones. Lessons learned are presented in the conclusion.

Keywords: Collaborative governance · Decision-making · Social Responsibility · Public-private partnerships · Post-conflict reconstruction

1 Introduction

As early as 1932 Dodd argued that "companies like individuals, should strive to be
good corporate citizens by contributing to the community to a greater extent as is
generally required". Since this first impetus, the maximization for private firm of
corporate social activities has been studied. It is recognized as key in improving the
value of the firm [1]. Such initiatives and activities promote societal peace and stability
and range from protection of human rights to education and public health [2–5]. Such
activities require investing in infrastructures such as school buildings, community
clinic, and Information Technology (IT).

Most definitions of Corporate Social Responsibility (CSR) relate directly to the
concept of Social Responsibility (SR). "Social Responsibility (…) implies a public
posture toward society's economic and human resources and a willingness to see that
those resources are used for broad social ends and not simply for the narrowly cir-
cumscribed interests of private persons and firms" [6]. The CSR debate focuses on the
entanglement of maximizing shareholder wealth with the requirements of involving a
range of stakeholders in the decision-making process. CSR initiatives lead to a better

© Springer International Publishing Switzerland 2015
B. Kamiński et al. (Eds.): GDN 2015, LNBIP 218, pp. 31–45, 2015.
DOI: 10.1007/978-3-319-19515-5_3

reputation for private firm loyalty and customer identification [7, 8] and increased human and capital resources [9]. However, other multinational engagements have contributed to public mistrust. For example, Shell's joint venture with the Nigerian government in 1995 illustrates corporate failure to protect and enable civil rights. Furthermore, CSR has failed as a peacebuilding tool in the Democratic Republic of Congo. This likely because the post conflict equation lacked numerous governance prerequisites. Enabling such CSR activities requires the ability to make sense of the environment and to maintain an ongoing dialogue amongst stakeholders. Improvements in corporate governance, accountability, and transparency are key to ensuring successful CSR initiatives. However, this is a challenge because stakeholders' motives are diverse, depending on the antecedents and consequences of engaging in these activities [10, 11] proposed focusing on sensemaking to better understand the institutional factors that led to CSR activities.

Since the end of World War II intergovernmental organizations (IGOs) such as the UN, the NATO, and the EU have been involved in the stabilization and reconstruction of post-conflict zones. Governments and organizations perceive these zones as highly complex environments. Reconstruction includes programs for disarmament, destruction of weapons, repatriating refugees, demining, training of police and other security personnel, election monitoring, human rights promotion, and "reforming or strengthening governmental institutions and promoting formal and informal processes of political participation" [12, 13]. Effective reconstruction of post-conflict zones requires collaboration, coherence, and coordination amongst stakeholders.

In the post-conflict zones, there is an absence of war, but not essentially real peace. Brahimi states that "the end of fighting does propose an opportunity to work towards lasting peace, but that requires the establishment of sustainable institutions, capable of ensuring long-term security" [14]. Examples of post-conflict zones in recent history are the former Socialist Federal Republic of Yugoslavia, Iraq and Afghanistan.

The post-conflict reconstruction, building peace, and bringing stability in these zones has always been a high priority for IGOs. Boutros-Ghali defined post-conflict reconstruction as the "comprehensive efforts to identify and support structures which will tend to consolidate peace and advance a sense of confidence and well-being among people" [15]. The post-conflict zones are characterized by a lack of cohesion and high levels of inequality amongst citizens, resulting in environmental and social instability [16, 17]. A main challenge is that each stakeholder has its own agenda, jurisdiction, and approaches. This including solutions from a military, civil society and organizational perspectives. It is a challenge to align these various approaches. The Comprehensive Approach (CA) is an interesting intent in doing so.

The Comprehensive Approach represents the governmental stakeholders involved in the decision-making process. CA is usually implemented in failing or failed states aiming at post-conflict reconstruction. Governance in the CA context is operationalized as collaboration, covering lasting and well-structured relationships, resources flow and other interactions between specific organizations seeking to attain both common as well as separate goals [18, 19]. Besides the participation of IGOs such as NATO and the UN, there is an increasing number of Non-Governmental Organizations (NGOs) and private firms who play an important role in the reconstruction process. These stakeholders deploy organizational resources and activities to the benefit of the society, government, and environment. Establishing joint goals in public-private partnerships [20], especially in an unstable context require governmental efforts and financial investments. Therefore the involvement of multiple stakeholders is central in the reconstruction process.

While the literature reports extensively on the benefits of engaging in CSR to firms, little is known about the added-value of CSR in public-private partnerships from a governmental perspective. To better understand the vision of the government on the role of business firm as enabling SR activities, we conducted a single case study in a unique context: the reconstruction mission of the Task Force Uruzgan (TFU) in Afghanistan that ran from 2006 to 2010. Post-conflict zones such as Afghanistan are interesting for studying the engagements of private firms and socially responsible activities. First, political circumstances are changing rapidly and require collaboration amongst stakeholders. Second, post-conflict zones are the center of governmental and financial efforts of reconstruction. In this article, we particularly look at the joint goal of reconstruction from the Comprehensive Approach perspective as a social construction [21].

We argue that the mental frameworks of stakeholders regarding CSR are underlying organizational sensemaking vis-à-vis the involvement of business firm in the CA perspective. This paper presents the preliminary results of a single, explanatory case study. The goal of the study is to first better grasp how mental frameworks of stakeholders affected their perceptions of the Comprehensive Approach. Second and therefore, this exploratory research aim at mapping the decision-making process when considering CSR engagement. Additionally, we investigated the stakeholders' perceptions on the role of private firms and SR as enabler of reconstruction in post-conflict zones.

The structure and mechanisms of the Comprehensive Approach are well-documented in the literature [22]. However, little is known on the process of decision-making, and to our actual knowledge, nothing has been reported on the participative role of private firms within the Comprehensive Approach. In this study we first concentrate on the decision-making process of each line of operation of the three main stakeholders of the Comprehensive Approach i.e., defense, diplomacy, and development during the post-conflict reconstruction. Second, we address to the public-private partnerships and to the role of SR as enablers of reconstruction.

The paper is organized as follows. In the theoretical section we provide an overview on CSR definitions and introduce the concept of Comprehensive Approach in detail. The case study is then presented in the next section. The results of the content analysis and narratives of the participants are used to illustrate and support the discussion. The conclusion reflects on the social consciousness and political roles of stakeholders, private firms, and governments in SR engagements in post-conflict reconstruction.

1.1 Corporate Social Responsibility: Definitions

Definitions of Corporate Social Responsibility have proliferated in the literature. These vary as a function of the diversity of the stakeholders involved, their values, and therefore associated goals. The definition of Carroll is one of the most commonly used: "Corporate Social Responsibility involves the conduct of a business so that it is economically profitable, law abiding, ethical and socially supportive" [23]. Epstein [24] also related CSR to organizational decision-making activities. He stated that the normative correctness of the outcome of corporate action relating to a specific problem, i.e. to be socially responsible, should have beneficial rather than adverse effects on stakeholders. That is to be socially responsible. Steiner stated that taking up "social responsibilities is more of an attitude, of the way a manager approaches his decision-making task, than a great shift in the economics of decision making. It is a philosophy that looks at the social interest and the enlightened self-interest of business over the long run as compared with the old, narrow, unrestrained short-run self-interest" [25].

Dahlsrud [26] identified thirty-seven definitions of CSR. The author categorized these definitions based on five dimensions: Environmental e.g., a cleaner environment, environmental concerns; Social e.g. relationship between business and society, contributing to a better society, integrating social concerns in their business operation; Economic e.g. financial aspects, describing CSR in terms of business operation;, Stakeholder e.g. treating the stakeholders, and Voluntariness e.g. action not prescribed by law. Aguilera et al. [10] combined four organizational levels individual, organizational, national, and transnational and their associated motives. They identified three types of associated motives. The first motive is instrumental. It posits that humans are searching for control. The second associated motive is relational. It concerns the quality of the relationships between individuals and groups in linkage to the psychological need for belongingness. The last associated motive is the type of relation. This third motive is moral and it is related to the need for meaningful existence. This need is based on the common idea that most human beings are sharing basic respect and human dignity with each other. Different types of pressure are placed on the firm to engage in CSR as function of the interaction between organizational levels and motives.

In the context of the following case study, we are focusing on the transnational levels. Particularly, we are looking at the engagement of private firms from a developed country i.e., the Netherlands, in CSR activities to support post-conflict reconstruction in a developing country, i.e., Afghanistan. We focused first on the decision-making process of the three inter-organizational pillars, namely the Dutch ministry of defense, diplomacy, and international trade and development. Second, we addressed the public-private partnership and the role of CSR in facilitating reconstruction based on the Comprehensive Approach.

1.2   A Comprehensive Approach to Post-conflict Reconstruction

The North Atlantic Treaty Organization (NATO) defines the Comprehensive Approach as: "the integration of military security efforts in diplomacy and development. The aim of the CA is to achieve greater harmonization and synchronization among the activities of the various international and local actors. This, across the analysis, planning, implementation, management and evaluation aspects of the program-cycle" [27]. The CA encompasses a wide range of security, governance, and development tasks, but little direction is given on how these are activities are integrated [28].

The Comprehensive Approach can be continued and interdepartmental policy development can be further stimulated for example in the areas of environment, health care, energy, water and agricultural production. In addition to the participation of IGOs, multiple stakeholders are involved in the Comprehensive Approach. NGOs and private firms play an important role in deploying resources and socially responsible activities from the "home country" in the "host country".

De Coning and Friis [18] identified six types of structured relationships within the Comprehensive Approach: unity integration, cooperation, coordination, coexistence, and competition. Depending on the level of analysis, those relationships may over-lap. They linked their framework to the structured relationships amongst the stake-holders e.g. centralized vs. decentralized leadership, and the flow of resources e.g. level of interdependence, required to enlighten the decision-making process within the CA. For example, they view cooperation as complementing deployment of resources and activities, allowing the stakeholders to operate jointly when resources are scarce. Coordination intents to prevent conflict or friction between the deployments of the resources and mostly consists of sharing information with partner organizations with "deconfliction" as a primary goal.

2   Case Study

The Dutch government started the mission of the Task Force Uruzgan (TFU) in Afghanistan in 2006 with a plan consisting of three main lines of operation. Each line had its own set of goals, but the interdependence between the lines symbolized the effective collaboration required to reach effective reconstruction. The TFU served as a case study to investigate how the mental frameworks of the stakeholders affected sensemaking. Henceforth, how these frameworks influenced the decision-making process regarding SR engagements. Eight members of the core components of the TFU, senior military personnel, senior diplomats, as well as civilian entrepreneurs, each representing each a unit of analysis of the Comprehensive Approach took part in this research: The TFU's Headquarters (HQ), the Battle Group (BG), and the Provincial Reconstruction Team (PRT). Yin [29] argued that "the essence of a case study (…) is that it tries to illuminate a decision or set of decisions: why they were taken how they were implemented, and with what results" [30]. This exploratory study undertook part of this challenge.

2.1   Case Description

The International Security Assistance Force (ISAF) mission was established in 2002 by NATO as part of a broader international strategy. The meta-goal of ISAF is to enable the Afghan government to guarantee security and stability within its own borders facilitating post-conflict reconstruction.

After the Dutch government contributed to the reconstruction of Baghlan Province (2002–2006) they shifted their focus to the southern Afghanistan. The Dutch gov-ernment agreed on contributing to the reconstruction of Uruzgan province as of 2006 by establishing the TFU. Beyond deploying their forces, they explicitly included requirements for governance and economic reconstruction in their planning.

In accordance with the ISAF mandate, the Dutch government focused on increasing the support of the local population to take support away from the Taliban and other insurgency groups. Although offensive military actions were needed in particular situations, the core task of the mission was to improve the efficiency of the Afghan government, stimulate good governance, uphold rule of law, and implement projects and reconstruction activities. From the start of the mission, all three ministries defense, diplomacy and development, were actively engaged in the mission. Its core components were the Battle Group (BG) and the Provincial Reconstruction Team (PRT). The main responsibility of the BG was to maintain security in the Province. The PRT was responsible for reconstruction efforts and maintaining contact with civilians. It also advised and supported activities of reconstruction.

2.2   Participants

This case study focuses on the decision-making process regarding development activities as well as the potential role of private firms and CSR engagement in the reconstruction efforts. To collect the data we conducted semi-structured interviews with members of the core components of the TFU, each represent a unit of analysis of the CA: TFU Headquarters (HQ), the Battle Group (BG) and the Provincial Reconstruction Team (PRT). The participants were selected on the basis of their seniority and the nature of their responsibilities in the decision-making process. Two out of four senior diplomats with the role of Civilian Representative (CivRep) and two out of four senior military commanders with the role of Commander Task Force Uruzgan (C-TFU) were selected from the HQ. Furthermore, two out of eight senior military commanders involved in the TFU with the role of respectively Commander Battle Group (C-BG) and Commander PRT (C-PRT) participated in this research. Finally, two civilian entrepreneurs from the PRT with the role of private sector developer were interviewed.

3   Results and Discussion

The narratives of the participants illustrate two levels of understanding: manifest and latent. The answers provided during the interview support the analysis. We first addressed how the participants made sense of the decision-making process during the time they were involved in the TFU mission. We found elements in the narrative addressing goal-setting, decision-making, organizational culture, motives for collaboration, stakeholder relationships, and CSR perception.

3.1   Goal-Setting

The Dutch government deployed troops in Uruzgan province with a master plan: support the local Afghan government in stabilizing and reconstructing Uruzgan province. A military commander stated that: "*the meta-goal of the mission was to kick start the local economy in a self-sustainable system with local means that will not require external financing or NGO involvemen*t". A diplomat perceived the goal as:

"an overall effort of stabilization, province return to prosperity, political transition, opening up the province, private sector to locally develop". Indeed, shaping the conditions for the reconstruction activities was one of the core aims according to another military commander: "we needed to integrate and coordinate the non-military effects into the mission". After one year the TFU learned an important lesson as a military commander stated: *"we learned the importance of having a global campaign plan. So no short-term vision but mid-term vision and also hopefully a long-term vision where all three stakeholders of defense, diplomacy and development of the Comprehensive Approach were addressed"*.

Interestingly, a civilian entrepreneur was more critical regarding the alignment of the goals. He emphasized that *"the meta-goal was of course ISAF as a Stabilization Force. All the things that ISAF wanted to achieve were in line with the Dutch International Security Strategy (IVS). They matched. The problem although was that the goals NL and ISAF had were not made SMART"*. The strategic direction of an organization requires a clear definition of its purpose. Still, this does not imply an agreement on so-called SMART goals. The Dutch government accepted a certain level of fuzziness in the strategy and therefore failed in describing its goals in detail, which has been recognized as key to better outcome. A military commander recognized the need for precise goals: *"being only there four 4 or 5 months makes it really important to have a broader structure in which your operations are tied into. If you do not have a far-stretching campaign plan for more than only your term of deployment, it is always opportunistic and short-term"*.

3.2 Decision-Making

The collaboration within TFU was based on hierarchical institutionalized structures i.e. asymmetrical influence. It included centralized planning and decision-making, while execution was decentralized. From 2006 until 2009 the decisions were taken by the military commander of the TFU (C-TFU). The Political Advisor (PolAd) of the ministry of foreign affairs functioned as a personal advisor to the C-TFU. According to a diplomat this part of the mission's operational design was identified as inefficient. Consequently, as of 2009 the mission came under dual-headed command. One of the diplomats described the leadership and decision-making process as follows: *"the TFU was set up in an integrated way. Two people were put at the head of the TFU; Defense and Diplomacy. The two had very different responsibilities, dual-lead in all decisions were taken by both"*. A military commander underlined the importance of "deconfliction" particularly when NGOs entered into the picture. He stated that: *"NGOs and foreign affairs were not often on the same page. There was a need to deconflict and sometimes to synchronize them"*.

The joint decision-making process was perceived as effective since they could easily consult each other. They converge in planning the actions core to the mission and agreed allocating the scarce resources. One of the mission's diplomats describes the dual-headed mission command as: *"creating an organizational climate in which both military and civilian personnel could collaborate efficiently"*. The mode of coordination was based on a strong instructional hierarchy and conveyed authoritative decision-making. A diplomat explained that in practice there was room for TFU personnel to engage in a form of mixed governance with space left to negotiate under the supportive shadow of hierarchy: *"the TFU was civilian-military led by both a CivRep and a military commander. Each had very different responsibilities. Needless to say it was not my task to lead the military operations but yet all decisions were taken by the two persons together. The different perspectives were compared and then on the basis of that integrated approach a decision was taken"*. We conclude that collaboration at the operational level – between civilians and the military – was a cooperative relational type before 2009 and an integrated relational type after 2009 [18].

3.3   Organizational Culture

Organizational culture surely plays an important role in such dual-headed situations. One military commander describes the collaboration as having: *"mutual understanding for each other's organizational cultures"*. Interestingly, the CivRep was not directly part of the ISAF structure. A diplomat reported that regarding the local command and control: *"it was not problematic for the diplomats"*. Overall, he insisted that: *"We were trying to create synergistic effects by collaborating intensively"*. Synergy between the different organizational cultures started from the home country prior to the deployment: The diplomat emphasized that: *"we come from different organizations, we have different working cultures. It is important to learn from each other"*.

The collaborative process generated tension. A military commander reported that: *"there was always tension, it never became smooth. It is not bad because if we have development and military going hand in hand in an operation and there is no struggle something is wrong. We need each other. You cannot create stability without security, you cannot create security without stability so it goes hand in hands"*.

Coming from different organizations means that integration needs to take place at an early stage as one of the diplomats emphasized: *"if you lose one or two months there in getting used to each other and understand what each other's plans are, then that is a time lost"*.

3.4   Motives for Collaboration

Sense of control can maximize the favorability of the outcomes [10]. Organizations have several instrumental motives for collaborating with other organizations. Time played an important role in the collaboration and decision-making process. The mandate of the mission was signed two years ago and there was a sense of urgency to collaborate. The Dutch government did not have endless time to participate in the reconstruction of Uruzgan province. Time has always been a critical factor. Especially in relation to the level of progress that have been made. Nowadays the sense of urgency for Western organizations is heightened by the financial recession they are facing in their home countries. The initial two-year mandate was perceived as a real limitation for credible goal-setting. A military commander stated that during his deployment he did not know whether the missions mandate was going to be extended: *"the question was if we still had to plan for long-term reconstruction or we had to shift to a possible re-deployment back to the Netherlands.* " A diplomat emphasized that *"reconstruction efforts are generally a long-term activity. The uncertainty about TFU's mandate made it really hard to plan for long-term development and allocate the necessary resources"*.

Furthermore, TFU had to deal with scarce resources for its operations, making efficiency a key motive for collaboration as a diplomat stated: *"we were operating with very scares resources. Whether an armored vehicle was used for a certain patrolling activity that day or for a visit to the governor. These type of choices were considered in terms of what was most necessary that day"*. Therefore, joining and coordinating deployment of resources were more efficient. Enhancing social cohesion is an important motive for collaboration. The OECD [31] stated that a cohesive society works towards the well-being of all its members, fights exclusion and marginalization, creates a sense of belonging, promotes trust, and offers its members the opportunity of upward mobility. While the notion of social cohesion is often used with different meanings, its constituent elements include concerns about social inclusion, social capital and social mobility. Social cohesion is key to reconstructing a well-functioning society. A military commander provided a good illustration of social cohesion: *"you basically need a social contract. If the economy is working, people will pay taxes to the government and then it will provide safety, security, and a healthcare system. You need to build government functionalities"*.

3.5   Stakeholder Relationships and Network Governance

Freeman defines stakeholders as "any group or individual who can affect or is affected by the achievement of an organization's purpose. For instance, some corporations must count 'terrorist groups' as stakeholders". The TFU mission required collaboration with several other stakeholders; IOs, NGOs, and private firms. Effective collaboration is a key element to success within the CA, because actions are highly interdependent. Having mutual understanding of each other's intentions is therefore important, as a military commander stated: *"if we deployed our forces into a certain area to provide security, it is very important to have a good understanding of what the other agencies, reconstruction teams or NGOs are doing during this military operation, but this also after securing part of the area. So the importance is not only of having a successful military operation, but also to stress the importance of the follow on"*.

The U.S. military units deployed into Uruzgan province under the mandate of Operation Enduring Freedom (OEF). The C-TFU had received orders from his leadership in the Netherlands not to collaborate, but to "only support in extremis". There was neither functional relationship nor institutionalized structure between them. In practice this meant that both C-TFU and the leadership of the OEF units in the Province informed each other about their intentions. The primary goal of this information sharing was, according to a military commander: *"not coordinating but deconflicting each other's planning of future activities. There was a sense of mutual recognition. "* In other words, creating *"shared awareness about each other's activities"*. This type of collaboration at the operational level between TFU and U.S. military units was coordination [18]. The motive behind this collaboration can be found in the motive for consistency; organizations seek for collaboration since they recognize the importance of having "shared awareness" of their environment [10].

Communication and collaboration with NGOs took place through the PRT. There was no functional relationship between the C-PRT and the NGOs. The institutionalized structure was non-hierarchical (i.e. mutual influence) and agreements were achieved through bargaining. A diplomat observed: *"there was a mutual tension between them and collaboration took place based on inter-personal relationships"*. Security was at stake and required careful coordination. A diplomat underlined that: *"some attacks showed how difficult it is for civilian players like the UN to play their coordinating role; Setting up the structure right before the beginning is required to make it to make it more solid"*. The motive to increase legitimacy drove the collaboration with other organizations. Interestingly, the diplomat underlined that: *"if you want the development activities to be sustainable, and to reach out to people, you need to do it with the NGO, military cannot do it alone. NGOs can do their work when the area is secure enough. Then you see the excellent collaboration between the civilian and the military that leads to concrete results"*. More actors working together will increase the political and moral legitimacy. Moreover, the relational motive of enhancing social cohesion is not only an important motivator for NGOs, but also for the collaboration with the TFU.

The collaboration with the Afghan local government also took place through the PRT. Both the C-TFU and CivRep had their own relationships with government officials such as the governor of Uruzgan province. The institutional structure for this collaboration was non-hierarchical (i.e. mutual influence). The mode of communication was agreement by bargaining or arguing, a form of intergovernmental cooperation. The underlying motives are both instrumental and relational. The meta-goal of the TFU was to support the local Afghan government with stabilizing and reconstructing Uruzgan province. As emphasized by a diplomat: *"it is all about the Afghan people"*.

Although the Dutch government intended to be in a supporting role, in practice however, due to the lack of a credible local Afghan government, the TFU had to take the lead instead. A military commander stated that: *"instead of building their own country the Afghans were helping us rebuild it"*. The role of experts and particularly of the tribal advisor was recognized as key to supporting the decision-making process. *"The tribal advisor was terribly good"*, said one of the diplomats.

Some of the private firms involved were actually an element of the TFU itself. There were for example private firms which contributed to the security of TFU's compound by deploying Unmanned Arial Vehicles. There were also firms responsible for part of the logistics such as food and water. The relationship between the C-TFU and the personnel of these private firms was functional, since they had established a contract. The collaboration motives were instrumental since the Dutch government hired them for several reasons. First, they had unique resources that could be deployed quickly. Second, their services were cheaper. The private firms most likely collaborated with the TFU to make profit. This typical entanglement of maximizing wealth of private firms and the requirements to involve a range of stakeholders from Uruzgan province as well as from the defense and diplomacy in the decision-making process, is challenging. A diplomat added: *"despite a lot was possible with securing the ground. UN and NGOs were the most involved in projects such as education, healthcare, and infrastructure"*.

Other private firms with whom the TFU interacted focused on deploying resources and activities into the actual province. One private firm supporting the "saffron project" - meant to support the Afghan farmers replacing their agriculture activities from poppy into saffron - was mentioned several times. This private firm positioned itself in Uruzgan province using the business concept of "People, Planet, Profit". Through this socially responsible business model they intended to make profit and at the same time create a better situation for the Afghan people. The socially responsible and economic vision of CSR was their main drive. A civilian entrepreneur specified the importance of *"promoting the growth of saffron instead of poppy. Saffron will ultimately provide more revenues for the farmers themselves. The starting cost was paid by the Dutch ministry of foreign affairs. For these projects the TFU used civilian entrepreneurs from the Netherlands to support the education of local Afghan entrepreneurs"*.

3.6   CSR Perception

Each participant gave a definition of CSR and tried to put it in context. The definitions were analyzed in accordance with the coding schemes [26]. The participants were mostly referring to the social and economic dimensions of CSR. Their main conclusion was that it was a difficult task to envisage the participation of private firms without proper planning and security. *"Not so many private companies were able to play their role, which will come about later on"*, said a diplomat. An important lesson the Dutch government learned is that the operational design of a mission, including all relevant stakeholders, is made at the strategic level before deploying into the post-conflict zone. A diplomat stated that: *"one of the key lessons learned is that we need to have the operational design ready up front. What happened at the first couple of Task Forces was that they did not have the comprehensive design as a whole before going to Uruzgan"*. At the operational level the collaboration of all stakeholders is the key to a successful CA as a military commander emphasized: *"we need to partner up. The true grasp of what is happening on the ground cannot be left to Defense, Diplomacy, or Development alone; it is their common picture that will show what is happening"*. The CA requires more than the participation of various government agencies, as a diplomat emphasized: *"a true Comprehensive Approach requires the participation of NGOs and private firms. Those stakeholders need to be involved in order to be successful on the long-term"*. Moreover, he emphasized that "government cannot take over private firm tasks while private firm can indeed take over government tasks".

Successful participation of private firms begins at the analysis, decision-making process and design of the mission at the strategic level, as a civilian entrepreneur stated: *"in order to successfully attract private firms for the CA they need to be engaged at the initial decision-making process. Collaborative goals need to be developed so those private firms can conduct their own long-term planning and allocation of resources"*. A diplomat mentioned that: *"talking about the analyses, what is the problem, what are we going to do about it? There you have the possibility to really include the options for the private sector to take their role"*. To design a true CA, governments have to include the possible participation of private firms in the operational design of the mission. However, private firm involvement depends on the security situation. One of the diplomats reported that: *"there is a potential to trigger the private sector and it was incredibly important we did not focus only on the NL consultancy firm but they were unable or unwilling to come because of insecurity."* Moreover, he emphasized: *"private firm involvement is very dependent on the security situation"*. Private firms, after their initial participation in the decision-making process at the strategic level, may be on stand-by with the deployment of their resources at the operational level until security conditions are favorable enough for them to actively engage in the mission. A diplomat said: *"it is quite possible that they actively engage at a later stage, but how to get there is something we should be talking to them about because that will help us in laying out our strategy and comprehensive design"*.

4 Conclusions

To conclude, we first underline the necessity to develop a case specific "working definition" of CSR. This is required to tackle the complexity and fuzziness of the concept. As Aguilera et al. [10] underlined, combining organizational levels and associated motives is key to providing a clear definition. In the case of the TFU, we are focusing on the stakeholders' perspectives during the decision-making process in a particularly unstable context. The organizational level is transnational and the motives are instrumental, relational and moral according to the narratives of the participants. We therefore propose a working definition of CSR as the "joint public-private decision-making process with respect to the local community, linked to compliance with legal and security requirements, ethical values which will contribute to stability and sustainable peacebuilding during post-conflict reconstruction rather than exclusive corporate profit".

Second, engaging in CSR activities toward reconstruction is a joint goal for the stakeholders engaged in the Comprehensive Approach. The stakeholders from defense, development, and diplomacy taking part of the Comprehensive Approach, all demonstrated a sense of social consciousness in their decision-making and approach to reconstruction. In that respect the CA meets the expectation of our society that organizations adapt proper responsible social values in their legal, ethical, and discretionary activities [10].

Third, the three stakeholders perceived the role of public-private partnerships to be potential enablers of stabilization and reconstruction in post-conflict zones. However, the role of the private sector has changed in global society, mainly the resulting from the proliferation of cross-border trading. Indeed, private firms engage more actively in corporate social initiatives, therefore taking up a more prominent political role.

Private sector engagement in CSR has been criticized for going beyond its economic role, assuming a state-like role protecting, enabling, and implementing citizenship rights. This shift mostly occurs when the state system fails, taking over functions which used to be the responsibility of the governmental agencies [32]. Since the growing of transnational interdependence, economic globalization is one of the factors that has accelerated CSR engagements. We therefore stress the importance of government involvement in the decision-making process when involving private firm participation during reconstruction. We state that a standardization would be sterile in such context. Standardization in CSR reporting has been manipulated in the past, accelerating a spiral of distrust [35]. Improvements in corporate governance, accountability, and transparency should therefore be a minimal requirement for ensuring successful CSR initiatives in the context of the Comprehensive Approach. We therefore propose adopting a particular governance approach. Indeed, the planning and management of the deployment of international resources is one of the key challenges in post-conflict reconstruction. Moreover, the complexity of the environment makes it difficult for post-conflict organizations to accurately gage effective allocation and reallocation of resources to high-return uses. Therefore, post-conflict reconstruction requires effective collaborative governance or network governance [33].

Overall, it is clear that with the development of the CA, the organization of post-conflict reconstruction is no longer solely focused on military superiority. Neither is it centered on the previously mentioned ministries, but also the ministries responsible for justice, police, correctional services, home affairs and finance. Governmental organizations are only one piece of the pie and this also applies for other IGOs, NGOs, and private firms. The analysis shows that to be successful it is necessary to have all relevant stakeholders integrated at the strategic level in the "home country"; that is where the decision-making process starts and the comprehensive design is developed.

Furthermore, IT resources are key to efficiently supporting collaborative and networked governance [34]. Nowadays, the stakeholders involved in post-conflict reconstruction use and also depend on IT resources in their activities. Therefore we propose to investigate the role played by IT in the Comprehensive Approach into more detail.

This paper reports the results of an exploratory research. We focused on the operational level. This explains the so far small sample size. More data are being actually collected. Future research should therefore address the tactical and strategic level of collaboration. The main limitation of this study is that we collected the data with a governmental approach. In future research we will collect the perspective of the NGOs and the private sector. Last, it will be interesting to investigate the perception of CSR engagements from an intercultural perspective.

Acknowledgments. We are grateful to all our interviewees featured in this paper. We thank in particular Anne van Bruggen for kindly editing this paper. This paper would not have been written without their generous assistance.

References

1. Fombrun, C.J.: The leadership challenge: building resilient corporate reputations. In: Doh, J.P., Stumpf, S.A. (eds.) Handbook on Responsible Leadership and Governance in Global Business, pp. 54–68. Edward Elgar, Cheltenham (2005)
2. Fort, T.L., Schipani, C.A.: The role of the corporation in fostering sustainable peace. Vanderbilt J. Transnatl. Law **35**, 389–436 (2002)
3. Kruk, M.E., Freedman, L.P., Anglin, G.A., Waldman, R.J.: Rebuilding health systems to improve health and promote statebuilding in post-conflict countries: a theoretical framework and research agenda. Soc. Sci. Med. **70**, 89–97 (2009)

4. Melandri, M.: The state, human rights and the ethics of war termination: what should a just peace look like? A critical appraisal. J. Glob. Ethics **7**, 241–249 (2011)
5. Quaynor, L.J.: Citizenship education in post-conflict contexts: a review of the literature. Educ. Citizsh. Soc. Justice **7**, 33–57 (2012)
6. Frederick, W.C.: The growing concern over business responsibility. Calif. Manag. Rev. **2**, 54–61 (1960)
7. Klein, B., Leffler, K.: The role of market forces in assuring contractual performance. J. Polit. Econ. **89**, 615–641 (1981)
8. Keh, H.T., Xie, Y.: Corporate reputation and customer behavioral intentions: the roles of trust, identification and commitment. Ind. Mark. Manag. **38**, 732–742 (2009)
9. Milgrom, P., Roberts, J.: Price and advertising signals of product quality. J. Polit. Econ. **94**, 796–821 (1986). University of Chicago Press
10. Aguilera, R.V., Rupp, D.E., Williams, C.A., Ganapathi, J.: Putting the S back in corporate social responsibility: a multilevel theory of social change in organizations. Acad. Manag. Rev. **32**, 836–863 (2007)
11. Basu, K., Palazzo, G.: Corporate social responsibility: a process model of sensemaking. Acad. Manag. Rev. **33**, 122–136 (2008)
12. Weinberger, N.: Civil-military coordination in peacebuilding: the challenge in Afghanistan. J. Int. Aff. **55**, 245–274 (2002)
13. Manning, C.: Local level challenges to post-conflict peacebuilding. Int. Peacekeeping **10**, 25–43 (2003)
14. Brahimin, L.: Report of the Panel on United Nation Peace Operations. United Nations, New York (2000). http://www.un.org/en/ga/search/view_doc.asp?symbol=A/55/305
15. Boutros-Ghali, B.: An Agenda for Peace. United Nations, New York (1997)
16. Bjarnegård, E., Melander, E.: Disentangling democratization, gender, and peace: the negative effects of militarized masculinity. J. Gend. Stud. **20**, 139–154 (2011)
17. Wallensteen, P.: The origins of contemporary peace research. In: Höglund, K., Öberg, M. (eds.) Understanding Peace Research: Methods and Challenges. Routledge, London (2011)
18. de Coning, C., Friis, K.: Coherence and coordination: the limits of the comprehensive approach. J. Int. Peacekeeping **15**, 243–272 (2011)
19. Damanpour, F., Schneider, M.: Phases of the adoption of innovation in organizations: effects of environment, organization and top managers. Br. J. Manag. **17**, 215–236 (2006)
20. Eden, C., Ackermann, F.: 'Joined-Up' policy-making: group decision and negotiation practice. Group Decis. Negot. **23**, 1385–1401 (2013)
21. Berger, P., Luckmann, T.: The Social Construction of Reality: A Treatise in the Sociology of Knowledge. Penguin, London (1966)
22. van der Lijn, J., van den Berg, P.: 3D 'The Next Generation' Lessons learned from Uruzgan for future operations. The Hague, The Clingendael Institute and Cordaid (2011)
23. Carroll, A.B.: the pyramid of corporate social responsibility: toward the moral management of organizational stakeholders. Bus. Horiz. **34**, 39–48 (1991)
24. Epstein, E.M.: The corporate social policy process: beyond business ethics, corporate social responsibility and corporate social responsiveness. Calif. Manag. Rev. **19**, 252–284 (1987)
25. Steiner, G.A.: Social policies for business. Calif. Manag. Rev. **15**, 17–24 (1972)
26. Dahlsrud, A.: How corporate social responsibility is defined: an analysis of 37 definitions. Corp. Soc. Responsib. Environ. Manag. **15**, 1–13 (2006)
27. Stavridis, J.G.: The comprehensive approach in Afghanistan. PRISM **2**, 65–76 (2011)
28. Travers, P., Owen T.: Peacebuilding While Peacemaking: The Merits of a 3D Approach in Afghanistan. UBC Center for International Relations Security and Defense Forum Working Paper, No.3 (2007)
29. Yin, R.: Case Study Research. Design and Methods. Sage, Thousand Oaks (1994)

30. Schramm, W.: Notes on Case Studies for Instructional Media Projects. Working paper for Academy of Educational Development, Washington DC (1971)
31. OECD: Perspectives on Global Development. Social Cohesion in a Shifting World, OECD Publishing (2012). http://dx.doi.org/10.1787/persp_glob_dev-2012-en
32. Matten, D., Crane, A.: Corporate citizenship: toward an extended theoretical conceptualization. Acad. Manag. Rev. **30**, 166–179 (2005)
33. Moynihan, D.P.: The network governance of crisis response: case studies of incident command systems. J. Public Adm. Res. Theory **19**, 895–915 (2009)
34. Nevo, S., Wade, M.R.: The formation and value of it-enabled resources: antecedents and consequences. Manag. Inf. Syst. Q. **34**, 163–184 (2010)
35. Sims, R.R., Brinkmann, J.: Enron Ethic Culture Matters more than Codes. J. Bus. Ethics **45**, 243-256 (2003)

How to Order the Alternatives, Rules, and the Rules to Choose Rules: When the Endogenous Procedural Choice Regresses

Takahiro Suzuki[(⊠)] and Masahide Horita

Department of International Studies, Graduate School of Frontier Sciences,
The University of Tokyo, Kashiwa, Japan
k147628@inter-k.u-tokyo.ac.jp

Abstract. A procedural choice problem occurs when there is no ex ante agreement on how to choose a decision rule nor an exogenous authority that is strong enough to single out a decision rule in a group. In this paper, we define the manner of procedural selection as a relation-valued procedural choice rule (PCR). Based on this definition, we then argue for some necessary conditions of a PCR. One of the main findings centers on the notion of consistency, which demands concordance between judged-better procedures and judged-better outcomes. Specifically, we found that the consistency principle and a modified version of the Pareto principle yield a simple impossibility result. We then show how the weakening of these conditions results to a degenerate PCR or the existence of a procedural veto. Finally, we show that the restriction of the preference domain to an extreme consequentialism can be seen as a positive result.

Keywords: Procedural choice · Infinite regress · Consequentialism

1 Introduction

Meta-level procedural choice, or 'how to decide how to decide (and so forth)', is a classical problem in collective decision-making. While the choice of voting rules has long been studied in social choice theory stating that a *good* rule is the one that satisfies widely accepted normative properties such as Pareto principle, there is yet another point of view that a *good* rule is one that is favored by the members of the group, even if it does not satisfy such normative properties. This view can be rephrased in terms of how the group members can find the best procedure 'locally' that suits the best with their procedural judgments. Some people might esteem anonymous and neutral procedures in purely public issues of decision-making. Others, however, might esteem a dictatorship by the most experienced engineer in terms of technical decision-making. These same people might even have procedural judgments over the rule to choose the rule. In other words, some people might hope that the chairperson should determine the decision rule, or there may be pros and cons concerning the rule to choose the rule for the choice of texts in the constitution. While these cases appeal to the necessity of procedural choice, they can yield an infinite regress of 'how to decide how to decide

© Springer International Publishing Switzerland 2015
B. Kamiński et al. (Eds.): GDN 2015, LNBIP 218, pp. 47–59, 2015.
DOI: 10.1007/978-3-319-19515-5_4

how to decide…'. Each procedural choice is no less important than the 1st level procedural choice, as we can see that the final choice depends on what procedure was used to arrive upon that choice. An instructive example is shown by Dyer and Miles [1], where scientists and engineers faced a collective choice problem as for the pair of trajectories for spacecraft Mariner Jupiter/Saturn 1977 Projects. They showed, using the submitted cardinal/ordinal preferences, the consequences totally varied among three well-known procedures: (1) Rank sum, (2) Additive form (weighted sum of utilities), and (3) Multiplicative form.

The purpose of our study is to search for rational ways of selecting decision-making procedures when each individual has procedural preferences and yet there has not been an ex ante agreement. It is only recently that some solution concepts have been proposed in the social choice theory. Koray [2], and Barbera et al. [3] have adopted the so-called fix point approach (Kultti and Miettinen [4]). The axioms they have proposed demand that a decision rule should select itself among a series of alternative rules. Much characterization or investigation of this methodology is now being done in subsequent research (Nicolas [5], Koray and Slinko [6], Diss and Merlin [7]). Above all, Kultti and Miettinen [8] extend these results to higher levels of procedural arguments, showing that the existence of self-stable rules with more than two levels and providing an explanation for why many of the real world principles stipulate only two levels of procedural choice.

These concepts are all very intuitive in that the use of self-stable rules (or self-selective social choice functions) does not allow for deviation from the status quo. Thus, we seemingly escape from the annoying regress of procedural choice. However, we can imagine a situation where all of the members do not favor self-stability for some reason, and they would instead prefer Borda count, for example. This view is even more convincing when we realize that different countries possessing different voting systems based on their own history and justice. Such systems might be judged based on their procedural cost, rapidity, or affinity with peace. Considering these cases, the assumption of the people's consequentialism, which is assumed in much of the fixed-point approach, is not well suited at all.

Dietrich [9], on the other hand, has constructed a rather different approach called procedural autonomy for the purposes of this article. This approach first defines the procedural autonomy premise, which says:

> The manner in which the profile is aggregated into a collective decision should be determined by the procedural judgments within the group. (Dietrich [9], pp. 364)

Dietrich's approach mainly focused on finding legitimate alternatives from the premise of procedural autonomy and does not base specific structures on people's preferences. As long as a society adopts the procedural autonomy premise, monarchies, non-unanimous rules, non-self-selective rules, non-self-stable rules, or every other social choice rule can (or should) be elected if the society favors it, no matter what the outside people think. This necessitates the need to consider the manner through which a society can choose legitimate alternatives, procedures, the procedures to choose procedures, and so on. By adopting this approach, we invent an order-valued procedural choice with more than one level of regress. We first define the procedural choice rule (PCR), which expresses a manner of procedural choice, and then discuss what kind of property it should satisfy and what manners have normative status. Section 2 will provide the notation for this system, followed by Sect. 3 where we will present the normative properties we consider necessary for PCRs. Section 4 provides the technical results and Sect. 5 presents conclusions.

2 Notation

Let $N = \{1, 2, ..., n\}$ denote a society with at least two individuals. Let X denote the set of alternatives, whose cardinality is $2 \leq |X| < \infty$. Assume that the society tries to make an endogenous decision over X.

A binary relation R over a non-empty set A is defined as a subset of $A \times A$. Binary relation R over A is said to be reflexive if for all $a \in A$, $(a, a) \in A$, R is transitive if for all $a, b, c \in A$, $(a, b) \in A$ and $(b, c) \in A$ imply $(a, c) \in A$, R is complete if for all $a, b \in A$, $(a, b) \in A$ or $(b, a) \in A$ and R is anti-symmetric if for all $(a, b) \in A$, $(a, b) \in R$ and $(b, a) \in R$ implies $a = b$. Binary relation R is called a weak ordering if it is reflexive, transitive and complete, and a linear ordering if it is an anti-symmetric weak ordering. Let $W(A)$ be the set of all weak orderings over A, and $L(A)$ be the set of all linear orderings over A. Let $P(R)$ and $I(R)$ denote the asymmetric and symmetric part of binary relation R:

$$P(R) := \{(a, b) \in A \times A | (a, b) \in R \text{ and } (b, a) \notin R\}$$
$$I(R) := \{(a, b) \in A \times A | (a, b) \in R \text{ and } (b, a) \in R\}$$

Given a binary relation R over A and a non-empty subset $B \subseteq A$, we denote by $G(R, B)$ the greatest element of B relative to R; $G(R, B) := \{x \in B | \forall y \in B, xRy\}$.

We call $R = (R_1, R_2, ..., R_n) \in W(A)^n$ as a preference profile over A, whose i^{th} element $R_i \in W(A)$ denotes the individual i's preference ordering. A social choice function f over A is a function that assigns an alternative to each preference profile over A, such that $f : W(A)^n \to A$. Let N denote the set of natural numbers. For all $k \in N \cup \{0\}$, we define the level-k procedural set F^k inductively:

1. $F^0 := X$.
2. For any $k \in N - \{0\}$, F^{k+1} is the set of all social choice functions over F^k.

We call an element of F^k as a level-k SCF, or level-k procedure (rule) interchangeably. A level-k SCF is a social choice function over F^{k-1} and a rule [to choose the rule] (k − 1 times) to choose an alternative. For all $k \in N \cup \{0\}$, we assume that each individual in the society N has a preference ordering R_i^k over F^k. A level-k preference profile $R^k = (R_1^k, R_2^k, ..., R_n^k)$ is a preference profile over F^k. Integrating the level-k (k = 0, 1, ..., K) preference profile $R^0, R^1, ..., R^K$, we call $R = (R^0, R^1, ..., R^K)$ as a level-K meta-profile. Next we define the manners of procedural choice in the society.

Definition 1. Procedural Choice Rule (PCR): Let $K \in N$ and $D \subseteq W(X)^n \times W(F^1)^n \times ... \times W(F^K)^n$. A level-K PCR (: Procedural Choice Rule) E of domain D is a function assigning a level-K social meta-preference $E = (E^0, E^1, ..., E^K)$ to each level-K meta-profile: $E : D \to W(X) \times W(F^1) ... \times W(F^K)$. For $K = \infty$, level-∞ PCR is a function $E : \prod_{k \in N \cup \{0\}} W(F^k)^n \to \prod_{k \in N \cup \{0\}} W(F^k)$.

Thus a PCR expresses a way of procedural choice in a society. Given a meta-profile, each individual's procedural judgment, a PCR returns a set of the procedural judgments of the society. Unlike usual social welfare functions, PCRs consider how they evaluate the (possible infinite) levels of procedures. This can be a counterpart of Dietrich [9] 's decision rule, which focuses only on the final choice over the set of alternatives. In order to get a clearer understanding, we provide two manners of procedural choice.

Definition 2. x-Supporting Rules $F^k[x]$: For all $k, l \in N \cup \{0\}$ with $k < l$, for all $x \in F^k$, we define the x-supporting rules of F^l (relative to a given meta-profile), $F^l[x]$, as

$$F^l[x] := \left\{ f^l \in F^l : f^l\left(R^{l-1}\right)\left(R^{l-2}\right) \ldots \left(R^k\right) = x \right\}$$

This is a notation to descript which rules in a certain level ultimately result in a certain alternative. As an immediate consequence, we have that

$$\forall k < l, F^l = \coprod_{x \in F^k} F^l[x]$$

Example 1. Level-1 Dictatorial PCR: Suppose a society where procedural choice is totally determined by individual $j \in N$. Social preference E^0 over the set of alternatives X is completely determined by j's level-1 preference R^1_j, in the following way.

For any $x, y \in X$, xE^0y if and only if $F^1[x]O\left(R^1_j\right)F^1[y]$. For any $f, g \in F^1, fE^1g$ if and only if fR^1g.

In this manner of procedural choice, individual j is "dictatorial" since his/her preference over the procedure is sufficient to determine the social preference over alternatives regardless of the other individuals' meta-preferences.

Example 2. Level-K Always-Majority Procedural Choice: Suppose a society where any agenda $(X, F^1, F^2, \ldots)$ are judged according to majority rule. For any level $k \in \{0, 1, \ldots, K\}$ and for any alternatives/procedures $x, y \in F^k$,

$$xP(E^k)y \text{ if and only if } \left| \{ i \in N | xP(R^k_i)y \} \right| > \left| \{ i \in N | yP(R^k_i)x \} \right|$$

This is not a PCR, since it can generate a cyclic social preference for some preference profiles such as $N = \{1, 2, 3\}, X = \{x, y, z\}, R^0_1 : xyz, R^0_2 : yzx, R^0_3 : zxy$ (a Condorcet profile). However, this always-majority procedural choice has another counterintuitive problem. Suppose the following preference profile.

$$N = \{1, 2, 3, 4, 5\} \text{ and } X = \{a, b, c\}$$
$$R^0_1 : abc, R^0_2 : abc, R^0_3 : abc$$
$$R^1_i : BG^1D_4(i = 1,2,3), R^1_k : G^1D_4B(k = 4, 5)$$
$$R^2_i : BG^2D_4(i = 1,2), R^2_3 : G^2(D \sim B), R_k : G^2D_4B$$

where B is the Borda count, $G^1 \in F^1$ and $G^2 \in F^2$ are both one of the generalized Borda counts with 5 points for top alternatives, 4 points for the second, 0 points for the others (just in order to distinguish, we denote G^1 for the one in F^1 and G^2 for the one in F^2).

D_4 is the dictatorship by individual 4 in the usual sense. Since each alternative/ procedural set has a Condorcet winner, the always-majority procedural choice admits them as the greatest element in the social preference. However, there lies a paradoxical result; while in the second level set we have $G^2 \in G(E^2, F^2)$, its outcome $G^2(R^1) = G^1$ is defeated by B^1 according to E^1. This is an inconsistency of judgment between the procedures and alternatives. E^2 and E^1 defined this way are not consistent. Procedures ranked higher by E^2 do not always output better alternatives.

While the dictatorial PCR does not look desirable in an intuitive sense, the always-majority manner of procedural choice in Example 2. can yield two unintuitive paradoxes: a well-known Condorcet paradox and an inconsistency paradox. The objective of our study is to design normative PCRs that avoid paradoxical outcomes. Since the PCR by definition expresses the manner of how the society ranks each alternative/ procedure facing the potential of opposing judgments by each individual, the main benefit of designing a normative PCR is therefore to propose how we can rationally stop the regress of procedural choice and make an endogenous and democratic procedural choice in our society.

In addition, we impose a consistency property upon our PCRs that rules out the inconsistency observed in Example 2. Having a consistent hierarchy of procedures can be a foundation of the social meta-preference.

3   Axioms for Procedural Choice Rule

Now we turn to discuss the properties of PCRs. After introducing a further definition, we examine each property.

Definition 3. Optimistically Induced Preference: For all non-empty set X and a binary relation R on X, we say $O(R)$ is an optimistically[1] induced preference over the set of non-empty subsets of X if

$$\forall S, T \subseteq X : (S, T) \in O(R) \Leftrightarrow \forall t \in T, \exists s \in S \; s.t. (s, t) \in R$$

$O(\cdot)$ is an operator that induces from the original preference a related preference over the power set. We say $O(R)$ just as "induced preference" if there is no fear of misleading. And we immediately get the following result.

---

[1] The manners to induce the preference over the power set, including optimistic manner, are very well studied in the strategy proof social choice rules, see [10] and [11].

Note 3.1. If R is a weak ordering over X, then $O(R)$ is a weak ordering over $2^{X-\{\phi\}}$. Although $O(\cdot)$ is a way to extend the original preference to the power set of the set of alternatives, we do not demand that each individual induces a related preference in this way. This notation is just defined in order to state formally the normative axioms of PCRs.

Definition 4. The Procedural Weak Pareto Principle (PWP): A level-$K(<\infty)$ PCR E satisfies PWP if and only if for all $R \in D, k \in \{0, 1, 2, \ldots, K-1\}, \forall x, y \in F^k$: if $\forall l \in \{k+1, k+2, \ldots, K\}, \forall i \in N, F^l[x]P\big(O(R_i^l)\big)F^l[y]$, then we have $xP(E^k)y$. For $K = \infty$, a level-∞ PCR E satisfies PWP if and only if for all $R \in D_\infty :=$ $\prod\limits_{k \in N \cup \{0\}} W(F^k), k \in N \cup \{0\}$, and $x, y \in F^k$: if $\forall l \in N - \{0, 1, \ldots, k\}, \forall i \in N, F^l[x]P\big(O(R_i^l)\big)F^l[y]$, then we have $xP(E^k)y$.

We give several comments on this property. First to note is that the PWP principle does not demand the usual Pareto principle. They are totally independent. While the latter demands that if everyone prefers alternative x to alternative y, so should do the society, our PWP principle demands that if everyone prefers x-supporting rules to y-supporting rules at every higher level, the society should rank x above y. Whether or not Pareto optimal alternatives are ranked high totally depends on the procedural judgments by the members of the society. A good ground for this is found in the law system of Sanhedrin:

> Unlike in contemporary US law, where capital cases require a unanimous jury decision (Mitchell and Eckstein, 2009) the Sanhedrin would automatically acquit a defendant if all members argued to convict in such a case (Talmud, Tractate Sanhedrin, 17a). While such a practice could seem counterintuitive, it may have been established as a last-ditch measure to prevent groupthink-like outcomes. If all 70 members vote unanimously, without any dissension at all, then there is reason to fear that groupthink conformity pressures may be to blame. (Schnall and Michael [12])

As long as we esteem the premise of procedural autonomy, and as long as they accept the unanimity-rejection principle, the procedure of Sanhedrin does not matter at all. Even if contemporary theorists *unanimously* favor unanimous procedures, the premise of procedural autonomy can acknowledge the use of non-unanimous procedures if the society members favor.

Second to note on the definition of our PWP is rather similar, but it is a direct application of the above discussion. As long as we evaluate alternatives/procedures on the basis of procedural judgments, we have no reason to stop meta-level reasoning at any finite level. Even if there is a unanimous agreement at level 1, that has no particular significance if the society members do not agree at level 2 on the use of unanimous procedure of level 1. The level 1 preference cannot be evaluated until we carefully investigate the preferences of higher levels.

Definition 5. Procedural Independence of Irrelevant Alternatives (PIIA): A level-K PCR E satisfies PIIA if and only if for all $R, \tilde{R} \in D, k \in \{0, 1, 2, \ldots, K-1\}$, and $x, y \in F^k$: if for all

$$l \in \{k+1, k+2, \ldots, K\}, and \ i \in N, O(R_i^k)|_{\{F^l[x] \cup F^l[y]\}} = O(\tilde{R}_i^k)|_{\{F^l[x] \cup F^l[y]\}},$$

then

$$O(E(R))|_{\{x,y\}} = O(E(\tilde{R}))|_{\{x,y\}}.$$

This is a modified version of Arrow's IIA condition. PIIA demands two main contents. One is that for any alternatives x and y, social ranking between x and y should completely depend on the individual's meta-profile over x-supporting rules and y-supporting rules. The other to note is that the set of x-supporting rules and y-supporting rules generally depends on the outcome of the procedures. Under the assumption of completeness of individuals' preferences, PIIA property does not demand anything if R and R are different in the eyes of the given procedures.

Definition 6. Inter-Level Consistency (ILC): A level-K PCR satisfies ILC if and only if the following holds. For all $R \in D, k \in \{0, 1, \ldots, K-1\}$ and $f, g \in F^{k+1}$: fE^k ^{+1}g **if and only if** $f(R^k)E^k g(R^k)$.

This consistency property rules out such inconsistent social meta-preferences found in **Example 2**. The 'if' part demands that if a procedure f is ranked above g, their outcome should be ranked the same. **Only if** part demands that if an alternative $x = f$ (R^k) above $y = g(R^k)$, then the society ranks procedure f above g. In other words, the society cannot rank an alternative x above another y without accepting the rule that supports x.

Definition 7. Procedural Vetoer: For any level $k \in \{0, 1, \ldots, K\}$ and alternatives/procedures $x, y \in F^k$, an individual $i \in N$ is a (procedural) vetoer over the pair (x, y) if and only if for all meta-profile $R = (R^0, R^1, \ldots, R^K) \in D$ and, if $F^K[x]P(O(R_i^K))F^K[y]$, then xE^ky. The individual $i \in N$ is a vetoer if and only if for any level $k \in \{0, 1, \ldots, K-1\}$ and for any $x, y \in F^k$, (s)he is a vetoer over the pair (x, y).

This is very similar to the concept of veto power developed in the Arrovian framework (Blair and Robert [13]). A procedural vetoer is an individual who can force x to be socially at least as good as y by presenting a preference whose induced preference strictly prefers x-supporting rules to y-supporting rules.

Definition 8. Arbitrary Focus (AF): A level-K PCR E satisfies AF if and only for all $j \in \{0, 1, 2, \ldots, K-1\}$, there exists a function

$$E_{[j]} : W(F^j)^n \times W(F^{j+1})^n \times \ldots \times W(F^K)^n \to W(F^j) \times W(F^{j+1}) \times \ldots \times W(F^K)$$

such that for all

$$R = (R^0, R^1, \ldots, R^K) \in D,$$

$$E^l\left(R^0, R^1, \ldots, R^j, \ldots, R^K\right) = E^{l-j}_{[j]}\left(R^j, R^{j+1}, \ldots, R^K\right) \text{ for all } l \in \{j, j+1, \ldots, K\}$$

The AF condition demands that for any meta-profile $R = (R^0, R^1, \ldots, R^K)$, the social meta-preference over R^j does not depend on R^k for $k < j$. More intuitively, the social preference over some procedures x and y should be totally determined by the preference on the rules to choose them, the rules to choose the rule to choose them, and so forth. And it should not depend on the preferences over their outcomes.

4 Results

4.1 Basic Impossibility Results

The majority of our results are focused around the following elementary impossibility. In the following part of Sects. 4.1 and 4.2, we fix $D = \prod_{k=0}^{K} W(F^k)^n$.

Proposition 1. [1] Let $2 \leq K$. There is no level-K PCR that satisfies PWP and ILC. [2] Let $K = \infty$. There is no level-∞ PCR that satisfies PWP and ILC. (All proofs are in the Appendix)

As can be seen in the proof, this is a direct consequence from PWP and 'if' part of ILC. However, this presents us with an elementary note to consider the procedural choice. When we regress in procedural choice, the outcome is expected to be consistent in the sense that each level of social meta-preference is well related to the given meta-profile. The proposition states, unfortunately, that we cannot expect consistency of the social meta-preference $E^0, E^1, \ldots, E^K$ and the Procedural Weak Pareto principle at the same time. This expresses the elementary impossibility in considering the procedural choice with consistency. Though we explicitly refer only to ILC and PWP condition, there are some other implicit conditions imposed on PCRs. The rest of the article is to search for plausible PCRs by weakening each of the axioms shown in Proposition 1. Some remarks on the remained axioms are in Sect. 5.

4.2 Weakening PWP and ILC

The 'if' part and PWP condition are both essential to derive the impossibility result in **Proposition 1**. In fact, as we will show later, there exists a PCR that satisfies ILC and there exists a PCR that satisfies the 'only if' part of the ILC and PWP. However, these apparently positive results are not fully satisfactory, for they immediately yield other negative results.

Proposition 2. [1] There exists a PCR E that satisfies the ILC and AF if and only if E is degenerated in the sense that for all meta-profile $R \in D$, for all $k \in \{0, 1, \ldots, K\}$ and for all alternatives/procedures $x, y \in F^k$, $xI(E^k)y$. [2] There exists a PCR E that satisfies the 'only if' part of the ILC, PWP, AF, and PIIA, but it yields at least one procedural voter.

4.3 Restricting the Preference Domain of PCRs

Definition 10. Consequentialist Preference Domain: The consequential preference domain $D_C \subseteq \prod_{k=0}^{K} W(F^k)^n$ is such that for all $R = (R_1, R_2, ..., R_n) \in D_C$, for all i $\in$ N, for all $k \in \{0, 1, ..., K-1\}$, and for all $f, g \in F^{k+1}$, $f(R^k)P(R_i^k)g(R^k)$ implies $fP(R_i^{k+1})y$. When an individual's meta-preference satisfies the underlined part, he/she is said to be a consequentialist.

Definition 11. Extremely Consequentialist Preference Domain: For any finite set X and the sets of procedures F^1, F^2, ..., F^K, the consequentialist preference domain $D_{EC} \subseteq \prod_{k=0}^{K} W(F^k)^n$ is such that for all $R = (R_1, R_2, ..., R_n) \in D_{EC}$, for all i $\in$ N, for all $k \in \{0, 1, ..., K-1\}$, and for all $f, g \in F^{k+1}$, $f(R^k)\boldsymbol{R_i^k}g(R^k)$ implies $f\boldsymbol{R_i^{k+1}}g$. When an individual's meta-preference satisfies the underlined part, he/she is said to be an extreme consequentialist.

The difference between the two is the bold style. An extremely consequentialist individual evaluates the rules simply by looking at their procedures. If two procedures' outcomes are different according to his/her measure, he/she chooses the procedure with the most preferable outcome. Otherwise, he/she is completely indifferent among the two.

On the other hand, a simple consequentialist individual evaluates the rules mainly on their outcomes, but not completely. If two procedures' outcomes are different according to his/her measure, he/she chooses the procedure with the most preferable outcome. Otherwise, it is possible that he/she has a strict preference over them according to their internal judgments.

Proposition 3. [1] Let K $\geq$ 2 be finite or infinite. There does not exist a level-K PCR of domain $D = D_C$ satisfying the PWP and ILC. [2] Let K $\geq$ 2 be finite or infinite. There exists level-K PCRs of domain $D = D_{EC}$ that satisfies the PWP and ILC.

This proposition gives an ironic solution to the impossibility proposed in **Proposition 1**. If we consider non-extremely-consequentialist individuals, we cannot order the social meta-preference to satisfy the consistency property and the procedural Pareto principle. However, if the society is extremely consequentialist, we do have potential to realize both the consistency and the PWP at the same time. The extremely consequentialist domain is at first sight hard to deal with since the opposition at level 0 remains the same no matter how high we take the levels. These people do not have any standardized concept of procedural justice in common, such as "a majority based SCF is better than dictatorial SCF," or "unanimity is not admissible at any level," and so on. All these people have is only the principle that the value of procedures resolves at their outcomes. All the other information has no importance.

5　Conclusion

When a society is going to make a collective decision but has no ex ante agreement or exogenous factors strong enough to stipulate the possible decision procedures, the choice of decision procedures is also a matter of endogenous decision making. We first defined a function of PCRs that express a way to make a procedural choice endogenously within a given society. Next, we investigated what kind of normative property we should impose on PCRs, and then we observed the performance and (im)possibility of the PCRs.

Our work centers around the basic impossibility result (**Proposition 1**) that describes the incompatibility between the ILC and PWP properties. The former demands for consistency between the judged-better procedures and judged-better alternatives whereas the latter is a derivation of the Pareto Principle modified for our PCRs. In Sects. 4.2 and 4.3 we searched for escape routes from this impossibility. Weakening of each conditions does yield a positive result, but only to yield another problems in its aftermath (**Proposition 2**). On the other hand, restriction of the preference domain to those that are extremely consequential can, in fact, be an ironic solution.

Finally, we make a few comments on the other implicit conditions imposed on PCRs. The first one is the set of procedures $F^1, F^2, \ldots, F^K$. When we consider decision makings very generally, our assumption of F^k has all the possible SCFs and indeed has some rationality. No social choice function should be deleted before the endogenous argument of which procedures are better than others. However, to look at practical cases we sometimes practice endogenous decision-making within the constraints of knowledge, time, or some other exogenous factors. Considering the referendum, it is unrealistic to collect all of the citizens' meta-preferences for all possible SCFs at each level. Though some of our results do not completely depend on the completeness of procedural sets, there is room to study PCRs under the restriction of procedural sets. The second point is the extreme richness of the preference domain.

While the extreme consequentialist domain is too small to deal with procedural satisfaction or the concept of justice, our preference domain allows for such peculiar meta-preferences, as "for any level $p \in \mathbf{N}$, if p is a prime number, I prefer my dictatorship to all the other SCFs. Otherwise, I am indifferent for all the SCFs." It is perhaps of less practical importance for a real-world procedure to be fully prepared to deal with such an implausible preference. There is room to determine practically what kind of meta-preferences people actually have in mind. We need to take into account the results of recently developed experimental approaches for endogenous procedural choice (Weber [14], Ertan, Page, and Putterman [15]) in future studies.

Appendix (Proofs of the Propositions)

Lemma 1. Let $K \geq 1$ be either finite or infinite. If a level-K PCR E satisfies the 'if' part of ILC, then for all $x \in X, k \in \{1, 2, \ldots, K\}$ and $f, g \in F^k[x]$, we have $fI(E^k)g$.

Proof. We show the lemma inductively. Take arbitrary $x \in X$ and $f, g \in F^1[x]$. Then, by reflexivity of E^0, we have xE^0x, or $f(R^0)E^0g(R^0)$. Therefore, by the 'if' part of the ILC, we have fE^1g. Since this argument is symmetric over f and g and does not depend on what x is, we have for all x and for all $f, g \in F^1[x], fI(E^1)g$.

Take any level $k \in \{1, \ldots, K-1\}$. Assume that for all $f, g \in F^k[x], fI(E^k)g$. Let $u, v \in F^{k+1}[x]$ be any x-supporting rules of level $(k+1)$. Then, by the completeness of E^{k+1}, we have either $uE^{k+1}v$ or $vE^{k+1}u$. Suppose one of these, for example $uE^{k+1}v$, does not hold.

Then from the contraposition of 'if' part of the ILC, we have $\neg(u(R^k)E^k v(R^k))$. By the completeness of E^k, it is equivalent to $v(R^k)P(E^k)u(R^k)$. This contradicts the assumption, since $u, v \in F^{k+1}[x]$ implies $u(R^k), v(R^k) \in F^k[x]$ and therefore the assumption demands $u(R^k)I(R^k)v(R^k)$. Therefore, we have inductively shown that $fI(E^k)g$ holds for all $x \in X, k \in \{1, 2, \ldots, K\}$, and $f, g \in F^k[x]$.   ∎

Proof of Proposition 1 [1]. $2 \leq K < \infty$: Take any $x \in X$. Consider a meta-profile $R = (R^0, R^1, \ldots, R^{K-1}, R^K)$ such that for all $i \in N$, $F^K[f]P(O(R_i^K))F^K[g]$ for some $f, g \in F^{K-1}[x]$. By PWP on f and g, we have $fP(E^{K-1})g$. This contradicts Lemma 1, which demands that $fI(E^{K-1})g$.   ∎

[2] $K = \infty$: Take any $x \in X$ and $k \in N$. Take any $R^j \in W(F^j)^n (j = 0, 1, \ldots, k-1)$ and let $f, g \in F^k[x]$. Consider a meta-profile such that for all $i \in N$ and for all $l \in \{k+1, k+2, \ldots\}$, $uP(R_i^l)v$ for all $u \in F^l[f], v \in F^l[g]$. Note that $u, v \in F^l[x]$. At this point the PWP condition demands $fP(E^k)g$ while the Lemma 1 demands $fI(E^k)g$. Contradiction.   ∎

Proof of Proposition 2 [1]. The 'if' part is trivial. We show the 'only if' part. Suppose PCR E satisfies ILC and AF. Take any meta-profile $R \in D$, level $k \in \{1, \ldots, K\}$ and procedures $f, g \in F^k$. There are two possibilities concerning the similarity of f and g as a function. (1) There exists a level-$k - 1$ preference profile $\tilde{R}^{k-1} \in W(F^{k-1})$ such that $f(\tilde{R}^{k-1}) = g(\tilde{R}^{k-1})$. Consider a meta-profile $\tilde{R} = (R^0, R^1, \ldots, \tilde{R}^{k-1}, R^k, \ldots, R^K)$. Then, by Lemma 1, we have $fI(\tilde{E}^k)g$. On the other hand, we have $E^k|_{\{f,g\}} = \tilde{E}^k|_{\{f,g\}}$. Therefore, we have $fI(E^k)g$. (2) Otherwise, we consider SCF h over F^{k-1} such that $h(R^{k-1}) = f(R^{k-1})$ and $h(R'^{k-1}) = g(R'^{k-1})$ for all $R'^{k-1} \in W(F^{k-1}) - \{R^{k-1}\}$,. Since F^k is the set of all possible SCFs over F^{k-1}, such a SCF h is in F^k. By applying (1) we have $fI(E^k)$ h and $gI(E^k)h$. Thus, we have $fI(E^k)g$.

Finally we must show that the PCR E is also indifferent for any alternatives $x, y \in X$. However, it is easy from the 'only if' part of the ILC and the above fact that $fI(E^1)g$ for any $f, g \in F^1$.   ∎

Lemma 2.(Arrow [16]). If a SWF $f : W(A)^n \to W(A)$ satisfies WP and IIA, then there exists a dictator, where:

WP: $\forall S = (S_1, \ldots, S_n) \in W(A)^n, \forall a, b \in A, [aP(S_i)b \, \forall i \in N] \to aP(f(S))b$

IIA : $\forall S, S' \in W(A)^n, \forall a, b, \in A, S_i|_{\{a,b\}} = S_i'|_{\{a,b\}} \to f(S)|_{\{a,b\}} = f(S')|_{\{a,b\}}$

A dictator is an individual $i \in N$ such that for all $S \in W(A)$ and for all $a, b \in A$, $aP(S_i)$ b implies $aP(f(S))b$.

Proof of Proposition 2 [2]. Let E be a PCR that satisfies the 'only if' part of ILC, PWP, AF, and PIIA. Fix $(R^0, R^1, \ldots, R^{K-1}) \in W(X) \times W(F^1) \times \ldots \times W(F^{K-1})$ and let A be a set such that $A := \{F^K[f] | f \in F^{K-1}\}$. By AF, we have a function G such that for all R^K, $E^{K-1}(R^0, ..., R^K) = G(R^K)$. Moreover, by PIIA, there exists a function $G':W$ $(A)^n \to W(F^{K-1})$ such that $G(R^K) = G'\left(O(R_1^K), O(R_2^K), ..., O(R_n^K)\right)$ for all $R^K \in W$ (F^K). Let us consider another function $\mu : W(F^{K-1}) \to W(A)$ such that for all $\tilde{R}^{K-1} \in$ $W(F^{K-1})$ and f, $g \in F^{K-1}$, $f\tilde{R}^{K-1}g$ if and only if $F^K[f]\mu(\tilde{R}^{K-1})F^K[g]$. Construct a composite function $v := \mu \bigcirc G' : W(A)^n \to W(A)$. This is a SWF for the set A, and it is easy to see that our PWP and PIIA condition demands the WP and IIA for SWF v. Therefore, by Lemma 2 we have a dictator $j \in N$ (of SWF v) such that for all $S \in W$ (A) and for all $F^K[f]$, $F^K[g] \in A$, if $F^K[f]P\left(O\left(R_j^K\right)\right)F^K[g]$, then $fP(v(S))g$. By the way we have constructed μ, we have $fP(E^{K-1})g$. Since this argument does not depend on the value of R^0, R^1, ..., R^{K-1} or what f and g are, we can conclude that the set of axioms yield a vetoer over any pair in F^{K-1}.

We must only show the level under $K - 1$. Take any level $l \in \{0, 1, ..., K - 2\}$ and any alternatives/procedures $x, y \in F^l$. Assume that $F^K[x]P\left(O\left(R_j^k\right)\right)F^K[y]$. Take $f \in F^K$ $^{-1}[x]$ and $g \in F^{K-1}[y]$ such that $F^K[f'] \in G\left(O\left(R_j^K\right), B_x\right)$ and $F^K[g'] \in$ $G\left(O\left(R_j^K\right), B_y\right)$, where $B_x := \{F^K[h]|f \in F^{K-1}[x]\}$ and $B_y := \{F^K[h]|f \in F^{K-1}[y]\}$. Since $O\left(R_j^K\right)$ is a weak ordering over 2^{F^K}, $G\left(O\left(R_j^K\right), B_w\right)(w = x, y)$ are non-empty and we can take such f' and g'. Now, the definition of the operator $O(\ )$ and the assumption of $F^K[x]P(O(R))F^K[y]$ together yield $F^K[f']P(O(R))F^K[g']$. From the above paragraph we get $f'P(E^{K-1})g'$. Finally, iterating the 'only if' part of ILC we get xE^ky. ∎

Proof of Proposition 3 [1]. The counterexample showed in the proof of Proposition 1 also applies under D_C. ∎

[2] Let us consider a SWF $S:W(X)^n \to W(X)$ which satisfies the Pareto principle: for all preference profile of level 0 $R^0 \in W(X)$, $[xP(R_i^0)y$ for all $i \in N]$ implies $xP(S(R^0))$ y. Now we define PCR E_S such that (1) for all $x, y \in X$, xE^0y if and only if $xS(R^0)y$ and (2) for all $k \in \{1, 2, ..., K\}$ and $f, g \in F^k$, $fE^{k+1}g$ if and only if $f(R^k)E^kg(R^k)$. We will show that this E_S is actually a PCR and satisfies the ILC and PWP. The completeness of each $E_S^k(k = 0, 1, ..., K)$ is obvious. To show they are transitive, suppose $E_S^k \in W(F^k)$. Take any procedures $f, g, h \in F^{k+1}$ and assume $fE^{k+1}g$ and $gE^{k+1}h$. By (2) we have $f(R^k)$ $E^kg(R^k)$ and $g(R^k)E^kh(R^k)$. This implies $f(R^k)E^kh(R^k)$ by the transitivity of E^k. By (2) once again we get $fE^{k+1}h$. Since $E^0 \equiv S(R^0)$ is transitive, we have inductively that $E^k \in W(F^k)$ for all $k \in \{0, 1, ..., K\}$. Now we show that E_S satisfies the ILC and PWP, but the former is obvious because of (2). So we show PWP. Take any $k \in \{0, 1, ..., K - 1\}$ and $f, g \in F^k$. Suppose $F^l[f]P\left(O(R_i^l)\right)F^l[g]$ for all $l \in \{k + 1, ..., K\}$. Iterating the condition of extremely consequentialist, we have for all $l \in \{k + 1, ..., K\}$. Iterating the condition of extremely consequentialist, we have for all $i \in N$ $f(R^{k-1})P(R_i^{k-1})g(R^{k-1}), f(R^{k-1})(R^{k-2})P(R_i^{k-2})g(R^{k-1})(R^{k-2}), ..., xP(R_i^0)y$, where $f \in F^k[x]$ and g $\in F^k[y]$. The Pareto prinicple of $E^0 \equiv S(R^0)$ implies $xP(E^0)y$. Iteration of the contraposition of the 'only if' part of the ILC gives $fP(E^k)g$.

References

1. Dyer, J.S., Miles Jr., F.: An actual application of collective choice theory to the selection of trajectories for the Mariner Jupiter/Saturn 1977 project. Oper. Res. **24**(2), 220–244 (1976)
2. Koray, S.: Self-selective social choice functions verify arrow and Gibbard-Satterthwaite theorems. Econometrica **68**(4), 981–996 (2000)
3. Barbera, S., Jackson, M.O.: Choosing how to choose: self-stable majority rules and constitutions. Q. J. Econ. **119**, 1011–1048 (2004)
4. Kultti, K., Miettinen, P.: Stable set and voting rules. Math. Soc. Sci. **53**(2), 164–171 (2007)
5. Houy, N: Dynamics of Stable Sets of Constitutions, Mimeo (2005)
6. Semih, K., Slinko, A.: Self-selective social choice functions. Soc. Choice Welfare **31**(1), 129–149 (2008)
7. Diss, M., Vincent, M.: On the stability of a triplet of scoring rules. Theory Decis. **69**(2), 289–316 (2010)
8. Kultti, K., Miettinen, P.: Stability of constitutions. J. Public Econ. Theory **11**(6), 891–896 (2009)
9. Dietrich, F.: How to reach legitimate decisions when the procedure is controversial. Soc. Choice Welfare **24**(2), 363–393 (2005)
10. Taylor, A.D.: Social choice and the mathematics of manipulation. Cambridge University Press, Cambridge (2005)
11. Endriss, U.: Sincerity and manipulation under approval voting. Theor. Decis. **74**(3), 335–355 (2013)
12. Schnall, E., Greenberg, M.J.: Groupthink and the Sanhedrin: an analysis of the ancient court of Israel through the lens of modern social psychology. J. Manag. Hist. **18**(3), 285–294 (2012)
13. Blair, D.H., Pollak, R.A.: Acyclic collective choice rules. Econometrica: J. Econometric Soc. **50**, 931–943 (1982)
14. Weber, M.: Choosing voting systems behind the veil of ignorance: A two-tier voting experiment. Tinbergen Institute, No. 14-042/I (2014)
15. Ertan, A., Talbot P., Putterman, L.: Can endogenously chosen institutions mitigate the free-rider problem and reduce perverse punishment? WP 2005–13. Brown University, Department of Economics (2005)
16. Arrow, K.J.: Social Choice and Individual Values. Monograph/Cowles Foundation for Research in Economics at Yale University, 12. Wiley, New York (1963)

The Hidden Costs of the Door-in-the-Face Tactic in Negotiations

Ricky S. Wong[(⊠)]

Department of Supply Chain Management, School of Decision Sciences,
Hang Seng Management College, Shatin, Hong Kong, China
rickywong@hsmc.edu.hk

Abstract. Past studies have shown that Door-in-the-face tactics can induce compliance from negotiators. This research examines the hidden costs of the use of the Door-in-the-face tactic in dyadic negotiations. It shows that learning about opponents' use of this tactic affects negotiators' feelings of mistreatment and their behaviours in the subsequent negotiation. It also induces negotiators' covert, retaliatory behaviour. The results showed that negotiators who had dealt with opponents using the Door-in-the-face tactic made larger demands and attained higher outcomes in the subsequent negotiation. It was also found that feelings of mistreatment by opponents tended to spread over into future negotiations. Feelings of mistreatment mediated the effect of opponents' use of Door-in-the-face tactics on covert retaliation. Implications of results are discussed and directions for future research are given.

Keywords: Door-in-the-face technique · Covert retaliation · Feelings of mistreatment · Negotiation

1 Introduction

Negotiation is a complex process in which negotiators' outcomes are interdependent: outcomes received by negotiators rely on their behaviours and decisions and those of their counterparts [1]. To increase negotiators' abilities to claim surplus, persuasion research has examined different strategic techniques [2, 3]. However, these tactics might become ineffective once they are revealed to negotiators [4] and the consequences beyond negotiated outcomes should be factored in. One of the pervasive compliance strategies that has attracted theoretical and empirical attention is the door-in-the-face (DITF) tactic [4–6]. For the sake of brevity, I referred to negotiators using DITF tactic as requesters and to their opponents as targets.

The DITF tactic has been shown to be beneficial in different contexts, such as marketing, retail, participation in health research and charity donation [6–9]. Requesters attempt to convince targets to comply by offering a large request initially that is likely to be rejected; they then offer a more realistic offer that appears more reasonable compared to the first offer [4–6, 10]. In retail business, the consumers making a purchase decision increased from 15 % to 40 %, when comparing the sellers who did not use DITF with those who adopted the DITF strategy [7]. Meta-analyses found that increases in compliance rates were between 15 % and 27 % over control groups when the DITF approach

© Springer International Publishing Switzerland 2015
B. Kamiński et al. (Eds.): GDN 2015, LNBIP 218, pp. 61–74, 2015.
DOI: 10.1007/978-3-319-19515-5_5

was used [10, 11]. Studies also demonstrated that the effect size of DITF technique was between $r = .15$ and $r = .25$ [9, 12]. Although the DITF technique is shown to be effective, studies examining DITF tactic focussed only on short-term benefit. Notably absent from the existing research are the longer-term, adverse effects of the DITF tactic on negotiation relationship and targets' reaction, once it is revealed.

The DITF tactic has been documented for over 30 years [9] and organisations spend a lot of resources on negotiation training [13]. A large body of research provides individuals with prescriptive and strategic advice about how to attain better outcomes [11]. It is therefore conceivable that more people have a better understanding of different negotiation tactics, and that the DITF tactic may incur a cost when targets are familiar with the underlying psychology: targets may be less likely to fall prey to DITF. At the very least, more people recognise how to use DITF technique and will not comply if their opponent uses it.

In this paper, I present an exploratory study of the longer-term effects and hidden costs of the DITF tactic in dyadic negotiations. Colleagues, friends and collaborative partners in different organisations often become involved in repetitive negotiations instead of a one-off negotiation. The major research questions addressed in this study are: If targets learn that they have been manipulated by their opponents using the DITF tactic, would they behave differently in future interactions? How does it change negotiators' feelings towards their opponents and would they retaliate covertly when opportunities come along? Does DITF tactic backfire in subsequent negotiations, placing the requesters in a disadvantageous position in the longer run? These questions are addressed in the remainder of this paper by deriving insights from experimental data.

The theoretical rationale for the current research stems from past findings that negotiators are concerned with more than negotiated outcomes [14–17]. For example, Curhan, Elfenbein and Xu [14] have shown that subjective value (e.g. feelings about the relationship and negotiation process) is as important as negotiated outcomes to negotiators. It is possible that negotiators' feelings about the negotiation process may be adversely affected when they know that they have been manipulated by their counterparts. If targets deem the use of DITF tactic inappropriate, their reactions need to be considered and their behaviour in future negotiations with the same partner may change. In other words, although the DITF tactic can increase the requesters' negotiated outcomes, it does not mean that the DITF tactic is costless. Before proceeding to the potential costs of DITF tactic explored in this research, a brief review of why this tactic works is useful.

1.1   The Psychology Behind the DITF Tactic

There are a few psychological mechanisms explaining the effectiveness of the DITF tactic. One such mechanism is the norm of reciprocity. If someone does you a favour, you need to return him or her a favour [18]. When targets face requesters using the DITF tactic, targets may perceive the seemingly smaller, second request as a concession. After rejecting the initial large request, targets may feel obliged to reciprocate by accepting the subsequent smaller request [5, 9, 19, 20]. Diekmann [19] provided evidence that negotiators who were strangers tended to 'reciprocate' even when the money at stake was high.

The second explanation of why the DITF tactic is effective is that rejecting an initially extreme offer induces some levels of guilt feelings from targets. Studies examined the relationship between the amount of guilt induced by the rejection of the first unreasonable request and the amount of compliance with the subsequent request [12]. Agreeing to the second request reduces the level of guilt that targets feel and compliance with the second request was more likely to be elicited when the rejection of the initial large request elicited high levels of guilt [21, 22].

The DITF tactic is effective for a third reason self-presentation. Past findings support the contention that self-presentation is a factor that influences targets to comply with the DITF request [9, 23–25]. These studies argue that targets become motivated to comply with the second request, because they do not want to be considered to be unhelpful and uncooperative people [23, 24]. Millar (2002) has extended this explanation by the importance of friend (vs. stranger) making DITF request in raising targets' concern about self-presentation. Next, I will discuss the potential costs of using the DITF tactic.

1.2   Feelings of Mistreatment and Covert Retaliation

Although many have examined the benefits of DITF tactic and how it works, this study explores the two potential costs that may be incurred. One is the behaviour of targets in subsequent negotiation with the requesters, and another is the potential covert retaliation from targets. After learning that the DITF technique has been used by their opponents, it is possible for targets to demand more in subsequent negotiations. Learning of being manipulated may also affect the targets' choice to 'get back at' the requesters.

The cognitive mechanism of targets' reactions to DITF tactic in negotiation is underdeveloped. To explore the potential costs of using DITC tactic, I take an interpersonal perspective by looking at how learning counterparts' use of DITF tactic in the previous negotiation affects targets' responses and feelings. I also draw on past studies on the relationship between emotional expressions and affective reactions in observers [26, 27]. It is speculated that in negotiations one's use of persuasive tactics cannot be directly observed and is likely to be subtle. However, the revelation of such tactics may also change targets' behaviours and feelings towards their counterparts. Does learning requesters' use of DITF tactic result in targets' feelings of mistreatment?

Extending this logic, knowing that one's opponent is using the DITF tactic may influence the way targets feel they have been treated, which may then influence how they behave and make decisions in future interactions with the requesters. Accordingly, targets who have knowingly dealt with requesters using DITF tactic may feel mistreated (or inappropriately treated). Although never examined, there may be a spill-over effect of DITF tactic on feelings of mistreatment. The speculation is that feelings of mistreatment may spread to future encounters, even though the requesters do not use the same tactic in the subsequent negotiation. Such feelings have strategic importance in negotiations. These feelings may generalise to inform targets' perceptions in subsequent negotiations with the same opponents. It is also expected that other negative reactions may be present after they have learnt the requesters' use of DITF tactic. In particular, targets may become less willing to engage in future interaction with the DITF requesters. To examine the speculation, the relationships among learning opponents' use of the DITF tactic, targets' feelings of mistreatment and willingness to engage with the same requesters in future are explored in this study.

If targets learn that they have been manipulated by their opponents using the DITF tactic, will they negotiate more aggressively in subsequent negotiations with the same requesters? In fact, one who has learnt that s/he has been manipulated previously may have an incentive to get back at the opponent. According to the retaliation hypothesis, targets may have a desire to get even and thus demand more in the later negotiation because of the negative impression that the targets have developed of their counterparts in the previous negotiation. Specifically, I expect that targets, who learn that requesters have used the DITF tactic, demand more and receive more surplus in the later negotiation than those who have not dealt with DITF requesters in previous negotiation.

Exploring another potential cost of using the DITF tactic, previous research on organisational retaliation behaviour and on negotiators' retaliatory reactions to counterparts' display of negative emotions is helpful. When employees (or third parties) perceive that fairness is violated in workplace, more retaliatory behaviours are observed [15, 28–30]. Similarly, when negotiators feel that they are appropriately treated during negotiation, the effects are usually positive; when they are mistreated, the negative effects are significant [16, 31, 32]. Literatures on workplace retaliation suggest that employees tend to comply with unfair treatment but secretly punish their employers, through acts such as theft and sabotage [29, 33].

Similar patterns of results have been found in negotiation settings. Negotiators dealing with angry counterparts are more likely to show overtly concessionary and covertly retaliatory responses [16]. When opportunities for such covert retaliation are available to negotiators, the current study investigates how targets react to the use of DITF tactic, to complement past studies on the DITF tactic. When retaliation is less risky, feelings of mistreatment may result in greater overt retaliatory behaviour. While an impasse in subsequent negotiations may be costly especially when targets do not have a good BATNA (Best Alternative to the Negotiated Agreement), it does not mean that targets will not pursue other means to punish requesters using DITF secretly. It begs a question: when there are opportunities for targets to get back at requesters anonymously, does learning of being manipulated by requesters induce covert retaliatory behaviours from targets? I expect that targets who have knowingly dealt with requesters using DITF tactic are more likely to retaliate covertly than those whose counterparts have not used the DITF tactic.

Finally, this study explores the possibility that targets' feelings of mistreatment play a mediating role. That is, a negative impression of an opponent mediates the relationship between revelation of opponents' DITF tactic and targets' covert retaliatory behaviour. Targets' decision to covertly retaliate depends on, at least partially, feelings of mistreatment.

2 Method

2.1 Participants and Procedure

One hundred and fifty participants (71 females, 79 males; M_{age} = 21.92 years, SD = 1.45) from Hang Seng Management College (Hong Kong) took part in two negotiation simulations that involved the sale of smart phones, for course credit. To explore the hidden costs of DITF tactic, I engaged participants in two subsequent face-to-face negotiations, Negotiation 1 and Negotiation 2. Participants learned that they would be assigned the role of either seller or the buyer of smart phones. They were randomly assigned into one of the two experimental conditions: Control and Door-in-the-face (DITF). Requesters (i.e. negotiators using the DITF tactic) were the sellers in the DITF condition.

To ensure that the DITF tactic was deployed effectively, thirty-eight students, who were blind to hypotheses, were told that they had been randomly assigned the role of seller. These students were enrolled in a 14-week negotiation course and they had learned about the rationale of DITF tactic in the last two weeks of the negotiation course. Before the start of this experimental study, one full training session (45 m) was used to train these students (the sellers in the DITF condition) how to use DITF strategy in a natural manner. During this training session, the experimenters explained to the sellers in the DITF condition how DITF technique worked. In the DITF condition, the requesters (i.e. sellers) were instructed to use the DITF tactic *only* in Negotiation 1. After completing Negotiation 1, buyers and sellers were sent to different rooms. Participants were asked to complete a questionnaire regarding their feelings towards their counterparts. Having finished the questionnaire, participants met with the same opponents and proceeded to Negotiation 2. After completing Negotiation 2, targets were asked to complete another questionnaire and to take part in an exercise that was described as "unrelated" to negotiation. The questionnaires were used to assess participants' appraisals of their feelings to be treated in Negotiations 1 and 2, examining whether feelings towards counterparts would spread over into future interaction, as implied by the spill-over hypothesis. They were shown short descriptions of four different tasks (i.e. two were positive and two were negative) and were asked to indicate the extent to which they would like their opponents to perform in each task. A final questionnaire was given to targets, which elicited their willingness to interact with the same opponents in future.

Negotiation 1. The first negotiation involved a single, distributive issue. Participants were either a buyer or seller of smart mobile phones. They were given 15 min to reach an agreement. Both buyers and sellers were provided with their own payoff charts (see Appendix).

Negotiation 2. Participants were paired with the same partner, who in this case did not use the DITF tactic in all experimental conditions. From this negotiation, participants could earn a maximum point of 8,500 points. Negotiation 2 involved four issues to be resolved, including delivery time, warranty, price and quantity. Participants were given 30 min to reach a deal and a disagreement would result in zero points. Again, participants were given their payoff schedules prior to Negotiation 2 (see Appendix). Pilot study found that 15 min and 30 min were more than ample for Negotiations 1 and 2.

2.2   Experimental Manipulation

There were two experimental conditions to which participants were randomly assigned. In the DITF condition, requesters (i.e. sellers) were instructed to initially make an extreme request to targets, the buyers, ($3,000), followed by a smaller request ($2,600). In the control group, requesters were not instructed to use the DITF tactic. In other words, the buyers were the targets in the DITF condition. After Negotiation 1, the experimenter told the targets (i.e. buyers) in the DITF condition that the sellers had used the DITF tactic technique. Prior to Negotiation 2, requesters in DITF condition were told not to use the DITF tactic in the upcoming negotiation.

2.3   Dependent Measures

Feelings of Mistreatment. Targets' feelings of mistreatment were assessed using an item (*"The seller treated me in an inappropriate manner in the previous negotiation"*). It was scored on a 5-point Likert scale (1 = *strongly disagree* to 5 = *strongly agree*). A higher score indicated greater mistreatment. Targets in both experimental conditions were asked to complete this measurement after Negotiations 1 and 2. A higher score indicated that targets felt mistreated during the negotiation.

Willingness to Engage in Future Interaction. Targets' willingness to interact with their counterparts in future were examined using an item ("*I would like to interact with the same opponent again in future*"). It was scored on a 5-point Likert scale (1 = strongly disagree to 5 = strongly agree). A higher score indicated that targets were more willing to engage in future interaction with requesters. Participants in both experimental conditions were asked to complete this measurement after Negotiation 2.

Negotiated Outcomes and Final Offers. Negotiated outcomes obtained in Negotiations 1 & 2 were measured. Participants were instructed to record the offers that they made during the negotiations. Targets' final offers and their attained outcomes in Negotiation 2 were used to examine if learning requesters' DITF tactic after Negotiation 1 made them demand more and obtain higher outcomes in the subsequent negotiation (Negotiation 2).

Covert Retaliation. I used an existing task assignment to measure the covert retaliatory behaviour of targets [16]. Wang et al. (2012) found that Tasks 1 and 3 were perceived highly attractive and appealing whereas Tasks 2 and 4 were considered highly unattractive and unappealing. Targets were told that their decisions would not be disclosed to their counterparts. The details of tasks were shown to targets, as in Wang et al. (2012)'s study:

"*Task 1:   This task studies positive emotions in the workplace. If you choose to perform this task, you would be induced to feel positive emotions. In particular, you are likely to experience a variety of positive feelings, such as satisfaction, happiness, respect, amusement, and enthusiasm*

Task 2:   *This task studies negative emotions in the workplace. If you choose to perform this task, you will be induced to feel negative emotions. In particular, you are likely to experience a variety of negative feelings, such as frustration, sadness, disrespect, guilt, and shame*

Task 3:   *This task studies investment strategies. In this task, you will be shown several investment tactics that have been proven to be successful. You will be asked to invest the $20 you earned for participating in this experiment using these strategies. Please be aware that you may win $10 or lose $1 if you choose to do this task. However, the chance of winning money is much higher than that of losing money. Based on previous research, the odds of winning $10 are 95 % and the odds of losing $1 are 5 %*

Task 4:   *This task studies the effects of gambling and risky behaviours. You will be asked to gamble with the $20 you earned for participating in this experiment. Please beware that you may win $1 or lose $10 if you choose to do this task. In addition, the chance of losing money is much higher than that of winning money. Based on previous research, the odds of winning $1 are 5 % and the odds of losing $10 are 95 %.*"

Participants were reminded that these four tasks were not related to the negotiation simulations. After Negotiation 2, targets were asked to indicate how much they wanted their opponents to perform in each of the four tasks (1 = *not at all* and 7 = *very much*). Higher scores for Tasks 2 and 4 reflected stronger tendencies to covertly retaliate. Participants who covertly retaliated against their counterparts would assign higher scores to Tasks 2 and 4 than those who did not. The reverse applied to Tasks 1 and 3.

3   Results

3.1   Manipulation Check

An independent t test showed that requesters in the DITF condition gained significantly higher outcomes those in the control condition ($M = 832$, $SD = 158$ vs. $M = 625$, $SD = 127$) in Negotiation 1, $t(71) = 6.16$, $p < 0.0005$. It suggests that DITF technique worked as intended. The result replicates previous findings that DITF technique helps requesters to obtain higher payoffs.

3.2   Feelings of Mistreatment

Did learning opponents' use of DITF lead to targets' feelings of mistreatment? Did targets' feelings of mistreatment spread over into future negotiations with the same requesters? An independent t test revealed that targets in the DITF condition after completing Negotiation 1 showed a higher level of mistreatment ($M_{DITF} = 3.58$) than those in the control group ($M_{Control} = 2.57$), $t(73) = -4.90$, $p < 0.0005$. To examine whether the effect of learning opponents' use of DITF tactic spilled over to subsequent negotiation, targets' feelings of mistreatment were measured again after completing

Table 1. Means (standard deviations) of dependent measures. Comparisons of means were made within each row. *Significant at $p < 0.01$; **significant at $p < 0.0005$.

Targets' responses	Control group	Door-in-the-face group
	$(n = 37)$	$(n = 38)$
Negotiation 1: Feelings of Mistreatment	2.57** (1.02)	3.58** (0.76)
Negotiation 2: Feelings of Mistreatment	2.32** (0.92)	3.16** (0.72)
Willingness to Engage in Future Interaction	3.16* (0.86)	3.62* (0.68)
Negotiation 2: Final Offers	5,664** (900)	6,522** (741)
Negotiation 2: Outcomes	5,609** (915)	6,362** (838)

Negotiation 2. Targets in the DITF condition indicated greater mistreatment ($M_{DITF} = 3.16$) than those in the control group ($M_{Control} = 2.32$), $t(73) = -4.38$, $p < 0.0005$. The results suggest that targets in the DITF condition thought that they were inappropriately treated even when opponents did not use DITF tactic in Negotiation 2 (Table 1).

3.3   Willingness to Engage in Future Interaction

Result revealed that targets in the DITF condition were less willing to engage in interaction with the requesters in future ($M_{Control} = 3.62$) than targets in the control condition ($M_{DITF} = 3.16$), $t(73) = 2.60$, $p < 0.01$. No difference in willingness to engage in future interaction between male targets and female targets was found, $t(73) = -0.48$, $p = ns$.

3.4   Targets' Final Offers and Outcomes in Negotiation 2

If targets learned that they had been manipulated by their opponents using the DITF tactic, would they demand more and obtain higher outcomes in the subsequent negotiation? In Negotiation 2, targets in the DITF condition made statistically higher final offers ($M_{DITF} = 6,522$) than control targets ($M_{Control} = 5,664$), $t(73) = 4.51$, $p < 0.0005$. The finding also showed that targets in the DITF condition attained higher outcomes ($M_{DITF} = 6,362$) than those in the control condition ($M_{Control} = 5,609$), $t(72) = 3.69$, $p < 0.0005$. No effects of gender were found on targets' final offers, $t(73) = 0.28$, $p = ns$, and final offers, $t(73) = 0.50$, $p = ns$.

3.5   Covert Retaliation

Would targets retaliate covertly after learning opponents' use of DITF tactic? Illustrated in Table 2 shows how much targets wanted their counterparts to perform in the four different tasks. Regarding the negative tasks (Tasks 2 & 4), targets in the DITF condition indicated that they wanted their opponents to take part in the negative tasks to a greater extent ($M_{DITF} = 5.50$; $M_{DITF} = 5.61$) than control targets ($M_{Control} = 2.11$; $M_{Control} = 2.27$), $t(73) = -16.33$, $p < 0.0005$; $t(73) = -12.70$, $p < 0.0005$. In contrast, targets in the DITF condition tended to be less likely to assign positive tasks to their opponents (M_{DITF} = 2.53; M_{DITF} = 2.39) than those in the control condition ($M_{Control}$ = 5.51; $M_{Control}$ = 6.08), $t(73)$ = 11.01, $p < 0.0005$; $t(73)$ = 14.08, $p < 0.0005$. The findings lend support to the speculation that learning opponents' use of DITF tactic would induce targets' covert, retaliatory behaviours. No effect of gender was found on targets' retaliatory behaviour, $t(73)$ = 0.42, $p = ns$.

Table 2. Means (standard deviations) of task assignments. Comparisons of means were made within each row. **Significant at $p<0.0005$.

	Control group (n = 37)	Door-in-the-face group (n = 38)
Task 1 - Positive task	5.51** (0.96)	2.53** (1.35)
Task 2 - Negative task	2.11** (0.70)	5.50** (1.06)
Task 3 - Positive task	6.08** (0.72)	2.39** (1.42)
Task 4 - Negative task	2.27** (0.96)	5.61** (1.29)

3.6 Mediation Analyses Between Covert Retaliation and Feelings of Mistreatment

Finally, I examined whether targets' feelings of mistreatment mediated the relationship between learning others' use of DITF tactic and their retaliatory behaviour. I demonstrated that learning counterparts' use of DITF tactic in the previous negotiation predicted targets' feelings of mistreatment, β = 1.01, $t(73)$ = 4.90, $p<.0005$. And, feelings of mistreatment were associated with targets' covert retaliatory behaviour, β = 0.42, $t(73)$ = 3.94, $p<.0005$. When controlling for feelings of mistreatment, the effect of DITF tactic on retaliatory behaviour was still statistically significant, β = 3.39, $t(73)$ = 3.94, $p<.0005$; however, this effect became somewhat lower, β = 2.96, $t(73)$ = 13.55, $p<.0005$. To test the significance of the indirect effect (i.e., the path through the mediator), I followed a bootstrapping procedure [34]. The result of 1,000 resamples demonstrated that zero fell outside of the 95 % CI of the indirect effect of feeling of mistreatment (95 % CI [0.18, 0.85]). Therefore, feelings of mistreatment partially mediated the relationship between learning others' use of DITF tactic and retaliatory behaviour.

4 Discussion

Although the DITF tactic appears to help negotiators get what they want at the negotiation table, the costs of using it may be substantial. The purpose of this paper is to examine whether the use of this tactic can have hidden costs beyond the immediate economic benefits. Results from the current study replicated previous findings that DITF technique would be beneficial to the requesters. This notion entails assumptions that negotiators are involved in a one-off negotiation and that the use of DITF is never revealed to targets. The results from the current research suggest that the hidden costs are targets' higher demands in the subsequent negotiation and also their covert retaliatory behaviour.

I found that learning one is being manipulated by a counterpart using DITF tactics induced targets' negative feelings towards counterparts, precisely, feelings of being inappropriately treated. These negative feelings remained significant even when requestors did not use the DITF tactic in the later negotiation. It was also found that targets who had faced DITF requesters in Negotiation 1 and learnt of the use of DITF tactics made higher demands and received more outcomes in Negotiation 2, than targets who had not. Consistent with the prediction in this research, when opportunities to covertly retaliate were available to targets, they tended to do so. In fact, feelings of mistreatment partially mediated the effect of learning others' use of DITF tactics on covert retaliation. Targets' willingness to engage in future interaction with the DITF requesters was also reduced after requesters' use of DITF tactic. A practical implication is that when targets and requesters are involved in similar business networks or work in the same organisation, it is believed that more opportunities are available for targets to retaliate in real-life situations.

Being the first to illustrate the effects of DITF tactics on targets' covert retaliation, willingness to co-operate in the future and feelings of mistreatment, the current findings makes important contributions to research on the benefits of using DITF technique in negotiations. First, this paper extends previous work on the DITF tactic and has taken an interpersonal perspective, by examining how learning counterparts' use of the DITF tactic in a negotiation influenced targets' feelings and behaviours in a subsequent negotiation.

Second, the results add to the growing research on negotiation persuasion techniques, by showing that the revelation of the DITF tactic may continue to shape targets' behaviour and decisions in future interactions, even if this tactic is not adopted in the subsequent negotiation. The implications of current findings may be beyond the use of DITC tactics and speak to other types of negotiation strategies. These tactics, such as foot-in-the-door compliance strategy [4] and the framing technique [35, 36] have been shown to affect negotiators' decisions. They might, however, yield similar adverse effects on targets' behaviours and feelings. Central to theoretical implications of the current findings is whether negative feelings towards counterparts may be elicited after learning that opponents have used other types of persuasion tactics. Egocentric interpretations of fairness are common in negotiation settings, which have drawn attention from negotiation scholars [37, 38]. In this respect, targets and requesters may have diverse interpretations of inappropriateness or perceptions of fairness when it comes to the use of persuasion tactics. If so, this may lead to more conflict and hostility between negotiators.

Although support was found for targets' covert retaliatory behaviour, one might wonder whether targets would still choose to retaliate if it comes with a price. Studies on retaliatory responses to others' expressions of anger may help shed light into this. They have shown that negotiators retaliate to angry counterparts in coalition negotiations and in ultimatum bargaining even when retaliation leads to lower outcomes or nothing [32, 39]. Future studies should consider situations where covert retaliation may cost. It is also important to note that all the participants involved in this study are Hong Kong Chinese. More research is necessary to examine if participants from different ethnic backgrounds react to the revelation of opponents' use of DITF tactic similarly. Another avenue for future research is to consider the changes in targets' behaviour when they have a strong best alternative to a negotiated agreement (BATNA). It means that the cost of walking away from subsequent negotiations is lower (note that targets did not have a BATNA in this research). Also, fairness may come to the fore once the DITF tactic is revealed to targets. Targets may attempt to bring DITF requesters back to a balanced level and distribute power evenly by retaliating in a punishing way.

This study is not free from limitation. It assumed that targets in the DITF condition did not realise the DITF tactic and that requesters' use of DITF tactic was revealed by the experimenter. Perhaps negotiation training (e.g. learning the logic of DITF tactics in a classroom) is sufficient for negotiators to notice others' use of such a tactic, without the revelation by a third party. Future research is needed to establish whether learning the DITF tactic or other persuasive tactics prior to negotiation leads to different results. And, what if targets are experienced negotiators? If they are already familiar with the logic behind the tactic(s) used by requesters, it is very likely that they do not comply with such requests. The advantages of DITF tactic may diminish. Another potential limitation is that no financial incentive was offered to the participants. Since the current study concerns the absolute differences across experimental conditions, any effect of financial incentive (versus no incentive) should not interfere with the validity of findings. Nevertheless, it is worth examining whether performance-based incentive will change the magnitude of the DITF effects. Finally, this study used single-item measures of negotiators' feelings towards opponents and their willingness to engage in future interaction. Future research should adopt multi-item measures to strengthen the validity and reliability of constructs.

This paper reveals that the DITF tactic may hurt negotiators using it by bearing substantial hidden costs of which they may be unaware (e.g. covert retaliation, targets' reduced willingness to interact again, etc.). As an employer negotiating with a valued new employee, where the DITF tactic might later be discovered, and where there are opportunities for retaliation, DITF carries risk. In a one-shot negotiation (e.g. the sale of a house), the hidden costs are likely to be fewer. Therefore, negotiators need to consider the context specific utility of such a tactic.

Acknowledgments. I thank Tony Chan, Racy Liu and KC Tam for their assistance in conducting this research.

Appendix

Negotiation 1 - Payoff schedules

Price	Buyer's payoff	Seller's payoff
$3000	100	1100
$2800	200	1000
$2600	300	900
$2400	400	800
$2200	500	700

(*Continued*)

Negotiation 1 - Payoff schedules (*Continued*)

Price	Buyer's payoff	Seller's payoff
$2000	600	600
$1800	700	500
$1600	800	400
$1400	900	300
$1200	1000	200
$1000	1100	100

Negotiation 2 - Payoff schedules

Price	Buyer's payoff	Seller's payoff	Quantity	Buyer's payoff	Seller's payoff
$2000	300	2500	100	2500	300
$1800	600	2250	200	2250	600
$1600	900	2000	300	2000	900
$1400	1200	1750	400	1750	1200
$1200	1500	1500	500	1500	1500
$1000	1800	1250	600	1250	1800
$800	2100	1000	700	1000	2100
$600	2400	750	800	750	2400
$400	2700	500	900	500	2700
$200	3000	250	1000	250	3000
Warranty (months)	Buyer's payoff	Seller's payoff	*Delivery Time (weeks)*	Buyer's payoff	Seller's payoff
12	100	2000	1	2000	100
13	200	1800	1.5	1800	200
14	300	1600	2	1600	300
15	400	1400	2.5	1400	400
16	500	1200	3	1200	500
17	600	1000	3.5	1000	600
18	700	800	4	800	700
19	800	600	4.5	600	800
20	900	400	5	400	900
21	1000	200	5.5	200	1000

References

1. Thompson, L., Wang, J., Gunia, B.C.: Negotiation. Annu. Rev. Psychol. **61**, 491–515 (2010)
2. Sinaceur, M., Neale, M.A.: Not all threats are created equal: how implicitness and timing affect the effectiveness of threats in negotiations. Group Decis. Negot. **14**(1), 63–85 (2005)

3. Van Kleef, G.A., De Dreu, C.K.W.: Longer-term consequences of anger expression in negotiation: retaliation or spillover? J. Exp. Soc. Psychol. **46**(5), 753–760 (2010)
4. Cialdini, R.B., Goldstein, N.J.: Social influence: compliance and conformity. Annu. Rev. Psychol. **55**, 591–621 (2004)
5. Cialdini, R.B., Vincent, J.E., Lewis, S.K., Catalan, J., Wheeler, D., Darby, B.L.: Reciprocal concessions procedure for inducing compliance: the door-in-the-face technique. J. Pers. Soc. Psychol. **32**(2), 206–215 (1975)
6. Feeley, T.H., Anker, A.E., Aloe, A.M.: The door-in-the-face persuasive message strategy: a meta-analysis of the first 35 years. Commun. Monogr. **79**(3), 316–343 (2012)
7. Ebster, C., Neumayr, B.: Applying the door-in-the-face compliance technique to retailing. Int. Rev. Retail Distrib. Consum. Res. **18**(1), 121–128 (2008)
8. Mowen, J.C., Cialdini, R.B.: On implementing the door-in-the-face compliance technique in a business context. J. Mark. Res. **17**, 253–258 (1980)
9. Turner, M.M., Tamborini, R., Limon, M.S., Zuckerman-Hyman, C.: The moderators and mediators of door-in-the-face requests: is it a negotiation or a helping experience? Commun. Monogr. **74**(3), 333–356 (2007)
10. Dillard, J.P.: The current status of research on sequential-request compliance techniques. Per. Soc. Psychol. B. **17**, 283–288 (1991)
11. Rodafinos, A., Vucevic, A., Sideridis, G.D.: The effectiveness of compliance techniques: foot in the door versus door in the face. J. Soc. Psychol. **145**(2), 237–239 (2005)
12. Hale, J.L., Laliker, M.: Explaining the door-in-the-face: is it really time to abandon reciprocal concessions? Commun. Stud. **50**(3), 203–210 (1999)
13. Movius, H.: The effectiveness of negotiation training. Negot. J. **24**(4), 509–531 (2008)
14. Curhan, J.R., Elfenbein, H.A., Xu, H.: What do people value when they negotiate? mapping the domain of subjective value in negotiation. J. Pers. Soc. Psychol. **91**, 493–512 (2006)
15. Ferguson, M., Moye, N., Friedman, R.: The lingering effects of the recruitment experience on the long-term employment relationship. Negot. Confl. Manage. Res. **1**(3), 246–262 (2008)
16. Wang, L., Northcraft, G.B., Van Kleef, G.A.: Beyond negotiated outcomes: the hidden costs of anger expression in dyadic negotiation. Organ. Behav. Hum. Decis. **119**(1), 54–63 (2012)
17. Welsh, N.A.: Perceptions of fairness in negotiation. Marquette Law Rev. **87**, 753–767 (2004)
18. Goulder, A.W.: The norm of reciprocity: a preliminary statement. Am. Sociol. Rev. **25**(2), 161–178 (1960)
19. Diekmann, A.: The power of reciprocity: fairness, reciprocity, and stakes in variants of the dictator game. J. Confl. Resolut. **48**, 487–505 (2004)
20. Reeves, R.A., Baker, G.A., Boyd, J.G., Cialdini, R.B.: The door-in-the-face technique: reciprocal concessions vs self-presentational explanations. J. Soc. Behav. Pers. **6**(3), 545–558 (1991)
21. O'Kleef, D.J., Figge, M.: A guilt-based explanation of the door-in-the-face influence strategy. Hum. Commun. Res. **24**, 64–81 (1997)
22. O'Kleef, D.J., Figge, M.: Guilt and expected guilt in the door-in-the-face technique. Commun. Monogr. **66**, 312–324 (1999)
23. Millar, M.G.: The effectiveness of the door-in-the-face compliance strategy on friends and strangers. J. Soc. Psychol. **142**(3), 295–304 (2002)
24. Pendleton, M., Batson, C.: Self-presentation and the door-in-the-face technique for inducing compliance. Pers. Soc. Psychol. B. **5**, 77–81 (1979)
25. Tusing, K.J., Dillard, J.P.: The psychological reality of the door-in-the-face: it's helping not bargaining. J. Lang. Soc. Psychol. **19**(1), 5–25 (2000)

26. Barsade, S.G.: The ripple effect: emotional contagion and its influence on group behavior. Admin. Sci. Q. **47**, 644–677 (2002)
27. Van Kleef, G.A.: How emotions regulate social life: the emotions as social information (EASI) model. Curr. Dir. Psychol. Sci. **18**, 184–188 (2009)
28. Barclay, L., Skarlicki, D., Pugh, S.: Exploring the role of emotions in injustice perceptions and retaliation. J. Appl. Psychol. **90**(4), 629–642 (2005)
29. Skarlicki, D.P., Folger, R.: Retaliation in the workplace: the role of distributive, procedural, and interactional justice. J. Appl. Psychol. **82**(3), 434–443 (1997)
30. Skarlicki, D.P., Rupp, D.E.: Dual processing and organizational justice: the role of rational versus experiential processing in third-party reactions to workplace mistreatment. J. Appl. Psychol. **95**(5), 944–952 (2010)
31. Kim, S.H., Smith, R.H.: Revenge and conflict escalation. Negot. J. **9**(1), 37–43 (1993)
32. Kopelman, S., Rosette, A.S., Thompson, L.: The three faces of Eve: an examination of the strategic display of positive, negative, and neutral emotions in negotiations. Organ. Behav. Hum. Dec. **99**, 81–101 (2006)
33. Baron, R.A., Neuman, J.H.: Workplace violence and workplace aggression: evidence on their relative frequency and potential causes. Aggressive Behav. **22**, 161–173 (1996)
34. Preacher, K.J., Hayes, A.F.: Asymptotic and resampling strategies for assessing and comparing indirect effects in multiple mediator models. Behav. Res. Methods **40**(3), 879–891 (2008)
35. Chang, L., Cheng, M., Trotman, K.T.: The effect of framing and negotiation partner's objective on judgments about negotiated transfer prices. Account. Org. Soc. **33**(7–8), 707–714 (2008)
36. Curseu, P.L., Schruijer, S.: The effects of framing on inter-group negotiation. Group Decis. Negot. **17**, 347–362 (2008)
37. Gelfand, M.J., Higgins, M., Nishii, L.H., Raver, J.L., Dominguez, A., Murakami, F.: Culture and egocentric perceptions of fairness in conflict and negotiation. J. Appl. Psychol. **87**(5), 833–845 (2002)
38. Thompson, L., Loewenstein, G.: Egocentric interpretations of fairness and interpersonal conflict. Organ. Behav. Hum. Dec. **51**, 176–197 (1992)
39. Van Beest, I., Van Kleef, G.A., Van Dijk, E.: Get angry, get out: the interpersonal effects of anger communication in multiparty negotiation. J. Exp. Soc. Psychol. **44**, 993–1002 (2008)

Preference Analysis
and Decision Support

Warsaw School of Economics

Understanding and Using the Group Decision Analysis Model

Ralph L. Keeney[(⊠)]

Fuqua School of Business, Duke University, Durham, NC 27708, USA
keeney@duke.edu

Abstract. Decision analysis is usually thought of as a model for decisions with a single decision-maker. Many attempts to extend decision analysis to group decisions have led to results indicating how it cannot be done. Other analyses, such as the well-known impossibility theorem of Arrow (1963) [1], have tried to combine rankings of alternatives by individual group members to produce a group ranking. As a result, there had been no logically consistent way to extend the principle of decision analysis to group decisions. A different approach was used in Keeney (2013), where each member of a decision-making group could have a different decision frame for their common decision. Using the assumptions of decision analysis for each member's analysis of their group decision and using an analogous set of decision analysis assumptions for the group decision to combine the member's decision analyses produced a group decision analysis model. This article discusses the concepts and intuitive logic for the model and practical aspects of applying it.

Keywords: Group decisions · Framing · Expected utility · Group utility · Group judgment principle

1 Introduction

The term group decision does not have a universally accepted meaning. It can refer to a collection of very broad classes of decisions that includes voting, negotiations, arbitration, mediation, organizational decisions, social planning, and decisions with stakeholders. Hence, any technical work on group decisions must carefully define the meaning of a group decision for that work. In this article, group decisions are decisions for which a group of two or more individuals must collectively select an alternative from a set of two or more alternatives. A typical decision of interest is where a group of 2 to 10 members in a company, organization, or family have the opportunity and responsibility for making a joint decision.

All of the original foundational work in developing decision analysis focuses on the case where the decision-maker is an individual [13]. Clearly, each individual faces many important decisions worthy of thought. In addition, this convenient assumption allowed one to avoid the additional complexities of group decisions and focus on the basic logic for how to model an important decision to best address its complexity and inform the decision-maker about the relative desirabilities of the alternatives and about the effects of various factors on those relative desirabilities.

© Springer International Publishing Switzerland 2015
B. Kamiński et al. (Eds.): GDN 2015, LNBIP 218, pp. 77–86, 2015.
DOI: 10.1007/978-3-319-19515-5_6

There have been several attempts to extend decision analysis to group decisions [5, 10, 14]. These attempts have all implicitly assumed that each of the members of the decision-making group accepts a commonly held frame of their decision problem. This means that each member of the group (i) is concerned with the same possible consequences of the alternatives and (ii) considers the same events to be those relevant to the decision. Each of these investigations then assumed that the group members needed to combine individual probabilities for common events into group probabilities for those events and individual utilities for common consequences into group utilities for those consequences. All of this work led to the general conclusion that the logic of decision analysis could not be extended to group decisions.

The previous attempts to extend decision analysis to group decisions indicated that for such an extension to be successful, the approach cannot begin with a single group decision frame analogous to an individual decision frame. Eliminating this assumption and addressing a broader, and often more realistic, situation where group members may be concerned about different events and different consequences of their common group decision led to a general group decision analysis result [7]. This result, which used foundational assumptions analogous to the individual decision analysis assumptions, can be applied to group decisions using decision analysis techniques and procedures.

This article elaborates on logical and practical aspects of this group decision analysis model. Section 2 clarifies the intuitive logic behind the framing of the group decision analysis model. Section 3 presents the group decision analysis model. Section 4 summarizes background information leading to and motivating the results, and Sect. 5 discusses the logical intuition embedded in that result. Practical aspects of implementing the results are in Sect. 6, followed by a short summary and comments in Sect. 7.

2   Framing the Group Decision Problem

To illustrate how reasonable it is for group members to have different frames for a common decision, the following example from [7] is useful. "Consider a relatively simple decision concerning two people planning to have dinner together at a restaurant. One individual wants to have a substantial discussion at a convenient location (e.g. not far away and with easy parking) and the other individual is interested in great food at an exciting location. As they have different objectives, they will be concerned about different consequences of their joint decision. This will also result in different events being relevant to each individual, as the first individual would be concerned with how quiet the restaurants are and about travel times to the restaurants, whereas the second individual may be concerned with whether the restaurants have exotic fusion dishes and a trendy clientele."

To solve such a decision, the process can be described as having two stages. In the first stage, the individuals will appraise the various possible restaurants. In the second stage, they will jointly share perspectives and reach a choice. Of course there would likely be discussion to communicate information to each other in both stages, and aspects of both stages may be occurring at the same time.

For more complex group decisions, the decision process may involve several group meetings over time with individual effort between meetings as each group member must evaluate the desirabilities of the alternatives for the group. In stage 1, each group member frames the group's common decision from his or her own perspective (i.e. decision frame) and evaluates alternatives, perhaps using an individual decision analysis model. In stage 2, the group uses the member's evaluations to collectively evaluate alternatives. This could be done using a discussion, a voting mechanism, or a systematic group decision analysis as described in [7].

It is useful to recognize that each group member's evaluation of alternatives in stage 1 should be viewed as an individual decision problem, where each individual is evaluating what he or she thinks is best for the group. These analyses can be done without any interaction with other group members, although information from others may be very useful in these individual analyses. Stage 2 can be considered to be the essence of the group decision, where the potentially different perspectives of the various decision-makers must be somehow balanced in an acceptable way to the members in order to evaluate alternatives from the collective group perspective.

3   The Group Decision Analysis Model

Before discussing the motivation for the group decision analysis model, it is useful to state the result. This both clarifies the subsequent discussion and succinctly defines the notation that is used in what follows.

In stage 1, each group member focuses on the following decision. Each member, denoted by I_m, $m = 1, \ldots, M$, must evaluate and choose among a set of alternatives A_n, $n = 1, \ldots, N$. He or she has specified a set of mutually exclusive and collectively exhaustive events E_{mj}, $j = 1, \ldots, J$, one of which will occur, and a set of consequences c_{nmj}, $n = 1, \ldots, N$, and $j = 1, \ldots, J$, that will result if alternative A_n is chosen and event E_{mj} then occurs. It is important to recognize that each consequence includes all things that matter to that member about the choice of an alternative.

The prescriptive assumptions that provide the logical foundation for the individual decision analysis model are listed in Table 1. Given this frame and assumptions, the individual decision analysis result that follows is that the member should evaluate alternative A_n using its expected utility

$$U_m(A_n) = {}_j\, p_m(E_{mj})\, u_m(c_{nmj}), \quad n = 1, \ldots, N, m = 1, \ldots, M, \qquad (1)$$

where:

$p_m(E_{mj})$ is member I_m's, $m = 1, \ldots, M$, probability for event E_{mj},
$u_m(c_{nmj})$ is member I_m's, $m = 1, \ldots, M$, utility for consequence c_{nmj},
$U_m(A_n)$ is member I_m's, $m = 1, \ldots, M$, expected utility for alternative A_n, and
Σ_j means to sum over all j.

It is generally accepted that the expected utility in (1) is an appropriate basis for analyzing individual decisions. There has been plenty of experience applying the individual decision analysis model, so we will not elaborate on that stage 1 topic.

Table 1. Decision analysis assumptions for individual decisions (from Pratt et al., 1964).

Principles of consistent behavior		
IT	Transitivity	As regards any set of lotteries among which the decision maker has evaluated his or her feelings of preference or indifference, these relations should be transitive.
IS	Substitutability	If some of the prizes in a lottery are replaced by other prizes such that the decision maker is indifferent between each new prize and the corresponding original prize, then the decision maker should be indifferent between the original and the modified lotteries.
Principles for scaling preferences for consequences and judgments concerning events		
IP	Preferences	The decision maker can scale his or her preference for any consequence c by specifying a number $\pi(c)$ such that he or she would be indifferent between (1) c for certain, and (2) a lottery giving a probability $\pi(c)$ chance at c^* and a complementary chance at c^o.
IJ	Judgments	The decision maker can scale his judgment concerning any possible event E_j by specifying a number $p(E_j)$ such that he or she would be indifferent between (1) a lottery with consequence c^* if E_j occurs, c^o if it does not, and (2) a lottery giving a probability $p(E_j)$ chance at c^* and a complementary chance at c^o.

In stage 2, the group collectively focuses on the following decision: the group of $M \geq 2$ members must evaluate and choose among a set of alternatives A_n, $n = 1, \ldots, N$. Assuming that

1. each member accepts the decision analysis assumptions for individual decision-making in Table 1 for his or her analysis of the group decision,
2. the decision-making group accepts the group decision analysis assumptions in Table 2 for their group decision, and
3. the decision-making group accepts the group identical indifference assumption below,

the group expected utility of any alternative A_n, denoted $U_G(A_n)$, is

$$U_G(A_n) \quad = \Sigma_m\, w_m U_m(A_n) \; = \Sigma_m\, w_m\big(\Sigma_j\, p_m(E_{mj})u_m(c_{nmj})\big), \tag{2}$$

where the w_m, $m = 1, \ldots, M$, are scaling factors that sum to one, and $0 \leq w_m \leq 1$.
Group member I_m has a utility function that can be scaled by

$$u_m(c_m{}^o) = 0 \text{ and } u_m(c_m{}^*) = 1, \; m = 1, \ldots, M, \tag{3}$$

where any two of the $c_m{}^o$ consequences need not, but may, be the same, and the same circumstance holds for the $c_m{}^*$.

The group identical indifference assumption is necessary to relate the individual's preferences to the group's preferences. This assumption is the following: If some of the consequences in a lottery facing the group are replaced by other consequences such that the probabilities of all consequences relevant to each group member in the original and

Table 2. Decision analysis assumptions for group decisions.

Principles of consistent behavior		
GT	Group Transitivity	As regards any set of lotteries among which the decision making group has evaluated its feelings of preference or indifference, these relations should be transitive.
GS	Group Substitutability	If some of the prizes in a lottery are replaced by other prizes such that the decision making group is indifferent between each new prize and the corresponding original prize, then the decision making group should be indifferent between the original and the modified lotteries.
Principles for scaling preferences for consequences and judgments concerning events		
GP	Group Preferences	The decision making group can represent its preferences over the consequences $(c_1, ..., c_M)$ in terms of a group utility function.
GJ	Group Judgments	The decision making group agrees that there exists a representation of its judgments about the possible occurrence of any combination of the events $(E_1, ..., E_M)$ in terms of a joint probability distribution function.

modified lotteries are identical, then the decision-making group should be indifferent between the original and modified lotteries.

This assumption is less restrictive than a Pareto assumption, which would require that the group must be indifferent between lotteries when each of the individual group members was indifferent.

4 Motivation for the Group Decision Analysis Model

The inherent problems in developing a complete group decision analysis model are mainly related to stage 2. Indeed, much of the work on group decisions is framed as identifying a logically sound way to combine individual group member's preferences for alternatives to produce a group preference. Notationally, one wants to obtain group preferences P for all A_n, $n = 1, ..., N$, given the group member's preferences P_m, $m = 1, ..., M$, so

$$P_G(A_n) = f(P_1(A_n), ..., P_M(A_n)), \qquad (4)$$

where f is a function

The most well-known result of this type is due to Arrow (1963) [1], who investigated a specific group decision formulation where the preferences P and P_m were rankings of the alternatives. He postulated five assumptions and proved that they were inconsistent. Hence, an impossibility theorem resulted, which has been very influential and continues to generate great interest.

Suppose one maintains the general formulation (4), but changes the preferences P and P_m of the group and group members to be ratings instead of rankings of the alternatives. Then, based on assumptions analogous to Arrow's using ratings, specifically the expected utilities of the alternatives, the impossibility theorem disappears.

Specifically, for this formulation it has been proven that result (2) follows based on reasonable assumptions different than the assumptions of decision analysis [6].

Two other well-known and substantial shortcomings of using rankings rather than ratings for group decisions can be easily illustrated. First, consider a two-person group decision with two alternatives A and B. Suppose individual 1 ranks A better than B and that individual 2 ranks B better than A. What should the group preference be? The key information to resolve this concerns strengths of preferences, how much does individual 1 prefer A over B and how much does individual 2 prefer B over A, and interpersonal comparison of preferences, how important should those differences be to the group decision. Rankings do not allow one to address these concerns, but ratings, specifically expected utilities, do.

Now consider a three-person group decision. Individual 1 ranks alternative A better than B, which is better than alternative C. Individual 2 ranks alternative B better than C, which is better than A, and individual 3 ranks alternative C better than A, which is better than B. Now suppose the group compares pairs of the alternatives. This will easily show that two individuals prefer alternative A to alternative B, two prefer alternative B to alternative C, and two prefer alternative C to alternative A. Using majority rule indicates that A is preferred to B which is preferred to C, which is preferred to A. Any evaluation method that leads to intransitivities such as this would not be appropriate for evaluating alternatives for important group decisions. The basis for this significant flaw is that rankings do not provide the needed information to make value judgments necessary for logically sound and justifiable group decisions.

In group decisions, you need three additional types of information: ratings (e.g. expected utilities) of the alternatives for each individual, interpersonal comparison of the differences in preferences of the individuals for those alternatives, and the relative importance of each individual's preferences to the decisions. The first item is addressed in stage 1 of the group decision model and the other two items are addressed in stage 2.

5   Logical Foundation for the Group Decision Analysis Result

The intuitive logical foundation for the group decision frame is based on two principles.

Origin of Judgments Principle: *All original judgments must be developed in the mind of an individual.*

This principle refers to all judgments such as recognizing or creating alternatives, describing consequences, identifying events, specifying probabilities for events, and constructing utilities for consequences. The logic for this principle is that judgments can only be developed in a mind, and only individuals have minds. Groups do not have a mind, so all original judgments must occur in the mind of an individual. However, a group of individuals can decide to express group judgments, which leads us to a related principle.

Group Judgments Principle: *Any group judgments must be constructed by the group based only on the judgments of the individuals in the group.*

The logic for this principle is as follows. If no group member recognizes some aspect that might be relevant to a group decision, then clearly the group cannot recognize and consider it. If a potentially relevant aspect is recognized by one group member, but neither that group member nor any other member cares about it, the group should not care about it. Succinctly, this principle implies that the group should not care about anything that no member in the group cares about and the group should care about anything that at least one group member does care about.

The group decision analysis result can be interpreted in terms of these two principles. When each group member evaluates alternatives at the end of stage 1, he or she can only use information that has been personally generated or communicated from others. The group member should include only and all of the information that he or she thinks is relevant to the decision. Hence, the resulting expected utilities of the alternatives are a succinct summary of all that matters from the perspective of that group member for the group decision.

If group member I_1 cares about how well the alternatives please group member I_2, then I_1's decision analysis should include this in his or her analysis. Similarly, if group member I_3 cares about the equity or fairness of the alternatives to other members of the group, then I_3's decision analysis needs to include this concern in his or her decision analysis. The point is that anything relevant to group member I_m about the decision should be included in that member's individual decision analysis of the group decision, it is not something additional that should be added to the results of the collective set of group member's decision analyses.

From the group judgments principle, the group evaluation of alternatives should be based only on the judgments of group members. Since the members expected utilities of alternatives include everything that each considers to be relevant to the decision, the set of the members expected utilities for alternatives is a complete set of everything relevant to evaluating alternatives from the group perspective. The additive formulation (2) weights each group member's perspective.

Other than the weights of these expected utilities, there are no other terms that should be included in the group evaluation, as everything relevant to any member is already included in his or her decision analysis. Having any other terms added to this weighted expected utility would be irrelevant and/or double counting.

6   Use of the Model

The individual member's analyses of what the group should do are each an individual decision analysis. For these, the procedures to do this are well developed and understood [2, 8, 12]. Each member's utility function over the possible consequences of interest to them provides a strength of preference scale necessary for assessing the weights w_m, $m = 1, \ldots, M$ in (2) in stage 2.

The group must collectively produce those weights. To do this requires two separate types of information. First the interpersonal comparison of utilities is essential which provides values of the relative importance to members of having their utilities change from its worst to its best levels (i.e., $u_m = 0$ to $u_m = 1$). Second, the relative importance of the individual's values for the group decision must be assessed. For many decisions, it is reasonable to assume that the values of each group member are equally important. There are other decisions where this assumption may not be appropriate. As an example, if a group of individuals jointly owned a company, they may decide that their relative importance should be proportional to their percentage of ownership in the company. The relative interpersonal comparison preference weights are multiplied by the relative importance of the member's values, and these results are normalized so that the w_m weights in the group decision analysis model (2) sum to one. Details about this procedure are in [7].

In practice, it may be very difficult to specify precise weights for two separate reasons. First, the interpersonal comparison of utilities is a complex contentious issue [3, 4, 9, 15]. Second, the larger the weight w_m is on for one group member's preferences, the smaller the weights must be for other member's preferences. This may result in an inability of the group members to agree on a set of weights w_m, $m = 1, ..., M$, and this is the only input needed for the group decision analysis model that requires agreement from the collective group.

Fortunately, two distinct strategies can be used, alone or together, to address this situation to resolve the group decision. One is a bounding strategy that puts constraints on the possible weights. The other is a sensitivity strategy that examines implications of various combinations of weights.

Regarding the bounding strategy, even if the group members do not agree on the weights w_m, they may agree on one component of the weights, the interpersonal comparison of preferences or the relative importance of the group member's values for the decision. Bounding can be used on either or both and the implications of the bounds on the weights can be calculated. The ideas are the same for both, so I will illustrate the approach using the relative importance component only.

Suppose there was no universal agreement by the members of the decision-making group about the relative importance of the group member's values for that decision. They may be able to agree that no member's important should be more than twice any other member's importance. For a group of two members, then the relative importance weights for each member would have to be between 1/3 and 2/3. For a group of seven members, no individual could have an importance weight more than 0.25, as this would force at least one other member to have an importance weight less than 0.125. There may be other unanimously agreed upon constraints, such as the importance weights for members I_1 and I_2 should be equal or that the importance weight for I_3 should be larger than that for I_4. The implications of all bounding constraints would collectively bound the acceptable range of the set of weights w_m, $m = 1, ..., M$.

A sensitivity analysis of the alternatives with all possible combinations of acceptable weights may conclude that one alternative is better for all cases. If not, the analysis would likely indicate that some alternatives are inferior to others in all cases and those inferior alternatives can be eliminated. This sensitivity analysis could be used to specify the ranges of acceptable weights that result in different alternatives being preferred and also indicate by how much each is preferred over which other alternatives in each of these situations. These insights may suggest which alternative should be chosen or at least some alternatives that should be dropped from further consideration.

For the remaining contending alternatives, it would indicate where additional thought about the importance weights, or the weights on the interpersonal comparisons of preferences, could be focused to decrease the acceptable ranges of the weights w_m and select an alternative among the remaining contenders.

Note that this process can begin with a sensitivity analysis using only constraints that $0 \leq w_m \leq 1$ and $w_1 + w_2 + \ldots + w_M = 1$, and then subsequently pursuing bounding strategies as necessary. Bounding the weights and sensitivity analysis complement each other and help identify and eliminate poor alternatives and converge to selecting the best, or at least a good, alternative.

7   Summary and Comments

Over several decades, there been numerous attempts to extend the concepts of decision analysis from individual decisions to group decisions. These attempts have invariably led to impossibility theorems, suggesting that it could not be done and that decision analysis is not relevant to group decisions. In retrospect, a key contributor to the difficulty in finding a positive solution to the group decision problem was the implicit assumption in all of these efforts that the decision frame for the group decision was the same for each of the group members. This implied that the group members were each concerned with the same set of consequences and with the same set of possible events. As a result, the search for a solution proceeded to specify group probabilities for these events and group utilities for these consequences to produce a group decision analysis. As group members could have different judgments about probabilities of events and different preferences for the consequences, simultaneously combining both different probabilities and different utilities turned out to be problematic.

The approach taken in [7] addresses the more realistic general decision problem where group members may have different view points, and therefore different decision frames, of their common group decision. The group decision frame explicitly incorporates each member's frame, so it is broader than any member's decision frame. This allows each member to incorporate his or her potentially different consequences and events of concern, as well as different probabilities of events and utilities of consequences into the group decision. For this group decision frame, using group decision analysis assumptions analogous to those for an individual decision, a decision analysis solution for group decisions is derived. This solution is that the group expected utility for an alternative is the weighted sum of the individual member's expected utilities for that alternative. This result incorporates and maintains the integrity of each member's decision analysis of what he or she feels is in the group's interest and, in addition, explicitly addresses how the evaluations of the group members should be combined. The result is a logically sound operational framework to conduct a decision analysis of any group decision.

References

1. Arrow, K.J.: Social Choice and Individual Values, 2nd edn. Wiley, New York (1963)
2. Clemen, R.T., Reilly, T.: Making Hard Decisions with Decision Tools. Duxbury, Pacific Grove (2001)
3. Dyer, J.S., Sarin, R.K.: Group preference aggregation rules based on strength of preference. Manage. Sci. **25**(9), 822–832 (1979)
4. Harsanyi, J.C.: Cardinal welfare, individualistic ethics, and interpersonal comparisons of utility. J. Polit. Econ. **63**(4), 309–321 (1955)
5. Hylland, A., Zeckhauser, R.: The impossibility of Bayesian group decision making with separate aggregation of the beliefs and values. Econom. **47**(6), 1321–1336 (1979)
6. Keeney, R.L.: A group preference axiomatization with cardinal utility. Manage. Sci. **23**(2), 140–143 (1976)
7. Keeney, R.L.: Foundations for group decision analysis. Decis. Anal. **10**(2), 103–120 (2013)
8. Kirkwood, C.W.: Strategic Decision Making. Wadsworth Publishing Company, Belmont (1997)
9. Luce, R.D., Raiffa, H.: Games and Decisions. Wiley, New York (1957)
10. Mongin, P.: Consistent Bayesian aggregation. J. Econ. Theory **66**(2), 313–351 (1995)
11. Pratt, J.W., Raiffa, H., Schlaifer, R.: The foundations of decision under uncertainty: an elementary exposition. Am. Stat. Assoc. J. **59**(306), 353–375 (1964)
12. Raiffa, H.: Decision Analysis. Addison-Wesley, Reading (1968)
13. Seidenfeld, T., Kadane, J.B., Schervish, M.J.: On the shared preferences of two Bayesian decision makers. J. Philos. **86**(5), 225–244 (1989)
14. Sen, A.: Collective Choice and Social Welfare. Holden-Day, San Francisco (1970)

Distributive Justice, Legitimizing Collective Choice Procedures, and the Production of Normative Equilibria in Social Groups: Towards a Theory of Social Order

Tom Burns[1,2], Nora Machado[3], and Ewa Roszkowska[4(✉)]

[1] Woods Institute for Environment Energy, Stanford University,
California, USA
tom.burns@soc.uu.se
[2] Department of Sociology, University of Uppsala, Box 821, 75108
Uppsala, Sweden
[3] Centre for Research and Studies of Sociology Lisbon,
Lisbon University Institute, Lisbon, Portugal
nora.machado@iscte.pt
[4] Faculty of Economy and Management, University of Bialystok,
Warszawska 63, 15-062 Bialystok, Poland
erosz@o2.pl

Abstract. This paper focuses on group normative procedures and distributional norms that are utilized in functioning groups in the production/generation of normative equilibria, that is, the major basis of social order in groups and communities. The group is an organizational arrangement with some degree of division of labor and characterized by group purposes and goals, a normative order and patterns of interaction and output. We identified three patterns of particular interest: (1) legitimation procedures in groups to resolve conflicts and make collective choices; (2) patterns of just outcomes satisfying the normatively prescribed group outcomes/outputs of a principle of distributive justice's; (3) normative equilibria, which are group patterns of interaction or collective decision that tend to stability because they satisfy or realize one or more key group norms.

Keywords: Collective choice procedure · Distributive justice · Norm · Normative equilibria · Fair procedure · Adjudication · Democratic vote · Negotiation

1 Introduction

Generally speaking, a group is an organizational arrangement with some degree of division of labor and characterized by group purposes and goals, a normative order and patterns of interaction and output [4, 5, 17]. In other words, a group, produces particular patterns of interaction and outputs/developments. A particular group in a given context will under the right internal and external conditions generate interaction patterns and perform outputs according more or less to the group's shared conceptions,

© Springer International Publishing Switzerland 2015
B. Kamiński et al. (Eds.): GDN 2015, LNBIP 218, pp. 87–98, 2015.
DOI: 10.1007/978-3-319-19515-5_7

values, and norms, in general, its rule regime. Although the group components are usually to a greater or lesser extent compatible, they entail to varying degrees gaps and inconsistencies, in part because the construction and development of groups are typically piecemeal, historical in character and the external and internal conditions of their activities shift.

Functioning groups are characterized by multiple norms in their rule regime that applies to shaping and regulating group activities. This paper focuses on group normative procedures and distributional norms that are utilized in functioning groups in the production/generation of normative equilibria, that is, the major basis of social order in groups and communities. It is precisely when they the directives of multiple norms converge that they are sources of normative equilibria. When they diverge or clash, group actors are faced with dilemmas, contradictions, and uncertainties; and, of course, consensus and normative equilibria are not immediately forthcoming concerning group issues and collective decisions.

Of particular attention in our earlier group research [3, 4, 8–11, 19] are normatively grounded patterns in group behavior. In this research, we identified three patterns of particular interest: (1) legitimation procedures in groups to resolve conflicts and make collective choices; the group procedures or algorithms, when performed properly, have the capacity to legitimize an outcome of the procedure as "fair" or "just"; (2) patterns of just outcomes satisfying the normatively prescribed group outcomes/outputs of a principle of distributive justice's; (3) normative equilibria, which are group patterns of interaction or collective decision that tend to stability because they satisfy or realize one or more key group norms. An individual member is disposed to contribute to and to accept the pattern because it is right and fair – by doing this, a member expresses a group norm and demonstrate her group belonging; (4) on the collective level in a functioning, stable group, such norms serve as "focal points" for group members to direct critical discourses and to mobilize in reacting to individual or sub-group deviance from the norms, that is, to violate the relevant norms.

In earlier papers we also formulated and applied the concepts of rule regime and normative equilibrium. (1) Groups and communities function are maintained on the basis of a rule regime specifying norms, roles, role relationships (including leadership and authority), and appropriate interaction patterns and goals/group performances. Most key or fundamental norms in a group are robust in that there are multiple pushes and pulls to adhere to them, and thus, maintain group social order; there are instrumental and non-instrumental motives at play. (2) States of the world, performance outcomes which satisfy one or more relevant group or community norms are typically normative equilibria. (3) Two of the major ways normative equilibria are produced is: (a) through the implementation of legitimating collective decision procedures such as democratic voting and adjudication by an appropriate agent; and bilateral and multilateral negotiation; and (b) realization of a relevant principle of distributive justice in group interactions and their outcomes.

That is, in one class of cases such equilibria can be generated/produced by a group following/applying established procedures for legitimating a collective decision (including resolving group conflicts). Among the major group procedures, which our research has identified, are negotiation (bi-lateral or multi-lateral), adjudication (judicial, administrative), and democratic voting or decision-making. The other basis of normative equilibrium is obtained through realizing a justice principle in group action and/or outcomes In other words, normative equilibria are realized by implementing in a right and proper way a normatively specified procedure and/or by realizing or satisfying a norm in a particular condition, state of the world, or process. Groups make collective decisions – and also resolve conflicts – by applying one or more legitimating procedures or production function(s) with reference to a norm that defines the proper dimensions of the choice process and the properties of just or fair outcomes.

In sum, this article focuses on particular types of universal judgments in human groups and communities, namely those concerning justice and fairness, which serve to legitimize decisions and resolutions. We apply several key elements of sociological game theory (SGT) in formulating particular conceptions of justice and their interplay.

2 The Structural Embeddedness of Social Interaction and Games

The Sociological Game Theory (SGT) approach stresses *the institutional and cultural embeddedness of games and other forms of social interaction.* It provides a cultural and institutional basis for defining and analyzing social interaction and group behavior in their social context—game is re-conceptualized as a *social and often institutionalized form* [3, 9–12]. In such approach human action and interaction are explained in a form of rule-application as well as rule-following action; this mechanism underlies diverse modalities such as instrumental-rational, normative, and expressive as well as "playful" modalities for determining choices and actions [3, 11, 12]. SGT entails the extension and generalization of classical game theory through the systematic development of the mathematical theory of rules and rule complexes (the particular mathematics is based on contemporary developments at the interface of mathematics, logic, and computer science) [3, 6, 16, 17]. The rule complex(es) of a game applied (and interpreted) in a particular social context guide and regulate the participants in their actions and interactions. What is important SGT introduces in its conceptual framework a sociologically important type of equilibrium, namely *normative equilibrium*, which is the basis of much social order [3, 10–12].

In SGT game theory, an activity, program, outcome, condition or state of the world is in a *normative equilibrium* if it is judged by participants to realize or satisfy appropriate norm(s) or value(s) relevant or applicable in the given interaction situation. Although the concept of normative equilibria may be applied to role performances and to individuals following norms, we are particularly interested here in normative equilibria related to collective choice in given institutionalized settings. This means that the participants judge interactions and/or outcomes in terms of the degree they realize or satisfy a collective norm, normative procedure, or outcome-legitimizing institutional arrangement. Examples of particular procedures that are capable of *producing normative equilibria are adjudication, democratic voting, and negotiation as well as the exercise of legitimate authority.*

3 Normative Legitimizing Procedures for Collective Choice and Conflict Resolution

Agents in a group – by virtue of their differing positions in the social order - may have different interpretations of reality, in particular their common social situations and hold and apply values from different perspectives and, consequently, there are potential contradictory value judgments and incompatible equilibria. Regulative procedures – or social algorithms –are typically applied to problems of conflict and suboptimality in a multiple value world such as Pareto envisioned. We can specify several models of institutionalized regulatory mechanisms that resolve conflicts, inefficient or non-optimal states, and disequilibria. Procedures such as democratic voting, administrative and adjudication decision-making, and multi-lateral negotiation are capable of producing outcomes that in many cases are widely accepted as legitimate and become social equilibria (at least within some range of conditions). The principle of Pareto optimality is replaced in SGT with the principle of legitimating collective decisions [8, 9]. A normative legitimating procedure entails making a collective choice or resolving a conflict by applying or following an institutionalized algorithm or procedure, at least in the judgment of many of the key actors (or groups of actors) involved. This can be accomplished through one of the established societal procedures [8, 9, 18]:

- *negotiations* and settlements that purchase a better result for those negotiating
- *a democratic vote* that enables citizens to settle a contentious issue through the *choice of one of* the options, e.g. who will lead the community or what policy or program will apply
- an *administrative agency or court* with legal responsibility and authority decides an issue, for instance that everyone at workplaces is expected to use special safety equipment.

Procedural and outcome equilibria concern states that agents judge to be fair or just according to shared or community norms and values including legitimating procedures to settle or resolve conflicts – included here are norms of distributive justice (see earlier). The application of a legitimating procedure is tantamount to realizing or satisfying a norm or principle. The procedure is judged by the community as the right and proper way to conduct collective decision, *whatever the outcome of the procedure* [5, 7–9]. This might be the case, for example, in utilizing a democratic procedure to make a collective choice (rather than utilizing bureaucratic or authoritarian procedures). Voting conducted according to a democratic rule is intended to legitimize the outcome of voting as just or right *per se* – whatever the outcome might be. Indeed, the correct following of such normatively prescribed procedures is expected to result in outcomes that are generally recognized as normative equilibria in a democratic society, other things being equal. Legitimate outcomes are normative equilibria, because they have been arrived at or constructed in *ways* that are to a great extent socially recognized as "fair," or "right and proper". In such ways, many institutional arrangements, laws, and constitutions may be constructed as normative equilibria in their own right.

The regimes of some governance procedures, above all administration, adjudication, and arbitration, include principles and norms not only regulating the procedure but guiding the proper content or distributional properties of the choice or imposition. Other regimes such as those of a democracy, but also those of market exchange and bargaining, specify more or less right and proper procedures but often leave the options "negotiated" to the discretion of the players (although, of course, social norms and laws put constraints on what may be negotiated and decided upon). Participants are free to define and adjust outcomes within the procedural framework. Under some conditions, the proper performance or realization of the procedure is necessary and sufficient for a normative equilibrium, for instance the outcome is judged "right and proper" because the procedure itself is judged as fair and legitimate, although it may not be sufficient. In many instances, the outcome(s) must be constructed in such a way as to satisfy justice principles [6, 7, 15, 20]. This entails structuring the procedure in a way that rules out or excludes certain actions, outcomes, and even particular evaluations and preferences that violate one or more of the multiple norms and values defined by the procedure. For instance, in adjudication, the court process is conceptually ordered into phases so as to frame or shape the deliberations, structuring the options as well as the legal concepts and norms that apply in the different phases [13, 14]. Among the principles and rules, which are to be realized or satisfied in the collective choice process, we emphasize the following:

1. There are principles and rules about the organization of the process, specifying which agents (who) are to be included (and by implication, who should be excluded). That is, "Whose voice counts" – at least in formal terms. Of course, there may be deviation in that those whose voices should count are excluded and a few (an oligarchy) or one (a tyranny) determine the decision.
2. There may be other rules specifying time and place for a collective decision to be conducted. Adherence to these rules contributes to legitimizing the collective decision. Deviation opens the way for challenges or de-legitimizing reactions in the group or organization.

In general, we are talking about a finite set of principles and rules whose concrete realization may vary to differing degrees. When such a rule regime [4, 5] is fully realized, then there is optimal legitimacy and an increased likelihood of acceptance of the results and the achievement of normative equilibrium. The violation of one or more rules of the regime implies movement away from optimal legitimacy and reduced likelihood of acceptance and accomplishment of normative equilibrium.

In general, the problem of fair division of goods is the subject of extensive literature in the social sciences, law, economics, game theory and other. Usually, the object or good (or bad) can be desirable, undesirable or a mixture of desirable and undesirable. It may be infinitely divisible (as we usually regard real cake, money) or only divisible into discrete pieces (such as house, car, furniture).

The Case of Adjudication. We are concerned here about a particular type of judgment, namely making a collective choice (or resolving a conflict). As indicated earlier, our models refer to the cultural, institutional and situational context $S(t)$ at time t. Also, the actors involved, their role and role relationships are indicated. In adjudicated procedures, the role of the legitimate authority or judge P is considered in relation to the N members of the group or community $N = \{A_1, A_2, ..., A_n\}$, the situational beliefs or model and the values of the actors in the situation. P may be a member of the group or an external judge or agent. Situational value may indicate either equality or differentiation in the division of the good (bad) – and the differences may range from a basic need to the level of professional credentials (and other measures of status), or performance/contribution to the group. Finally, a procedure or algorithm is used to allocate in accordance with the value applying in the situation.

In adjudication, the judge **J** may be appointed by the group leadership, or elected by members of the group, or imposed by a larger community (in the latter case, with authorization presumably to address the particular group in question). Burns and Flam [5] refer to cases of allocation by government agents in different sectors to subgroups of citizens (for instance, housing, health care, and educational subsidies). Some firms and organizations conduct such distribution among its employees, and many voluntary organizations provide its members with access to training and other educational opportunities, health care, possibly pensions.

In adjudication, the group or its leadership accept and adhere to a procedure or algorithm for determining allocation, including possibly the application of a formula to realize a norm of distributive justice. But the arrangement may enable a judge/leader to allocate *arbitrarily*, or to do it on the basis of a *lottery*. If the allocation is arbitrary, the leader or judge typically requires a great deal of authority or power to overcome any disappointment and envy arising among the members. Often a judge deciding arbitrarily would try to conceal the amounts of allocation to different members (that is, through the use of secrecy or a "veil of ignorance").

The legitimizing procedure such as negotiation or democratic vote may be applied by the group or by a significant part of it. As in the case of adjudication, the group may make use of a distributive justice formula in collectively deciding a distribution of a good (or bads) among members.

4   Distributive Justice Formulas

A legitimate authority P has a right and the knowledge to apply an allocation or distribution procedure concerning a good (or goods) G to be divided a group or population of N actors, $N = \{A_1, A_2, ..., A_n\}$. The good G may be divisible into N parts, not necessarily equal parts; or divisible into just K parts; or is non-divisible. The basis of the allocation is a principle (an **allocation rule**) for example, to allocate according to equal shares, or status, according to "merit" (performance or credentials), according to need, or according to property rights, etc. *The formula used, if any, is embedded in the social process of adjudication or in the negotiation or democratic vote.*

Several justice formulas can be considered [1, 2, 20, 21]. These range from consideration of equal division in a group or community to consideration of cases of inequality on the basis of status or authority differences, performance or contribution differences, or differences in need. We continue to *make a distinction between judgments of allocation that are adjudicated and those that are self-organized or managed by members of the group or community itself.* In the first case, a judge, arbitrator, political agent (they may be elected, appointed, or of traditional origin) applies a principle or procedure in the distribution process, whether it concerns equality, differentiation according to status, responsibility, performance or need. In self-organized fair distribution those participating apply division procedures or algorithms that divide fairly (although the outcomes may differ in several ways from equal division).

Two key characterizations of groups are egalitarian, on the one hand, and differentiated (typically hierarchical), on the other. For instance, a group may have been established and functions on egalitarian principles, so allocation of an available goods would be done according to a principle of equality (this may take different forms). Individuals in a particular group context may operate from the beginning on an assumption that they are equal to one another others in the situation; or, they may operate on an assumption that they are either superior or inferior to others in an hierarchical group. One may distinguish absolute equality, for instance equal rights to vote, or each member receiving an absolute equal amount. But some allocations are designed to accomplish relative equality, in that each member receives the same percentage increase (or decrease) over their current salary. But the amounts differ according to their different wage baselines.

In a group with prominent status differences, which are considered highly central and relevant for the distribution of a resource (such as the results of a business or professional venture), distribution would be uneven. Some procedures or rules may be recognized by the group as accomplishing unequal distribution. The levels differentiated may be 2, 3,...,k. As mentioned above, the differentiation can be based on status or authority in the community or group, performance for, or contribution to, the group may be a basis for such differentiation. Those performing at the highest level or contributing the most to the group or its output would receive the most, and those performing or contributing less, would receive less, and so on. Finally, differentiation may be based on need, the needier receiving more than those needing less.

In general, one may distinguish distributions that are intended to accomplish or reflect equality and those that accomplish or reflect differentiation. And, as suggested above, the bases of differentiation may vary from group status, performance or contribution to the group, to neediness. But groups may choose to ignore differences in performance or contribution or even need and allocate to every member the same good, or amount of a good, regardless. Or, they may act arbitrarily not following any particular rule or formula.

Description of the Problem.[1] Let $N = \{1, ..., n\}$ be a set of agents (or players, or individuals) in a group or community who are to share several goods (or resources, items, objects). An allocation A is a mapping of agents to bundles of goods. Most criteria will not be specific to allocation problems, so we also speak of agreements (or outcomes, solutions, alternatives, states). A group of agents each have individual preferences over a collective agreement or the allocation of goods to be found. The problem is the definition of the notion and understanding of *"fairness"* allocation. In other words, is there a formula which satisfies distributive justice or *just or fairness*

---

[1] The theory of *fair division* among a group of individuals was initiated in the 1940s by three Polish mathematicians: Hugo Steinhaus, Bronisław Knaster and Stefan Banach. Brams and Taylor [2] give a historical introduction of fair division and provide a detailed discussion of many procedures.

conditions? The theory of fair division provides explicit criteria for various different types of distributive justice or fairness [1, 2, 7, 15, 21]. A distributive justice principles specifies the division of one or several goods amongst two or more agents in a way that satisfies a suitable property. There are several possible definitions of what constitutes a just or "fair" division, where proportionality, envy freeness and equitability being several of the major fairness criteria considered [2, 21]. Such a formula give the instruction how to partition, the *number of (non-intersecting) cuts* allowed for partitioning the good and what certain properties are satisfied by the allocation. Procedures can be distinguished on the basis of certain technical aspects. One distinction is between discrete and moving-knife procedures. In *discrete procedures*, the players' moves are in a sequence of steps, whereas in *moving-knife procedures*, there is a continuous evaluation of pieces of cakes by the single players. Another further essential distinction is based on the *number of (non-intersecting) cuts* allowed for partitioning the good.

Algorithms satisfying desirable *fair* properties, or showing the limits of doing so, can be referred as institutionalized legitimizing procedures. We can analyze step-by-step rules or algorithms to implement the fair division of goods in normative aspects and study their distributional consequences. By making precise criteria that one wishes the fair-division procedure to satisfy, and clarifying relationships among these criteria, we can analyze the distributional effects of those procedures in a social context to resolve conflict. In such a conceptualization a Fair Division Procedure is a complex of rules that, when properly applied, produces a fair division of the objects to be divided. It is a program of actions or operations to be performed by an agent of the group (adjudicators) or by group themselves in their negotiations in terms of the visible data and their valuations. Usually a fair division procedure is expected to be decisive, so if the rules are followed, a fair division is assured, internal to the players, so *no outside intervention is required to obtain a fair division, legitimize (players agree for it)*. The agent or agents has to agree on criteria for a fair division, select a valid procedure and follow its rules. Beyond fairness, additional desirable properties of procedure include simplicity, self-implementation, minimal number of cuts, and applicability to any number of participants in real-life conflicts.

5   Extensions and Discussion

As stressed earlier, not only does a group or community have multiple norms, but these may or may not be applied (or applicable) together, indeed, they may not be fully compatible in legitimizing collective decisions and producing normative equilibria. Here we want to briefly consider cases of normative convergence and coherency, as compared to cases where only one of the normative rules is applied. We hypothesize that when a legitimizing procedure is applied properly together with the proper application of a distributive justice principle, the likelihood of producing or achieving normative equilibria increases. This contributes to social stability in the group or community.

On the other hand if the institutionalized procedure is performed on its own without distributive justice considerations, the outcomes are likely to be challenged and instability occurs – particularly if the outcomes are visible and monitored by some or all of the participants. If collective decisions are made (or conflict resolution measures

Table 1. Convergence and divergence of legitimizing normative procedures and principles of distributive justice

	Relevant distributive justice formula applied properly	Relevant distributive justice formula not applied or not applied properly
Legitimating choice procedure applied properly	(++): Normative equilibrium likely	(+-): Partial normative equilibrium with respect to applying legitimizing procedure (most stable when outcome not visible or deferred)
Legitimating choice procedure not applied or not applied properly	(-+): Partial normative equilibrium with respect to allocative justice	(–): No normative equilibrium. Equilibrium may be established by force or by remunerating acceptance of the decision.

are attempted) without following legitimizing procedures, the likelihood of challenge would be high, and normative equilibria unachievable. We use this simple scheme (see Table 1) to analyze diverse cases presented in earlier work.

Let us observe that legitimizing procedures may or may not be combined with the application of a distributive justice principle. In the former case, the legitimizing procedures are adjusted and adapted to construct and take into account justice norms: (1) in self-organizing applications, a group shares and applies normative rules legitimating a group decision procedure and a relevant principle of distributive justice; (2) in adjudication, a leader, judge, administrator of the group, knowledgeable in the rules, applies a relevant distributive justice norm in carrying out the adjudication procedure. Consider the following normatively complex situations:

(1) *Negotiation.* Group actors who follow a negotiation procedure legitimize their settlements to a greater or lesser extent – simply by following the procedure properly. Those participating as well as those observing recognize that any given settlement is collectively or mutually agreed to. There is consensus. Nevertheless, there may be differing levels of satisfaction (or dissatisfaction) with the settlement. As Roszkowska and Burns [19] show, mutually or collectively agreed settlements or agreements may be perceived or experienced by the participating negotiators as unfair according to a relevant or applicable rule of distributive justice such as equal division (or in some cases unequal proportions). A *bona fide* normative equilibria fails to be accomplished, at least as viewed by one or more of the participants. Consequently, there will be ("justified") seeking for alternatives or "outs", and the settlement is, therefore, potentially unstable (not in full equilibrium).

Thus, negotiated agreements reached may or may not satisfy one or more expectation levels based on a relevant distributive justice rule so that the agreement, although negotiated, is not a normative equilibrium.

(2) *Adjudication.* Adjudication can often legitimize a collective decision outcome, to a greater or lesser extent. In other words, the procedure may produce normative

equilibrium, other things being equal, since it legitimizes to a greater or lesser extent the decision. However, if the outcomes are observable and measurable, members may apply a relevant principle of distributive justice in making judgments or assessments about the outcome(s). If the adjudicator has effectively applied the norm, then the actors will be "satisfied" (feel that the outcome(s) is justified), and a normative equilibrium obtains. On the other hand, if the judge has ignored the norm or is ineffective in his application and interpretation, the outcome (s) will be judged unjustified, and a disequilibrium obtains. In sum, when an adjudicated decision or outcome clashes with one or more norms of distributive justice, one or more of those affected will feel unjustly treated and be predisposed to reject the judgment, introducing uncertainty and instability into the social situation. Thus, adjudicators often try to take into account as much as is feasible the norm(s) of justice applying in the situation and realize them in their decision (and in the discourses associated with the decision). In other words, in some instances such norms are taken into account in the process of formulating a proposal or settlement, thus serving to generate a multi-facetet normative equilibrium.

(3) *Democratic voting.* Such a procedure is a common basis of legitimizing a collective decision. However, it may fail to satisfy one or more norms of justice. In order to avoid outcomes which might be perceived or experienced as highly unjust, relevant norms may be taken into account in formulating alternatives on which to vote. For instance, a decision about the tax system that would focus largely on one segment of the tax payer population (dividend recipients or young people or retired people) is likely to produce feelings of unjust treatment among some of the population and would provoke them to oppose the decision and to try to rescind or avoid it. Thus, alternative proposals are formulated which avoid violating the relevant distributive applying to major groups of voters.

The legitimacy accomplished by following a legitimizing procedure may take precedence over the *actual outcomes, which in fact may be experienced as highly unjust and, therefore, unsatisfactory in themselves,* to one or several or, indeed most parties involved. One of the following holds: (1) The agents assign more value to the procedure (to its inherent procedural qualities) than to any of the consequences embodied in a given or expected outcome. That is, a "negative outcome," while obviously not preferred, would be more acceptable than devaluing or rejecting the properly performed procedure. (2) The agents believe that, on the average or over the long-run, the procedure gives positive results, possibly not always the best result but at least commonly agreed "not worst" results, assessed in terms of norms or values, r_{just} and r_{eff}, that specify degrees of justice and efficiency, respectively. Results include the avoidance of escalating conflict, disruption, or chaos.

Our research has shown that the normative equilibria produced through institutionalized procedures to deal with disagreement or allocation issues need not depend ostensibly on actual outcomes – or assessments of actual outcomes – but on following the procedures in a right and proper way, whether a democratic vote, adjudication, or negotiation procedure. Outcomes generated by such institutional procedures are normative equilibria to the degree that the procedures are judged to be fair and to be implemented in a right and proper way according to established social norms and laws.

However, the outcomes of the procedure may or may not satisfy a relevant distributive norm adhered to by one or more involved groups, organizations, or movements.

In general, the likelihood of group opposition to and rejection of the outcomes of a legitimizing procedure may increases as a result of, on the hand, improper performance of the procedure or, on the other hand, outcomes of the procedure violating an established, core distributive justice norm – whether it concerns basic need, position of authority or status, or other differentiating norm. Future research will concern the differentiated judgments about procedural fairness and/or distributive justice among members of groups and their consequences.

6 Conclusion

This article focuses on particular types of universal judgments in human groups and communities, namely those concerning procedural fairness and justice outcomes, which serve to legitimize group decisions and outcomes and to generate group equilibria. Conflict and destabilizing conditions in groups are also identified. The normatively grounded outcome equilibria may be obtainable through the utilization of societal procedures that settle (not necessarily resolving) conflicting views, interpretations (of a law or contract), values or evaluations, proposals, etc. Institutionalized procedures such as voting, adjudication, negotiation with the participation of opposing agents not only partially *legitimizes resulting outcomes but gives them a normative force.* An outcome is collectively defined or understood as right and proper by virtue of having resulted from application of the right and proper procedure. In other words, the procedure itself generates outcomes that *derive normative force from the proper application of the procedure.* This is a type of *institutional alchemy.* In societies with such social processes, participants (and other societal agents) are likely to sanction negatively those who refuse to accept the outcomes of a right and proper procedure, for the refusal is tantamount to criticizing or denigrating the procedure and its normative context.

There are limits, however, to such legitimation of outcomes. Violation of relevant distributive justice principles – whether egalitarian or differentiated – makes for opposition and challenge to outcomes, and thus disequilibration. This is clear, above all, in the case of substantive matters of a sacred character to one or more participants, for instance, issues such as abortion, physician assisted suicide, violation or desecration of sacred objects or places, etc., but also other issues that touch on deep cleavages between sub-groups, classes, and sub-communities.

Acknowledgements. This research was supported by the grant from Polish National Science Centre (DEC-2011/03/B/HS4/03857).

References

1. Brams, S.J., Edelman, P.H., Fishburn, P.C.: Paradoxes of Fair Division. Economic Research Reports C.V. Starr Center for Applied Economics Department of Economics Faculty of Arts and Science New York University. New York University, New York (2000)

2. Brams, S.J., Taylor, A.D.: Fair Division: From Cake-cutting to Dispute Resolution. Cambridge University Press, New York (1996)
3. Burns, T.R., Caldas, J.C., Roszkowska, E.: Generalized game theory's contribution to multi-agent modelling: addressing problems of social regulation, social order, and effective security. In: Dunin-Keplicz, B., et al. (eds.) Monitoring, Security and Rescue Techniques in Multiagent Systems. Springer, Berlin (2005)
4. Burns, T.R., Corte, U., Machado, N.: Toward a Universal Theory of the Human Group: Sociological Systems Framework Applied to the Comparative Analysis of Groups and Organizations. CIES/ISCTE WP, Lisbon (2014)
5. Burns, T.R., Flam, H.: The Shaping of Social Organization: Social Rule System Theory with Applications. Sage Publications, London (1987)
6. Burns, T.R., Gomolińska, A.: The theory of socially embedded games: the mathematics of social relationships, rule complexes, and action modalities. Qual. Quant. Int. J. Methodol. **34** (4), 379–406 (2000)
7. Burns, T., Machado, N., Roszkowska, E.: Distributive justice: from steinhaus, knaster, and banach to elster and rawls: the perspective of sociological game theory. Stud. Log. Gramm. Rhetoric **37**(50), 11–38 (2014)
8. Burns, T., Roszkowska, E.: A social procedural approach to the pareto optimization problematique: part I: pareto optimization and its limitations versus the GGT conception of the solution of multi-value conflict problems through societal procedures. Qual. Quant. **43**(5), 781–803 (2009)
9. Burns, T., Roszkowska, E.: A social procedural approach to the pareto optimization problematique: part II. institutionalized procedures and their limitation. Qual. Quant. **43**(5), 805–832 (2009)
10. Burns, T.R., Roszkowska, E.: Fuzzy games and equilibria: the perspective of the general theory of games on nash and normative equilibria. In: Pal, S.K., Polkowski, L., Skowron, A. (eds.) Rough-Neural Computing. Techniques for Computing with Words, pp. 435–470. Springer-Verlag, Berlin (2004)
11. Burns, T.R., Roszkowska, E.: Social judgment in multi-agent systems: the perspective of generalized game theory. In: Sun, Ron (ed.) Cognition and Multi-agent Interaction. Cambridge University Press, Cambridge (2005)
12. Burns, T.R., Roszkowska, E.: Multi-value decision-making and games: the perspective of generalized game theory on social and psychological complexity, contradiction, and equilibrium. In: Shi, Y. (ed.) Advances in Multiple Criteria Decision Making and Human Systems Management. IOS Press, Amsterdam (2007)
13. Chapman, B.: Law games: defeasible rules and revisable rationality. Law Philos. **17**, 443–480 (1998)
14. Chapman, B.: More easily done than said: rules, reasons and rational social choice. Oxf. J. Leg. Stud. **18**, 293–329 (1998)
15. Elster, J.: Fairness and norms. Soc. Res. **73**(2), 365–376 (2006)
16. Gomolińska, A.: Rule complexes for representing social actors and interactions. Stud. Logic Gramm. Rhetoric **3**(16), 95–108 (1999)
17. Gomoliska, A.: Fundamental mathematical notions of the theory of socially embedded games: a granular computing perspective. In: Pal, S.K., Polkowski, L., Skowron, A. (eds.) Rough-Neural Computing: Techniques for Computing with Words, pp. 411–434. Springer-Verlag, Berlin (2004)
18. Keeney, R.L.: Foundations for group decision analysis. Decis. Anal. **10**(2), 103–120 (2014)
19. Roszkowska, E., Burns, T.: Fuzzy bargaining games: conditions of agreement satisfaction equilibrium. Group Decis. Negot. **19**(5), 421–440 (2010)
20. Rawls, J.: A Theory of Justice. Belknap Press, Cambridge (1971)
21. Steinhaus, H.: The problem of fair division. Econometrica **16**, 101–104 (1948)

A Multiple Criteria Model for Comparison of Subjective-Objective Evaluations and Its Application

Ye Chen[1(✉)], Yao Li[1], Wangqun Sun[2], and Haiyan Xu[1]

[1] College of Economics and Management, Nanjing University of Aeronautics and Astronautics, Nanjing, Jiangsu, China
chenye@nuaa.edu.cn
[2] College of Logistics, Linyi University, Linyi, Shandong, China

Abstract. A multiple criteria decision model addressing the comparison of both subjective and objective evaluation results is proposed in this paper. Firstly, based on cluster analysis, a method to select representative sample data set from all alternatives under evaluation is designed; next, experts are invited to review these sample data and dominance-based rough set theory is used to analyze expert decisions in format of a set of decision rules; then, these trained decision rules are applied to all alternatives and hence, an objective-oriented results can be obtained and used to compare with the alternatives' self-evaluation results which contains subjective orientation; finally, the method is applied to analyze the graduate's leaning ability to demonstrate its feasibility.

Keywords: Multiple criteria decision analysis · Comparison of subjective-objective evaluation results · Cluster analysis · Dominance-based rough set theory · Graduate's learning ability

1 Introduction

Due to ever increasing complexity of human society, people often need to consider multiple criteria (attributes, factors, objectives) to make decisions. The research area of multiple criteria decision analysis (MCDA) is developed to provide decision aid for complex decision situations. MCDA aims to furnish a set of decision analysis techniques to help decision-makers (DMs) logically identify, compare, and evaluate alternatives according to diverse, usually conflicting, criteria arising from societal, economic, and environmental considerations. This body of literature has also been referred to as multiple criteria (attribute) decision aid and support [1].

Early classic MCDA methods include AHP [2], MAUT [3], OUTRANKING [4]. With the relevant research making processes in both depth and breadth, various hybrid approaches to MCDA have been developed to address different decision scenarios, which contain Fuzzy logic-based MCDA [5], dominance-based rough set approach [6], grey system theory-based MCDA [7], and many others. The research on MCDA becomes a hot topic internationally, for example, in China, through searching Chinese Hownet (CNKI, the most popular academic database [8]) with relevant keywords, alone in 2014 over 600 journal papers have been listed.

© Springer International Publishing Switzerland 2015
B. Kamiński et al. (Eds.): GDN 2015, LNBIP 218, pp. 99–105, 2015.
DOI: 10.1007/978-3-319-19515-5_8

An MCDA problem can be summarized using the following three steps, which are also illustrated in Fig. 1:

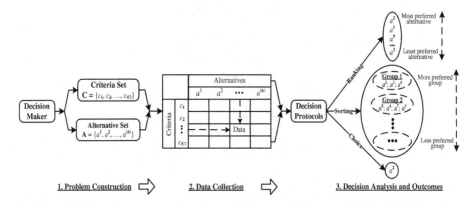

Fig. 1. Steps in multiple criteria decision analysis

1. Problem Construction: Define objectives, arrange them into criteria and identify all possible alternatives.
2. Data Collection: Obtain an information matrix, in which each column represents an alternative and each row provides evaluations of the performance of the alternatives over that criterion.
3. Decision Analysis: Analyze the data to obtain *ranking*, *sorting* or *choice* results for the alternatives as an aid to decision-making.

In this paper a specific decision scenario for sorting problem in MCDA is addressed:

1. There are a large number of alternatives needed to be sorted.
2. The alternatives are interactive agents such as people instead of inanimate objects.

Often, DMs cannot just make final decisions purely depending on their own preferences. Usually DMs will anticipate the potential feedback from such alternatives and make possible adjustments to induce positive incentives to alternatives. Practical examples include student scholarship evaluation and employee annual performance evaluation.

Here, a framework based on the concept from both statistics and rough set theory is designed to address evaluation problem with a large size, and provide a fine-tune mechanism.

2   The Framework of Subjective-Objective Evaluation Result

2.1   The Overall Comparison Procedure

To clarify the relevant concepts, the following seven concepts are defined:

1. *Subjective Evaluation Results*: alternatives' evaluation results are obtained by themselves.
2. *Objective Evaluation Results*: alternatives' evaluation results are obtained by the DM's evaluation without the considering of alternatives' feedbacks.

Next, a framework of subjective-objective evaluation result comparison is constructed, as shown in Fig. 2. The framework comprises the following five steps:

1. *All Alternatives' Information Collection*: the criteria and alternatives are identified and information on all alternatives is collected (see Fig. 2).

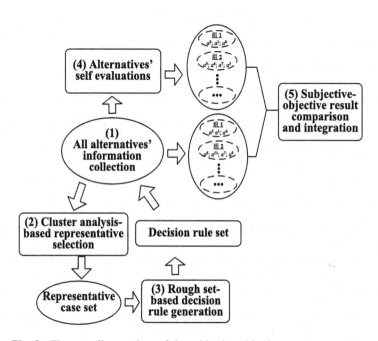

Fig. 2. The overall procedure of the subjective-objective result comparison

2. *Cluster Analysis-Based Representative Selection*: the application of cluster analysis [9] in order to identify representative alternatives.
3. *Rough Set-Based Decision Rule Generation*: the invitation of experts to provide overall evaluations on representatives and apply rough set theory [10] to generate decision rules.
4. *Alternatives' Self-Evaluations*: Request for alternatives' self-evaluation.
5. *Subjective-Objective Result Comparison and Integration*: the undertaking of the comparison of two kinds of evaluation results and making the final decision integration.

2.2   Basic Notation

The following notations are used to describe the aforementioned procedure:

Alternative set A:

$$A = \{a^1, a^2, \ldots, a^i, \ldots, a^n\}$$

where a^i is a specific alternative and n is the total number of alternatives.

Criteria set C:

$$C = \{c_1, c_2, \ldots, c_j, \ldots, c_q\}$$

where c_j is a specific criterion and q is the total number of criteria.

Classification set G:

$$G = \{g^1, g^2, \ldots, g^m\},$$

where $g^1 \succ g^2 \ldots \succ g^m$ ($\succ$ reads "preferred to"), and m is the total number of group, for which A is assigned into G according to C. Hence, it satisfies: $m \leq n$.

Representative case set, P:

$$P = \{p^1, p^2, \ldots, p^d, \ldots, p^h\},$$

where P^d is a specific case, and h is the total number of representative cases. Here, P is obtained by cluster analysis of A.

We define decision rule set $R = \{r_1, r_2, \ldots r_l\}$, where l is the total number of decision rule. Here, R is obtained through the application of rough set theory on P.

Subjective evaluation result set A^s:

$$A^s = \{g^s(a^1), g^s(a^2), \ldots, g^s(a^i), \ldots, g^s(a^n)\},$$

where subscript s represents 'subjective'.

Objective evaluation result set A^o:

$$A^o = \{g^o(a^1), g^o(a^2), \ldots, g^o(a^i), \ldots, g^o(a^n)\}$$

where subscript o represents 'objective'. DMs apply R in order to obtain A^o.

3   The Analysis of Result Comparison

3.1   Result Comparison Modeling

To carry out the subjective-objective evaluation result comparison we denote:

$I(a^i)$ – the difference between subjective-objective evaluation result for a^i, i.e.,

$$I(a^i) = g^s(a^i) \; g^o(a^i).$$

$I(A)$ – the overall difference between subjective-objective evaluation result for A, i.e.,

$$I(A) = \frac{\sum_{i=1}^{n} I(a^i)}{n}.$$

Note that $1 - m \leq I(A) \leq m - 1$.

The standard deviation σ ($0 \leq \sigma \leq m - 1$) of the difference between the subjective-objective evaluation result for A, is given by:

$$\sigma = \sqrt{\frac{\sum_{i=1}^{n} [I(a^i) - I(A)]^2}{n}}.$$

We also denote A^{so} to represent the integration of the subjective-objective evaluation result for A:

$$A^{so} = \left\{ w_s g^s(a^1) + w_o g^o(a^1), w_s g^s(a^2) + w_o g^o(a^2), \ldots, w_s g^s(a^n) + w_o g^o(a^n) \right\}$$

where w_s and w_o are the weight of subjective and objective evaluation result, respectively. We assume further that $0 \leq \{w_s \text{ and } w_o\} \leq 1$, $w_s + w_o = 1$.

3.2 Comparison Analysis

To effectively compare the subjective-objective results, two thresholds are set.

1. Threshold α represents the overall indifference between subjective and objective results for A. A suggested initial value is $(m - 1)/n$. Based on the available information of α, three types of subjective-objective result comparison can be achieved:
   a. If $-\alpha \leq I(A) \leq \alpha$, then subjective and objective results are consistent.
   b. If $I(A) > \alpha$, then subjective results are better than objective results.
   c. If $I(A) > -\alpha$, then objective results are better than subjective results.
2. Threshold β represents the average indifference between subjective and objective results for A. A suggested initial value is $1/m$. Based upon the available information of β, two types of subjective-objective result comparison can be set:
   a. If $\sigma > \beta$, then there exists significant deviation between subjective and objective evaluation results.
   b. If $\sigma \leq \beta$, then there are no significant deviation between subjective and objective evaluation results.

3.3   Fine-Tuning and Integration Strategy

Based upon the above definition of α and β, the following fine-tune and integration strategy can be designed, as shown in Table 1.

Table 1. The fine-tune and integration strategy of subjective-objective results

	No significant deviation between subjective and objective results	Significant deviation between subjective and objective results
Subjective-objective result are consistent	Set equal weights for both subjective and objective results for final integration, hence, $w_s = w_o = 0.5$	Set $w_s = w_o = 0.5$ for final integration and check with alternatives with $I(a^i)$ being very large or small.
Subjective results are better than objective results	Set $w_s \geq w_s$ for better representation of subjective results	Explain the evaluation procedure clearly to alternatives and re-take alternatives' self-revaluation
Objective results are better than subjective results	Set $w_o \geq w_s$ for better representation of objective results	Explain the evaluation procedure clearly to alternatives and re-take alternatives' self-revaluation

4   Case Study

A case study of graduate student's learning ability survey data [11] is used to demonstrate the proposed method. The survey identifies graduate's learning ability according to three dimensions including self-learning ability, research ability and innovation ability based on 17 criteria. Then, 248 copies of data were collected from four universities' graduate students in Nanjing, China. The following steps are carried out the analysis:

Table 2. The fine-tune and integration strategy for the survey

	$I(A)$	σ
Self-learning ability	Since $I(A) = 0.014 > 0.008$, students' self-evaluation results are better than experts' evaluation results. Set $w_s = 0.6$ and $w_0 = 0.4$ satisfying $w_s \geq w_0$.	Since $\sigma = 0.24 < 0.33$, there are no significant deviation between subjective and objective results.
Innovation ability	Since $I(A) = 0.006 > 0.008$, students' self-evaluation results are consistent with experts' evaluation results. Set $w_s = w_0 = 0.5$.	Since $\sigma = 0.17 < 0.33$, there are no significant deviation between subjective and objective results.
Research ability	Since $I(A) = 0.007 < 0.008$, students' self-evaluation results are consistent with experts' evaluation results. Set $w_s = w_0 = 0.5$.	Since $\sigma = 0.21 < 0.33$, there are no significant deviation between subjective and objective results.

1. Apply Minitab software and use K-means algorithm to identify 30 representative sample data. Then invite experts to evaluate these 30 data.
2. Apply 4eMka2 software to analyze experts' judgment and use domain-based rough set theory to identify 10 key decision rules.
3. Set the parameters and carry out the comparison of subjective and objective results.

Let $\alpha = (m-1)/n$ and $\beta = 1/m$. Since $m = 3$ and $n = 248$ then $\alpha = 0.008$ and $\beta = 0.33$. Based on the results we calculate $I(A)$ and σ. The following results are obtained (Table 2):

Apparently, according to the above analysis, students' self-evaluation results on self-learning ability are better than experts' evaluation results while the other two assessments are roughly consistent from both students and experts' perspectives.

Acknowledgments. This work was supported in part by the National Natural Sciences Foundation of China through the project entitled "Research for the inverse problem based on the graph model for conflict resolution (71471087) and by the Chinese Society of Academic Degrees and Graduate Education through the project "The empirical study of the graduates' learning ability in research-oriented university" (B2-2013Y13-176). Also, the authors wish to express their sincere appreciation to anonymous referees whose comments helped us to improve the paper significantly.

References

1. Belton, V., Stewart, T.J.: Multiple Criteria Decision Analysis: An Integrated Approach. Kluwer, Dordrecht (2002)
2. Saaty, T.L.: Analytic Hierarchy Process. McGraw Hill, New York (1980)
3. Keeney, R.L., Raiffa, H.: Decision with Multiple Objectives: Preferences and Value Tradeoffs. Wiley, New York (1976)
4. Roy, B.: Multicriteria Methodology for Decision Aiding. Kluwer, Dordrecht (1996)
5. Lootsma, F.A.: Fuzzy Logic for Planning and Decision Making. Kluwer, Dordrecht (1997)
6. Slowiński, R.: Rough set theory for multicriteria decision analysis. Eur. J. Oper. Res. **129**, 1–47 (2001)
7. Liu, S., Yang, Y., Wu, L.: Grey System Theory and Its Application. Science Press, Beijing (2014)
8. China Knowledge Resource Integrated Database. http://www.cnki.net
9. Cluster analysis. http://en.wikipedia.org/wiki/Cluster_analysis
10. Graduate learning ability survey. http://www.diaochapai.com/survey/93e2c4a9-e15e-4114-a27d-7c9aabfd3308
11. Minitab software. http://www.minitab.com/
12. 4eMka2 software. Institute of Computing Science, Poznan University of Technology, Poland. http://idss.cs.put.poznan.pl/site/4emka.html

Using Surrogate Weights for Handling Preference Strength in Multi-criteria Decisions

Mats Danielson[1] and Love Ekenberg[1,2(✉)]

[1] Department of Computer and Systems Sciences, Stockholm University,
Forum 100, SE-164 40 Kista, Sweden
mad@dsv.su.se
[2] International Institute of Applied Systems Analysis, IIASA, Schlossplatz 1,
A-2361 Laxenburg, Austria
ekenberg@iiasa.ac.at

Abstract. Various proposals for how to eliminate some of the obstacles in multi-criteria decision making exist and methods for introducing so called surrogate weights have proliferated for some time in the form of ordinal ranking methods for the criteria weights. Considering the decision quality, one main problem is that the input information to ordinal methods is often too restricted. At the same time, decision-makers often possess more background information, for example regarding the relative strengths of the criteria, and might want to use that. Thus, some form of strength relation often exists that can be utilised when transforming orderings into weights. In this article, using a quite extensive simulation approach, we suggest a thorough testing methodology and analyse the relevance of a set of ordering methods including to what extent these improve the efficacy of rank order weights and provide a reasonable base for decision making.

Keywords: Multi-criteria decision analysis · Criteria weights · Criteria ranking · Rank order

1 Introduction

Despite having been around for some decades and been used in various applications, tools for supporting decision problems under multiple criteria (MCDA) are nevertheless still far too seldom utilised in real-life decision problems. This situation seems to be partly due to the lack of convergence between time constraints and cognitive abilities of decision-makers versus decision aid requirements. For instance, a vast number of methods have been suggested for assessing criteria weights using exact numbers. These range from relatively simple ones, like commonly used direct rating and point allocation methods, to more advanced procedures. Generally, a precise numerical weight is assigned to each criterion to represent the information extracted from the user. Despite the straightforward mathematics, a problematic issue is that the selection of adequate criteria weights is difficult and the time spent on determining and assigning precise numbers with enough accuracy seems to be a problem for actual decision-makers. Barron and Barrett [1] argue that the elicitation of exact weights demands an exactness which does not exist in the mind of the decision-maker, and already von Winterfeldt and

© Springer International Publishing Switzerland 2015
B. Kamiński et al. (Eds.): GDN 2015, LNBIP 218, pp. 107–118, 2015.
DOI: 10.1007/978-3-319-19515-5_9

Edwards [2] claim that "*the precision of numbers is illusory*". And, for example, ratio weight procedures can be difficult to accurately employ due to response errors [3]. The usual lack of reasonably complete information increases this problem significantly.

Several attempts have been made to resolve this issue. Methods allowing for less demanding ways of ordering the criteria, such as ordinal rankings or interval approaches for determining criteria weights and values of alternatives, have been suggested. The idea is, as far as possible, not to force decision-makers to express unrealistic, misleading, or meaningless statements, but at the same time being able to utilise the information the decision-maker is able to supply. An approach of this type is to use surrogate weights, which are derived from ordinal importance information [1, 4, 5]. In such methods, the decision-maker provides information on the rank order of the criteria, i.e. supplies ordinal information on importance. Thereafter, this information is converted into numerical weights consistent with the extracted ordinal information. Several proposals on how to convert the rankings into numerical weights exist, e.g., rank sum (RS) weights and rank reciprocal (RR) weights [6], and centroid (ROC) weights [7]. However, the use of only ordinal information is often perceived as being too vague or imprecise, resulting in a lack of confidence in the alternatives' final rankings.

Furthermore, it is not obvious how "correct" a surrogate weight method is, since the "real" weights are unknown or even inexistent (in some objective sense). The decision quality of a method was at first mostly assessed in case studies until Barron and Barrett [1] introduced a process utilising systematic simulations. The basic idea is to generate surrogate weights as well as "true" reference weights from some underlying distribution and investigate how well the result of using surrogate numbers match the result of using the "true" numbers. The idea is good, but is nevertheless vulnerable since the validation result is heavily dependent on the distribution used for generating the weight vectors.

In this article, we propose a set of methods for increasing the expressive power of user statements, with a particular aim at how the weight function(s) still can be reasonably elicited while preserving the comparative simplicity and correctness of ranking approaches. Below we discuss and compare some important aspects of a number of extensions to a set of ranking methods for weights as well as their relevance and correctness. After having briefly recapitulated some *ordinal* ranking methods in the following section, we continue with ranking methods taking strength into account and discuss a set of interesting candidates. Thereafter, using simulations, we investigate some properties of the methods and conclude with pointing out, according to the results, a particularly attractive method for weight elicitation.

2 Ordinal Ranking Methods

In multi-criteria decision making (MCDM), different elicitation formalisms have been proposed by which a decision-maker can express preferences. Such formalisms are sometimes based on scoring points, as in point allocation (PA) or direct rating (DR) methods.[1] In PA, the decision-maker is given a point sum, e.g. 100, to distribute among the criteria. Sometimes, it is pictured as putty with the total mass of 100 being divided

---

[1] PA and DR are akin to elements of the SAW approach [8].

and put on the criteria. The more mass, the larger weight on a criterion, and the more important it is. In PA, there is consequently $N-1$ degrees of freedom (DoF) for N criteria. DR, on the other hand, puts no limit to the number of points to be allocated.[2] The decision-maker allocates as many points as desired to each criterion. The points are subsequently normalized by dividing by the sum of points allocated. Thus, in DR, there are N degrees of freedom for N criteria. Regardless of elicitation method, the assumption is that all elicitation is made relative to a weight distribution held by the decision-maker.[3]

One very early idea in MCDM was to just skip the criteria elicitation and assign equal weights to every criterion, but the information loss is then very large. It is therefore worthwhile to at least rank the criteria when applicable, since rankings are normally easier to provide than precise numbers. From the ranking, so called surrogate weights can then be derived. This technique is utilised in [1, 4, 5], and many others. Needless to say, for practical decision making, surrogate weights can sometimes be perceived as a peculiar way of motivating a method. Nevertheless, validation in this field is very difficult, due to difficulties regarding elicitation, and the surrogate methods are quite widely used and can be considered as attempts of trying to motivate the various generation methods. The crucial issue is then rather how to assign surrogate weights while losing as little information as possible and preserving the "correctness" when assigning the weights. Stillwell et al. [6] discuss the weight approximation techniques rank sum (RS) and rank reciprocal (RR) weights. They are suggested in the context of maximum discrimination power, and are both alternatives to ratio based weight schemes. The rank sum is based on the idea that the rank order should be reflected directly in the weights. Assume a simplex S_w generated by $w_1 > w_2 > ... > w_N$, $\Sigma w_i = 1$ and $0 \leq w_i$.[4] Assign an ordinal number to each item ranked, starting with the highest ranked item as number 1. Denote the ranking number i among N items to rank. Then the RS weight for all $i = 1,..., N$ becomes

$$w_i^{RS} = \frac{N + 1 - i}{\sum_{j=1}^{N}(N + 1 - j)}$$

Another idea, also discussed in [6] is rank reciprocal weights. They have a similar origin as RS weights, but are based on the reciprocals (inverted numbers) of the rank order for each item ranked. These are obtained by assigning an ordinal number to each item ranked, starting with the highest ranked item as number 1. Denote the ranking number i among N items to rank. Then the rank reciprocal (RR) weight becomes

---

[2] Sometimes there is a limit to the individual numbers but not a limit to the sum of the numbers.

[3] For various cognitive and methodological aspects of imprecision in decision making, see, e.g., [8, 9] and other papers by the same authors.

[4] We will henceforth, unless otherwise stated, presume that decision problems are modelled as simplexes S_w generated by $w_1 > w_2 > ... > w_N$, $\Sigma w_i = 1$, and $0 \leq w_i$.

$$w_i^{RR} = \frac{1/i}{\sum_{j=1}^{N} \frac{1}{j}}$$

A decade later, Barron [7] suggested a weight method based on vertices of the simplex of the feasible weight space. The ROC (rank order centroid) weights are the centroid vector components of the simplex S_w. That is, ROC is a function based on the average of the corners in the polytope defined by the simplex $S_w = w_1 > w_2 > \ldots > w_N$, $\Sigma w_i = 1$, and $0 \leq w_i$. The weights then become the centroid (mass point) of S_w. The ROC weights for the ranking number i among N items to rank are given by

$$w_i^{ROC} = 1/N \sum_{j=i}^{N} \frac{1}{j}$$

Examining the weights, ROC resembles RR more than RS but is, particularly for lower dimensions, more extreme than both in the sense of weight distribution, especially for the largest and smallest weights.

As discussed in [10], RS, RR, and ROC perform well only for specific assumptions on decision-maker behaviour. If we assume that the decision-maker in his/her mind stores his/her criteria preferences in a way similar to a given point sum, for example pictured as putty with the fixed total mass, there are consequently $N-1$ degrees of freedom (DoF) for N criteria. On the other hand, if we assume that the decision-maker stores his/her criteria preferences in a way that puts no limit to the total number of points (or mass) allocated, then there are N degrees of freedom for N criteria. Those two models of decision-maker behaviour yield very different results in assessing surrogate weights. The RS weight model is tailored to the assumption of N degrees of freedom and the RR and ROC models are tailored to the $N-1$ DoF assumption. Since the models RS and RR are in this sense opposites, and in reality the preferences are reasonably stored in either one of the above ways or somewhere in between, a weight function combining the properties of RS and RR was proposed in [10]. The SR weight method is an additive combination of Sum and Reciprocal weight functions:

$$w_i^{SR} = \frac{1/i + \frac{N+1-i}{N}}{\sum_{j=1}^{N} \left(1/j + \frac{N+1-j}{N} \right)}$$

In [10], we carried out a large set of simulations of the above ordinal methods and confirmed some previous results as well as discussed some new results regarding a mixed model of decision-maker behaviour that takes into account the different possible degrees of freedom available. A crucial issue here is whether a decision-maker does keep the linear dependence of the criteria weights in mind when allocating values ($N-1$ DoF) or whether this is not an actual constraint since normalisation can be perceived to be possible to carry out after the criteria assessments (N DoF). Of course, various linear combinations of these would be possible, but the important observation is made using SR and comparing it with the others. The actual linear combination between the methods would affect the result according to its proportions. The details there are not crucial for our main point that all these results are a sensitive result of the underlying assumptions regarding the mind-settings of decision-makers. If this nevertheless would be important, the reasonable proportions must be elicited and fine-tuned with respect to the individual in question. Thus, in reality, we cannot absolutely know which DoF model a decision-maker employs, or if it is a mixed model somewhere in between. We concluded that to be *robust*, a rank ordering method should fare well under both of these assumptions and others. Of the above methods, SR was found to be the most robust and will, together with ROC, be used as references in the following comparative study.

3 General Ranking Methods

Providing ordinal rankings of criteria seems to avoid some of the difficulties associated with the elicitation of exact numbers. It puts fewer demands on decision-makers and is thus, in a sense, effort-saving. Furthermore, there are techniques such as those above for handling ordinal rankings with some success. However, decision-makers might in many cases have more knowledge of the decision situation, even if the information is not precise. For instance, importance relation information containing strengths may implicitly exist.[5] However, these cannot be taken into account in the transformation of an ordinal rank order into weights. This entails that the surrogate weights may not really reflect what the decision-maker actually means by his/her ranking. Some form of strengths often exist and this information should reasonably be used when transforming orderings into weights to utilise all the information the decision-maker is able to supply. Below, we will therefore investigate whether the above (ordinal) methods can be successfully extended to accommodate some information regarding relational strengths as well, i.e. to handle ordinal information together with strength relations information, while still preserving the property of being less demanding and more practically useful than other types of methods. The idea is that instead of using a predetermined conversion method (as in, e.g., ROC weights) to obtain surrogate weights from an ordinal criteria ranking, the decision-maker will be able to express and utilise known differences in importance between the criteria.

3.1 Preference Strength

Assume that there exists an ordinal ranking of N criteria. In order to make this order into a stronger ranking, information should be given about how much more or less important the criteria are compared to each other. Such rankings also take care of the problem with ordinal methods of handling criteria that are found to be equally important, i.e. resisting pure ordinal ranking. In this paper, we will use the following notations for the strength of the rankings between criteria as well as some suggestions for a verbal interpretation of these:

---

[5] For example: "A is slightly more important than B while B is vastly more important than C" must, in an ordinal ranking, be expressed as "A is more important than B which is more important than C".

$>_0$ equally important
$>_1$ slightly more important
$>_2$ more important (clearly more important)
$>_3$ much more important

While being more cognitively demanding than ordinal weights, they are still less demanding than, for example, AHP weight ratios (usually employing nine ratios, i.e. 1/9, 1/7, 1/5, 1/3, 1, 3, 5, 7, and 9) or point scores like SMART (usually employing several integers). In an analogous manner as for ordinal rankings, the decision-maker statements can be converted into weights.

3.2   Weights of Preference Strength

In analogy with the ordinal weight functions above, counterparts can straightforwardly be derived.

1. Assign an ordinal number to each importance scale position, starting with the most important position as number 1 (see Fig. 1, where A–F are criteria and where A $>_2$ B; B $>_1$ C; C $>_2$ D; D $>_0$ E; and E $>_3$ F).
2. Let the total number of importance scale positions be Q ($\Sigma s_i + 1$, where $i = 1,\ldots, N -1$ for N criteria; e.g. 9 positions in Fig. 1). Each criterion i has the position $p(i) \in \{1,\ldots,Q\}$ on this importance scale, such that for every two adjacent criteria c_i and c_{i+1}, whenever $c_i >_{s_i} c_{i+1}$, $s_i = |\,p(i+1) - p(i)\,|$. The position $p(i)$ then denotes the importance as stated by the decision-maker.
3. Then the counterparts to the ordinal ranking methods above can be found as follows.

To begin with, we consider the counterpart to RS weights. The concept of CRS weights is based on the idea that the rank order strength should be reflected directly in the weights. Then the corresponding counterpart to the ordinal rank sum weight (CRS) is obtained as

$$w_i^{CRS} = \frac{Q+1-p(i)}{\sum_{j=1}^{N}(Q+1-p(j))'}$$

based on the importance positions $p(i)$ as stated by the decision-maker. The counterpart to ordinal rank reciprocal weights is analogously defined. According to step 2, let the total number of importance scale positions be Q. Each criterion i has the position $p(i)$ on the importance scale such that $p(i) \leq p(j)$ if $i < j$. Then the corresponding rank reciprocal (CRR) weights are obtained as

Fig. 1. Importance scale positions

$$w_i^{CRR} = \frac{\frac{1}{p(i)}}{\sum_{j=1}^{N} \frac{1}{p(j)}}$$

with the usual property that a higher weight is assigned to lower ranking numbers. ROC weights are generalised in the same way. The ordinal weights are given by the formula

$$w_i^{ROC} = \frac{1}{N} \sum_{j=i}^{N} \frac{1}{j}$$

which could be interpreted as candidate weights for positions on the importance scale. Then the corresponding preference strength rank order centroid weights (CRC) are obtained as

$$w_i^{CRC} = \frac{\sum_{j=p(i)}^{Q} \frac{1}{j}}{\sum_{k=1}^{N} \left(\sum_{j=p(k)}^{Q} \frac{1}{j} \right)}$$

Finally, generalising SR weights is done in the same way. The ordinal SR weights are given by the formula

$$w_i^{SR} = \frac{\frac{1}{i} + \frac{N+1-i}{N}}{\sum_{j=1}^{N} w_j^{SR}}$$

which could now be interpreted as candidate weights for positions on the importance scale. Using steps 1–3 above, the corresponding preference strength SR weights (CSR) are obtained as

$$w_i^{CSR} = \frac{\frac{1}{p(i)} + \frac{Q+1-p(i)}{Q}}{\sum_{j=1}^{N} \left(\frac{1}{p(j)} + \frac{Q+1-p(j)}{Q} \right)}$$

which is a similar generalisation as the other weights. Thus, using the idea of importance steps, ordinal weight methods are easily generalised to their respective counterparts. Having obtained weights for preference strength relationships, we now proceed by assessing them together with ordinal weights.

4   Generalised Assessment of Models for Weights

Given that we have a set of methods as in the previous section, how can they be validated? For ordinal weights, simulation studies similar to [1, 11–13], and others have become a kind of de facto standard for comparing multi-criteria surrogate weight methods. The underlying assumption of most studies is that there exist a set of 'true' weights in the decision-maker's mind which are inaccessible in its pure form by any elicitation method. We will utilise the same technique for determining the efficacy, in this sense, of the ranking approaches suggested above. The modelling assumptions regarding decision-makers' mind-sets we discussed above are mirrored in the generation of decision problem vectors by a random generator. Thus, following an $N-1$ DoF model, a vector is generated in which the components sum to 100 %, i.e., a process with $N-1$ degrees of freedom. Following an N DoF model, a vector is generated keeping components within [0 %, 100 %] and subsequently normalising, i.e., a process with N degrees of freedom. Other distributions modelling actual decision makers would of course be possible, and could maybe be elicited in one way or another. However, this is not the main point herein. The important observation is that these validation methods are highly dependent of the model of decision makers and this yields significant effects on the reliability of the validations. The degree of freedom is only one type of dichotomy, but one actually expressing a meaningful semantics for discriminating cognitive models in this respect.

When following an $N-1$ DoF model, a vector is generated in which the components sum to 100 %. This simulation is based on a homogenous N-variate Dirichlet distribution generator. Details on this kind of simulation can be found, e.g., in [14]. On the other hand, following an N DoF model, a vector is generated without an initial joint restriction, only keeping components within [0 %,100 %] yielding a process with N degrees of freedom. Subsequently, they are normalised so that their sum is 100 %. Details on this kind of simulation can be found, e.g., in [15]. We will call the $N-1$ DoF model type of generator an $N-1$-generator and the N DoF model type an N-generator. Depending of the simulation model used (and consequently the background assumption of how decision-makers assess weights), the results become very different. For instance, ROC weights in N dimensions coincide with the mass point for the vectors of the $N-1$-generator over the polytope S_w. In [10], the close relationships between ROC weights and the $N-1$-generator as well as between RS weights and the N-generator were discussed, and we concluded that the choice of degrees of freedom for the random number generator significantly affects the results. Thus, when using $N-1$ DoF generated random vectors, ROC will always outperform all other surrogate weights in a simulation study. This is not a measure of ROC's superiority but of its match to the random generating function. Similarly, since RS weights are very close to the mass point of an N-generator over the polytope S_w, it is likewise not a measure of RS's superiority that it outperforms other surrogate weights when an N DoF simulator is employed.

In reality, though, we cannot know whether a specific decision-maker (or even decision-makers in general) adhere more to $N-1$ or N DoF representations of their knowledge. Both as individuals and as a group, they might use either or be anywhere in between. A, in a reasonable sense, *robust* rank ordering mechanism must therefore employ a surrogate weight function that handles both styles of representation and anything in between. Thus, the evaluation of surrogate weights in this paper will use both types of generators and combinations thereof to find the most efficient and robust weights.

Barron and Barrett [1] compared RS, RR, and ROC, where the idea was to measure the validity of the method by simulating a large set of scenarios utilising surrogate weights and see how well different methods provided results similar to scenarios utilising "true" weights. Again, note that the notion of a "true" weight is dependent on the decision-maker model. The Barron and Barrett study obviously assumes an $N-1$ DoF model and presents a computer simulation consisting of four steps, assuming the problem is modelled as the simplex S_w.

Generation Procedure

1. For an N-dimensional problem, generate a random weight vector with N components. This is called the TRUE weight vector. Determine the order between the weights in the vector. For each method $\mathbf{X'}$, use the order to generate a weight vector $w^{\mathbf{X'}}$.
2. Given M alternatives, generate $M \times N$ random values with value v_{ij} belonging to alternative j under criterion i.
3. Let $w_i^{\mathbf{X}}$ be the weight from weighting method $\mathbf{X}$ for criterion i (where $\mathbf{X}$ is either $\mathbf{X'}$ or TRUE). For each method $\mathbf{X}$, calculate $V_j^{\mathbf{X}} = \sum_i w_i^{\mathbf{X}} v_{ij}$. Each method produces a preferred alternative, i.e. the one with the highest $V_j^{\mathbf{X}}$.
4. For each method $\mathbf{X'}$, assess whether $\mathbf{X'}$ yielded the same decision (i.e. the same preferred alternative) as TRUE. If so, record a hit.

This is repeated a large number of times (simulation rounds). The hit rate (or frequency) is defined as the number of times a weighting method made the same decision as TRUE. The study also used two other measures of efficacy, average value loss and average proportion of maximum value range achieved. The two latter measures are strongly correlated to the hit ratio and do not add much insight into method performance. The results of the original study in [1] were that ROC outperformed the other two weighting methods. Of the two other, RR was slightly superior to RS. Since the three methods require equally much input from the decision-maker, the conclusion was made that ROC was to be preferred among the surrogate weights. Using an $N-1$-generator simulation model over the simplex S_w, the results of the Barron and Barrett study can easily be verified. However, note again that this distribution favours the ROC method since the centroid of the generated "true" weights is the same as the vector of the corresponding ROC weights.

It should also be noted that most simulation studies to date arrive at the same conclusions regarding ROC, RS, and RR. A study by Roberts and Goodwin [15], though, came up with a different result where RS performed better than ROC with RR in third place. The random weight distribution is in most other simulations (in step 1 of the generation procedure above) generated by an $N-1$ procedure, thus generating a vector with $N-1$ DoF. Instead, Roberts and Goodwin [15] employ a different distribution generating function where a fixed number, say 100, is given to the most important criterion and the others are uniformly generated as U[0, 100], i.e. an N-generator. As explained above, this N-generator is not the same as $N-1$-generators based on a Dirichlet distribution and thus, their simulation study instead yields the result that RS outperforms ROC with RR in third place. This is also confirmed in [10], i.e. given an N-generator RS outperforms ROC and RR while ROC is marginally better than RR. While yielding a different "best" weighting method, this result is consistent with the other study results considering it is merely a consequence of choice of DoF in the simulator generator.

4.1   Comparing weight methods

Our comparative simulations were carried out with a varying number of criteria and alternatives. There were four numbers of criteria $N = \{3, 6, 9, 12\}$ and five numbers of alternatives $M = \{3, 6, 9, 12, 15\}$ creating a total of 20 simulation scenarios. Each scenario was run 10 times, each time with 10,000 trials, yielding a total of 2,000,000 decision situations generated. An N-variate joint Dirichlet distribution was employed to generate the random weight vectors for the N–1 DoF simulations and a standard round-robin normalised random weight generator for the N DoF simulations. Similar to [1], unscaled value vectors were generated uniformly, and no significant differences were observed with other value distributions.[6]

Table 1. The winner frequency for the methods using an N-1 generator

N–1 DoF		ROC	SR	CRC	CRS	CRR	CSR
3 criteria	3 alternatives	90.2	89.3	93.6	93.0	94.1	94.7
3 criteria	15 alternatives	79.1	76.9	84.7	84.7	85.7	87.3
6 criteria	6 alternatives	84.8	83.1	92.1	81.8	88.6	89.1
6 criteria	12 alternatives	81.3	78.9	89.9	77.9	85.6	86.0
9 criteria	9 alternatives	83.5	81.2	89.9	73.9	85.6	83.3
12 criteria	6 alternatives	86.4	84.1	89.9	73.4	87.2	82.9
12 criteria	12 alternatives	83.4	80.2	86.8	68.1	83.7	78.6

Table 2. The winner frequency for the methods using an N generator

N DoF		ROC	SR	CRC	CRS	CRR	CSR
3 criteria	3 alternatives	87.3	89.1	91.7	93.7	93.1	94.4
3 criteria	15 alternatives	77.9	80.6	83.0	88.1	87.1	89.5
6 criteria	6 alternatives	80.1	85.1	90.5	88.8	89.0	92.4
6 criteria	12 alternatives	76.4	82.0	88.3	86.0	86.2	90.4
9 criteria	9 alternatives	76.3	83.0	90.0	84.3	86.8	89.9
12 criteria	6 alternatives	77.5	84.6	90.9	83.9	88.5	89.6
12 criteria	12 alternatives	73.4	81.7	88.8	80.2	85.5	87.0

---

[6] Success measures we used were (a) "winner", having the same preferred alternative, (b) matching of the three highest ranked alternatives ("podium"), and (c) matching of all ranked alternatives ("overall"), the number of times all evaluated alternatives using a particular method coincide with the true ranking of the alternatives. The two latter sets correlated strongly with the first and are not shown in this paper.

Table 3. The winner frequency for the methods using a combined generator

Combined		ROC	SR	CRC	CRS	CRR	CSR
3 criteria	3 alternatives	88.8	89.2	92.7	93.4	93.6	94.6
3 criteria	15 alternatives	78.5	78.8	83.9	86.4	86.4	88.4
6 criteria	6 alternatives	82.5	84.1	91.3	85.3	88.8	90.8
6 criteria	12 alternatives	78.9	80.5	89.1	82.0	85.9	88.2
9 criteria	9 alternatives	79.9	82.1	90.0	79.1	86.2	86.6
12 criteria	6 alternatives	82.0	84.4	90.4	78.7	87.9	86.3
12 criteria	12 alternatives	78.4	81.0	87.8	74.2	84.6	82.8

In Table 1, using an $N-1$-generator, it can be seen that all four preference strength methods generally outperform the ordinal ones and in most cases CRC is the best one, except for the case with three criteria, when CSR is the best. The frequencies have changed in Table 2, according to expectations, since we employ a model with N degrees of freedom. Still the preference strength methods perform better than the ordinal ones. CRS improves and CRC generally fares a bit worse. In general, strength methods perform clearly better than ordinal ones. In Table 3, in looking for a robust method, the N and $N-1$ DoF models are combined with equal emphasis on both. Still the latter methods perform better than the ordinal ones and we can see that in total CSR performs the best for three criteria. Above three, CRC is the best performer. From this it is clear that the CRC, CRR, and CSR methods outperform the best ordinal methods under varying assumptions of decision-maker weight generation.

5   Concluding Remarks

There is obviously a need of weighting and elicitation methods that neither require formal decision analysis knowledge, nor is too cognitively demanding by forcing people to express unrealistic precision or to state more than they are able to. Furthermore, it should not require too much time and must be able to actually make use of the information the decision-maker is able to supply. However, elicitation methods available today are often either cognitively demanding and require too much time and effort or unable to use available information. We have therefore focused pragmatically on providing a method trying to balance between the need of simplicity and the requirement for accuracy, i.e. providing a more useful MCDA weighting method with reasonable elicitation components. We have extended some well-known ordinal scale approaches as well as a newly proposed one with the possibility to supply information regarding preference strength as well. We have furthermore compared the approaches in various ways by carrying out a large set of simulations of the methods and have found some interesting simulation results. We have also discussed some new results regarding a mixed model of decision-maker behaviour and which degree of freedom that is adequate. From the above, it is clear that imprecise information should definitely be used when available. Furthermore, the proposed CSR method has turned out to be particularly appealing. CSR extends the rank order weighting procedure SR from [10] by also taking strength preference into account in a more straightforward way than previously suggested in [16].

Acknowledgments. This research was funded by the Swedish Research Council FORMAS, project number 2011-3313-20412-31, as well as by Strategic funds from the Swedish government within ICT – The Next Generation.

References

1. Barron, F., Barrett, B.: Decision quality using ranked attribute weights. Manag. Sci. **42**(11), 1515–1523 (1996)
2. von Winterfeldt, D., Edwards, W.: Decision Analysis and Behavioural Research. Cambridge University Press, New York (1986)
3. Jia, J., Fischer, G.W., Dyer, J.: Attribute weighting methods and decision quality in the presence of response error: a simulation study. J. Behav. Decis. Making **11**(2), 85–105 (1998)
4. Barron, F., Barrett, B.: The efficacy of SMARTER: simple multi-attribute rating technique extended to ranking. Acta Psych. **93**(1–3), 23–36 (1996)
5. Katsikopoulos, K., Fasolo, B.: New tools for decision analysis. IEEE Trans. Syst. Man Cybern. – Part A: Syst. Hum. **36**(5), 960–967 (2006)
6. Stillwell, W., Seaver, D., Edwards, W.: A comparison of weight approximation techniques in multiattribute utility decision making. Org. Behav. Hum. Perform. **28**(1), 62–77 (1981)
7. Barron, F.H.: Selecting a best multiattribute alternative with partial information about attribute weights. Acta Psych. **80**(1–3), 91–103 (1992)
8. Danielson, M., Ekenberg, L.: Computing upper and lower bounds in interval decision trees. Eur. J. Oper. Res. **181**(2), 808–816 (2007)
9. Danielson, M., Ekenberg, L., Larsson, A., Riabacke, M.: Weighting under ambiguous preferences and imprecise differences in a cardinal rank ordering process. Int. J. Comp. Int. Syst. **7**, 105–112 (2013)
10. Danielson, M., Ekenberg, L.: Rank ordering methods for multi-criteria decisions. In: Zaraté, P., Kersten, G.E., Hernández, J.E. (eds.) GDN 2014. LNBIP, vol. 180, pp. 128–135. Springer, Heidelberg (2014)
11. Arbel, A., Vargas, L.G.: Preference simulation and preference programming: robustness issues in priority derivation. European J. Operational Res. **69**, 200–209 (1993)
12. Stewart, T.J.: Use of piecewise linear value functions in interactive multicriteria decision support: a monte carlo study. Manage. Sci. **39**(11), 1369–1381 (1993)
13. Ahn, B.S., Park, K.S.: Comparing methods for multiattribute decision making with ordinal weights. Comput. Oper. Res. **35**(5), 1660–1670 (2008)
14. Rao, J.S., Sobel, M.: Incomplete dirichlet integrals with applications to ordered uniform spacing. J. Multivar. Anal. **10**, 603–610 (1980)
15. Roberts, R., Goodwin, P.: Weight approximations in multi–attribute decision models. J. Multi-criteria Decis. Anal. **11**, 291–303 (2002)
16. Danielson, M., Ekenberg, L., He, Y.: Augmenting ordinal methods of attribute weight approximation. Decis. Anal. **11**(1), 21–26 (2014)

Veto Values Within MAUT for Group Decision Making on the Basis of Dominance Measuring Methods with Fuzzy Weights

Pilar Sabio, Antonio Jiménez-Martín[✉], and Alfonso Mateos

Universidad Politécnica de Madrid, 28660 Boadilla del Monte, Spain
{p.sabio,antonio.jimenez,alfonso.mateos}@upm.es
http://www.dia.fi.upm.es/grupos/dasg/index.htm

Abstract. In this paper we extend the additive multi-attribute utility model to incorporate the concept of veto in a group decision-making context. Moreover, trapezoidal fuzzy numbers are used to represent the relative importance of criteria for each DM, and uncertainty about the alternative performances is considered by means of intervals. Although all DMs are allowed to provide veto values, the corresponding vetoes are effective for only the most important DMs. They are used to define veto ranges. Veto values corresponding to the other less important DMs are partially taken into account, leading to the construction of adjust ranges. Veto and an adjust function are then incorporated into the additive model, and a fuzzy dominance matrix is computed. A dominance measuring method is then used to derive a ranking of alternatives for each DM, which are then aggregated to account for the relative importance of DMs.

Keywords: Group decision-making · Multi-attribute utility theory · Veto · Dominance measuring methods · Fuzzy weights

1 Introduction

The *veto* concept is considered as a real-world approach for representing the limits of decision-maker (DM) preferences, and is thus an important tool in multicriteria and group decision-making.

For example, let us consider a couple who decide to buy a home, where both have veto power. They identify several criteria for selecting the house, like price, location, size or age. One of the two might rule out any house smaller than $80\,\text{m}^2$, regardless of house price, location and age, whereas the other might rule out any more expensive than 250,000 euros and smaller than $60\,\text{m}^2$.

The concept of veto was originally justified in social theory by the *prudence axiom* enunciated by Arrow and Raynaud [1]. The main idea behind this axiom is that it is not prudent to accept highly conflicting alternatives that may result in vulnerable decisions. Regarding the prudence axiom, Moulin defines the *principle of proportional veto* in a group of DMs [2], according to which any coalition of

© Springer International Publishing Switzerland 2015
B. Kamiński et al. (Eds.): GDN 2015, LNBIP 218, pp. 119–130, 2015.
DOI: 10.1007/978-3-319-19515-5_10

DMs should be given the right to veto a certain number of alternatives, which is approximately proportional to the size of this coalition. In addition, the fact of allocating veto power across the various groups of social participants has ethical implications, since it entails attaching different weights to different groups.

In MCDM problems the concept of veto has been used to manage *non-compensatory methods*. In outranking methods the use of veto usually represents the intensity of preference of the minority [3]. For instance, Nowak [4] used ELECTRE-III to build a multi-attribute ranking using preference thresholds to distinguish situations of strict and weak preference in stochastic dominance approaches.

Later, Munda [5] implemented a veto-based threshold to deal with environmental and resource management and policies aimed at sustainable development. A fuzzy set theory framework was used to represent *qualitative information* by means of the concept of *linguistic variable*. Ranking policy options were derived by means of the majority principle implemented by Concordet, whereas the power of a subgroup of DMs to veto some alternatives was accounted for by means of Moulin's proportional veto function.

On the other hand, additive *compensatory methods* have also incorporated the concept of veto. For instance, Bana e Costa et al. [6] define a multi-criteria approach for prohibiting alternatives by the *measuring attractiveness by a categorical based evaluation technique* (MACBETH) for facilitating bid evaluation processes, such as interventions in an international public call for tenders. The result is a procedure called the *determinants technique*, whose groundwork is aligned, albeit not directly, with the notion of *veto power* used to model non-compensatory situations.

In connection with research based on the power of veto, Marichal [7] proposes to axiomatize individual indices to rate whether each criterion behaves as a veto or an aggregator using the Choquet integral. These indices for measuring the degree to which each criterion behaves like a blocker or a pusher, make it possible to identify and measure the *dictatorial* tendency of criteria, which is a particular interaction phenomenon. Here, the veto is not a preference parameter given by the DM but an effect phenomenon when aggregating criteria. Therefore, the veto concept is related to the impact caused by a criterion on the global evaluation of alternatives.

Liginlala and Ow [8] use the same idea of veto effects, expressing degrees of conjunction, disjunction, veto and approval given by the indices through fuzzy analysis measures, which represent a risk tolerance measure of the DM. The veto power examines how tolerant DMs are about accepting or rejecting evaluations of alternatives associated with specific actions on a given attribute.

More recently, Daher and de Almeida [9] developed an additive group preference model that incorporates a utility reduction factor. DMs express their preferences in terms of a ranking of alternatives and are able to make an informed veto by providing information about the undesirable or unacceptable ranking of some alternatives. The ranking veto is achieved by using a reduction factor on the global utility of the alternatives.

In this paper, we consider the classical additive multi-attribute utility model (MAUT) in a group decision-making situation. Here, the concept of veto threshold is related to the definition of each attribute's preference bounds, whereby alternatives whose criteria are rated above or below these bounds are rejected by DMs depending on whether their utility function is increasing or decreasing and irrespective of the value that they take for other attributes. At the same time, this use of the veto concept is an attempt to account for the flawedness or ambiguity of the evaluation of alternatives in order to reach a consensus.

Moreover, uncertainty about the alternative performances is accounted for by means of intervals, the relative importance of DMs is known and the attribute weights are represented by trapezoidal fuzzy numbers. A dominance measuring method is used to derive a complete fuzzy ranking of alternatives for each DM accounting for imprecise information. These are then aggregated to reach a consensus ranking.

In Sect. 2 we introduce an extension of the additive multi-attribute value model to account for veto and adjust ranges. First, veto and adjust functions are defined from the veto values provided by DMs. Then, a dominance measuring method accounting for fuzzy weights and performance intervals is introduced in Sect. 2.1 to derive a ranking of alternatives for each DM. Finally, we aggregate the rankings from the different DMs to reach a consensus ranking in Sect. 2.2. An example to illustrate the proposed methodology is given in Sect. 3. Finally, some conclusions are provided in Sect. 4.

2   Group Decision-Making Within MAUT Accounting for Veto

We consider a group decision-making problem with m alternatives $\{A_1, ..., A_m\}$ and n attributes $\{X_1, ..., X_n\}$. DM preferences are modeled by an additive multi-attribute utility function, which is a compensatory model considered to be a valid approach in most practical situations [10,11]. Its functional form is

$$u(A_i) = \sum_{j=1}^{n} w_j u_j(x_{ij}), \tag{1}$$

where x_{ij} is the performance of alternative A_i with respect to attribute X_j, $u_j(.)$ is the component utility function representing DM preferences for the values of attribute X_j and w_j is the weight representing the relative importance of each attribute. Note that $\sum_j w_j = 1$ and $w_j \geq 0$.

We account for uncertainty about the alternative performances by means of uniformly distributed intervals. We denote by $[x_{ij}^L, x_{ij}^U]$ the performance of alternative A_i with respect to attribute X_j.

We also consider a set of k DMs, denoted by DM_l, $l = 1, ..., k$, whose relative importance is known and denoted by w_{DM_l}. Without loss of generality we assume that the most important DM is DM_1, followed by DM_2, and so on, until DM_k. Consequently, $w_{DM_1} \geq w_{DM_2} \geq ... \geq w_{DM_k}$, and $\sum_l w_{DM_l} = 1$.

The question of how to measure the weights of DMs in a group decision-making context is an interesting research topic. Yue [12] provides a brief overview of approaches proposed by different authors to determine the weights of DMs. Morever, a new approach based on an extended TOPSIS method [13,14] is also proposed.

All DMs are allowed to provide veto values, but the corresponding veto will be effective for only the r most important DMs, $r \leq k$. Veto values corresponding to the $k - r$ remaining DMs will be partially taken into account, as described later.

We denote by v_j^l the veto threshold provided by the l-th DM for the attribute X_j, i.e. the l-th DM wants the alternative performances to be equal to or greater (lower) than v_j^l if an increasing (decreasing) component utility function is associated with attribute X_j. Consequently, the veto interval for the l-th DM is $(0, v_j^l]$ in attribute X_j. For simplicity's sake, we will consider from now on that component utility functions are increasing.

A *veto range* can then be identified for each attribute $[v_j^L, v_j^U]$, where $v_j^L = r_j^L$, $[r_j^L, r_j^U]$ being the attribute range, and $v_j^U = max_{l=1,...,r}\{v_j^l\}$, i.e. the highest veto value for attribute X_j for the r most important DMs.

We build an *adjust range* for each attribute X_j, $(a_j^L, a_j^U]$, with $a_j^L = v_j^U = max_{l=1,...,r}\{v_j^l\}$, i.e. the highest veto value for attribute X_j for the r most important DMs, and $a_j^U = max_{l=1,...,k}\{v_j^l\}$, i.e. the highest veto value for attribute X_j considering all DMs.

We add the above information to the additive multi-attribute utility function by means of the following functions:

- $v(A_i)$ is the *veto function* that checks if the performances for a given A_i are within the respective veto intervals:

$$v(A_i) = \prod_{j=1}^n v_j(A_i), \quad \text{with } v_j(A_i) = \begin{cases} 1, \text{ if } x_{ij} > v_j^U \\ 0, \text{ if } x_{ij} \leq v_j^U \end{cases}. \tag{2}$$

Note that $v(A_i) = 0$ if at least one performance is within the veto interval for the corresponding attribute.

- $d_j(A_i)$ is the *adjust function* that decreases the utility associated with the alternative performances within the corresponding adjust range. A first possible approach is to apply a linear adjust function. However, we believe that the veto values for the $k - r$ less important DMs should be added by means of this adjust function. Veto values provided by the $k - r$ DMs may be within the adjust interval. In this case, we use this information to build a piecewise linear function.

The adaptation of the additive multi-attribute utility function to account for the veto and adjust functions would be as follows:

$$u^l(A_i) = [\sum_{j=1}^n u_j^l(x_{ij})w_j^l d_j(A_i)] \times v(A_i). \tag{3}$$

This expression would then be used to derive a ranking of the alternative under consideration for each DM that should be aggregated taking into account the relative importance of DMs to reach a consensus ranking.

Note, however, that in the decision-making scenario under consideration, we have assumed trapezoidal fuzzy numbers to represent the relative importance of criteria, $\widetilde{w}_j^l$, which have to be added to Eq. (3), leading to $\widetilde{u^l(A_i)}$.

The DMs can select linguistic terms from a linguistic term scale ([15,16]) to represent these weights. Direct assessment on the basis of this kind of scales is much faster and more commonplace in decision-making processes involving fuzzy logic. A more efficient way of allocating the weights without the biases inherent in the use of linguistic scales is reported in [17].

In the next section we describe how to derive a ranking of alternatives using *dominance measuring methods*, which are based on the notion of pairwise dominance. Then, we aggregate the rankings for each DM taking into account their relative importance to reach a consensus ranking.

2.1 A Dominance Measuring Method for Deriving a Fuzzy Ranking of Alternatives for Each DM

A recent approach for dealing with imprecise information is to compute different measures of dominance to derive a ranking of alternatives ([18–20]). They are known as *dominance measuring methods (DMMs)*. DMMs are based on the computation of a *dominance matrix*, D, including pairwise dominance values, which are exploited in different ways to derive measures of dominance to rank the alternatives under consideration

In our decision-making scenario with performance intervals and fuzzy weights the corresponding fuzzy dominance matrix for the l–th DM is:

$$\widetilde{D}^l = \begin{pmatrix} - & \widetilde{D}_{12}^l & \cdots & \widetilde{D}_{1m}^l \\ \widetilde{D}_{21}^l & - & \cdots & \widetilde{D}_{2m}^l \\ \widetilde{D}_{31}^l & \widetilde{D}_{32}^l & - & \widetilde{D}_{3m}^l \\ \vdots & \vdots & \ddots & \vdots \\ \widetilde{D}_{m1}^l & \widetilde{D}_{m2}^l & \cdots & - \end{pmatrix}, \text{ where } \begin{array}{l} \widetilde{D}_{ks}^l = \min\{\widetilde{u^l(A_k)} - \widetilde{u^l(A_s)}\} \\ s.t. \\ x_{kj}^L \leq x_{kj} \leq x_{kj}^U, \ j = 1,\ldots,n \\ x_{sj}^L \leq x_{sj} \leq x_{sj}^U, \ j = 1,\ldots,n. \end{array} \quad (4)$$

Note that is not necessary to solve the above fuzzy linear optimization problems since $\widetilde{D}_{ks}^l$ can be computed as follows (for increasing component utility functions):

$$\widetilde{D}_{ks}^l = [\sum_{j=1}^n u_j^l(x_{kj}^L)\widetilde{w_j^l}d_j(x_{kj}^L)] \times v(x_{kj}^L) - [\sum_{j=1}^n u_j^l(x_{sj}^U)\widetilde{w_j^l}d_j(x_{sj}^U)] \times v(x_{sj}^U). \quad (5)$$

We propose using the *DMM* introduced in [18] in the group decision-making scenario under consideration to derive a ranking of alternatives for each DM. Consequently, we first compute the strength of dominance of alternative A_k by adding the trapezoidal fuzzy numbers in the kth row of $\widetilde{D}^l$,

$$\tilde{d}^l_k = (d^l_{k1}, d^l_{k2}, d^l_{k3}, d^l_{k4}) = \left(\sum_{s=1\,s\neq k}^{m} D^l_{ks1}, \sum_{s=1\,s\neq k}^{m} D^l_{ks2}, \sum_{s=1\,s\neq k}^{m} D^l_{ks3}, \sum_{s=1\,s\neq k}^{m} D^l_{ks4} \right). \tag{6}$$

Then, a *dominance intensity*, DI^l_k, is computed for each alternative A_k by multiplying the proportion of the positive part of the fuzzy number $\tilde{d}^l_k$ by the distance of the fuzzy number to zero. Specifically, the dominance intensity for alternative A_k is computed according to the location of $\tilde{d}^l_k$ as follows:

1. If $\tilde{d}^l_k$ is completely located on the left of zero, then DI^l_k is minus the distance of $\tilde{d}^l_k$ to zero, because there is no positive part in $\tilde{d}^l_k$.
2. If $\tilde{d}^l_k$ is completely located on the right of zero, then DI^l_k is the distance of $\tilde{d}^l_k$ to zero, because there is no negative part in $\tilde{d}^l_k$.
3. If $\tilde{d}^l_k$ includes the zero in its base, then the fuzzy number will have a part on the right of zero that we denote $\tilde{d}^{lR}_k$ and another part on the left of zero that we denote $\tilde{d}^{lL}_k$. DI^l_k is the proportion that represents $\tilde{d}^{lR}_k$ with respect to $\tilde{d}^l_k$ multiplied by the distance of $\tilde{d}^l_k$ to zero less the proportion that represents $\tilde{d}^{lR}_k$ with respect to $\tilde{d}^l_k$ multiplied by the distance of $\tilde{d}^l_k$ to zero. Specifically:
   - If $d^l_{k3} < 0$ and $d^l_{k4} > 0$, Fig. (1a), the corresponding trapezoidal fuzzy number is again divided by the vertical axis into two parts, and the dominance intensity of alternative A_k is

$$DI^l_k = \frac{\left(d^l_{k4}\right)^2}{\left(d^l_{k4} - d^l_{k3}\right)\left(d^l_{k4} + d^l_{k3} - d^l_{k2} - d^l_{k1}\right)} D(\tilde{d}^l_k, 0, f) -$$
$$\frac{d^l_{k4}\left(d^l_{k4} - d^l_{k2} - d^l_{k1}\right) - d^l_{k3}\left(d^l_{k3} - d^l_{k2} - d^l_{k1}\right)}{\left(d^l_{k4} + d^l_{k3} - d^l_{k2} - d^l_{k1}\right)\left(d^l_{k4} - d^l_{k3}\right)} D(\tilde{d}^l_k, 0, f).$$

   - If $d^l_{k1} < 0$ and $d^l_{k2} > 0$, see Fig. (1b), the corresponding trapezoidal fuzzy number is divided by the vertical axis (at zero) into two parts. The dominance intensity of alternative A_k is defined as

$$DI^l_k = \frac{d^l_{k2}(-d^l_{k2} + d^l_{k3} + d^l_{k4}) - d^l_{k1}(d^l_{k3} + d^l_{k4})}{\left(d^l_{k2} - d^l_{k1}\right)\left(d^l_{k4} + d^l_{k3} - d^l_{k2} - d^l_{k1}\right)} D(\tilde{d}^l_k, 0, f) -$$
$$\frac{\left(d^l_{k1}\right)^2}{\left(d^l_{k4} + d^l_{k3} - d^l_{k2} - d^l_{k1}\right)\left(d^l_{k2} - d^l_{k1}\right)} D(\tilde{d}^l_k, 0, f).$$

   - If $d^l_{k2} < 0$ and $d^l_{k3} > 0$, see Fig. (1c), the dominance intensity of alternative A_k is

$$DI^l_k = \frac{-d^l_{k2} - d^l_{k1}}{d^l_{k4} + d^l_{k3} - d^l_{k2} - d^l_{k1}} D(\tilde{d}^l_k, 0, f) - \frac{d^l_{k4} + d^l_{k3}}{d^l_{k4} + d^l_{k3} - d^l_{k2} - d^l_{k1}} D(\tilde{d}^l_k, 0, f). \tag{7}$$

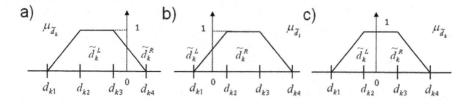

Fig. 1. Building DI_k^l

Note that an adaptation of Tran and Duckstein's distance for the *generalization of the left and right fuzzy numbers* (GLFRN) is used to account for the distance between a trapezoidal fuzzy number and a constant (specifically 0).

Moreover, Tran and Duckstein's distance [21] incorporates a function f that makes DM participation flexible. For example, if a risk-neutral DM is considered $(f(\alpha) = \alpha)$, then

$$D^2(\tilde{a}, 0, f) = \left(\frac{a_2 + a_3}{2}\right)^2 + \frac{1}{3}\left(\frac{a_2 + a_3}{2}\right)\left[(a_4 - a_3) - (a_2 - a_1)\right] + \frac{2}{3}\left(\frac{a_3 - a_2}{2}\right)^2$$
$$+ \frac{1}{9}\left(\frac{a_3 - a_2}{2}\right)\left[(a_4 - a_3) + (a_2 - a_1)\right] + \frac{1}{18}\left[(a_4 - a_3)^2 + (a_2 - a_1)^2\right]$$
$$- \frac{1}{18}\left[(a_2 - a_1)(a_4 - a_3)\right].$$

$$(8)$$

Expressions for a risk-prone and a risk-averse DM can be found in Aguayo et al. [19].

Once the dominance intensity has been computed for each alternative A_k, the alternatives are ranked accordingly, where the best (rank 1) is the alternative with the greatest DI_k^l and the worst is the alternative with the least DI_k^l.

We proceed analogously for all DMs, applying the above *DMM* from the corresponding dominance matrix, D^l, $l = 1, ..., k$, leading to k rankings of alternatives.

2.2   Aggregating Alternative Rankings

Different methods for aggregating rankings by different authors can be found in the literature. Lin [22] discusses three classes of methods, namely *distribution-based methods*, for instance, the original Thurstone scaling and its extensions [23]; *heuristic methods*, ranging from simple arithmetic averages of ranks (Borda's methods, [24]) to Markov chains and stationary distributions [25]; and *stochastic optimization search methods*, such as the Kemeny optimal aggregation.

In our decision-making scenario, complete rankings and the relative importance of such rankings (relative importance of DMs) are available. Moreover, the values that lead to the corresponding rankings (global dominance intensities) are also available. The only aggregation methods that exploit all the above information is the Kemeny method [26] and its extensions.

Kemeny optimal aggregation optimizes the average Kendall distances between a candidate aggregate ranking and each of the input rankings. As computing the Kemeny optimal aggregate is NP-hard even when the number of ranked lists to be aggregated is small, we have used the *order explicit algorithm* (OEA)[27] to solve the combinatorial problem under consideration.

OEA uses a global optimization technique, called the *cross-entropy Monte Carlo method*, which searches iteratively for an optimal list that minimizes a criterion, the sum of weighted distances between the candidate (aggregate) list and each of the input ranked lists. The method is, however, general and amenable to any other optimization criterion. A modified Kendall's tau measure and the *Spearman's footrule* are used to measure the distance between two ranked lists.

3   An Illustrative Example

We consider five DMs whose relative importance is $w_{DM_1} = 0.35 \geq w_{DM_2} = 0.25 \geq w_{DM_3} = 0.2 \geq w_{DM_4} = 0.1 = w_{DM_5} = 0.1$. Seven alternatives $\{A_1, ..., A_7\}$ will be analyzed on the basis of four attributes $\{X_1, ..., X_4\}$, whose ranges are [0,100] in all cases.

The corresponding veto will be effective for the three most important DMs only. Table 1 shows the veto values provided by the DMs. Note that, except for DM_1, DMs do not provide veto values for all attributes, and the only veto for attribute X_4 is provided by DM_1.

Table 1. Veto values for DMs.

	X_1	X_2	X_3	X_4
DM_1	20	15	10	20
DM_2	15	10	5	-
DM_3	25	10	-	-
DM_4	30	25	-	-
DM_5	27	10	-	-

Each DM expresses the relative importance of the attributes under consideration by means of trapezoidal fuzzy numbers, see Table 2. We assume that the four component utility functions are linear and increasing in the attribute ranges, [0, 100], for the five DMs.

Table 3 shows the alternative performances for the four attributes under consideration as well as the veto and adjust ranges for each attribute. The vetoed performance endpoints are marked in bold.

The adjust functions for attributes X_1 and X_2 are shown in Fig. 2. Note that the adjust function for attribute X_2 is a linear function since none of the DMs provided a veto value within the adjust range, whereas the adjust function for attribute X_1 is a piecewise utility function since the veto value 27 corresponding to DM_5 is within the adjust range and assigned a value of 0.5.

Table 4 shows the values output by the veto and adjust function for the endpoints of the performance intervals included in Table 3. All values in the

Table 2. Relative importance of attributes for DMs.

	w_1	w_2	w_3	w_4
DM_1	(0.30,0.33,0.37,0.40)	(0.25,0.28,0.32,0.35)	(0.15,0.18,0.22,0.25)	(0.10,0.13,0.17,0.20)
DM_2	(0.05,0.08,0.12,0.15)	(0.15,0.18,0.22,0.25)	(0.15,0.18,0.22,0.25)	(0.45,0.48,0.52,0.55)
DM_3	(0.35,0.38,0.42,0.45)	(0.20,0.23,0.27,0.30)	(0.15,0.18,0.22,0.25)	(0.10,0.13,0.17,0.20)
DM_4	(0.05,0.08,0.12,0.15)	(0.30,0.33,0.37,0.40)	(0.10,0.13,0.17,0.20)	(0.35,0.38,0.42,0.45)
DM_5	(0.15,0.18,0.22,0.25)	(0.10,0.13,0.17,0.20)	(0.15,0.18,0.22,0.25)	(0.40,0.43,0.47,0.50)

Table 3. Alternative performances and veto and adjust ranges.

	X_1	X_2	X_3	X_4
A_1	[32,38]	[56,65]	[37,43]	[32,38]
A_2	**[19,22]**	[19,22]	[37,43]	[60,70]
A_3	**[24,**28]	[25,29]	[28,32]	[87,100]
A_4	[37,43]	[32,38]	[46,54]	[58,68]
A_5	[32,38]	[79,91]	**[9,**11]	[31,35]
A_6	[56,65]	[16,18]	[19,22]	[42,48]
A_7	**[25,**29]	[60,70]	[19,22]	[42,48]
Veto range	[0, 25]	[0, 15]	[0, 10]	[0, 20]
Adjust range	(25, 30]	(15, 25]	-	-

performance interval of alternative A_2 are vetoed for attribute X_1. Consequently, the utility for that alternative will be 0, see Eq. (2), and it will always be the worst-ranked alternative. Thus, we have omitted this alternative from further analyses. Besides, some values in the performance interval of alternatives A_3 and A_7 are vetoed for attribute X_1, i.e. the corresponding lower endpoints are vetoed while the upper ones are not. Finally, some values in A_5 are vetoed for attribute X_3.

Adjust functions decrease the upper component utility associated with alternatives A_3 and A_7 for attribute X_1 and both the lower and upper component

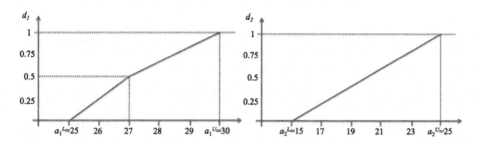

Fig. 2. Adjust functions for attributes X_1 and X_2

Table 4. Veto and adjust values.

	A_1	A_2	A_3	A_4	A_5	A_6	A_7
$\{v(A_i^L), v(AU_i)\}$	{1,1}	{0,0}	{0,1}	{1,1}	{0,1}	{1,1}	{0,1}
$\{d_1(x_{i1}^L), d_1(x_{i1}^U)\}$	{1,1}	{1,1}	{0,0.658}	{0.997,1}	{1,1}	{1,1}	{0,0.837}
$\{d_2(x_{i2}^L), d_1(x_{i2}^U)\}$	{1,1}	{0.35,0.65}	{1,1}	{1,1}	{1,1}	{0.072,0.327}	{1,1}

Table 5. Imprecise component utilities accounting for the veto and adjust functions.

	A_1	A_3	A_4	A_5	A_6	A_7
X_1	{0.32,0.38}	{0,0.18}	{0.37,0.43}	{0,0.38}	{0.56,0.61}	{0,0.24}
X_2	{0.56,0.65}	{0,0.29}	{0.32,0.38}	{0,0.91}	{0.01,0.06}	{0,0.7}
X_3	{0.37,0.43}	{0,0.32}	{0.46,0.54}	{0,0.11}	{0.19,0.22}	{0,0.22}
X_4	{0.32,0.38}	{0,1}	{0.58,0.68}	{0,0.31}	{0.42,0.48}	{0,0.48}

utilities associated with alternatives A_2 and A_6 for attribute X_2. Finally, the lower component utility associated with alternative A_3 is decreased for X_2.

Table 5 shows the component utility associated with the performance interval endpoints for each alternative accounting for adjust and veto functions shown in Table 4, $\{u_j(x_{ij}^L)d_j(x_{ij}^L)v(A_i^L), u_j(x_{ij}^U)d_j(x_{ij}^U)v(A_i^U)\}$.

The dominance matrices corresponding to the five DMs can be computed by solving the problem in Eq. (5) with the information included in Tables 2 and 5. For instance, the pairwise dominance between alternatives A_1 and A_3 for DM_1, D_{13}^1, is the trapezoidal fuzzy number $(-0.0215, 0.0124, 0.0576, 0.0915)$.

The dominance measuring method described in Sect. 2 is then applied to compute global dominance intensities (GDIs) for each DM on the basis of which to derive the corresponding ranking of the considered alternatives. Table 6 shows the GDIs associated with each alternative and the resulting rankings for the DMs under consideration.

Finally, OEA method is used to aggregate the rankings in Table 6, also taking into account the relative importance of such rankings (relative DM importance),

Table 6. Global dominance intensities and alternative rankings for DMs.

	DM_1	DM_2	DM_3	DM_4	DM_5	Consensus
1st	$A_1(0.057)$	$A_4(0.113)$	$A_4(0.002)$	$A_4(-0.118)$	$A_4(0.168)$	A_4
2nd	$A_4(0.038)$	$A_1(-0.544)$	$A_1(-0.121)$	$A_1(-0.367)$	$A_1(-0.523)$	A_1
3rd	$A_6(-0.719)$	$A_6(-1.006)$	$A_6(-0.522)$	$A_6(-1.317)$	$A_6(-0.708)$	A_6
4th	$A_5(-2.089)$	$A_3(-2.244)$	$A_5(-2.079)$	$A_3(-2.336)$	$A_3(-2.199)$	A_5
5th	$A_7(-2.159)$	$A_7(-2.436)$	$A_7(-2.146)$	$A_5(-2.391)$	$A_7(-2.379)$	A_7
6th	$A_3(-2.205)$	$A_5(-2.467)$	$A_3(-2.174)$	$A_7(-2.409)$	$A_5(-2.400)$	A_3

$w_{DM_1} = 0.35 \geq w_{DM_2} = 0.25 \geq w_{DM_3} = 0.2 \geq w_{DM_4} = 0.1 = w_{DM_5} = 0.1$, to derive the consensus ranking shown in the last column of Table 6.

Note that alternative A_4 is best ranked in the consensus ranking, followed by A_1 and A_6. Alternative A_4 was best ranked by all but the most important DM (DM_1), who placed it second. Moreover, the same ranking was derived for risk-prone and risk-averse DMs, which implied different expressions for Tran and Duckstein's distance in the application of the dominance measuring method.

4   Conclusions

In this paper we have extended the additive multiattribute utility model to incorporate the concept of veto in a group decision-making context as an approximation to real situations to represent DM constraints. Moreover, trapezoidal fuzzy numbers are used to represent the relative importance of criteria.

Although all DMs are allowed to provide veto values, the corresponding vetoes are effective for only the most important DMs. They are used to define veto ranges. Veto values corresponding to the other less important DMs are partially taken into account, leading to the construction of adjust functions that decrease the utility associated with the alternative performances.

A dominance measuring method is then used to derive a fuzzy ranking of alternatives for each DM. Finally, we have used the *order explicit algorithm* to aggregate the individual fuzzy alternative rankings accounting for the relative importance of DMs.

Moreover, the cases of risk-prone, neutral and risk-averse DMs have been considered by means of Tran and Duckstein's distance in the application of the dominance measuring method to rank the alternatives under consideration.

References

1. Arrow, K.J., Raynaud, H.: Social Choice and Multicriterion Decision Making. MIT Press, Cambridge (1986)
2. Moulin, H.: The proportional veto principle. Rev. Econ. Stud. **48**, 407–416 (1981)
3. Roy, B., Slowinski, R.: Handling effects of reinforced preference and counter-veto in credibility of outranking. Eur. J. Oper. Res. **188**, 185–190 (2008)
4. Nowak, M.: Preference and veto threshold in multicriteria analysis based on stochastic dominance. Eur. J. Oper. Res. **158**, 339–350 (2004)
5. Munda, G.: A conflict analysis approach for illuminating distributional issues in sustainability. Eur. J. Oper. Res. **194**, 307–322 (2009)
6. Bana e Costa, C.A., Corra, E., De Corte, J.M., Vansnick, J.C.: Faciling bid evalutation in public call for tenders: a social-technical approach. Omega **30**, 227–242 (2002)
7. Marichal, J.L.: Tolerant or intolerant character of interacting criteria in aggregation by the choquet integral. Eur. J. Oper. Res. **155**, 771–791 (2004)
8. Liginlala, D., Ow, T.T.: Modeling attitude to risk in human decision processes: an application of fuzzy measures. Fuzzy Sets Syst. **157**, 3040–3054 (2006)

9. Daher, S.S.D., de Almeida, A.T.: The use of ranking veto concept to mitigate the compensatory effects of additive aggregation in group decisions on a water utility automation investment. Group Decis. Negot. **21**, 185–204 (2012)
10. Raiffa, H.: The Art and Science of Negotiation. Harvard University Press, Cambridge (1982)
11. Stewart, T.: Robustness of additive value function method in MCDM. J. Multicriteria Decis. Anal. **5**, 301–309 (1996)
12. Yue, Z.: A method for group decision-making based on determining weights of decision makers using TOPSIS. Appl. Math. Model. **35**, 1926–1936 (2011)
13. Yoon, K.: System selection by multiple attribute decision making. PhD thesis. Kansas State University Press, Manhattan (1980)
14. Hwang, C.L., Yoon, K.: Multiple Attribute Decision-making: Methods and Applications. Springer, Berlin (1981)
15. Dokas, I.M., Nordlander, T.E., Wallace, R.J.: Fuzzy fault tree representation and maintenance based on frames and constraint technologies: a case study. In: Workshop on Knowledge Capture and Constraint Programming. Whistler, British Columbia, Canada (2007)
16. Vicente, E., Mateos, A., Jiménez-Martín, A.: Risk analysis in information systems: a fuzzification of the MAGERIT methodology. Knowl.-Based Syst. **66**, 1–12 (2014)
17. Vicente, E., Jiménez, A., Mateos, A.: An interactive method of fuzzy probability elicitation in risk analysis. In: Intelligent Systems and Decision Making for Risk Analysis and Crisis Response, pp. 223–228. CRC Press, New York (2013)
18. Jiménez, A., Mateos, A., Sabio, P.: Dominance intensity measure within Fuzzy weight oriented MAUT: an application. Omega **41**, 397–405 (2013)
19. Aguayo, E., Mateos, A., Jiménez-Martín, A.: A new dominance intensity method to deal with ordinal information about a DM's preferences within MAVT. Knowl.-Based Syst. **69**, 159–169 (2014)
20. Mateos, A., Jiménez-Martín, A., Aguayo, E., Sabio, P.: Dominance intensity measuring methods in MCDM with ordinal relations regarding weights. Knowl.-Based Syst. **70**, 26–32 (2014)
21. Tran, L., Duckstein, L.: Comparison of fuzzy numbers using a Fuzzy number measure. Fuzzy Sets Syst. **130**, 331–341 (2002)
22. Lin, S.: Rank aggregation methods. WIREs Comput. Stat. **2**, 555–570 (2010)
23. Green, P.: Research for Marketing Decisions. Prentice-Hall, New Jersey (1978)
24. Borda, J.: Memoire Sur les Elections au Scrutin. Histoire de l'Academie des Sciences, Paris (1981)
25. DeConde, R., Hawley, S., Falcon, S., Clegg, N., Knudsen, B., Etzioni, R.: Combining results of microarray experiments: a rank aggregation approach. Stat. Appl. Genet. Mol. Biol. **5**, 5–15 (2006)
26. Kemeny, J.: Mathematics without numbers. Daedalus **88**, 577–591 (1959)
27. Lin, S., Ding, J.: Integration of ranked lists via cross entropy monte carlo with applications to mRNA and microRNA studies. Biometrics **65**, 9–18 (2009)

Inaccuracy in Defining Preferences
by the Electronic Negotiation System Users

Ewa Roszkowska[1] and Tomasz Wachowicz[2(✉)]

[1] Faculty of Economics and Management, University of Bialystok,
ul. Warszawska 63, 15-062 Bialystok, Poland
erosz@o2.pl
[2] Department of Operations Research, University of Economics in Katowice,
ul. 1 Maja 50, 40-287 Katowice, Poland
tomasz.wachowicz@ue.katowice.pl

Abstract. In this paper we analyze how preferences are defined by negotiators in electronic negotiations if a SAW-based negotiation offer scoring system is used. We analyze a dataset of the Inspire electronic negotiation system, containing the transcripts of bilateral negotiation experiments and study how the negotiators use the preferential information provided in the case description and map it into a system of issues and options ratings in the discrete negotiation problem. We measure the accuracy of the preference systems by comparing the user-defined scoring systems with the reference ideal ones that stem directly from precise initial graphical information. Two notions of accuracy are used: (1) ordinal accuracy which measures if the negotiators followed the ranking order only; and (2) cardinal accuracy, defined by means of an original formula that takes into account weighted normalized distances between the negotiator's own system and the reference scoring one.

Keywords: Preferences · Preference elicitation · Negotiation issue and option ratings · Negotiation offer scoring systems · SAW

1 Introduction

Since negotiation is a complex decision making process involving two or more parties discussing many issues in an effort to reconcile their opposing interests [9], it may require support and facilitation to avoid impasses, deadlocks or stalemates. Therefore a number of support methods and software tools have been recently developed to facilitate negotiations. From the methodological viewpoint, various multiple criteria decision making (MCDM) methods [8, 12, 16] are applied to help negotiators at the prenegotiation phase in constructing their own negotiation offer scoring systems. Such systems measure the scales of concessions and visualize the negotiation progress and therefore are of use in quantitative evaluation of the negotiation offers. Various formal decision support models are implemented in the negotiation support systems (NSS) or electronic negotiation systems (eNS) used in business research and training, such as OpenNexus (http://en.opennexus.pl/), Inspire [5] or Negoisst [13]. Decision support provided by the vast majority of NSS/eNSs is based on the simple additive weighting

© Springer International Publishing Switzerland 2015
B. Kamiński et al. (Eds.): GDN 2015, LNBIP 218, pp. 131–143, 2015.
DOI: 10.1007/978-3-319-19515-5_11

(SAW) method [4]. For discrete negotiation problems, SAW requires assigning rating points to each element of the negotiation template assuming that more preferable issues and options obtain higher ratings. A SAW-based negotiation offer scoring system allows to evaluate any offer built with the options defined within the template by adding up the ratings of these options.

Even though SAW seems easy, intuitive and technically uncomplicated, there is some empirical evidence of its drawbacks and of problems with using SAW-based scoring systems. Interestingly, it has been observed [10] that a majority (57 %) of decision makers, when given a choice of the method for defining their preferences, express them qualitatively using linguistic or descriptive labels. If quantitative scores are used, they are usually of ordinal nature. Thus, it should not be surprising that some earlier electronic negotiation experiments showed that negotiators do not precisely know how to interpret SAW-based ratings and therefore misuse the scoring systems and incorrectly interpret the final scores of offers [17]. Furthermore, laboratory experiments performed with groups of students of economics asked to rank the negotiation offers and to compare them with other predefined rankings determined automatically by means of various versions of SAW, revealed many problems with comparing and selecting the predefined ranking that best fits the students' intrinsic preferences [11]. Most frequently, the negotiators evaluated as more useful (better) a predefined ranking that differed more from their own subjectively defined one. These are, however, interpretative problems that can be reduced or alleviated, as we believe, by implementing appropriate visualization techniques and tools [2].

In this paper we focus on the prenegotiation process of building a negotiation offer scoring system by means of SAW to find out whether the negotiators are able to construct systems that reflect their preferences in an accurate and reliable way. In our research we analyze a dataset of electronic negotiation experiments conducted in the Inspire system, with a predefined multi-issue bilateral business negotiation case. We study the ability of the negotiators to transform correctly the preferential information included in the case description into a system of ratings to be used later to evaluate complete packages exchanged by the parties during the actual negotiation. We measure the scale of potential inaccuracy in determining the negotiation offer scoring systems. Inspired by earlier research by Vetschera [15], we use a negotiation case with precise graphical information about the parties' preferences and therefore are able to introduce two separate measures of accuracy: a more general ordinal accuracy and a detailed cardinal accuracy measure. Finally we analyze the influence of the negotiators' correctness in defining the scoring systems on the negotiation results obtained as well as the difference between the objective quality of such compromises and the subjective perception of their quality resulting from inaccurate rating systems.

The paper consists of four more sections. In Sect. 2 we describe briefly the Inspire system and its protocol for defining the negotiators' preferences, as well as the case used in our experiment including details of the preference representation used. In Sect. 3 we discuss two notions of accuracy of preference definition that we use to measure the quality of the scoring systems built by the negotiators. In Sect. 4 we analyze the experimental results, while in Sect. 5 some future work is suggested.

2 Inspire

2.1 The System and Its General Functionalities

Inspire [5] is an eNS that supports bilateral negotiations conducted via the Web. It has been used for teaching and training, simulations and research in negotiations since the late 1990s. Data from the Inspire experiments have been widely used by a number of researchers investigating, among other things, cross-cultural aspects of electronic negotiations [7], the process of strategy formulation and communication [17], negotiators' behavior and motivations [6]; and decision aspects of negotiations [15].

Inspire supports negotiators throughout the whole negotiation process; however, for our experiment the most important are its decision support facilities implemented in the prenegotiation phase. As regards decision support Inspire offers a SAW-based tool that helps negotiators to analyze their preferences and set up priorities regarding different elements of the negotiation template. This tool is implemented as an element of the prenegotiation preparation check-list imposed on the users by the Inspire protocol. The process of building a negotiation offer scoring system consists of three steps which follow the general SAW requirements [4]. In the first stage a pool of 100 rating points is distributed among all the negotiation issues to define their weights. In the second stage the negotiator rates the options within each issue assigning the maximum score, equal to the issue weight, to the best (most preferred) option, and 0 to the least preferred one. All the intermediate options obtain scores greater than 0 but lower than the issue weight. In the third stage Inspire displays a list of selected complete packages with global scores determined as the sums of the ratings of options that comprise these packages. If the user changes the global scores of selected packages, Inspire, by applying elements of conjoint analysis [1], recalculates the ratings of issues and options in the initial scoring system.

2.2 The Negotiation Case and the Preferential Information

Various negotiation cases may be used for experiments with Inspire. In our experiment a Mosico-Fado bilateral negotiation case was implemented, in which a musician and a broadcasting company discuss the terms of a potential contract. In this case the negotiation template is defined by means of four issues, each with a predefined list of salient options, which allows to build 240 various offers (see Table 1).

Table 1. Mosico-Fado negotiation template.

Issues to negotiate	Issue options
Number of new songs (introduced and performed each year)	11; 12; 13; 14 or 15 songs
Royalties for CDs (in percent)	1.5; 2; 2.5 or 3 %
Contract signing bonus (in dollars)	$125,000; $150,000; $200,000
Number of promotional concerts (per year)	5; 6; 7 or 8 concerts

In the Mosico-Fado case each negotiator, representing either the musician or the broadcasting company, is provided with private information containing a detailed description of their preferences that should be used in building a negotiation offer scoring system. The structure of preferences of the parties is described both verbally and graphically. An example of preference description is presented in Fig. 1.

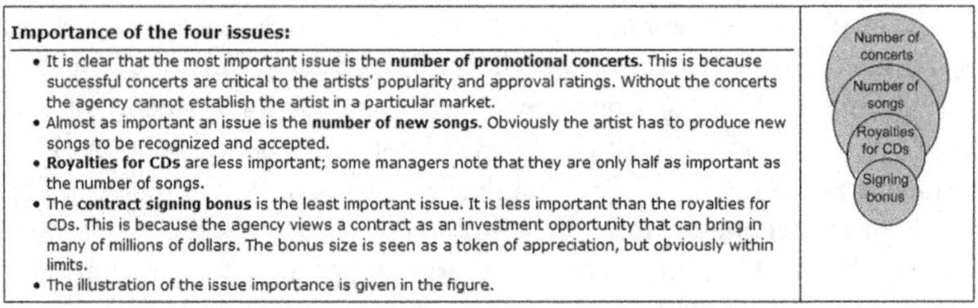

Importance of the four issues:
- It is clear that the most important issue is the **number of promotional concerts.** This is because successful concerts are critical to the artists' popularity and approval ratings. Without the concerts the agency cannot establish the artist in a particular market.
- Almost as important an issue is the **number of new songs.** Obviously the artist has to produce new songs to be recognized and accepted.
- **Royalties for CDs** are less important; some managers note that they are only half as important as the number of songs.
- The **contract signing bonus** is the least important issue. It is less important than the royalties for CDs. This is because the agency views a contract as an investment opportunity that can bring in many of millions of dollars. The bonus size is seen as a token of appreciation, but obviously within limits.
- The illustration of the issue importance is given in the figure.

Fig. 1. Verbal and graphical representation of preferences in Inspire

As noted in the case description, the graphical representation of the preferences was elaborated by the negotiating parties and accepted by their supervisors. The circle sizes indicate the importance of each issue and option. However, what was also emphasized in private information, the circles were drawn casually, so their radiuses do not necessarily reflect the preferences very precisely and accurately. Note that in the description of the circles mention was made of both the circle sizes (areas) and their radiuses, which may be confusing, since this indicates different reference points in the process of building a formal scoring system of offers. Complete graphical information about the preferences of both parties is presented in Appendix.

3   Measuring the Accuracy of the Negotiation Offer Scoring Systems

Inspire does not verify the correctness or accuracy of the scoring systems built individually by the negotiators; it allows them to rate the issues and options at their own discretion and according to their own understanding and interpretation of verbal and graphical preference information. Thus, a fundamental research question arises: if and to what extent the negotiators adhere to the preference description while building their SAW-based negotiation offers scoring system. The negotiators' accuracy can be measured with two different statistical concepts: (1) by analyzing the relationship between the scoring system and determining the correlation coefficients; (2) by analyzing the similarities of the scoring systems and measuring the distances between the negotiator's own system and the reference one. The first of these approaches could be implemented if the relationships between the rankings of full packages were to be studied, each represented by a single frequency distribution. In our problem, each scoring system is represented by a series of five frequency distributions (one representing issue weights and four representing option ratings within each issue) with some elements of these distributions being strongly mutually dependent. This would require a thorough reconsideration and modification of the correlation-based approach.

Therefore, the second approach will be applied here, which is easier to modify and interpret in the analytical context of our problem.

3.1 Ordinal Accuracy

Before measuring the similarities of the scoring systems, basic information about preserving the general preference information can be verified. It can be checked whether the negotiators follow the order of preference represented by the circle sizes for the ratings of both issues and options. This notion of agreement in defining preferences will be called *ordinal accuracy*. Formally, if n issues (or options) $A_1, \ldots, A_n$ are ordered according to decreasing preferences (the circle sizes representing these issues decrease while moving from A_1 to A_n), the ratings $u(A_i)$ of the issues are accurate if they satisfy the following condition

$$u(A_1) > u(A_2) > \ldots > u(A_n).$$ (1)

For instance, if the preferences regarding the negotiation issues presented in Fig. 1 are analyzed and scored, the ordinal accuracy requires that u("Number of concerts") > u ("Number of songs") > u("Royalties for CDs") > u("Signing bonus"). The ordinal accuracy index of the scoring system built by the ith negotiator can be represented as a ratio of the number of correct rankings (n_i^{cor}), i.e., subjective rankings that are in agreement with rankings in the reference order, to the total number (n) of all the rankings that have to be built for the negotiation template.

$$OA_i = \frac{n_i^{\mathrm{cor}}}{n}.$$ (2)

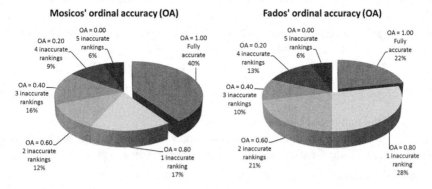

Fig. 2. The structure of globally ordinally accurate negotiators

In our problem, $n = 5$, since there is one ranking representing the importance of the issues and four others, reflecting the orders of salient options for each issue respectively. Note that ordinal accuracy can also be measured, for instance, by means of the Kendall tau rank correlation coefficient. However, as mentioned before, this would require a modification of the original formula since not every pair of elements of the negotiation template can be compared (e.g. rankings of options of different issues cannot be compared).

3.2   Cardinal Accuracy

By determining the global deviations (distances) between the ratings subjectively assigned by the negotiators and the ideal ratings which follow from the corresponding circles (areas or radiuses), **cardinal accuracy** of the negotiation offer scoring system can be measured. However, the specificity of the SAW algorithm, in which the option ratings of one issue depend on the ratings assigned previously to this issue (see Sect. 2.1), requires a different approach to measuring cardinal accuracy for issues and options. Cardinal inaccuracy of issue ratings (II_i) for the negotiator i is measured as a sum of differences in ratings for each issue j with respect to the reference ideal ratings:

$$II_i = \sum_j \left| u_j^{\text{ref}} - u_j^i \right|, \tag{3}$$

where: u_j^{ref} is the reference rating (radius-based or area-based) of the jth issue, and u_j^i is the subjective rating of the jth issue defined by the ith negotiator.

While determining the cardinal inaccuracy of option ratings we need only to verify if the proportions of the circle sizes (radiuses) are preserved by the negotiators regardless of the rating of the issue under consideration. This way we will avoid double-counting of the deviations resulting from the issue ratings incorrectly assigned. Thus, we will determine the normalized reference ratings for options of each issue separately ($\bar{u}_{jk}^{\text{ref}}$) and compare them with the normalized subjective ratings ($\bar{u}_{jk}^i$) of the negotiator to determine the normalized deviations. The normalized deviation for each option will be multiplied by the reference issue rating (u_j^{ref}) resulting in the option inaccuracy rate. Formally, cardinal inaccuracy of option ratings of the jth issue for the ith negotiator can be measured by the following formula:

$$OI_{ij} = u_j^{\text{ref}} \cdot \sum_{K=1,...,N_j} \left| \bar{u}_{jk}^{ref} - \bar{u}_{jk}^i \right|, \tag{4}$$

where N_j is the number of options of the issue j.

A simple example of measuring the inaccuracy of option ratings assigned by a representative of Fado for the issue of the number of concerts is presented in Table 2. The normalized deviations $\left| \bar{u}_{jk}^{ref} - \bar{u}_{jk}^i \right|$ are then aggregated according to formula (4) and the ordinal inaccuracy index is determined as $OI = 32.(0 + 0.2 + 0.37 + 0) = 18.24$.

To determine the global cardinal inaccuracy rate for the whole scoring system of the i-th negotiator, the issue inaccuracy rate and the option inaccuracy rates for all issues need to be aggregated:

Table 2. Normalized inaccuracy rates for option ratings.

Options	Reference ratings	Normalized reference ratings	User ratings	Normalized user ratings	Normalized deviation
5	32	1.00	17	1.00	0.00
6	25	0.78	10	0.58	0.20
7	21	0.66	5	0.29	0.37
8	0	0.00	0	0.00	0.00

$$CI_i = II_i + \sum_j OI_{ij}. \tag{5}$$

4 Online Experiment and Results

We analyzed the results of a negotiation experiment conducted in Inspire in the spring of 2014. The participants of this experiment were 378 students from Poland, Austria, China, Taiwan, Great Britain, Ukraine and Canada, paired into 189 active instances. Once the incomplete records had been eliminated, 176 representatives of the Mosico party and 174 representatives of the Fado party have been considered to analyze the accuracy of building a negotiation offer scoring system and its impact on the negotiation outcome.

4.1 Ordinal Accuracy in Building the Scoring Systems

Analyzing the Inspire's dataset we were surprised to find that 52 representatives of the Mosico party (32 %) and as many as 114 of the Fado party (66 %) party were inaccurate from the viewpoint of ordinal inaccuracy ($OA < 1$). Such a high percentage of inaccurate Fados may be caused by the peculiar structure of preferences defined for their party, with the first two issues equally important and represented by circles of the same size. However, due to some optical illusions (see [14]), for some of them those two circles might have looked different. Therefore we eliminated from the list of inaccurate Fados those who claimed that the number of concerts is more important than the number of songs (and vice versa), but by no more than 5 rating points, and were accurate for other issues. This still left as many as 81 of them (46 %) inaccurate. The situation looked similar if ordinal inaccuracy was determined for the ratings of options within each negotiation issue (see Table 3). It is surprising that Mosicos, who were more accurate in defining the issue ratings, are now more inaccurate than Fados in building their individual option ratings for the successive issues.

Based on the information regarding the inaccuracy of the issue and option ratings, we determined the global ordinal accuracy index according to formula (2). Thus, we counted for each negotiator the number of accurate rankings out of five different rankings they were ask to build. The results, determined separately for the Mosico and Fado parties, are shown in Fig. 2.

Table 3. Ordinal accuracy in option ratings.

Party	Number (%) of inaccurate negotiators while defining option ratings for issue of:			
	No. of concerts	No. of songs	Royalties for CDs	Contract bonus
Moscio	43 (24 %)	57 (32 %)	64 (36 %)	56 (32 %)
Fado	39 (22 %)	40 (23 %)	49 (28 %)	37 (21 %)

Even though the percentage of fully inaccurate negotiators is the same for the Mosico and Fado parties, the numbers of fully accurate ones differ significantly. In the Mosico group there were 69 negotiators (39 %) who built their negotiation offer scoring systems preserving the ordinal preferential information for both issue and option ratings ($OA = 1$). Among Fado's representatives the group of fully accurate negotiators was 17 pp smaller than among the Mosico's ones. These relatively small percentages of accurate negotiators are intriguing and thought-provoking, since we did not expect the negotiators to map the preferential information into the system of ratings precisely, but only to follow the order of preferences visualized by the circle sizes. This did not require any sophisticated calculations or analysis but only a thorough glance.

4.2 Cardinal Accuracy in Building the Scoring Systems

Next we analyzed the negotiators' scale of cardinal accuracy of issue ratings using formula (3). We used two reference ratings: area-based and radius-based (see Appendix). When analyzing the cardinal inaccuracy of issue ratings for ordinally accurate and inaccurate negotiators we found that the results differ depending on the reference rating used (see Table 4).

Table 4. Mosicos' and Fados' cardinal inaccuracy for issue ratings (*II*).

Group of the negotiators	Mosico's average cardinal inaccuracy		Fado's average cardinal inaccuracy	
	Radius-based	Area-based	Radius-based	Area-based
Ordinally accurate	29.37	15.10	19.824	22.122
Ordinally inaccurate	33.80	35.66	24.790	31.930
p (one-tailed test)	0.091	0.000	0.026	0.000

No matter which reference rating is applied, the representatives of the Fado party who are ordinally accurate are, on average, more cardinally accurate than the ordinally inaccurate ones. The same margin of five rating points in differences between the scores of the first two issues was applied, as in the ordinal accuracy analysis. However, Fados seem to refer to radiuses rather than the areas of circles. For the Mosicos, there is no significant difference in cardinal inaccuracy if a radius-based reference system is used ($p = 0.091$). However, the ordinally accurate Mosicos seemed focused more on circle sizes (areas) than on radiuses. If we compare them with ordinally inaccurate Mosicos, the difference in the cardinal accuracy is significant ($p = 0.000$). Comparing the Mosicos and the Fados, applying the same notion of ordinal accuracy, we see that the Mosicos are cardinally more accurate than the Fados.

Next, based on formula (5), we determined the global cardinal inaccuracy rates for Mosicos and Fados in our experiments (see Table 5). The *CI* rates prove once again that the ordinally accurate negotiators are also far more cardinally accurate (for both reference ratings the differences are statistically significant for $p = 0.000$) than those who did not preserve even the order of preferences. Therefore we can reject the conjecture formulated at the beginning of Sect. 4.2, that the ordinally inaccurate negotiators might have built rating systems that are relatively close to the reference ones (ideally accurate).

Table 5. Mosicos' and Fados' global cardinal inaccuracy (*CI*).

Group of the negotiators	Mosico's average cardinal inaccuracy		Fado's average cardinal inaccuracy	
	Radius-based	Area-based	Radius-based	Area-based
Ordinally accurate	58.539	47.093	43.556	52.179
Ordinally inaccurate	111.869	103.020	74.070	84.663
p (one-tailed test)	0.000	0.000	0.000	0.000

4.3 Accuracy of Scoring Systems and the Negotiation Outcomes

Knowing the scale of negotiators' inaccuracy in defining the scoring systems we aimed at verifying its potential impact on the negotiation agreement. The inaccurate negotiators, if they rely on their incorrect scoring systems, may have a false impression of the negotiation reality, may interpret the negotiation progress and concessions incorrectly and, consequently, may accept mediocre or weak agreements. Therefore we analyzed the percentage of agreements reached by accurate and inaccurate negotiators and scored the agreements reached using the negotiators' subjective scoring systems as well as the reference ones. The results for Mosicos and Fados are presented in Tables 6 and 7, respectively.

Table 6. Rates of agreements reached by the Mosico group.

Group of the negotiators	Agreements reached (%)	Average rating of an agreement for			
		Individual SS	Radius-based SS	Area-based SS	Significance
Ordinally accurate	83 %	78.0	75.8	75.5	ISS-to-RSS: $p = 0.301$
					ISS-to-ASS: $p = 0.211$
Ordinally inaccurate	80 %	77.7	72.9	72.0	ISS-to-RSS: $p = \mathbf{0.033}$
					ISS-to-ASS: $p = \mathbf{0.010}$
p (for one-tailed test)		0.973	0.151	0.086	

Table 7. Rates of agreements reached by the Fado group.

Group of the negotiators	Agreements reached (%)	Average rating of an agreement for			
		Individual SS	Radius-based SS	Area-based SS	Significance
Ordinally accurate	81 %	82.9	81.3	79.9	ISS-to-RSS: $p = 0.158$
					ISS-to-ASS: $p = 0.012$
Ordinally inaccurate	85 %	80.8	76.5	74.9	ISS-to-RSS: $p = 0.028$
					ISS-to-ASS: $p = 0.001$
p (for one-tailed test)		0.192	0.000	0.000	

In the Mosico group, if the results of accurate and inaccurate negotiators are compared within each type of the scoring system (individual, radius-based and area-based), no significant differences are observed. However, from the viewpoint of the external observer, both the accurate and inaccurate Mosicos reached agreements of similar quality. On the other hand, if we compare the outcomes for accurate and inaccurate Mosicos separately, we will see that the accurate negotiators, by relying on their accurate scoring system, had a correct perception of reality and were able to interpret the negotiation progress and history correctly. The differences in the ratings of agreements between the individual, radius- and area-based scoring systems are not significant. Yet, the inaccurate Mosicos had, on average, a false impression of their efficiency and of the quality of their performance. They thought they had reached quite profitable agreements (77.7 rating points on average), while objectively their agreements were significantly worse, i.e. 72.9 if measured by the radius-based scoring system, and 72.0, if by the area-based one. We may presume that they may similarly incorrectly interpret the whole negotiation process. The question is: if they had known the real value of the offers submitted and the potential agreement, would they have negotiated differently and obtained better results?

The situation is a little more evident if we analyze the results for the Fado group. Here, from the viewpoint of the external observer, the results obtained by the negotiators are objectively worse in the group of the inaccurate negotiators than in the group of the accurate ones (81.3 vs. 76.5 for the radius-based and 79.9 vs. 74.9 for the area-based scoring systems). Similarly, the inaccurate Fados interpreted their agreements to be significantly better (80.8 on average) than they actually were, when scored by means of the reference ratings (76.5 and 74.9 respectively).

5   Conclusions and Future Work

In our research we tried to check whether the negotiators build their negotiation offer scoring systems in accordance with their intrinsic preferences (or the ones that were imposed on them). We realized that in a vast majority of situations SAW-based scoring systems are inaccurate and give the negotiators a false perception of the negotiation progress and of the results they obtain. Unfortunately, we are not able to answer unambiguously the question: what (if anything) would have changed in the negotiation style, concession strategy or the results if the inaccurate negotiators had built their scoring system correctly and had had a correct perception of the negotiation situation throughout the whole negotiation process. The results obtained for the Mosico and Fado groups (see Tables 6 and 7) are ambiguous, and confirm that the accurate Fados performed significantly better, while Mosicos' results are even better, but are not confirmed by statistical significance tests.

There is, however, another question that was not answered here, mainly due to the lack of adequate data, and which is of a more fundamental nature: what is the cause for building such inaccurate scoring systems and how to help the negotiators to avoid making errors in rating the issues and options. To answer the first question, in-depth research is required that will examine the occurrence of a syndrome of fast thinking and various heuristics [3] in the analytical process of building negotiation offers scoring systems. It will also require experimenting with different methods of visualizing the preferences (e.g. using bars instead of circles) and different algorithms for eliciting the negotiators' preferences. Hence, our future research will consist in designing and performing new electronic negotiation experiments investigating in detail the causes of inconsistencies in the preference elicitation processes in electronic negotiations and producing prescriptive conclusions on the methodological solutions that would elimi-nate potential behavioral and technical errors made by the negotiators or caused by the support algorithm of too high cognitive demand.

Acknowledgements. This research was supported by the grant from Polish National Science Centre (DEC-2011/03/B/HS4/03857). We thank Professor Gregory Kersten for his valuable comments and suggestions that helped us to improve our research.

Appendix

Table A.1. Graphical representation of preferences for Mosico and Fado in Inspire.

Table A.2. The reference ratings.

Party	Reference rates															
	No. of concerts				No. of songs					Royalties for CDs				Contract bonus		
	5	6	7	8	11	12	13	14	15	1.5	2.0	2.5	3.0	125	150	200
Radius-based ratings																
Mosico	0	21	26	32	0	7	16	28	21	13	23	16	0	17	10	0
Fado	32	25	21	0	0	8	20	32	24	0	7	12	16	0	15	20
Area-based ratings																
Mosico	0	22	30	39	0	5	15	30	20	10	20	13	0	11	6	0
Fado	38	27	22	0	0	6	20	38	26	0	4	7	9	0	10	15

References

1. Angur, M.G., Lotfi, V., Sarkis, J.: A hybrid conjoint measurement and bi-criteria model for a two group negotiation problem. Socioecon. Plann. Sci. **30**(3), 195–206 (1996)
2. Blasco, X., Herrero, J., Sanchis, J., Martínez, M.: A new graphical visualization of n-dimensional Pareto front for decision-making in multiobjective optimization. Inf. Sci. **178** (20), 3908–3924 (2008)
3. Kahneman, D.: Thinking Fast and Slow. Macmillan, New York (2011)
4. Keeney, R.L., Raiffa, H.: Decisions with Multiple Objectives: Preferences and Value Trade-offs. Wiley, New York (1976)
5. Kersten, G.E., Noronha, S.J.: WWW-based negotiation support: design, implementation, and use. Decis. Support. Sys. **25**(2), 135–154 (1999)
6. Kersten, G.E., Wu, S., Wachowicz, T.: Why Do students Negotiate? The Impact of Objectives on Behavior, Process & Outcomes. Research Paper INR02 10 (2010)

7. Koeszegi, S., Vetschera, R., Kersten, G.: National cultural differences in the use and perception of internet-based nss: does high or low context matter? Int. Negot. **9**(1), 79–109 (2004)
8. Mustajoki, J., Hamalainen, R.P.: Web-HIPRE: global decision support by value tree and AHP analysis. **38**(3), 208–220 (2000)
9. Pruitt, D.G.: Negotiation behavior. Academic Press, New York (1981)
10. Roszkowska, E., Wachowicz, T.: Defining preferences and reference points – a multiple criteria decision making experiment. In: Zaraté, P., Kersten, G.E., Hernández, J.E. (eds.) GDN 2014. LNBIP, vol. 180, pp. 136–143. Springer, Heidelberg (2014)
11. Roszkowska, E., Wachowicz, T.: SAW-Based Rankings vs. Intrinsic Evaluations of the Negotiation Offers – An Experimental Study. In: Zaraté, P., Kersten, G.E., Hernández, J.E. (eds.) GDN 2014. LNBIP, vol. 180, pp. 176–183. Springer, Heidelberg (2014)
12. Salo, A., Hämäläinen, R.P.: Multicriteria decision analysis in group decision processes. In: Kilgour, D.M., Eden, C. (eds.) Handbook of Group Decision and Negotiation, pp. 269–283. Springer, The Netherlands (2010)
13. Schoop, M., Jertila, A., List, T.: Negoisst: a negotiation support system for electronic business-to-business negotiations in e-commerce. Data Knowl. Eng. **47**(3), 371–401 (2003)
14. Tversky, A.: Contrasting rational and psychological principles of choice. In: Zeckhauser, R.J., Keeney, R.L., Sebenius, J.K. (eds.) Wise Choices Decisions, Games, and Negotiations, pp. 5–21. Harvard Business School Press, Boston (1996)
15. Vetschera, R.: Preference structures and negotiator behavior in electronic negotiations. Decis. Support Sys. **44**(1), 135–146 (2007)
16. Wachowicz, T.: Decision support in software supported negotiations. J. Bus. Econ. **11**(4), 576–597 (2010)
17. Wachowicz, T., Wu, S.: Negotiators' strategies and their concessions. In: de Vreede, G.J. (ed.) Proceedings of The Conference on Group Decision and Negotiation 2010, pp. 254–259. University of Nebraska at Omaha, The Center for Collaboration Science (2010)

A Multi-criteria Group Decision-Making Approach for Facility Location Selection Using PROMETHEE Under a Fuzzy Environment

Reza Tavakkoli-Moghaddam[1]([✉]), Alireza Sotoudeh-Anvari[2],
and Ali Siadat[3]

[1] School of Industrial Engineering, College of Engineering,
University of Tehran, Tehran, Iran
tavakoli@ut.ac.ir
[2] Department of Industrial Engineering, Science and Research Branch,
Islamic Azad University, Tehran, Iran
ar.sotudeh@yahoo.com
[3] LCFC, Arts et Métier Paris Tech, Centre de Metz, Metz, France
ali.siadat@ensam.eu

Abstract. This paper presents a Z-PROMETHEE with Z-numbers as a new representation of vague information for a facility location selection (FLS) problem. The selection of a facility location, which is a kind of a multi-criteria decision making (MCDM) problem, should be considered from a strategic point of view. In a real-world situation, MCDM problems are generally under uncertainty. In order to overcome such a problem, fuzzy sets can be applied with the PROMETHEE to allow experts to combine inadequate information into the decision method. However, the fuzzy PROMETHEE also has some defects. The main problem is that the certainty of information is not taking into account. For explanation of real-life information, fuzziness and degree of the certainty of information are indispensable. In the proposed method, Z-numbers are used to evaluate the weights of the criteria. Hence, in comparison with the fuzzy model, the PROMETHEE with a Z-number (i.e., Z-PROMETHEE) can symbolize real life problems more realistically.

Keywords: PROMETHEE · Fuzzy set theory · MCDM · Facility location selection

1 Introduction

Facility location selection (FLS) is one of the most significant decisions at the strategic management level [28]. Various factors should be considered in the location selection process [13, 28]. According to Chou et al. [11], these factors can be categorized into three groups: (1) critical factors (e.g., accessibility of utilities) decide whether an option is checked for more assessment, (2) objective factors (e.g., investment costs) are defined in quantitative values, and (3) subjective factors (e.g., political stability) are qualitative. As a result, the essence of facility location selection is a multi-criteria decision making (MCDM) problem, which includes qualitative and quantitative factors. Most of these factors can be evaluated by human judgment. Hence, facility location selection processes involve the ambiguity inherent in linguistic terms [11].

© Springer International Publishing Switzerland 2015
B. Kamiński et al. (Eds.): GDN 2015, LNBIP 218, pp. 145–156, 2015.
DOI: 10.1007/978-3-319-19515-5_12

The theory of fuzzy sets developed by Zadeh [36] is used to model uncertainty in decision making models happening owing to lack of perfect information. Liang and Wang [25] proposed a fuzzy multi-criteria decision making (FMCDM) approach for facility site selection, on the basis of fuzzy set theory and hierarchical structure analysis. Chu [12] developed a fuzzy TOPSIS under group decisions to solve the FLS problem. Kahraman et al. [19] applied four fuzzy multi-attribute group decision making methods for selection of facility locations. Yong [35] developed a new fuzzy TOPSIS for selecting a plant location under linguistic terms. Ertuğrul and Karakaşoğlu [13] presented a comparison of AHP and TOPSIS for FLS under a fuzzy environment in a textile company.

Several approaches have been proposed for MCDM problems. There are no better methods and different MCDM approaches may give contradictory results when used to the same problem [17, 26]. Voogd [32] explained that at least 40 % of the time, each method generated a different outcome from any other approach. Among various approaches of MCDM, the PROMETHEE (Preference Ranking Organization METhod for Enrichment Evaluation) is appreciably appropriate for ranking applications. PROMETEE was introduced by Brans [4] and more developed by Vincke and Brans [31]. Al-Shemmeri et al. [2] illustrated that PROMETHEE is a little easier than ELECTRE to apply. Furthermore, Brans et al. [6] showed PROMETHEE is more stable than ELECTRE III. Goumas and Lygerou [15] explained that PROMETHEE is a reasonably easy ranking approach in idea and use compared with the other MCDM techniques. The achievement of this method in several applications is attributed to firm mathematical properties and simplicity [5]. However, a key drawback of the PROMETHEE, like other conventional MCDM approaches, is the need for accurate measurement of the performance values and criteria weights [33].

The criteria weights in real-life applications are frequently imprecise and subjective. The PROMETHEE does not offer a detailed strategy for determining these weights. Various techniques can be employed to establish the weights (e.g., fuzzy AHP, entropy analysis and Z-numbers) [29].

Incorporation of fuzzy sets and the PROMETHEE was primarily introduced by Le Teno and Mareschal [23]. Goumas and Lygerou [15] extended the fuzzy PROMETHEE to consider fuzzy inputs (performance of the alternative) and crisp weights for the ranking of alternative energy utilization projects. Geldermann et al. [14] applied fuzzy preference and fuzzy weights to gain fuzzy scores. They used trapezoidal fuzzy numbers to symbolize the ambiguities in iron and steel industry. Other fuzzy PROMETHEE are studied [7, 9, 10, 16, 18, 22, 24, 30, 34].

Although, during the last decades, conventional fuzzy set has been broadly used in the different fields and a lot of fruits have been attained [21], however, fuzzy sets face with the fundamental limitation. According to Aliev et al. [1], when dealing with real life information, it is not satisfactory to take into consideration only uncertainty. Another critical property of information is its level of reliability. In order to take into account this reality, Zadeh [38] introduced the idea of a Z-number as a more efficient notion for explanation of real world information. Kang et al. [21] suggested a new MCDM approach on the basis of Z-number to cope with linguistic terms. Azadeh et al. [3] proposed a novel AHP on the basis of Z-number. The key problem that occurs in processing Z-numbers-based method is computation with Z-numbers. According to

Zadeh [38], problem involving calculation with Z-numbers is straightforward to state but very complicated to solve. Kang et al. [20] proposed an efficient technique for transforming a Z-number into a fuzzy number based on a fuzzy expectation. In this paper, we extend PROMETHEE under a fuzzy environment to solve MCDM problems in which the criteria weights are Z-numbers, which can be transformed into traditional fuzzy numbers on the base of fuzzy expectation [20]. It is essential to state that transforming Z-numbers into conventional fuzzy numbers leads to loss of information. However, According to Aliev et al. [1] the key benefit of this method is low computational complexity, which allows for an extensive range of its use.

The rest of this study is ordered as follows. Section 2 contains the basic definitions are applied in the remaining parts of this study. Section 3 concentrates on the proposed approach. Section 4 provides an instance for illustrating the applicability of the proposed method. Section 5 presents conclusions.

2 Preliminaries

In this section, various fundamental definitions of a fuzzy set theory and the PROMETHEE are reviewed.

2.1 Fuzzy Set Theory

A fuzzy set is characterized with a membership function, which allocates to each element a degree of membership ranging between zero and one [27].

Definition 1 (*Linguistic Variables*): A linguistic variable is a variable whose values are linguistic term i.e., word or sentence [37]. These linguistic values can be represented by fuzzy numbers (see Table 1). In FMCDM problems, the ratings and weights of the criteria are expressed in linguistic variables and then transformed into triangular fuzzy numbers.

Table 1. Linguistic terms and their corresponding triangular fuzzy numbers

Triangle fuzzy numbers	Ratings of alternatives
(0,0,0.25)	Worst
(0,0.25,0.5)	Poor
(0.25,0.5,0.75)	Fair
(0.5,0.75,1)	Good
(0.75,1,1)	Best

Definition 2 (*Z- number*): The Z-number is a new fuzzy concept, relates to the topic of certainty of information. A Z-number has two components, $Z = (A, B)$, used to explain a value of a random variable X, where A is an estimation of a value of X and B is a measure of confidence of A [38]. For example, suppose a researcher gives the prediction of a condition of economy as follows [1]: *Prediction of a condition of economy*

for the next year = (sturdy growth, sure). This forecast can be expressed as a Z-number evaluation, where *X* is the variable *state of economy*, *A* is a fuzzy number applied to explain the constraint *"sturdy growth"* and *B* is a fuzzy number to describe the degree of certainty of *A*.

In this paper, *A* is a triangular fuzzy number and *B* is a linguistic terms (Table 2).

Table 2. Fuzzy numbers for each linguistic term

Criteria weights	Triangle fuzzy numbers
Very low (VL)	(0,0,0.25)
Low (L)	(0,0.25,0.5)
Medium (M)	(0.25,0.5,0.75)
High (H)	(0.5,0.75,1)
Very high (VH)	(0.75,1,1)

2.2   PROMETHEE and Fuzzy PROMETHEE

The fuzzy PROMETHEE is a mixture of the fuzzy logic and PROMETHEE, which is more applicable. The fuzzy PROMETHEE [10, 33] consists of the following steps.

Step 1: Determine alternatives, criteria and establish a group of experts. Assume *n* decision-makers (experts), *m* alternatives (options) and *k* criteria (factors).

Step 2: Characterize linguistic terms and their corresponding triangular fuzzy number. Linguistic values were applied to assess the criteria weights and performance ratings (see Tables 1 and 2).

Step 3: Aggregate expert's valuations. A result is concluded by aggregating the fuzzy criteria weights and fuzzy rating of alternatives (1). The preferences of experts of the alternative *i* under the criterion j can be calculated using (2).

$$\tilde{w}_j = \frac{1}{n} \left[\sum_{e=1}^{n} \tilde{w}_j^e \right] = \frac{1}{n} \left[\tilde{w}_j^1 + \tilde{w}_j^2 + \ldots + \tilde{w}_j^n \right] \tag{1}$$

$$\tilde{x}_{ij} = \frac{1}{n} \left[\sum_{e=1}^{n} \tilde{x}_{ij}^e \right] = \frac{1}{n} \left[\tilde{x}_{ij}^1 + \tilde{x}_{ij}^2 + \ldots + \tilde{x}_{ij}^n \right] \tag{2}$$

Step 4: Make a fuzzy decision matrix and calculate the average fuzzy weight of criterion, where $\tilde{x}_{ij}$ indicates the rating of the alternative *i* under the criterion *j* and $\tilde{w}_j$ is the weight of the criterion *j*.

$$D = \begin{array}{c} \\ A_1 \\ A_2 \\ . \\ A_m \end{array} \begin{array}{cccc} C_1 & C_2 & \cdots & C_k \\ \left[\begin{array}{cccc} \tilde{x}_{11} & \tilde{x}_{12} & \cdots & \tilde{x}_{1k} \\ \tilde{x}_{21} & \tilde{x}_{22} & \cdots & \tilde{x}_{2k} \\ \cdots & \cdots & \cdots & \cdots \\ \tilde{x}_{m1} & \tilde{x}_{m2} & \cdots & \tilde{x}_{mk} \end{array} \right] \end{array} \quad i = 1, 2, \ldots, m; j = 1, 2, \ldots, k \qquad (3)$$

$$\tilde{W} = [\tilde{w}_1, \tilde{w}_2, \ldots, \tilde{w}_n]$$

Step 5: Create the fuzzy preference function. Suppose A be a collection of alternatives. a and b are two alternatives of A. Preference function $\tilde{P}_j(a, b)$ can be determined as follows:

$$\tilde{P}_j(a, b) = \begin{cases} 0 & , \tilde{x}_{aj} \leq \tilde{x}_{bj} \\ \tilde{x}_{aj} - \tilde{x}_{bj} & , \tilde{x}_{aj} > \tilde{x}_{bj} \end{cases} \quad j = 1, 2, \ldots, k \qquad (4)$$

where $\tilde{P}_j(a, b)$ means the outranking severity that a is premier to b.

A preference function $\tilde{P}_j(a, b)$ is a function of the discrepancy between the ratings of two alternatives for every criterion. See to [5, 6] for more details. The following preference function is applied here:

$$\begin{cases} \tilde{x}_{aj} > \tilde{x}_{bj} \Leftrightarrow aPb \\ \tilde{x}_{aj} = \tilde{x}_{bj} \Leftrightarrow \tilde{x}_{aj} I \tilde{x}_{bj} \end{cases} \qquad (5)$$

Step 6: Determine the multi-criteria preference index to choose the rate of the outranking relation. This index $\tilde{\pi}(a, b)$ is calculated by:

$$\tilde{\pi}(a, b) = \frac{\sum_{j=1}^{k} [\tilde{w}_j \tilde{P}_j(a, b)]}{\sum_{j=1}^{k} [\tilde{w}_j]} \qquad (6)$$

Step 7: Compute the flow to preorder the options. Fuzzy PROMETHEE I: show a number of alternatives, which are incapable to compare together using partial preorder. The leaving flow is as follows:

$$\tilde{\phi}^+(a) = \sum_{y \neq a} \tilde{\pi}(a, y), \forall a, y \in A \qquad (7)$$

where $\tilde{\phi}^+(a)$ demonstrates the sum of preference that a is better another options. The entering flow is as follows:

$$\tilde{\phi}^-(a) = \sum_{y \neq a} \tilde{\pi}(a, y), \forall a, y \in A \qquad (8)$$

where $\tilde{\phi}^-(a)$ demonstrates the sum of preference that other options are superior to a. The further the leaving flow and the smaller the entering flow, the superior the alternative.

This stage applies the maximize set and minimize set [8] to defuzzification. Maximize set $R = \{(x, f_R(x))|x \in R\}$ and

$$f_R(x) = \begin{cases} \dfrac{x - x_1}{x_2 - x_1} & x_1 \leq x \leq x_2 \\ 0 & \text{otherwise} \end{cases} \qquad \begin{array}{l} \text{Minimize set} \\ L = \{(x, f_L(x))|x \in R\} \end{array} \tag{9}$$

$$f_L(x) = \begin{cases} \frac{x - x_2}{x_1 - x_2} & x_1 \leq x \leq x_2 \\ 0 & \text{otherwise} \end{cases} \tag{10}$$

Right Utility

$$U_R(\tilde{\phi}^+(i)) = \sup(f_{\tilde{\phi}^+(i)}(x) \wedge f_R(x)) \tag{11}$$

Left Utility

$$U_L(\tilde{\phi}^+(i)) = \sup(f_{\tilde{\phi}^+(i)}(x) \wedge f_L(x)) \tag{12}$$

Total preference rate is as follows:

$$U_T(\tilde{\phi}^+(i)) = \tilde{\phi}^+(i) = \frac{U_R(\tilde{\phi}^+(i)) + 1 - U_L(\tilde{\phi}^+(i))}{2} \qquad i = 1, 2, \ldots, m \tag{13}$$

The preference relation and the partial preorder $(P^{(I)}, I^{(I)}, R)$ as follows:

$$\begin{aligned} aP^+b : \begin{cases} P & \text{iff } \phi^+(a) > \phi^+(b), \forall a, b \in A \\ I & \text{iff } \phi^+(a) = \phi^+(b), \forall a, b \in A \end{cases} \\ aP^-b : \begin{cases} P & \text{iff } \phi^-(a) > \phi^-(b), \forall a, b \in A \\ I & \text{iff } \phi^-(a) = \phi^-(b), \forall a, b \in A \end{cases} \end{aligned} \tag{14}$$

According to (14), we find the outranking relation and the partial preorder as follows:

$$aP^{(I)}b \ (a \text{ outranks } b), \begin{cases} aP^+b : P & \text{and} & aP^-b : P \\ aP^+b : P & \text{and} & aP^-b : I \\ aP^+b : I & \text{and} & aP^-b : P \end{cases} \tag{15}$$

$aI^{(I)}b$ (a is indifferent to b), $aP^+b : I$ and $\quad aP^-b : I$

aRb (a and b are incomparable, otherwise)

Fuzzy PROMETHEE II: order all options by using the full preorder. This method ranks alternatives by their net flows. The net flow is calculated using the following equations.

$$\phi(a) = \phi^+(a) - \phi^-(a), \forall a \in A \tag{16}$$

A higher value shows a higher suitability of alternative. The preference relation is calculated as follows:

$$\begin{cases} aP^{(II)}b \text{ (a outranks b)} & \text{iff } \phi(a) > \phi(b), \forall a, b \in A \\ aP^{(II)}b \text{ (a is indifferent to b)} & \text{iff } \phi(a) = \phi(b), \forall a, b \in A \end{cases} \tag{17}$$

Step 8: Make a value outranking diagram to estimate the preference rank of each option.

3 Proposed Method

In this work, we extend the PROMETHEE to solve MCDM problem with Z-numbers. In our approach, we initially organize a committee of decision makers and establish our criteria and alternatives. Then by Z-numbers, we determine the weights of criteria. After that, a technique of transforming a Z-number into a traditional fuzzy number is used. The rating of each alternative is articulated in triangular fuzzy numbers. After that, we used these fuzzy values in the fuzzy PROMETHEE. According to Z-numbers and PROMETHEE, Z-PROMETHEE can be described as follows:

Step 1: Specify the factors that are the most considerable for the experts.

Step 2: Assign the criteria weights by applying Z-numbers. This step involves appropriation of Z-numbers to the criteria weights by the decision maker. The level of reliability $(\tilde{B})$ is prepared from Table 2. After that, a technique for transmuting a Z-number into a classical fuzzy number is used.

Kang et al. [20] presented an efficient approach of turning a Z-number into a fuzzy number on the base of the fuzzy expectation. This procedure is given as follows:

Step 2.1: Change the reliability $(\tilde{B})$ into a crisp value. This computation is made by:

$$\alpha = \frac{\int x\mu_{\tilde{B}}(x)dx}{\int \mu_{\tilde{B}}(x)dx} \tag{18}$$

As mentioned earlier, triangular fuzzy number is applied in this paper to state the degree of reliability. When $\tilde{B} = (b_1, b_2, b_3)$, the above formula becomes as follows:

$$\alpha = \frac{b_1 + b_2 + b_3}{3} \tag{19}$$

Step 2.2: Add the weight of the certainty $(\tilde{B})$ to the constraint$(\tilde{A})$. Weighted Z-number can be explained by:

$$\tilde{Z}^{\alpha} = \left\{ (x, \mu_{\tilde{A}^{\alpha}}) \middle| \mu_{\tilde{A}^{\alpha}} = \alpha\mu_{\tilde{A}}(x), x \in [0, 1] \right\} \tag{20}$$

Step 2.3: Transform the weighted Z-number into a fuzzy number by multiplying $\sqrt{\alpha}$ by:

$$\tilde{Z}' = \sqrt{\alpha} \times \tilde{A}^{\alpha} = (\sqrt{\alpha} \times a, \sqrt{\alpha} \times b, \sqrt{\alpha} \times c, \sqrt{\alpha} \times d) \tag{21}$$

The proofs of these theorems are omitted [20]. After this alteration, the Z-number model can be changed to the standard fuzzy form.

Step 3: Assign the suitable fuzzy numbers or linguistic terms for the rating of each alternative.

Step 4: Conducting fuzzy PROMETHEE to attain the final ordering results.

4 Numerical Example

In this part, we give an example to show how the proposed method can be used. This example is taken from [10, 33]. A firm desires to choose an appropriate location for establishing a new facility. The assessment is done by a group of four decision-makers. After introductory screening, four candidates stay for more assessment. This firm considers seven factors to select the most correct option. The committee used Z-numbers to rate the weight of each criterion. A is a triangular fuzzy number and B is stated by linguistic terms (Table 2). The result is shown in Table 3. The information in this table should be converted into a triangular fuzzy number in order to make computation possible. Table 4 demonstrates the outcomes of conversion according to [20]. Note that from this step, the results of [10] and [33] are exactly repeated.

Table 3. Weight of each criteria using the Z-number

	D_1	D_2	D_3	D_4
C_1	((0.86,1.15,1.15),H)	((0.86,1.15,1.15),H)	((0.707,1.06,1.41),M)	((0.57,0.866,1.15),H)
C_2	((0.522,0.783,1.04),VH)	((1.06,1.41,1.41),M)	((0.288,0.577,0.866),H)	((0.577,0.866,1.15,)H)
C_3	((0.866,1.154,1.154),H)	((0.866,1.154,1.154),H)	((0.577,0.866,1.54),H)	((0.707,1.06,1.414),M)
C_4	((1.06,1.414,1.414),M)	((0.86,1.154,1.154),H)	((0.522,0.783,1.04),VH)	((0.866,1.154,1.154),H)
C_5	((0.866,1.154,1.154),H)	((0.783,1.044,1.044),VH)	((0.577,0.866,1.154)),H)	((1.5,2,2),L)
C_6	((0.707,1.06,1.414),M)	((1,1.5,2),L)	((0261,0.522,0.783),VH)	((0.577,0.866,1.154),H)
C_7	((0.866,1.154,1.154),H)	((0.522,0.783,1.044),VH)	((0,0.288,0.577),H)	((0.5,1,1.5),L)

The committee applied linguistic terms (Table 1) to rate the four alternatives. The results are shown in Table 5. According to (2), the fuzzy preference function can be worked out. See an example shown in Table 6. After that, we can find the multi-criteria preference index $\tilde{\pi}(a, b)$. The result is shown in Table 7.

Table 4. Transformation from Z-numbers into fuzzy numbers

	D_1	D_2	D_3	D_4
C_1	(0.75,1,1)	(0.75,1,1)	(0.5,0.75,1)	(0.5,0.75,1)
C_2	(0.5,0.75,1)	(0.75,1,1)	(0.25,0.5,0.75)	(0.5,0.75,1)
C_3	(0.75,1,1)	(0.75,1,1)	(0.5,0.75,1)	(0.5,0.75,1)
C_4	(0.75,1,1)	(0.75,1,1)	(0.5,0.75,1)	(0.75,1,1)
C_5	(0.75,1,1)	(0.75,1,1)	(0.5,0.75,1)	(0.75,1,1)
C_6	(0.5,0.75,1)	(0.5,0.75,1)	(0.25,0.5,0.75)	(0.5,0.75,1)
C_7	(0.75,1,1)	(0.5,0.75,1)	(0.00,0.25,0.5)	(0.25,0.5,0.75)

Table 5. Rating of alternatives [10]

Criteria	Supplier	Decision makers			
		D_1	D_2	D_3	D_4
C_1	A_1	G	G	F	F
	A_2	G	G	G	F
	A_3	F	F	F	G
	A_4	F	F	F	F
C_2	A_1	G	G	G	G
	A_2	G	G	G	G
	A_3	F	G	F	F
	A_4	F	G	F	F
C_3	A_1	G	G	G	G
	A_2	G	B	G	G
	A_3	F	F	F	F
	A_4	F	F	F	F
C_4	A_1	G	F	F	F
	A_2	F	G	G	G
	A_3	F	F	F	F
	A_4	F	F	F	F
C_5	A_1	G	G	G	F
	A_2	F	G	G	G
	A_3	F	F	F	F
	A_4	F	F	F	F
C_6	A_1	G	F	F	G
	A_2	F	G	F	F
	A_3	F	F	F	F
	A_4	F	F	F	F
C_7	A_1	G	G	G	G
	A_2	F	G	F	G
	A_3	P	F	F	F
	A_4	P	F	F	F

Table 6. Fuzzy preference function [10]

	$\tilde{P}_j(A_1, A_2)$	$\tilde{P}_j(A_1, A_3)$
C_1	(0.250,0.750,1.25)	(0.375,0.875,1.375)
C_2	(0.250,0.750,1.25)	(0.5,1,1.5)
C_3	(0.312,0.750,1.25)	(0.562,1.062,1.250)
C_4	(0.187,0.750,1.25)	(0.375,0.750,1.375)
C_5	(0.312,0.812,1.312)	(0.5,1,1.5)
C_6	(0.375,0.875,1.375)	(0.437,0.937,1.437)
C_7	(0.437,0.937,1.437)	(0.625,1.125,1.625)

Table 7. The $\tilde{\pi}(a, b)$ index [10]

a	b	$\tilde{\pi}(a, b)$
A_1	A_2	(0.176,0.794,2.206)
	A_3	(0.278,0.976,2.511)
	A_4	(0.284,0.985,2.527)
A_2	A_1	(0.197,0.831,2.236)
	A_3	(0.292,0.994,2.518)
	A_4	(0.298,1.004,2.533)

According to (5) and (6), the fuzzy leaving flow and fuzzy entering flow are calculated. The next stage is the defuzzification. According to Table 8, A_2 is recognized as the best option. As shown here, the results generated are similar to [10] results. Note that this case is used just to explain the computational process of the proposed method and such a comparison may be worthless (Table 8).

Table 8. Ranking [10]

a	$\phi^+(a)$	$\phi^-(a)$	$\phi(a)$	Order
A_1	0.436	0.365	0.07	2
A_2	0.443	0.358	0.085	1
A_3	0.365	0.437	−0.072	3
A_4	0.36	0.441	−0.081	4

5   Conclusion

Selecting the proper facility location from a set of alternatives has been an intricate multi-criteria problem and several quantitative and qualitative factors should have been considered during this process. Due to the fact that determining the crisp values of the attributes is very difficult, it is more realistic to consider them as Z-numbers. In this paper, a new PROMETHEE with a Z-number called Z-PROMETHEE has been proposed to solve the facility location selection (FLS) problem by using Z-numbers to extend the traditional PROMETHEE. For explanation of real-life information, fuzziness and degree of certainty has been indispensable. The Z-numbers not only maintain the benefit of the fuzzy numbers, but also can handle the level of reliability of information. In the proposed method, Z-number has been applied to state the weight of each criterion and the criteria weights have been determined by transforming Z-number weights into triangular fuzzy numbers on the base of fuzzy expectation. This framework is very simple and flexible and can be applied in various other fields.

Acknowledgments. This work has been supported financially by the Center for International Scientific Studies & Collaboration (CISSC) and the French Embassy in Tehran as well as the Partenariats Hubert Curien (PHC) program in France. Additionally, the authors would like thank anonymous reviewers for their valuable comments.

References

1. Aliev, R.A., Alizadeh, A.D., Huseynov, O.H.: The arithmetic of discrete Z-numbers. Inform. Sci. **290**, 134–155 (2015)
2. Al-Shemmeri, T., Al-Kloub, B., Pearman, A.: Model choice in multicriteria decision aid. Eur. J. Oper. Res. **97**, 550–560 (1997)
3. Azadeh, A., Saberi, M., Atashbar, N.Z., Chang, E., Pazhoheshfar, P.: Z-AHP: a Z-number extension of fuzzy analytical hierarchy process. In: International Conference on Digital Ecosystems and Technologies (DEST), pp. 141–147 (2013)
4. Brans, J.P.: L'ingenierie de la decision; Elaboration d'instruments d'aide a la decision. Lamethode PROMETHEE. In: Nadeau, R., Landry, M. (Eds.) L'aide a la decision: Nature, Instruments et Perspectives d'Avenir. Presses de l'Univ. Laval, Canada, pp. 183–213 (1982)
5. Figueira, J., Greco, S., Ehrgott, M.: Multiple Criteria Decision Analysis: State of the Art Surveys, pp. 163–196. Springer Science + Business Media Inc, New York (2005)
6. Brans, J.P., Vincke, P., Mareschal, B.: How to select and how to rank projects: the PROMETHEE method. Eur. J. Oper. Res. **24**, 228–238 (1986)
7. Campos, A.C.S.M., Mareschal, M., Almeida, A.T.: Fuzzy FlowSort: an integration of the FlowSort method and fuzzy set theory for decision making on the basis of inaccurate quantitative data. Inform. Sci. **293**, 115–124 (2015)
8. Chen, S.H.: Ranking fuzzy numbers with maximizing set and minimizing set. Fuzzy Sets Syst. **17**, 113–129 (1985)
9. Chen, T.Y.: Multiple criteria decision analysis using a likelihood-based outranking method based on interval-valued intuitionistic fuzzy sets. Inform. Sci. **286**, 188–208 (2014)
10. Chen, Y.H., Chang, Y.H., Wu, C.Y.: Strategic decisions using the fuzzy PROMETHEE for IS outsourcing. Expert Syst. Appl. **38**, 13216–13222 (2011)
11. Chou, S.Y., Chauhan, S.S., Shen, C.Y.: A fuzzy simple additive weighting system under group decision-making for facility location selection with objective/subjective attributes. Eur. J. Oper. Res. **189**, 132–145 (2008)
12. Chu, T.C.: Selecting plant location via a fuzzy TOPSIS approach. Int. J. Adv. Manuf. Technol. **20**, 859–864 (2002)
13. Ertuğrul, I., Karakaşoğlu, N.: Comparison of fuzzy AHP and fuzzy TOPSIS methods for facility location selection. Int. J. Adv. Manuf. Technol. **39**, 783–795 (2008)
14. Geldermann, J., Spengler, T., Rentz, O.: Fuzzy outranking for environmental assessment. Case study: iron and steel making industry. Fuzzy Sets Syst. **115**, 45–65 (2000)
15. Goumas, M., Lygerou, V.: An extension of the PROMETHEE method for decision making in fuzzy environment: ranking of alternative energy exploitation projects. Eur. J. Oper. Res. **123**, 606–613 (2000)

16. Halouani, N., Chabchoub, H., Martel, J.M.: PROMETHEE-MD-2T for project selection. Eur. J. Oper. Res. **195**, 841–849 (2009)

17. Hatami-Marbini, A., Tavana, M.: An extension of the Electre I method for group decision-making under a fuzzy environment. Omega **39**, 373–386 (2011)

18. Hsu, T.H., Lin, L.Z.: Using fuzzy preference method for group package tour based on the risk perception. Group Decis. Negot. **23**, 299–323 (2014)

19. Kahraman, C., Ruan, D., Dogan, I.: Fuzzy group decision-making for facility location selection. Inform. Sci. **157**, 135–153 (2003)

20. Kang, B., Wei, D., Li, Y., Deng, Y.: A method of converting z-number to classical fuzzy number. J. Inform. Comput. Sci. **9**, 703–709 (2012)

21. Kang, B., Wei, D., Li, Y., Deng, Y.: Decision making using Z-numbers under uncertain environment. J. Inform. Comput. Sci. **8**, 2807–2814 (2012)

22. Kuang, H., Kilgour, M., Hipel, K.W.: Grey-based PROMETHEE II with application to evaluation of source water protection strategies. Inform. Sci. **294**, 376–389 (2015)

23. Le Teno, J.F., Mareschal, B.: An interval version of PROMETHEE for the comparison of building products' design with ill-defined data on environmentalquality. Eur. J. Oper. Res. **109**, 522–529 (1998)

24. Li, W., Li, B.: An extension of the PROMETHEE II method based on generalized fuzzy numbers. Expert Syst. Appl. **37**, 5314–5319 (2010)

25. Liang, G.S., Wang, M.J.J.: A fuzzy multi-criteria decision-making method for facilitysite selection. Int. J. Prod. Res. **29**, 2313–2330 (1991)

26. Løken, E.: Use of multicriteria decision analysis methods for energy planning problems. Renew. Sustain. Energy Rev. **11**, 1584–1595 (2007)

27. Rostamzadeh, R., Sofian, S.: Prioritizing effective 7Ms to improve production systems performance using fuzzy AHP and fuzzy TOPSIS (case study). Expert Syst. Appl. **38**, 5166–5177 (2011)

28. Shen, C.Y., Yu, K.T.: A generalized fuzzy approach for strategic problems: the empirical study on facility location selection of authors' management consultation client as an example. Expert Syst. Appl. **36**, 4709–4716 (2009)

29. Sotoudeh-Anvari, A., Sadi-Nezhad, S.: A new approach based on the level of reliability of information to determine the relative weights of criteria in fuzzy TOPSIS. Int. J. Appl. Decis. Sci. (2014, in press)

30. Taha, Z., Rostam, S.: A hybrid fuzzy AHP-PROMETHEE decision support system for machine tool selection in flexible manufacturing cell. J. Intell. Manuf. **23**, 2137–2149 (2011)

31. Vincke, J.P., Brans, P.: A preference ranking organization method. The PROMETHEE method for MCDM. Manag. Sci. **31**, 641–656 (1985)

32. Voogd, H.: Multicriteria Evaluation for Urban and Regional Planning. Pion, London (1983)

33. Wang, T.C., Chen, L.Y., Chen, Y.H.: Applying fuzzy PROMETHEE method orevaluating IS outsourcing suppliers. Paper presented at the 5th International Conference on Fuzzy Systems and Knowledge Discovery, China (2008)

34. Yilmaz, B., Dagdeviren, M.: A combined approach for equipment selection: F-PROMETHEE method and zero–one goal programming. Expert Syst. Appl. **38**, 11641–11650 (2011)

35. Yong, D.: Plant location selection based on fuzzy TOPSIS. Int. J. Adv. Manuf. Technol. **28**, 839–844 (2006)

36. Zadeh, L.A.: Fuzzy sets. Inf. Control **8**, 338–353 (1965)

37. Zadeh, L.A.: The concept of a linguistic variable and its application to approximate reasoning–I. Inform. Sci. **8**, 199–249 (1975)

38. Zadeh, L.A.: A note on Z-numbers. Inform. Sci. **181**, 2923–2932 (2011)

An Interval-Valued Hesitant Fuzzy TOPSIS Method to Determine the Criteria Weights

Reza Tavakkoli-Moghaddam[1](✉), Hossein Gitinavard[2],
Seyed Meysam Mousavi[3], and Ali Siadat[4]

[1] School of Industrial Engineering, College of Engineering,
University of Tehran, Tehran, Iran
tavakoli@ut.ac.ir
[2] School of Industrial Engineering,
Iran University of Science and Technology, Tehran, Iran
h.gitinavard@ut.ac.ir
[3] Industrial Engineering Department, Faculty of Engineering,
Shahed University, Tehran, Iran
sm.mousavi@shahed.ac.ir
[4] LCFC, Arts et Métier Paris Tech, Centre de Metz, Metz, France
ali.siadat@ensam.eu

Abstract. In a multi-criteria group decision analysis, numerous methods have been developed and proposed to determine the weight of each criterion; however, the group decision methods, except AHP, have rarely considered for obtaining the criteria weights. This study presents a new TOPSIS method based on interval-valued hesitant fuzzy information to compute the criteria weights. In this respect, the weight of each expert and the experts' judgments about the criteria weights are considered in the proposed procedure. In addition, an application example about the location problem is provided to show the capability of the proposed weighting method. Finally, results of the proposed method are compared with some methods from the related literature in the presented illustrative example to show the validation of the proposed interval-valued hesitant fuzzy TOPSIS method.

Keywords: Criteria weights · Group decision making · Interval-valued hesitant fuzzy set · Utility degree · Individual regret

1 Introduction

In modern group decision analysis, multi-criteria group decision making (MCGDM) problems are important part of operations research, which can rank the candidate potential alternative regarding to experts' judgments. One of the main factors that can affect ranking results is the criteria weight. In some decision problems, the authors focused on determining the criteria weights based on subjective, objective and integrated methods.

The subjective methods compute the criteria weights based on preferences and judgments of experts, such as ranking ordering method of criteria [1–3], direct rating method [4, 5], Delphi method [6], eigenvector method [7], point allocation method

© Springer International Publishing Switzerland 2015
B. Kamiński et al. (Eds.): GDN 2015, LNBIP 218, pp. 157–169, 2015.
DOI: 10.1007/978-3-319-19515-5_13

[8, 9], linear programming of preference comparisons [10], and linear programming techniques for multidimensional analysis of preferences [11]. In addition, the objective methods calculate the criteria weights by utilizing the information of objective decision matrix, such as criteria' importance through inter-criteria correlation method [12], entropy method [6, 12], maximizing deviation method [13, 14], and standard deviation method [12, 15]. Thus, in integrated methods, the weight of each criterion is determined based on considering the objective decision matrix and experts' subjective judgments [16, 17].

In real-world, the natures of the objects have been uncertain and imprecise, because the preferences and judgments of experts are hesitant or vague. Therefore, the criteria of group decision-making problems in an uncertain condition should be expressed by fuzzy values [18, 19], such as interval values [20, 21], linguistic variables [21, 22], intuitionistic fuzzy values [23, 24], hesitant fuzzy sets [25, 26], and interval-valued hesitant fuzzy sets [27, 28].

Fuzzy sets theory and its extensions have widely utilized in imprecise situations for evaluating the problems in many fields, such as artificial intelligence [29], management [30], pattern recognition [31] and group decision making [32, 33]. Hence, in fuzzy group decision-making problems the criteria weights is an important issue to provide the best solution. Therefore, some researchers have studied on determining the criteria weights regarding to the uncertain environment. In this respect, Fan et al. [34] proposed an optimization model to determine the criteria weights by according to the experts' fuzzy judgments and objective fuzzy decision matrices. Wang and Parkan [35] presented a general multi-attribute decision making framework by considering the objective information and subjective preferences to determine the criteria weights under fuzzy environment. Chen and Lee [36] proposed a fuzzy AHP method regarding to triangular fuzzy numbers for determining the criteria weights of professional conference organizer.

In some complex situations, the experts have defined their preferences and judgments by assigning some interval-values membership degrees for an object under a set to decrease the uncertainty risk and margin of errors. Therefore, the interval-valued hesitant fuzzy set (IVHFS), which first introduced by Chen et al. [27] is a powerful tool to deal with these situations. Thus, each criterion can be defined based on IVHFS and expressed in terms of experts' preferences. In this case, Zhang et al. [37] proposed an objective weighting approach by utilizing the Shannon information entropy under a hesitant fuzzy environment. Xu and Zhang [38] developed an optimization model regarding to the maximizing deviation method to determine the criteria weights under hesitant fuzzy and interval-valued hesitant fuzzy-environments. They proposed a hybridized group decision making method under some steps to specify the criteria weighs which led to be more easy to use versus the optimization model. Beg and Rashid [39] proposed a method to aggregate the preferences expert's judgments among the different criteria, in which the experts' opinions are expressed based on the hesitant fuzzy linguistic variables sets. Zhang et al. [40] constructed a hesitant fuzzy multiple attribute group decision making approach based on the distance measure to avoid the aggregation complexity of the hesitant fuzzy information. Feng et al. [41] utilized the TOPSIS (technique for order performance by similarity to ideal solution) method to solve the hesitant fuzzy multiple attribute decision making problems, in which the weight information are completely known. The literature review shows that determining the criteria weights based on ranking methods and especially under the IVHF-environments is still an open problem. In this paper, a hybridized group decision making approach is proposed based on the TOPSIS method and preferences experts' judgments about the criteria weights to determine the weight of each criterion under hesitant fuzzy environment. In addition, a group of experts is established to assess the problem based on the linguistic variables that indicate their subjective preferences. In sums, some merits and advantages of this study, which provide the proposed method to be more precise are expressed as follows: (1) Proposing a new TOPSIS method in an interval-valued hesitant fuzzy setting; (2) a group of experts is established to evaluate the problem by assigning their opinions by linguistic terms based on the interval-valued hesitant fuzzy information, which converted to interval-valued hesitant fuzzy elements; and (3) proposing a new relative closeness index to obtain the criteria weights. In addition, the weight of experts is applied in procedure of the proposed method. The validation of the proposed approach is obtained by comparing with other weighting methods for determining criteria weights.

The rest of this paper organized as follows. In Sect. 2, some methods to determine the criteria weights are explained and the interval-valued hesitant fuzzy TOPSIS (IVHF-TOPSIS) method for estimating the criteria weights are elaborated. In Sect. 3, an illustrative example in the selection of the best site for building a new factory provided to show the implementation process of the proposed approach. Finally, in Sect. 4, the paper is concluded.

2 Proposed Method

In this section, three techniques for computing the weight of criteria under the interval-valued hesitant fuzzy environment are extended to compare the computational results of criteria weights with the proposed approach.

2.1 Methods of Determining the Criteria Weights

In this subsection, these three approaches are considered to compute the criteria weights. Firstly, the criteria weights can be computed based on maximizing a deviation method introduced by Xu and Zhang [38] when the information completely unknown. In this paper, the method is extended based on the interval-valued hesitant fuzzy Hamming distance measure to determine the optimal weight vector as follows; let $h = \left[h_{ij}^L, h_{ij}^U \right]_{m \times n}$ is an interval-valued hesitant fuzzy element:

$$w_j = \frac{\sum_{i=1}^{m} \sum_{r=1}^{m} \left(\frac{1}{2l} \sum_{\lambda=1}^{l} \left(\left| h_{ij}^{\sigma(\lambda)^L} - h_{rj}^{\sigma(\lambda)^L} \right| + \left| h_{ij}^{\sigma(\lambda)^U} - h_{rj}^{\sigma(\lambda)^U} \right| \right) \right)}{\sqrt{\sum_{j=1}^{n} \left(\sum_{i=1}^{m} \sum_{r=1}^{m} \left(\frac{1}{2l} \sum_{\lambda=1}^{l} \left(\left| h_{ij}^{\sigma(\lambda)^L} - h_{rj}^{\sigma(\lambda)^L} \right| + \left| h_{ij}^{\sigma(\lambda)^U} - h_{rj}^{\sigma(\lambda)^U} \right| \right) \right) \right)^2}} \tag{1}$$

$$w_j^* = \frac{w_j}{\sum\limits_{j=1}^{n} w_j}, \quad j = 1, 2, \ldots, n \tag{2}$$

where the $h_{ij}^{\sigma(\lambda)^L}, h_{ij}^{\sigma(\lambda)^U}, h_{rj}^{\sigma(\lambda)^L}$ and $h_{rj}^{\sigma(\lambda)^U}$ are the λth largest value in $h_{ij}^L, h_{ij}^U, h_{rj}^L$ and h_{rj}^U, respectively and also the w_j^* is the normalized criteria weight vector.

Secondly, the decision makers (DMs) specify the relative importance of criteria weights by linguistic variables that can be converted to the IVHFS and denoted by $\upsilon_j = \left[\mu_j^L, \mu_j^U \right]$. In this respect, the final aggregated criteria weights regarding to DMs' judgments are obtained by the HIVFG operator are as follows:

$$\upsilon_j = HIVFG(\tilde{h}_1, \tilde{h}_2, \ldots, \tilde{h}_K) = \left(\overset{K}{\underset{k=1}{\oplus}} \left(\lambda_k^f \tilde{h}_k \right)^{\frac{1}{k}} \right)$$

$$= \bigcup_{\tilde{\gamma}_1 \in \tilde{h}_1, \tilde{\gamma}_2 \in \tilde{h}_2, \ldots, \tilde{\gamma}_k \in \tilde{h}_k} \left\{ \left[\prod_{j=1}^{K} \left(\lambda_k^L \gamma_k^L \right)^{\frac{1}{k}}, \prod_{j=1}^{K} \left(\lambda_k^U \gamma_k^U \right)^{\frac{1}{k}} \right] \right\} \tag{3}$$

where the weight of each DM is represented as $\lambda_k^f = \left[\lambda_k^L, \lambda_k^U \right]$ and is considered in the computational process of the criteria weights to decrease the errors.

Thirdly, the above-mentioned methods can be hybridized, and thus the following relation for computing the criteria weights can be proposed by:

$$\omega_j = \bar{\upsilon}_j . w_j \tag{4}$$

$$\bar{\upsilon}_j = \frac{\mu_j^L + \mu_j^U}{2} \quad \forall j \tag{5}$$

$$\omega_j^* = \frac{\omega_j}{\sum\limits_{j=1}^{n} \omega_j}, \quad j = 1, 2, \ldots . n \tag{6}$$

where ω_j^* is the normalized criteria weight vector.

2.2 TOPSIS Method with IVHFS

In this subsection, the proposed novel TOPSIS method is introduced based on the IVHFSs. In this respect, the DMs' opinions about the relative importance of each criterion are considered in process of the proposed method. Therefore, the procedure of the proposed method is defined based on the following steps:

Step 1. Construct an interval-valued hesitant fuzzy decision matrix (IVHF-decision matrix) for each criterion (C_j; 1,2,…,n) regarding to the possible alternatives (A_i; 1,2, …,m) and opinions of each DM (k; 1,2,…,K).

$$
G_j = \begin{array}{c} \\ A_1 \\ \vdots \\ A_m \end{array}
\begin{array}{cccc} k_1 & k_2 & \cdots & k_K \end{array}
\left(
\begin{array}{cccc}
\left[\mu_{1j}^{L1}, \mu_{1j}^{U1}\right] & \left[\mu_{12}^{L2}, \mu_{12}^{U2}\right] & \cdots & \left[\mu_{1j}^{Lk}, \mu_{1j}^{Uk}\right] \\
\vdots & \vdots & \ddots & \vdots \\
\left[\mu_{mj}^{L1}, \mu_{mj}^{U1}\right] & \left[\mu_{mj}^{L2}, \mu_{mj}^{U2}\right] & \cdots & \left[\mu_{mj}^{Lk}, \mu_{mj}^{Uk}\right]
\end{array}
\right)_{m \times k} \quad \forall j
\tag{7}
$$

where $\left[\mu_{mj}^{Lk}, \mu_{mj}^{Uk}\right]$ is the interval-value membership degree for the m-th alternative that expressed by the k-th expert to construct the j-th IVHF-decision matrix.

Step 2. Normalize each IVHF-decision matrix $\left(G_j^N = \left[\eta_{ij}^{Lk}, \eta_{ij}^{Lk}\right]_{m \times k}\right)$ based on the following relations [42]:

$$
\eta_{ij}^{Lk} = \frac{\mu_{ij}^{Lk}}{\sqrt{\sum\limits_{i=1}^{m} \sum\limits_{k=1}^{K} \left[(\mu_{ij}^{Lk})^2 + (\mu_{ij}^{Uk})^2\right]}} \quad \forall i, j, k
\tag{8}
$$

$$
\eta_{ij}^{Uk} = \frac{\mu_{ij}^{Uk}}{\sqrt{\sum\limits_{i=1}^{m} \sum\limits_{k=1}^{K} \left[(\mu_{ij}^{Lk})^2 + (\mu_{ij}^{Uk})^2\right]}} \quad \forall i, j, k
\tag{9}
$$

Step 3. Determine the interval-valued hesitant fuzzy positive ideal solution matrix (IVHF-PIS) and the interval-valued hesitant fuzzy negative ideal solution matrix (IVHF-NIS) by regarding to normalize IVHF-decision matrix as follows:

$$
A^* = \left(\left[\mu_i^{*Lk}, \mu_i^{*Uk}\right]\right)_{m \times k} = \begin{array}{c} \\ A_1 \\ \vdots \\ A_m \end{array}
\begin{array}{cccc} k_1 & k_2 & \cdots & k_K \end{array}
\left(
\begin{array}{cccc}
\left[\mu_1^{*L1}, \mu_1^{*U1}\right] & \left[\mu_1^{*L2}, \mu_1^{*U2}\right] & \cdots & \left[\mu_1^{*LK}, \mu_1^{*UK}\right] \\
\vdots & \vdots & \ddots & \vdots \\
\left[\mu_m^{*L1}, \mu_m^{*U1}\right] & \left[\mu_m^{*L2}, \mu_m^{*U2}\right] & \cdots & \left[\mu_m^{*LK}, \mu_m^{*UK}\right]
\end{array}
\right)_{m \times k}
\tag{10}
$$

$$
A^- = \left(\left[\mu_i^{-Lk}, \mu_i^{-Uk}\right]\right)_{m \times k} = \begin{array}{c} \\ A_1 \\ \vdots \\ A_m \end{array}
\begin{array}{cccc} k_1 & k_2 & \cdots & k_K \end{array}
\left(
\begin{array}{cccc}
\left[\mu_1^{-L1}, \mu_1^{-U1}\right] & \left[\mu_1^{-L2}, \mu_1^{-U2}\right] & \cdots & \left[\mu_1^{-LK}, \mu_1^{-UK}\right] \\
\vdots & \vdots & \ddots & \vdots \\
\left[\mu_m^{-L1}, \mu_m^{-U1}\right] & \left[\mu_m^{-L2}, \mu_m^{-U2}\right] & \cdots & \left[\mu_m^{-LK}, \mu_m^{-UK}\right]
\end{array}
\right)_{m \times k}
\tag{11}
$$

where the average of the group decision matrix is calculated by the following relations:

$$\mu_i^{*Lk} = \frac{1}{n}\sum_{j=1}^{n}\mu_{ij}^{Lk} \quad \forall i,k \tag{12}$$

$$\mu_i^{*Uk} = \frac{1}{n}\sum_{j=1}^{n}\mu_{ij}^{Uk} \quad \forall i,k \tag{13}$$

$$\mu_i^{-Lk} = \min_j\left\{\mu_{ij}^{Lk}\right\} \quad \forall i,k \tag{14}$$

$$\mu_i^{-Uk} = \max_j\left\{\mu_{ij}^{Uk}\right\} \quad \forall i,k \tag{15}$$

Step 4. Compute the separation measure for each normalized IVHF-decision matrix from the IVHF-PIS matrix and the IVHF-NIS matrix by using the IVHF-Euclidean distance measure, which indicates by S_j^* and S_j^-, respectively.

$$S_j^* = \sqrt{\frac{1}{2l_{x_i}}\sum_{i=1}^{m}\sum_{k=1}^{K}\sum_{\lambda=1}^{l_{x_i}}\left(\left|\mu_i^{Lk\sigma(\lambda)}(x_i) - A_i^{*Lk\sigma(\lambda)}(x_i)\right|^2 + \left|\mu_i^{Uk\sigma(\lambda)}(x_i) - A_i^{*kU\sigma(\lambda)}(x_i)\right|^2\right)} \quad \forall j \tag{16}$$

$$S_j^- = \sqrt{\frac{1}{2l_{x_i}}\sum_{i=1}^{m}\sum_{k=1}^{K}\sum_{\lambda=1}^{l_{x_i}}\left(\left|\mu_i^{Lk\sigma(\lambda)}(x_i) - A_i^{-Lk\sigma(\lambda)}(x_i)\right|^2 + \left|\mu_i^{Uk\sigma(\lambda)}(x_i) - A_i^{-Uk\sigma(\lambda)}(x_i)\right|^2\right)} \quad \forall j. \tag{17}$$

Step 5. Specify the relative closeness (C_j) for determining the most important criterion.

$$C_j = \frac{d\left(S_j^-,\bar{S}^-\right)}{d\left(S_j^-,\bar{S}^-\right) + d\left(S_j^*,\bar{S}^*\right)} \quad \forall j \tag{18}$$

$$\bar{S}^* = \frac{1}{n}\sum_{j=1}^{n}S_j^* \tag{19}$$

$$\bar{S}^- = \frac{1}{n}\sum_{j=1}^{n}S_j^- \tag{20}$$

where $\bar{S}^*$ the average of $S_j^*(j = 1,2,..,n)$, and also the average of $S_j^-(j = 1,2,..,n)$ represented as $\bar{S}^-$. The DM often specifies their opinion by linguistic variables, which applied in our proposed method and established a new hybrid approach. In this respect, the hybrid relative closeness $\left(C_j^h\right)$ is defined as follows:

$$C_j^h = \frac{\bar{\omega}_j . d\left(S_j^-, \bar{S}^-\right)}{d\left(S_j^-, \bar{S}^-\right) + d\left(S_j^*, \bar{S}^*\right)} \quad \forall j \tag{21}$$

$$\omega_j = HIVFG\left(\tilde{h}_1, \tilde{h}_2, \ldots, \tilde{h}_K\right) = \left(\overset{K}{\underset{k=1}{\oplus}}\left(\lambda_k^f \tilde{h}_k\right)^{\frac{1}{k}}\right)$$

$$= \underset{\tilde{\gamma}_1 \in \tilde{h}_1, \tilde{\gamma}_2 \in \tilde{h}_2, \ldots, \tilde{\gamma}_k \in \tilde{h}_k}{\cup}\left\{\left[\prod_{j=1}^{K}\left(\lambda_k^L \gamma_k^L\right)^{\frac{1}{k}}, \prod_{j=1}^{K}\left(\lambda_k^U \gamma_k^U\right)^{\frac{1}{k}}\right]\right\} \tag{22}$$

$$\bar{\omega}_j = \frac{\mu_j^L + \mu_j^U}{2} \quad \forall j \tag{23}$$

where the final weight of each DM is indicated by $\lambda_k = \left[\lambda_k^L, \lambda_k^U\right]$ that is computed by:

$$\lambda_k^L = \frac{\sum_i^m \sum_j^n \mu_{ij}^{Lk}}{\sum_k^K \sum_i^m \sum_j^n \mu_{ij}^{Lk}} \tag{24}$$

$$\lambda_k^U = \frac{\sum_i^m \sum_j^n \mu_{ij}^{Uk}}{\sum_k^K \sum_i^m \sum_j^n \mu_{ij}^{Uk}} \tag{25}$$

$$\lambda_k^f = \frac{\sum_i^m \sum_j^n \left(\mu_{ij}^{Uk} + \mu_{ij}^{Lk}\right)}{\sum_k^K \sum_i^m \sum_j^n \left(\mu_{ij}^{Uk} + \mu_{ij}^{Lk}\right)} \tag{26}$$

where, the final DMs' weights $\left(\lambda_k^f\right)$ is $\sum_{k=1}^{K} \lambda_k^f = 1$.

Step 6. Estimate the weight of each criterion w_j according to the relative closeness.

$$w_j = \frac{C_j}{\sum_{j=1}^{n} C_j} \quad \forall j \tag{27}$$

3   Illustrative Example

In this section, an illustrative example about the location problem from the literature [28] is considered to show the capability of the proposed method for determining the weight of each criterion. In this respect, for showing the verification of the proposed method, the application example is solved under three decision approaches, and then we compare them with our decision method. The application example is about the best site selection for building a new factory that is established by four DMs ($k = 1, 2, ..., 4$), three potential alternative (A_1, A_2, A_3) and the following six criteria: C_1: Climate, Condition; C_2: Regional demand; C_3: Expansion possibility; C_4: Transportation availability; C_5: Labor force; and C_6: Investment cost.

As represented in Tables 1 and 2, the relative importance of each hesitant fuzzy linguistic term for rating the importance of each criteria and rating of potential alternatives are defined, respectively. In addition, the evaluation of alternatives expressed by DMs' opinions with hesitant fuzzy linguistic terms that shown in Table 3. Similarly, the DMs' judgments about relative importance of each criterion are expressed as hesitant fuzzy linguistic terms in Table 4. Then, these linguistic variables are converted to interval-valued hesitant fuzzy elements (IVHFEs).

Table 1. Hesitant fuzzy variables for rating the importance of criteria and the DMs

Hesitant fuzzy linguistic variables	IVHFE
Very important (VI)	[0.90,0.90]
Important (I)	[0.75, 0.80]
Medium (M)	[0.50, 0.55]
Unimportant (UI)	[0.35, 0.40]
Very unimportant (VUI)	[0.10,0.10]

Table 2. Hesitant variables for the rating of possible alternatives

Hesitant fuzzy linguistic variables	IVHFE
Extremely good (EG)	[1.00,1.00]
Very very good (VVG)	[0.90,0.90]
Very good (VG)	[0.80, 0.90]
Good (G)	[0.70, 0.80]
Medium good (MG)	[0.60, 0.70]
Fair (F)	[0.50, 0.60]
Medium bad (MB)	[0.40, 0.50]
Bad (B)	[0.25, 0.40]
Very bad (VB)	[0.10, 0.25]
Very very bad (VVB)	[0.10,0.10]

The IVHF-decision matrix is normalized by regarding Eqs. (8) and (9); then, the IVHF-PIS matrix and the IVHF-NIS matrix are constructed by considering the relations Eqs. (10)−(15). In addition, the separation measure for each criterion is calculated by applying Eqs. (16) and (17). Then, the relative closeness is specified by Eqs. (18)−(20). Also, in hybrid decision approach, the DMs' judgments about the relative importance of each criterion are considered and computed by Eqs. (21)−(26). The Eq. (27) is utilized to compute the weight of each criterion in the proposed approach and in the

Table 3. Linguistic evaluations by the decision makers

Main criteria	Alternatives	k_1	k_2	k_3	k_4
C_1	A_1	MG	MG	G	VG
	A_2	VG	G	G	G
	A_3	F	MG	F	MG
C_2	A_1	G	VG	VG	MG
	A_2	F	MG	F	F
	A_3	VG	VG	VG	G
C_3	A_1	F	MG	MG	MG
	A_2	VG	G	VG	VG
	A_3	MG	MG	G	MG
C_4	A_1	VG	VG	G	G
	A_2	G	G	G	MG
	A_3	G	G	VG	F
C_5	A_1	MG	G	G	VG
	A_2	MG	MG	VG	G
	A_3	VG	G	G	G
C_6	A_1	VVG	VG	VVG	EG
	A_2	F	B	B	MG
	A_3	VVG	VVG	VG	VVG

Table 4. Linguistic evaluations for weights of criteria assigned by the decision makers

	k_1	k_2	k_3	k_4
C_1	I	VI	M	I
C_2	M	I	M	I
C_3	I	M	I	VI
C_4	I	UI	M	M
C_5	M	I	I	M
C_6	M	M	I	I

proposed hybrid approach. The above-mentioned results have shown in Tables 5 and 6. Other weighting techniques for determining the relative importance of criteria are extended and illustrated in Subsect. 2.1. In this respect, each extended technique in our application example is applied and for showing the low difference between them and proposed approaches, the mean value and the variance of criteria weights are provided.

Utilizing different techniques commonly leads to different results. It is unsuitable to say which method is powerful and capable because every method has various results underlying assertion or theory. However, the extended TOPSIS for the criteria weights is more capable for compromise of nearby to the ideal and farther from the negative ideal. Since the criteria weights of the proposed approaches are generated from the DMs' judgments, "biased" or "false" judgments lead to a low weight [43].

Table 5. Computational results of S_j^*, S_j^- and C_j

S_j^*		S_j^-		C_j	
S_1^*	0.111190	S_1^-	0.200000	C_1	0.034653
S_2^*	0.098327	S_2^-	0.191599	C_2	0.514718
S_3^*	0.119171	S_3^-	0.208143	C_3	0.38869
S_4^*	0.073633	S_4^-	0.186870	C_4	0.285241
S_5^*	0.060673	S_5^-	0.182734	C_5	0.273389
S_6^*	0.173451	S_6^-	0.229551	C_6	0.306194
$\bar{S}^*$	0.1060746	$\bar{S}^-$	0.199816		

Table 6. Final criteria weight by the proposed method and hybridized method and comparative analysis

w_j	Proposed approach	Proposed hybridized approach	Maximizing deviation method	Linguistic variables	Hybridized maximizing deviation method and linguistic variables
w_1	0.019221	0.021825	0.146788	0.182418	0.163515
w_2	0.285496	0.283351	0.183486	0.159441	0.178648
w_3	0.215593	0.244808	0.146788	0.182418	0.163515
w_4	0.158213	0.130957	0.091743	0.132973	0.074495
w_5	0.151639	0.150499	0.091743	0.159441	0.089324
w_6	0.169835	0.168558	0.339449	0.159441	0.330500
$\bar{X}$	0.166667	0.166666667	0.166666667	0.1626893	0.16666667
σ^2	0.0077124	0.008448048	0.008430828	0.00033864	0.00829643

In this regard, the proposed approach versus the classical/modern methods is more precise, because of two main features as considering the IVHFSs and the DMs' opinions about the criteria weights. The IVHFSs aid to DMs by assigning some interval-values membership degrees for an element under a set to margin of errors. In addition, the preferences DMs' judgments about the relative significance of the criteria are provided in procedure of the proposed method to decrease the errors.

4   Conclusions and Future Direction

In group decision making problems, determining the relative importance of each criterion has been very important issue. In this regard, a novel TOPSIS method has been proposed by utilizing the IVHFS regarding to the experts weights and their opinions about the criteria weights. In the proposed approach, the preferences experts' judgments have been expressed by linguistic variables which transformed to interval-valued hesitant fuzzy element. Hence, an illustrative example about the location problem has been considered to illustrate the steps of the proposed decision method. Finally, the results have been compared with several approaches implemented in a practical example to show the validation of the proposed approach. For future direction, the proposed method can be enhanced by considering the hierarchical structure of the criteria.

Acknowledgments. This work has been supported financially by the Center for International Scientific Studies & Collaboration (CISSC) and the French Embassy in Tehran as well as the Partenariats Hubert Curien (PHC) program in France. Additionally, the authors would like thank anonymous reviewers for their valuable comments.

References

1. Ahn, B.S., Park, K.S.: Comparing methods for multiattribute decision making with ordinal weights. Comput. Oper. Res. **35**(5), 1660–1670 (2008)
2. Barron, F.H., Barrett, B.E.: Decision quality using ranked attribute weights. Manage. Sci. **42**(11), 1515–1523 (1996)
3. Solymosi, T., Dombi, J.: A method for determining the weights of criteria: the centralized weights. Eur. J. Oper. Res. **26**(1), 35–41 (1986)
4. Bottomley, P.A., Doyle, J.R.: A comparison of three weight elicitation methods: good, better, and best. Omega **29**(6), 553–560 (2001)
5. Goodwin, P., Wright, G., Phillips, L.D.: Decision Analysis for Management Judgment. Wiley, London (2004)
6. Tzeng, G.-H., Huang, J.-J.: Multiple Attribute Decision Making: Methods and Applications. CRC Press, New York (2011)
7. Takeda, E., Cogger, K., Yu, P.: Estimating criterion weights using eigenvectors: a comparative study. Eur. J. Oper. Res. **29**(3), 360–369 (1987)
8. Roberts, R., Goodwin, P.: Weight approximations in multi-attribute decision models. J. Multi-Criteria Decis. Anal. **11**(6), 291–303 (2002)
9. Doyle, J.R., Green, R.H., Bottomley, P.A.: Judging relative importance: direct rating and point allocation are not equivalent. Organ. Behav. Hum. Decis. Process. **70**(1), 65–72 (1997)
10. Horsky, D., Rao, M.: Estimation of attribute weights from preference comparisons. Manage. Sci. **30**(7), 801–822 (1984)
11. Srinivasan, V., Shocker, A.D.: Linear programming techniques for multidimensional analysis of preferences. Psychometrika **38**(3), 337–369 (1973)
12. Xu, X.: A note on the subjective and objective integrated approach to determine attribute weights. Eur. J. Oper. Res. **156**(2), 530–532 (2004)
13. Wu, Z., Chen, Y.: The maximizing deviation method for group multiple attribute decision making under linguistic environment. Fuzzy Sets Syst. **158**(14), 1608–1617 (2007)
14. Wei, G.-W.: Maximizing deviation method for multiple attribute decision making in intuitionistic fuzzy setting. Knowl.-Based Syst. **21**(8), 833–836 (2008)
15. Deng, H., Yeh, C.-H., Willis, R.J.: Inter-company comparison using modified TOPSIS with objective weights. Comput. Oper. Res. **27**(10), 963–973 (2000)
16. Wang, Y.-M., Luo, Y.: Integration of correlations with standard deviations for determining attribute weights in multiple attribute decision making. Math. Comput. Modell. **51**(1), 1–12 (2010)
17. Ma, J., Fan, Z.-P., Huang, L.-H.: A subjective and objective integrated approach to determine attribute weights. Eur. J. Oper. Res. **112**(2), 397–404 (1999)

18. Mousavi, S.M., Torabi, S.A., Tavakkoli-Moghaddam, R.: A hierarchical group decision-making approach for new product selection in a fuzzy environment. Arab. J. Sci. Eng. **38** (11), 3233–3248 (2013)
19. Mousavi, S.M., Jolai, F., Tavakkoli-Moghaddam, R.: A fuzzy stochastic multi-attribute group decision-making approach for selection problems. Group Decis. Negot. **22**(2), 207–233 (2013)
20. Vahdani, B., Tavakkoli-Moghaddam, R., Mousavi, S.M., Ghodratnama, A.: Soft computing based on new interval-valued fuzzy modified multi-criteria decision-making method. Appl. Soft Comput. **13**(1), 165–172 (2013)
21. Vahdani, B., Zandieh, M.: Selecting suppliers using a new fuzzy multiple criteria decision model: the fuzzy balancing and ranking method. Int. J. Prod. Res. **48**(18), 5307–5326 (2010)
22. Parreiras, R., et al.: A flexible consensus scheme for multicriteria group decision making under linguistic assessments. Inf. Sci. **180**(7), 1075–1089 (2010)
23. Xu, Z.: Intuitionistic preference relations and their application in group decision making. Inf. Sci. **177**(11), 2363–2379 (2007)
24. Hashemi, H., Bazargan, J., Mousavi, S.M.: A compromise ratio method with an application to water resources management: an intuitionistic fuzzy set. Water Resour. Manage **27**(7), 2029–2051 (2013)
25. Torra, V., Narukawa, Y.: On hesitant fuzzy sets and decision. In: IEEE International Conference on Fuzzy Systems, 2009, FUZZ-IEEE 2009. IEEE (2009)
26. Torra, V.: Hesitant fuzzy sets. Int. J. Int. Syst. **25**(6), 529–539 (2010)
27. Chen, N., Xu, Z., Xia, M.: Interval-valued hesitant preference relations and their applications to group decision making. Knowl.-Based Syst. **37**, 528–540 (2013)
28. Wang, J.-q: Interval-valued hesitant fuzzy linguistic sets and their applications in multi-criteria decision-making problems. Knowl.-Based Syst. **288**, 55–72 (2014)
29. Greco, S., Matarazzo, B., Giove, S.: The Choquet integral with respect to a level dependent capacity. Fuzzy Sets Syst. **175**(1), 1–35 (2011)
30. Doria, S.: Characterization of a coherent upper conditional prevision as the Choquet integral with respect to its associated Hausdorff outer measure. Ann. Oper. Res. **195**(1), 33–48 (2012)
31. Demirel, T., Demirel, N.Ç., Kahraman, C.: Multi-criteria warehouse location selection using Choquet integral. Expert Syst. Appl. **37**(5), 3943–3952 (2010)
32. Wang, J.Q.: Multi-criteria outranking approach with hesitant fuzzy sets. OR Spectr. **36**(4), 1–19 (2013)
33. Qin, J., Liu, X.: Study on interval intuitionistic fuzzy multi-attribute group decision making method based on Choquet integral. Procedia Comput. Sci. **17**, 465–472 (2013)
34. Fan, Z.-P., Ma, J., Zhang, Q.: An approach to multiple attribute decision making based on fuzzy preference information on alternatives. Fuzzy Sets Syst. **131**(1), 101–106 (2002)
35. Wang, Y.-M., Parkan, C.: A general multiple attribute decision-making approach for integrating subjective preferences and objective information. Fuzzy Sets Syst. **157**(10), 1333–1345 (2006)
36. Chen, C.-F., Lee, C.-L.: Determining the attribute weights of professional conference organizer selection: an application of the fuzzy AHP approach. Tourism Econ. **17**(5), 1129–1139 (2011)
37. Zhang, Y., Wang, Y., Wang, J.: Objective attributes weights determining based on shannon information entropy in hesitant fuzzy multiple attribute decision making. Math. Probl. Eng. **2014**, 7 (2014)
38. Xu, Z., Zhang, X.: Hesitant fuzzy multi-attribute decision making based on TOPSIS with incomplete weight information. Knowl.-Based Syst. **52**, 53–64 (2013)

39. Beg, I., Rashid, T.: TOPSIS for hesitant fuzzy linguistic term sets. Int. J. Intell. Syst. **28**(12), 1162–1171 (2013)
40. Zhang, J.L., Qi, X.W., Huang, H.B.: A hesitant fuzzy multiple attribute group decision making approach based on TOPSIS for parts supplier selection. Appl. Mech. Mater. **357**, 2730–2737 (2013)
41. Feng, X.: TOPSIS method for hesitant fuzzy multiple attribute decision making. J. Intell. Fuzzy Syst. **26**(5), 2263–2269 (2014)
42. Jahanshahloo, G.R., Lotfi, F.H., Davoodi, A.: Extension of TOPSIS for decision-making problems with interval data: Interval efficiency. Math. Comput. Modell. **49**(5), 1137–1142 (2009)
43. Yue, Z.: An extended TOPSIS for determining weights of decision makers with interval numbers. Knowl.-Based Syst. **24**(1), 146–153 (2011)

Multiple Attribute Group Decision Making Under Hesitant Fuzzy Environment

Weize Wang [(⊠)], Qi-An Lu, and Li Yang

School of Economics and Management,
Guangxi Normal University, Guilin 541004, China
weizew@gmail.com

Abstract. Hesitant fuzzy set is a very useful means to depict the decision information in the process of decision making. In this paper, motivated by the extension principle of hesitant fuzzy sets, we export Einstein operations on fuzzy sets to hesitant fuzzy sets, and develop some new arithmetic averaging aggregation operators, such as the hesitant fuzzy Einstein weighted averaging (HFWA$^\varepsilon$) operator, hesitant fuzzy Einstein ordered weighted averaging (HFOWA$^\varepsilon$) operator, and hesitant fuzzy Einstein hybrid weighted averaging (HFHWA$^\varepsilon$) operator, for aggregating hesitant fuzzy elements. Finally, we apply the proposed operators to multiple attribute group decision making with hesitant fuzzy information.

Keywords: Hesitant fuzzy set · Einstein operation · Hesitant fuzzy Einstein arithmetic averaging operator · Multiple attribute group decision making (MAGDM)

1 Introduction

Aggregation operators, usually taking the forms of mathematical functions, are common techniques to fuse all the input individual data into a single one [1–4]. Three of the most common arithmetic operators for aggregating arguments are the weighted averaging (WA) operator [5], the ordered weighted averaging (OWA) operators [6] and the hybrid weighted averaging operator [3,7]. In the real-life world, due to the increasing complexity of the socioeconomic environment and the lack of knowledge or data about the problem domain, crisp data are sometimes unavailable. Thus, the input arguments may be vague or fuzzy in nature. Besides fuzzy sets (FSs) by Zadeh [8], several extensions of this concept have been introduced in the literature, for example, intuitionistic fuzzy sets (IFSs [9], interval-valued fuzzy sets [10], type 2 fuzzy sets [11,12], fuzzy multisets [13,14] and hesitant fuzzy sets (HFSs) [15]. IFSs are equivalent to interval-valued fuzzy sets [16,17], and the prominent characteristic of IFS is that it assigns to each element a membership degree and a nonmembership degree. The membership of an element to a type 2 fuzzy set is defined in terms of a fuzzy set on the domain of memberships. IFSs can be seen, from a mathematical point of view,

© Springer International Publishing Switzerland 2015
B. Kamiński et al. (Eds.): GDN 2015, LNBIP 218, pp. 171–182, 2015.
DOI: 10.1007/978-3-319-19515-5_14

as a particular case of type 2 fuzzy sets [18]. Fuzzy multisets, or fuzzy bags, permit us to have multiple occurrences of the elements. Recently, Torra [15] defined the notion of a HFS, whose basic elements are hesitant fuzzy elements (HFEs) [19], each of which is characterized by a membership degree consisting of a set of possible values. Although all HFSs can be represented as fuzzy multisets, the operations on fuzzy multisets do not apply properly on HFSs. In addition, it can be proved that HFSs can also be represented as type 2 fuzzy sets and IFS is a particular case of HFS [15].

In many practical situations, particularly in the process of group decision making under uncertainty and anonymity, the experts may come from different research areas and thus have different backgrounds and levels of knowledge, skills, experience, and personality, the experts may not have enough expertise or possess a sufficient level of knowledge to precisely express their preferences over the objects, and then, they usually have some uncertainty in providing their preferences; Moreover, the experts have only assigned a small and finite set in providing their preferences, where the difficulty may be caused by a doubt between a few different values. In such cases, the data or preferences given by the experts may be appropriately expressed in HFEs. For example, in multiattribute decision-making problems, such as presidential election, blind peer review of thesis, etc., each HFE provided by the experts can be used to express the degree that an alternative should satisfy a attribute. Up to now, some authors have paid attention to the HFS theory. Torra [15] proposed the concept of hesitant fuzzy sets, which is different from other extensions exist for fuzzy sets, and also introduce some basic operations on HFSs. Torra and Narukawa [20] present an extension principle of HFSs, which permits to generalize existing operations on fuzzy sets to HFSs, and also discuss their use in decision making. Xia and Xu [19] and Xu [21] developed a series of aggregation operators for hesitant fuzzy information, and then give their application in solving decision making problems. Based on the extension principle of HFSs, it is clear that the hesitant fuzzy aggregation operators proposed by Xia and Xu [19] are the corresponding extensions of the fuzzy aggregation operators [22–24]. Moreover, the extension principle of HFSs permits us to export operations on FSs to HFSs. The basic algebraic operations on FSs include Algebraic product and Algebraic sum, which are not the only operations that can be chosen to model the intersection and union on FSs, but they are the most commonly used ones in decision making applications [25,26]. For an intersection, a good alternative to the algebraic product is the Einstein product, which typically gives the same smooth approximations as the algebraic product. Equivalently, for an intersection, a good alternative to the algebraic sum is the Einstein sum. Therefore, how to extend the Einstein operations on FSs to aggregate these hesitant fuzzy elements is a meaningful work, which is also the focus of this paper.

The paper is structured as follows. In Sect. 1 we give an introduction of the research background. In Sect. 2 we briefly reviews some basic concepts related to the HFSs and the exsiting arithmetic averaging operators for aggregating HFEs. In Sect. 3 we introduce some Einstein operations on HFSs, and then develop some novel arithmetic aggregation operators for aggregating a collection of HFEs. In Sect. 4 we apply the hesitant fuzzy Einstein weighted averaging operator to MAGDM with hesitant fuzzy information. In Sect. 5 we have a conclusion.

2   Preliminary

The FS, an extension of the classical notion of set, was introduced by Zadeh [8] as follows

Definition 1. *Let a set X be fixed, a FS F on X is defined as:*

$$F = \{\langle x, \mu_F(x) \rangle | x \in X\}, \tag{1}$$

where μ_F is a mapping from X to the closed interval [0,1], and for each $x \in X$, $\mu_F(x)$ is called the degree of membership of x in X.

However, when giving the membership degree of an element on FS, the difficulty of establishing the membership degree is not because we have a margin of error, or some possibility distribution on the possibility values, but because we have several possible values. For such cases, Torra [15,20] proposed another generation of FS as follows:

Definition 2. *Let X be a reference set, then hesitant fuzzy set on X is defined in terms of a function h that when applied to X returns a subset of $[0,1]$.*

To be easily understood, Xia and Xu [19] express the HFS as follows:

Definition 3. *Let X be a fixed set, a HFS E on X is defined as:*

$$E = \{\langle x, h_E(x) \rangle | x \in X\}, \tag{2}$$

where $h_E(x)$ is a set-valued function from X to the power set of the unit interval (i.e., $2^{[0,1]}$), and denotes the possible membership degrees of the element $x \in X$ to the set E. For convenience, let Ω be the set of all HFSs on X.

Given $x \in X$, $h_E(x)$ is called as a hesitant fuzzy element (HFE) [19], which simply denoted as $h = h(x)$. For convenience, let H be the set of all HFEs on X.

To compare the HFEs, Xia and Xu [19] define the following comparison laws:

Definition 4. *For a HFE h, $s(h) = \frac{1}{\#h}\sum_{\gamma \in h} \gamma$ is called the score function of h, where $\#h$ is the number of the elements in h. For two HFEs h_1 and h_2, if $s(h_1) > s(h_2)$, then $h_1 > h_2$; If $s(h_1) = s(h_2)$, then $h_1 = h_2$.*

Torra and Narukawa [20] introduced the extension principle for extending functions to HFEs as follows.

Definition 5. *Let $\mathbb{O}$ be a function $\mathbb{O} : [0,1]^n \to [0,1]$, and $\hbar = \{h_1, h_2, \ldots, h_n\}$ be a set of n HFEs, then the extension of $\mathbb{O}$ on $\hbar$ is a function $\mathbb{O}_\hbar : H^n \to H$,*

$$\mathbb{O}_\hbar = \cup_{\gamma \in \{h_1 \times h_2 \times \ldots \times h_n\}} \{\mathbb{O}(\gamma)\}, \tag{3}$$

where $\mathbb{O}_\hbar$, the extension of an operator $\mathbb{O}$ on a set of HFEs $\hbar$, considers all the values in such sets and the application of $\mathbb{O}$ on them.

3   Hesitant Fuzzy Einstein Arithmetic Averaging Aggregation Operators

Let $\mathbb{O}$ be the Einstein sum and Einstein product on FSs respectively, then the extensions of the Einstein sum and Einstein product on h_1 and h_2, denoted by $h_1 \oplus_\varepsilon h_2$ $h_1 \otimes_\varepsilon h_2$, are defined as follows respectively:

$$h_1 \oplus_\varepsilon h_2 = \cup_{\gamma_1 \in h_1, \gamma_2 \in h_2} \left\{ \frac{\gamma_1 + \gamma_2}{1 + \gamma_1 \gamma_2} \right\} \tag{4}$$

$$h_1 \otimes_\varepsilon h_2 = \cup_{\gamma_1 \in h_1, \gamma_2 \in h_2} \left\{ \frac{\gamma_1 \gamma_2}{1 + (1 - \gamma_1)(1 - \gamma_2)} \right\}. \tag{5}$$

Theorem 1. *If n is any a positive integer and h is a HFE of H, then the scale multiplication operation $n \cdot_\varepsilon h$ is a mapping from $Z^+ \times H$ to H:*

$$n \cdot_\varepsilon h = \cup_{\gamma \in h} \left\{ \frac{(1+\gamma)^n - (1-\gamma)^n}{(1+\gamma)^n + (1-\gamma)^n} \right\}, \tag{6}$$

where $n \cdot_\varepsilon h = \overbrace{h \oplus_\varepsilon h \oplus_\varepsilon \ldots \oplus_\varepsilon h}^{n}$.

Proof. Mathematical induction can be used to prove that (6) holds for all positive integers n. (6) is called $P(n)$.

Basis: Show that the statement $P(n)$ holds for $n = 1$. The statement $P(n)$ amounts to the statement $P(1)$:

$$1 \cdot_\varepsilon h = \cup_{\gamma \in h} \left\{ \frac{(1+\gamma) - (1-\gamma)}{(1+\gamma) + (1-\gamma)} \right\}$$

In the left-hand side of the equation, $1 \cdot_\varepsilon h = h = \cup_{\gamma \in h} \{\gamma\}$; In the right-hand side of the equation, $\cup_{\gamma \in h} \left\{ \frac{(1+\gamma)-(1-\gamma)}{(1+\gamma)+(1-\gamma)} \right\} = \cup_{\gamma \in h} \{\gamma\}$. The two sides are equal, so the statement $P(n)$ is true for $n = 1$. Thus it has been shown the statement $P(1)$ holds.

Inductive Step: Show that if $P(n)$ holds, then also $P(n+1)$holds. Assume $P(n)$ holds (for some unspecified value of n). It must then be shown that $P(n+1)$ holds, that is:

$$(n+1) \cdot_\varepsilon h = \cup_{\gamma \in h} \left\{ \frac{(1+\gamma)^{n+1} - (1-\gamma)^{n+1}}{(1+\gamma)^{n+1} + (1-\gamma)^{n+1}} \right\}$$

Using the induction hypothesis that $P(n)$ holds, the left-hand side can be rewritten to $n \cdot_\varepsilon h \oplus_\varepsilon h$ and based on the Einstein sum operation of two HFEs (i.e., (4)), we have

$$n \cdot_\varepsilon h(x) \oplus_\varepsilon h = \cup_{\gamma \in h} \left\{ \frac{\frac{(1+\gamma)^n - (1-\gamma)^n}{(1+\gamma)^n + (1-\gamma)^n} + \gamma}{1 + \frac{(1+\gamma)^n - (1-\gamma)^n}{(1+\gamma)^n + (1-\gamma)^n} \cdot \gamma} \right\} = \cup_{\gamma \in h} \left\{ \frac{(1+\gamma)^{n+1} - (1-\gamma)^{n+1}}{(1+\gamma)^{n+1} + (1-\gamma)^{n+1}} \right\}.$$

Thereby showing that indeed $P(n+1)$ holds. Since both the basis and the inductive step have been proved, it has now been proved by mathematical induction that $P(n)$ holds for any positive integer n.

Theorem 2. *Let h, h_1 and h_2 be three HFEs and let $h_3 = h_1 + h_2$ and $h_4 = \lambda \cdot_\varepsilon h$ $\lambda > 0$, then both h_3 and h_4 are also HFEs.*

Proof. It is trivial.

In the following, let us look at $\lambda \cdot_\varepsilon h$ for some special cases of λ and h.

Proposition 1. *Let h, h_1 and h_2 be three HFEs, $\lambda, \lambda_1, \lambda_2 > 0$, then*

(i) $h_1 \oplus_\varepsilon h_2 = h_2 \oplus_\varepsilon h_1$,
(ii) $\lambda \cdot_\varepsilon (h_1 \oplus_\varepsilon h_2) = \lambda \cdot_\varepsilon h_1 \oplus_\varepsilon \lambda \cdot_\varepsilon h_2$,
(iii) $\lambda_1 \cdot_\varepsilon h \oplus_\varepsilon \lambda_2 \cdot_\varepsilon h = (\lambda_1 + \lambda_2) \cdot_\varepsilon h$,
(iv) $(\lambda_1 \lambda_2) \cdot_\varepsilon h = \lambda_1 \cdot_\varepsilon (\lambda_2 \cdot_\varepsilon h)$.

Proof.

(i) It is trivial.
(ii) Since $h_1 \oplus_\varepsilon h_2 = \cup_{\gamma_1 \in h_1, \gamma_2 \in h_2} \{ \frac{\gamma_1 + \gamma_2}{1 + \gamma_1 \gamma_2} \}$, which can be transformed into the following form:

$$h_1 \oplus_\varepsilon h_2 = \cup_{\gamma_1 \in h_1, \gamma_2 \in h_2} \left\{ \frac{(1 + \gamma_1)(1 + \gamma_2) - (1 - \gamma_1)(1 - \gamma_2)}{(1 + \gamma_1)(1 + \gamma_2) + (1 - \gamma_1)(1 - \gamma_2)} \right\}. \quad (7)$$

Let

$$a = (1 + \gamma_1)(1 + \gamma_2), \quad b = (1 - \gamma_1)(1 - \gamma_2), \quad (8)$$

then $h_1 \oplus_\varepsilon h_2 = \cup_{\gamma_1 \in h_1, \gamma_2 \in h_2} \{ \frac{a-b}{a+b} \}$, by Theorem 1, it follows that

$$\lambda \cdot_\varepsilon (h_1 \oplus_\varepsilon h_2) = \cup_{\gamma_1 \in h_1, \gamma_2 \in h_2} \left\{ \frac{a^\lambda - b^\lambda}{a^\lambda + b^\lambda} \right\}$$

based on (8), we have

$$\lambda \cdot_\varepsilon (h_1 \oplus_\varepsilon h_2) = \cup_{\gamma_1 \in h_1, \gamma_2 \in h_2} \left\{ \frac{(1 + \gamma_1)^\lambda (1 + \gamma_2)^\lambda - (1 - \gamma_1)^\lambda (1 - \gamma_2)^\lambda}{(1 + \gamma_1)^\lambda (1 + \gamma_2)^\lambda - (1 - \gamma_1)^\lambda (1 - \gamma_2)^\lambda} \right\}$$

Also since

$$\lambda \cdot_\varepsilon h_i = \cup_{\gamma_i \in h_i} \left\{ \frac{(1 + \gamma_i)^\lambda - (1 - \gamma_i)^\lambda}{(1 + \gamma_i)^\lambda + (1 - \gamma_i)^\lambda} \right\}, \quad i = 1, 2.$$

Let

$$a_1 = (1+\gamma_1)^\lambda, \; b_1 = (1-\gamma_1)^\lambda, \; a_2 = (1+\gamma_2)^\lambda, \; b_2 = (1-\gamma_2)^\lambda \qquad (9)$$

then, by the Einstein operation (8), it follows that

$$\lambda \cdot_\varepsilon h_1 \oplus_\varepsilon \lambda \cdot_\varepsilon h_2 = \cup_{\gamma_1 \in h_1} \left\{ \frac{a_1 - b_1}{a_1 + b_1} \right\} \oplus_\varepsilon \cup_{\gamma_2 \in h_2} \left\{ \frac{a_2 - b_2}{a_2 + b_2} \right\}$$

$$= \cup_{\gamma_1 \in h_1, \gamma_2 \in h_2} \left\{ \frac{a_1 a_2 - b_1 b_2}{a_1 a_2 + b_1 b_2} \right\}$$

based on the substitutes (9), we have

$$\lambda \cdot_\varepsilon h_1 \oplus_\varepsilon \lambda \cdot_\varepsilon h_2 = \cup_{\gamma_1 \in h_1, \gamma_2 \in h_2} \left\{ \frac{(1+\gamma_1)^\lambda (1+\gamma_2)^\lambda - (1-\gamma_1)^\lambda (1-\gamma_2)^\lambda}{(1+\gamma_1)^\lambda (1+\gamma_2)^\lambda - (1-\gamma_1)^\lambda (1-\gamma_2)^\lambda} \right\}.$$

Hence $\lambda \cdot_\varepsilon (h_1 \oplus_\varepsilon h_2) = \lambda \cdot_\varepsilon h_1 \oplus_\varepsilon \lambda \cdot_\varepsilon h_2$.

(iii) Since

$$\lambda_i \cdot_\varepsilon h = \cup_{\gamma \in h} \left\{ \frac{(1+\gamma)^{\lambda_i} - (1-\gamma)^{\lambda_i}}{(1+\gamma)^{\lambda_i} + (1-\gamma)^{\lambda_i}} \right\}, \lambda_i > 0 \quad (i=1,2)$$

Let

$$a_1 = (1+\gamma)^{\lambda_1}, b_1 = (1-\gamma)^{\lambda_1}, a_2 = (1+\gamma)^{\lambda_2}, b_2 = (1-\gamma)^{\lambda_2}. \qquad (10)$$

then, by the Einstein sum (4), it follows that

$$\lambda_1 \cdot_\varepsilon h \oplus_\varepsilon \lambda_2 \cdot_\varepsilon h = \cup_{\gamma \in h} \left\{ \frac{a_1 - b_1}{a_1 + b_1} \right\} \oplus_\varepsilon \cup_{\gamma \in h} \left\{ \frac{a_2 - b_2}{a_2 + b_2} \right\}$$

$$= \cup_{\gamma \in h} \left\{ \frac{\frac{a_1-b_1}{a_1+b_1} + \frac{a_2-b_2}{a_2+b_2}}{1 + \frac{a_1-b_1}{a_1+b_1} \cdot \frac{a_2-b_2}{a_2+b_2}} \right\}$$

$$= \cup_{\gamma \in h} \left\{ \frac{a_1 a_2 - b_1 b_2}{a_1 a_2 + b_1 b_2} \right\}$$

based on the substitutes (10), we have

$$\lambda_1 \cdot_\varepsilon h \oplus_\varepsilon \lambda_2 \cdot_\varepsilon h$$

$$= \cup_{\gamma \in h} \left\{ \frac{(1+\gamma)^{\lambda_1}(1+\gamma)^{\lambda_2} - (1-\gamma)^{\lambda_1}(1-\gamma)^{\lambda_2}}{(1+\gamma)^{\lambda_1}(1+\gamma)^{\lambda_2} - (1-\gamma)^{\lambda_1}(1-\gamma)^{\lambda_2}} \right\}$$

$$= \cup_{\gamma \in h} \left\{ \frac{(1+\gamma)^{\lambda_1+\lambda_2} - (1-\gamma)^{\lambda_1+\lambda_2}}{(1+\gamma)^{\lambda_1+\lambda_2} - (1-\gamma)^{\lambda_1+\lambda_2}} \right\} = (\lambda_1 + \lambda_2) \cdot_\varepsilon h$$

i.e., $\lambda_1 \cdot_\varepsilon h \oplus_\varepsilon \lambda_2 \cdot_\varepsilon h = (\lambda_1 + \lambda_2) \cdot_\varepsilon h$.

(iv) Since

$$\lambda_2 \cdot_\varepsilon h = \cup_{\gamma \in h} \left\{ \frac{(1+\gamma)^{\lambda_2} - (1-\gamma)^{\lambda_2}}{(1+\gamma)^{\lambda_2} + (1-\gamma)^{\lambda_2}} \right\}, \lambda_2 > 0.$$

Let

$$a = (1+\gamma)^{\lambda_2}, b = (1-\gamma)^{\lambda_2}, \tag{11}$$

then $\lambda_2 \cdot_\varepsilon h = \cup_{\gamma \in h} \{ \frac{a-b}{a+b} \}$, by Theorem 1, it follows that

$$\lambda_1 \cdot_\varepsilon (\lambda_2 \cdot_\varepsilon h) = \cup_{\gamma \in h} \left\{ \frac{a^{\lambda_1} - b^{\lambda_1}}{a^{\lambda_1} + b^{\lambda_1}} \right\}$$

based on the substitutes (11), we have

$$\lambda_1 \cdot_\varepsilon (\lambda_2 \cdot_\varepsilon h) = \cup_{\gamma \in h} \left\{ \frac{(1+\gamma)^{\lambda_1 \lambda_2} - (1-\gamma)^{\lambda_1 \lambda_2}}{(1+\gamma)^{\lambda_1 \lambda_2} + (1-\gamma)^{\lambda_1 \lambda_2}} \right\} = (\lambda_1 \lambda_2) \cdot_\varepsilon h$$

which completes the proof of Proposition 1.

Based on the above Einstein operational laws of HFEs and Definition 4, we will propose some arithmetic averaging aggregation operators for aggregating HFEs as listed below:

For a collection of n HFEs $h_j (j = 1, 2, \ldots, n)$, the hesitant fuzzy weighted averaging (HFWA$^\varepsilon$) operator is defined as

$$\text{HFWA}_{\boldsymbol{\omega}}^\varepsilon (h_1, h_2, \ldots, h_n)$$

$$= \cup_{\gamma_1 \in h_1, \gamma_2 \in h_2, \ldots, \gamma_n \in h_n} \left\{ \frac{\prod\limits_{j=1}^{n} (1+\gamma_j)^{\omega_j} - \prod\limits_{j=1}^{n} (1-\gamma_j)^{\omega_j}}{\prod\limits_{j=1}^{n} (1+\gamma_j)^{\omega_j} + \prod\limits_{j=1}^{n} (1-\gamma_j)^{\omega_j}} \right\} \tag{12}$$

where $\boldsymbol{\omega} = (\omega_1, \omega_2, \ldots, \omega_n)^T$ is the weight vector of $h_j (j = 1, 2, \ldots, n)$ with $\omega_j \in [0,1]$ and $\sum_{j=1}^{n} \omega_j = 1$.

Corollary 1. *The* HFWA *operator* [19] *and the* HFWA$^\varepsilon$ *operator have the following relation*

$$\text{HFWA}_{\boldsymbol{\omega}}^\varepsilon (h_1, h_2, \ldots, h_n) \leqslant \text{HFWA}_{\boldsymbol{\omega}} (h_1, h_2, \ldots, h_n),$$

where $h_j (j = 1, 2, \ldots, n)$ be a collection of HFEs and $\boldsymbol{\omega} = (\omega_1, \omega_2, \ldots, \omega_n)^T$ is the weight vector of $h_j (j = 1, 2, \ldots, n)$, with $\omega_j \in [0,1], (j = 1, 2, \ldots, n)$ and $\sum_{j=1}^{n} \omega_j = 1$,

$$\text{HFWA}_{\boldsymbol{\omega}} (h_1, h_2, \ldots, h_n) = \cup_{\gamma_1 \in h_1, \gamma_2 \in h_2, \ldots, \gamma_n \in h_n} \left\{ 1 - \prod\limits_{j=1}^{n} (1-\gamma_j)^{\omega_j} \right\} \tag{13}$$

Corollary 1 shows that the values obtained by the HFWA$^\varepsilon$ operator are smaller than the ones obtained by the HFWA operator.

Based on Eq. (12), we have some properties of the HFWA$^\varepsilon$ operator.

Proposition 2. *Let $h_j(j = 1, 2, \ldots, n)$ be a collection of HFEs, $\boldsymbol{\omega} = (\omega_1, \omega_2, \ldots, \omega_n)^T$ is the weight vector of $h_j(j = 1, 2, \ldots, n)$, with $\omega_j \in [0, 1], (j = 1, 2, \ldots, n)$ and $\sum_{j=1}^n \omega_j = 1$, then, we have the following properties.*

(i) *Idempotency: If all $h_j(j = 1, 2, \ldots, n)$ are equal, i.e., $h_j = h$, for all $j = 1, 2, \ldots, n$, then*

$$\mathrm{HFWA}_{\boldsymbol{\omega}}^\varepsilon(h_1, h_2, \ldots, h_n) = h;$$

(ii) *Boundary:*

$$h_{\min} \leqslant \mathrm{HFWA}_{\boldsymbol{\omega}}^\varepsilon(h_1, h_2, \ldots, h_n) \leqslant h_{\max}$$

where $h_{\min} = \min\{h_1, h_2, \ldots, h_n\}$ and $h_{\max} = \max\{h_1, h_2, \ldots, h_n\}$

(iii) *Monotonicity: Let h_j^1 and h_j^2, $(j = 1, 2, \ldots, n)$ be two collection of HFEs, and $h_j^1 \leqslant h_j^2$, for all j, then*

$$\mathrm{HFWA}_{\boldsymbol{\omega}}^\varepsilon(h_1^1, h_2^1, \ldots, h_n^1) \leqslant \mathrm{HFWA}_{\boldsymbol{\omega}}^\varepsilon(h_1^2, h_2^2, \ldots, h_n^2).$$

The hesitant fuzzy ordered weighted averaging (HFOWA$^\varepsilon$) operator is defined as

$$\mathrm{HFOWA}_{\mathbf{w}}^\varepsilon(h_1, h_2, \ldots, h_n) =$$

$$\bigcup_{\gamma_{\sigma(1)} \in h_{\sigma(1)}, \gamma_{\sigma(2)} \in h_{\sigma(2)}, \ldots, \gamma_{\sigma(n)} \in h_{\sigma(n)}} \left\{ \frac{\prod_{j=1}^n \left(1 + \gamma_{\sigma(j)}\right)^{\omega_j} - \prod_{j=1}^n \left(1 - \gamma_{\sigma(j)}\right)^{\omega_j}}{\prod_{j=1}^n \left(1 + \gamma_{\sigma(j)}\right)^{\omega_j} + \prod_{j=1}^n \left(1 - \gamma_{\sigma(j)}\right)^{\omega_j}} \right\}$$

$$(14)$$

where $h_{\sigma(j)}$ is the jth largest of $h_k(k = 1, 2, \ldots, n)$, and $\mathbf{w} = (w_1, w_2, \ldots, w_n)^T$ is the aggregation-associated vector with $w_j \in [0, 1]$ and $\sum_{j=1}^n w_j = 1$.

Corollary 2. *The HFOWA operator [19] and the HFOWA$^\varepsilon$ operator have the following relation*

$$\mathrm{HFOWA}_{\mathbf{w}}^\varepsilon(h_1, h_2, \ldots, h_n) \leqslant \mathrm{HFOWA}_{\mathbf{w}}(h_1, h_2, \ldots, h_n),$$

where $h_{\sigma(j)}$ is the jth largest of $h_k(k = 1, 2, \ldots, n)$, and $\boldsymbol{w} = (w_1, w_2, \ldots, w_n)^T$ is the aggregation-associated vector with $w_j \in [0, 1]$ and $\sum_{j=1}^n w_j = 1$.

$$\mathrm{HFOWA}_{\mathbf{w}}(h_1, h_2, \ldots, h_n)$$

$$= \bigcup_{\gamma_{\sigma(1)} \in h_{\sigma(1)}, \gamma_{\sigma(2)} \in h_{\sigma(2)}, \ldots, \gamma_{\sigma(n)} \in h_{\sigma(n)}} \left\{ 1 - \prod_{j=1}^n \left(1 - \gamma_{\sigma(j)}\right)^{w_j} \right\} \quad (15)$$

Corollary 2 shows that the values obtained by the HFOWA$^\varepsilon$ operator are smaller than the ones obtained by the HFOWA operator.

Similar to the HFWA$^\varepsilon$ operator, the HFOWA$^\varepsilon$ operator satisfies the properties of idempotency, boundary and monotonicity.

The hesitant fuzzy hybrid weighted averaging (HFHWA$^\varepsilon$) operator is defined as

$$\text{HFHWA}^\varepsilon_{\boldsymbol{\omega},\mathbf{w}}(h_1, h_2, \ldots, h_n) =$$

$$\bigcup_{\dot{\gamma}_{\sigma(1)}\in\dot{h}_{\sigma(1)},\dot{\gamma}_{\sigma(2)}\in\dot{h}_{\sigma(2)},\ldots,\dot{\gamma}_{\sigma(n)}\in\dot{h}_{\sigma(n)}} \left\{ \frac{\prod_{j=1}^{n}\left(1+\dot{\gamma}_{\sigma(j)}\right)^{\omega_j} - \prod_{j=1}^{n}\left(1-\dot{\gamma}_{\sigma(j)}\right)^{\omega_j}}{\prod_{j=1}^{n}\left(1+\dot{\gamma}_{\sigma(j)}\right)^{\omega_j} + \prod_{j=1}^{n}\left(1-\dot{\gamma}_{\sigma(j)}\right)^{\omega_j}} \right\}$$

(16)

where $\dot{h}_{\sigma(j)}$ is the jth largest of $\dot{h}_k (\dot{h}_k = h_k^{(n\omega_k)}, k = 1, 2, \ldots, n)$, and $\boldsymbol{\omega} = (\omega_1, \omega_2, \ldots, \omega_n)^T$ is the weight vector of $h_j (j = 1, 2, \ldots, n)$ with $\omega_j \in [0, 1]$ and $\sum_{j=1}^{n}\omega_j = 1$, $\mathbf{w} = (w_1, w_2, \ldots, w_n)^T$ is the aggregation-associated vector such that $w_j \in [0, 1]$ and $\sum_{j=1}^{n} w_j = 1$.

Corollary 3. *The* HFWA$^\varepsilon$ *and* HFOWA$^\varepsilon$ *operators are two special cases of the* HFHWA$^\varepsilon$ *operator.*

Similarly, the HFHWA$^\varepsilon$ operator satisfies the properties of idempotency, boundary and monotonicity.

4   Multiple Attribute Group Decision Making Based on Hesitant Fuzzy Information

Multiple attribute group decision making problems are widespread in real life decision situations. In some practical problems, such as presidential election or the blind peer review of thesis, the experts propose the preferences or opinions for the alternatives with anonymous in order to protect their privacy or avoid influencing each other. In such situations, hesitant fuzzy element permits us to represent the rating of the alternative on the attribute given by several experts, so we use a hesitant fuzzy decision matrix to describe the group decision making problems. Let $E = \{e_1, e_2, \ldots, e_l\}$ be the set of experts and $\lambda = \{\lambda_1, \lambda_2, \ldots, \lambda_l\}$ be the weight vector of experts, where $\lambda_k \in [0, 1], (k = 1, 2, \ldots, l)$ and $\sum_{k=1}^{l}\lambda_j = 1$, and suppose that there are n alternatives $X = \{x_1, x_2, \ldots, x_n\}$ and m attributes $G = \{g_1, g_2, \ldots, g_m\}$ with the attribute weight vector $\boldsymbol{\omega} = (\omega_1, \omega_2, \ldots, \omega_m)^T$ such that $\omega_j \in [0, 1], (j = 1, 2, \ldots, m)$ and $\sum_{j=1}^{m}\omega_j = 1$. The rating of alternative x_i $(i = 1, 2, \ldots, n)$ on attribute g_j $(j = 1, 2, \ldots, m)$ given by the expert e_k, $(k = 1, 2, \ldots, l)$ is HFE h_{ij}^k $(i = 1, 2, \ldots, n; j = 1, 2, \ldots, m; k = 1, 2, \ldots, l)$, where h_{ij}^k indicates the set of the degrees that the alternative x_i satisfies the attribute g_j given by the expert e_k. In the case where more than one experts provide the same value, then the value emerges only once in h_{ij}^k. Hence, a fuzzy multi-attribute group decision making problem can be concisely expressed in matrix format as follows: $D^k = (h_{ij}^k)_{n \times m}(k = 1, 2, \ldots, l)$.

In what follows, we propose an approach to hesitant fuzzy multiple attribute group decision making, which involves the following steps.

Step 1. Obtain the normalized hesitant fuzzy decision matrix. In general, attributes can be classified into two types: benefit attributes and cost attributes. In other words, the attribute set G can be divided into two subsets: G_1 and G_2, where G_t ($t = 1, 2$) is the subset of benefit attributes and cost attributes, respectively. Furthermore, $G_1 \cup G_2 = G$ and $G_1 \cap G_2 = \emptyset$, where $\emptyset$ is empty set. Since the m objectives may be measured in different ways, the decision matrix D_k needs to be normalized besides all the attributes $g_j(j = 1, 2, \ldots, m)$ are of the same type. In this paper we choose the following normalization formula to update the hesitant fuzzy decision matrices $D_k, k = 1, 2, \ldots, l$.

$$r_{ij}^k = \begin{cases} h_{ij}^k & j \in G_1 \\ \bar{h}_{ij}^k & j \in G_2 \end{cases} \tag{17}$$

where $\bar{h}_{ij}^k$ is the complement of h_{ij}^k. Hence, we obtain l normalized hesitant fuzzy decision matrices $R_k = (r_{ij}^k)_{n \times m}$, ($k = 1, 2, \ldots, l$), (see Table 1).

Table 1. The normalized hesitant fuzzy decision matrix R_k

	g_1	g_2	$\cdots$	g_m
x_1	r_{11}^k	r_{12}^k	$\cdots$	r_{1m}^k
x_2	r_{21}^k	r_{22}^k	$\cdots$	r_{2m}^k
$\vdots$	$\vdots$	$\vdots$	$\ddots$	$\vdots$
x_n	r_{n1}^k	r_{n2}^k	$\cdots$	r_{nm}^k

Step 2. Fuse all the individual decision opinion into a group opinion so as to make a final decision. Utilize the $\mathrm{HFHWA}^\varepsilon$ operator to aggregate all individual normalized decision matrices $R_k, k = 1, 2, \ldots, l$ into the collective normalized decision matrix $R = (r_{ij})_{n \times m}$, where

$$r_{ij} = \mathrm{HFHWA}_{\lambda, \mathbf{w}}^\varepsilon (r_{ij}^1, r_{ij}^2, \ldots, r_{ij}^l) =$$

$$\bigcup_{\dot{\gamma}_{ij}^{\sigma(1)} \in \dot{r}_{ij}^{\sigma(1)}, \dot{\gamma}_{ij}^{\sigma(2)} \in \dot{r}_{ij}^{\sigma(2)}, \ldots, \dot{\gamma}_{ij}^{\sigma(l)} \in \dot{r}_{ij}^{\sigma(l)}} \left\{ \frac{\prod\limits_{k=1}^{l} \left(1 + \dot{\gamma}_{ij}^{\sigma(k)}\right)^{w_k} - \prod\limits_{k=1}^{l} \left(1 - \dot{\gamma}_{ij}^{\sigma(k)}\right)^{w_k}}{\prod\limits_{k=1}^{l} \left(1 + \dot{\gamma}_{ij}^{\sigma(k)}\right)^{w_k} + \prod\limits_{k=1}^{l} \left(1 - \dot{\gamma}_{ij}^{\sigma(k)}\right)^{w_k}} \right\} \tag{18}$$

where $\dot{r}_{ij}^{\sigma(k)}$ is the kth largest of $\dot{r}_{ij}^t(\dot{r}_{ij}^t = (r_{ij}^t)^{l\lambda_t}, t = 1, 2, \ldots, l)$, and $\boldsymbol{\lambda} = (\lambda_1, \lambda_2, \ldots, \lambda_l)^T$ is the weight vector of $r_{ij}^t(t = 1, 2, \ldots, l)$ with $\lambda_t \in [0, 1]$ and $\sum_{t=1}^{l} \lambda_t = 1$, $\mathbf{w} = (w_1, w_2, \ldots, w_l)^T$ is the aggregation-associated vector such that $w_k \in [0, 1]$ and $\sum_{k=1}^{l} w_k = 1$.

Step 3. Compute the overall ratings of alternatives. Utilize the HFWA$^\varepsilon$ operator to aggregate all the rating values r_{ij} ($j = 1, 2, \ldots, m$) of the ith line and get the overall rating value r_i corresponding to the alternative x_i ($i = 1, 2, \ldots, n$), i.e.,

$$r_i = \mathrm{HFWA}_\omega^\varepsilon(r_{i1}, r_{i2}, \ldots, r_{im}), (i = 1, 2, \ldots, n). \tag{19}$$

Step 4. Rank the order of all alternatives. Utilize the method in Definition 4 to compute the scores of the overall rating values r_i ($i = 1, 2, \ldots, n$), and rank all the alternatives x_i ($i = 1, 2, \ldots, n$) in accordance with r_i ($i = 1, 2, \ldots, n$) in descending order, finally select the most desirable alternative(s) with the largest overall rating value.

Note that if we change the aggregation operators in hesitant fuzzy multi attribute group decision making, in general, the method returns different ranking orders and different winners. These hesitant fuzzy aggregation operators are based on different t-norms and their associated t-conorms. Each t-norm offers a kind of conjunction tool, and it represents the situation under a different perspective, i.e., some t-norms can be characterized as optimistic, whereas others as pessimistic in the aggregation process of decision making.

5 Conclusion

The Einstein operations on FSs typically give the same smooth approximations as the algebraic operations. In this paper, motivated by the extension principle of hesitant fuzzy sets, we have extended the Einstein operations on FSs to HFSs. We have developed some new hesitant fuzzy aggregation operators, including the HFWA$^\varepsilon$ operator, HFOWA$^\varepsilon$ operator, and HFHA$^\varepsilon$ operator. Then, we have applied the HFWA$^\varepsilon$ to the decision making problems with anonymity. It is worth point out that these aggregation operators are the same effective tools as the aggregation operators proposed by Xia and Xu [19], for aggregating hesitant fuzzy information. Correspondingly, using different hesitant fuzzy aggregation operators reflects the decision makers optimistic (or pessimistic) attitude. For example, the proposed HFWA$^\varepsilon$ operator shows the decision makers more pessimistic attitude than the HFWA operator proposed by Xu [19] in aggregation process.

Acknowledgements. This work is supported by Natural Science Foundation of Guangxi Province (2014jjAA10065), Scientific Research Foundation of Higher Education of Guangxi Province (KY2015YB050) and the 2014 Doctoral Scientific Research Foundation of Guangxi Normal University.

References

1. Yager, R.R., Kacprzyk, J.: The Ordered Weighted Averaging Operator: Theory and Applications. Kluwer, Boston (1997)
2. Calvo, T., Mayor, G., Mesiar, R.: Aggregation Operators: New Trends and Applications. Physica-Verlag, Heidelberg (2002)
3. Xu, Z.S., Da, Q.L.: An overview of operators for aggregating information. Int. J. Intell. Syst. **18**(9), 953–969 (2003)
4. Torra, V., Narukawa, Y.: Modeling Decisions: Information Fusion and Aggregation Operators. Springer, Berlin (2007)
5. Harsanyi, J.C.: Cardinal welfare, individualistic ethics, and interpersonal comparisons of utility. J. Polit. Econ. **63**(4), 309–321 (1955)
6. Yager, R.R.: On ordered weighted averaging aggregation operators in multicriteria decision-making. IEEE Trans. Syst. Man Cybern. Cybern. **18**(1), 183–190 (1988)
7. Torra, V.: The weighted OWA operator. Int. J. Intell. Syst. **12**(2), 153–166 (1997)
8. Zadeh, L.A.: Fuzzy sets. Inf. Control **8**(3), 338–353 (1965)
9. Atanassov, K.T.: Intuitionistic fuzzy sets. Fuzzy Sets Syst. **20**(1), 87–96 (1986)
10. Zadeh, L.A.: Outline of a new approach to analysis of complex systems and decision processes interval-valued fuzzy sets. IEEE Trans. Syst. Man Cybern. SMC **3**(1), 28–44 (1973)
11. Mizumoto, M., Tanaka, K.: Some properties of fuzzy sets of type 2. Inf. Control **31**(4), 312–340 (1976)
12. Dubois, D., Prade, H.M.: Fuzzy Sets and Systems: Theory and Applications. Academic Press, New York (1980)
13. Yager, R.R.: On the theory of bags. Int. J. Gen. Syst. **13**(1), 23–37 (1986)
14. Chakrabarty, K., Despi, I.: n^k-bags. Int. J. Intell. Syst. **22**(2), 223–236 (2007)
15. Torra, V.: Hesitant fuzzy sets. Int. J. Intell. Syst. **25**(6), 529–539 (2010)
16. Atanassov, K., Gargov, G.: Interval valued intuitionistic fuzzy sets. Fuzzy Sets Syst. **31**(3), 343–349 (1989)
17. Cornelis, C., Deschrijver, G., Kerre, E.E.: Implication in intuitionistic fuzzy and interval-valued fuzzy set theory: construction, classification, application. Int. J. Approx. Reason. **35**(1), 55–95 (2004)
18. Dubois, D., Gottwald, S., Hajek, P., Kacprzyk, J., Prade, H.: Terminological difficulties in fuzzy set theory - the case of "intuitionistic fuzzy sets". Fuzzy Sets Syst. **156**(3), 485–491 (2005)
19. Xia, M., Xu, Z.S.: Hesitant fuzzy information aggregation in decision making. Int. J. Approx. Reason. **52**(3), 395–407 (2011)
20. Torra, V., Narukawa, Y.: On hesitant fuzzy sets and decision. In: IEEE International Conference on Fuzzy Systems, FUZZ-IEEE 2009, pp. 1378–1382 (2009)
21. Xu, Z.S.: Hesitant Fuzzy Sets Theory. Springer International Publishing, Heidelberg (2014)
22. Xu, Z.S.: Intuitionistic fuzzy aggregation operators. IEEE Trans. Fuzzy Syst. **15**(6), 1179–1187 (2007)
23. Xu, Z.S., Yager, R.R.: Some geometric aggregation operators based on intuitionistic fuzzy sets. Int. J. Gen. Syst. **35**(4), 417–433 (2006)
24. Zhao, H., Xu, Z.S., Ni, M., Liu, S.: Generalized aggregation operators for intuitionistic fuzzy sets. Int. J. Intell. Syst. **25**(1), 1–30 (2010)
25. Schweizer, B., Sklar, A.: Probabilistic Metric Spaces. North Holland, New York (1983)
26. Hájek, P.: Metamathematics of Fuzzy Logic. Kluwer Academic Publishers, Dordrecht (1998)

Formal Models

Warsaw School of Economics

Using Ordinal Regression for Interactive Evolutionary Multiple Objective Optimization with Multiple Decision Makers

Miłosz Kadziński$^{(\boxtimes)}$ and Michał Tomczyk

Institute of Computing Science, Poznań University of Technology, Poznań, Poland
milosz.kadzinski@cs.put.poznan.pl, michal.k.tomczyk@gmail.com

Abstract. We present an interactive evolutionary multiple objective optimization (MOO) method incorporating preference information of several decision makers into the evolutionary search. It combines NSGA-II, a well-known evolutionary MOO method, with some interactive value-based approaches based on the principle of ordinal regression. We introduce several variants of the method distinguished by an elitist function indicating a comprehensive value that each solution represents to the group members. The experimental results confirm that all proposed approaches are able to focus the search on the group-preferred solutions, differing, however, with respect to both part of the Pareto front to which they converge as well as the convergence speed measured in terms of a change of utilitarian value of the returned solutions.

Keywords: Evolutionary multiple objective optimization · Interactive method · Group decision · Additive value function · Preference disaggregation · NEMO

1 Introduction

In Multiple Objective Optimization (MOO), several objectives are optimized simultaneously. As goals to be attained usually represent conflicting viewpoints, it is impossible to find a solution for which all objectives reach their individual optima [1]. Instead, we can identify a set of Pareto-optimal (non-dominated) solutions which are considered equivalent in case no additional information is available. A solution is called Pareto-optimal if none of the objective functions can be improved in value without deteriorating some of the other objective values. A possibly infinite set of such solutions forms a Pareto front in the objective space.

Traditionally, in MOO, two separate methodological streams have been developed: evolutionary and interactive ones [1]. On the one hand, the role of Evolutionary MOO (EMO) is to approximate the entire Pareto front. On the other hand, Interactive MOO (IMO) deals with identification of the most preferred solution. IMO techniques require participation of a Decision Maker (DM) who

© Springer International Publishing Switzerland 2015
B. Kamiński et al. (Eds.): GDN 2015, LNBIP 218, pp. 185–198, 2015.
DOI: 10.1007/978-3-319-19515-5_15

is expected to provide her/his subjective preference information. By developing a comprehensive model of such preferences, IMO makes the Pareto optimal solutions more comparable.

The recent trend in MOO consists in merging the interactive and evolutionary approaches (for a review, see [1,2]). This is achieved by integrating preference information into the EMO algorithms already during their optimization runs. The appealing effects of such integration consist in focusing the search on the area of the Pareto front which is most suitable to the DM and in speeding the convergence towards the most preferred region of the objective space.

The existing interactive EMO methods incorporate user preferences into evolutionary algorithms in different ways. In this paper, we focus on Necessary-preference-enhanced Evolutionary Multiobjective Optimizer (NEMO) [2] which combines the evolutionary method, called Non-dominated Sorting Genetic Algorithm II(NSGA-II) [3], with interactive ordinal regression approaches [4,5]. NEMO requires the DM to compare some pairs of solutions from the current population, and constructs a set of compatible general additive value functions. Then, it exploits these functions so that to guide the evolutionary search into regions of the Pareto front which are more desirable from the DM's point of view.

Our interest in NEMO comes from its favorable characteristics in terms of both preference information and preference model it employs. When it comes to pairwise comparisons, they represent indirect preference information whose elicitation is less demanding in terms of cognitive effort of the DM than direct specification of values for some preference model parameters. As far as general additive value functions are concerned, they can be computed efficiently with Linear Programming (LP), at the same time being flexible enough to handle preference information provided by the DMs with different value systems. Moreover, they do not require a pre-defined scaling of the objectives [2].

NEMO, alike other existing interactive EMO methods, was originally designed to deal with preferences expressed by a single DM. However, it is group decision making that is among the most important and frequently encountered processes within companies and organizations. When dealing with multiple DMs in the context of MOO, the main challenge consists in designing the algorithms so that they are able to focus the search on the group consensus solutions.

In this paper, we extend NEMO so that it is capable of dealing with MOO group decision problems. In particular, we propose a few variants of NEMO-GROUP that incorporate preference information of several DMs. Each of these variants is distinguished by a unique elitist function which indicates a comprehensive value that each solution represents to the group members. These values are employed to properly modify NSGA-II so that it promotes the group-preferred solutions in the optimization run. The use of proposed methods is illustrated by examples and experiments revealing the differences with respect to the regions of the Pareto front to which the methods converge, the quality of constructed solutions measured in terms of their utilitarian value, and the convergence speed.

The paper is organized as follows. The next section provides a brief reminder of ordinal regression methods. Section 3 describes the basic concepts of NSGA-II and NEMO. Section 4 presents different variants of our method, NEMO-GROUP. The experimental results are discussed in Sect. 5. The last section concludes.

2   Ordinal Regression

We are considering a multiple criteria decision problem where a set of solutions $A = \{a_1, a_2, \ldots\}$ is evaluated on a family $F = \{g_1, g_2, \ldots, g_n\}$ of n criteria. We assume, without loss of generality, that the smaller $g_j(a)$, $j = 1, \ldots, n$, the better solution a on criterion g_j, for all $a \in A$. Let G_j denote the value set of criterion g_j. Following [2], we assume that $G_j \subseteq \mathbb{R}$, and that the value space on each criterion g_j is bounded, such that $G_j = [\alpha_j, \beta_j]$, $\alpha_j < \beta_j$, where α_j and β_j are, respectively, the worst and the best evaluations. Consequently, $G = G_1 \times G_2 \times \ldots G_n$ represents the evaluation space, and each solution $a \in A$ is associated with an evaluation vector denoted by $g(a) = (g_1(a), g_2(a), \ldots, g_n(a)) \in G$.

Preference Model. To model the preferences provided by the DM and evaluate a set of solutions, we use an additive value function. It is defined on A as follows [4]:

$$U(a) = \sum_{j=1}^{n} u_j(g_j(a)) = \sum_{j=1}^{n} u_j(a), \tag{1}$$

where $u_j : G_j \rightarrow \mathbb{R}$, $j = 1, \ldots, n$, are subject to monotonicity and normalization constraints:

$$\left. \begin{array}{l} u_j(g_j(a)) \geq u_j(g_j(b)), \text{ if } g_j(a) < g_j(b), \\ u_j(\alpha_j) = 0, \ \sum_{j=1}^{n} u_j(\beta_j) = 1. \end{array} \right\} E^{\mathcal{U}} \tag{2}$$

Group Preference Model. We consider a set of DMs (let us denote it by $\mathcal{D} = \{DM_1, \ldots, DM_k, \ldots, DM_s\}$, where s is the number of DMs) cooperating to find a subset of the best consensus solutions. We assume that each DM plays the same role in the committee, so we do not differentiate their weights. Each $DM_k \in \mathcal{D}$ evaluates solutions with her/his individual "true" value function U_k^{TRUE}. The collective utilitarian preference model combines these evaluations into a comprehensive value that solution $a \in A$ represents to the whole committee:

$$U_{\mathcal{D}}(a) = 1/s \sum_{k=1}^{s} U_k^{TRUE}(a). \tag{3}$$

Preference Information. Each $DM_k \in \mathcal{D}$ offers individual preference information which is a set B_k of pairwise comparisons of some reference solutions in A_k^{REF}. In the considered setting, either each DM is allowed to choose the solutions (s)he wishes to compare on her/his own or the pairs to be compared by each DM are drawn randomly. Thus, in general, $A_k^{REF} \neq A_l^{REF}$ for $DM_k, DM_l \in \mathcal{D}$. The comparison of a pair $(a^*, b^*) \in B_k \subseteq A_k^{REF} \times A_k^{REF}$ provided by DM_k

states the strict preference, weak preference, or indifference. These relations are denoted by, $a^* \succ_k b^*$, $a^* \succsim_k b^*$, and $a^* \sim_k b^*$, respectively.

Let each pairwise comparison from B_k be denoted by B_k^t, $t = 1, \ldots, p_k$, where p_k is the number of comparisons contained in B_k. The set of constraints E^k given below translates such a reference pre-order provided by DM_k to a value function:

$$\left. \begin{array}{l} U(a^*) \geq U(b^*) + \varepsilon, \quad \text{for } B_k^t = (a^* \succ_k b^*) \\ U(a^*) \geq U(b^*), \quad \text{for } B_k^t = (a^* \succsim_k b^*) \\ U(a^*) = U(b^*), \quad \text{for } B_k^t = (a^* \sim_k b^*) \end{array} \right\} \text{ for } t = 1, \ldots, p_k \left. \begin{array}{l} \\ \\ \end{array} \right\} E^k \qquad (4)$$

where ε is an arbitrarily small positive value.

The pairwise comparisons provided by each $DM_k \in \mathcal{D}$ form the input data for the ordinal regression [4] that finds the whole set of value functions $\mathcal{U}_k$ being able to reconstruct these judgments. It is defined by a set of constraints $E^{\mathcal{U}_k} = E^{\mathcal{U}} \cup E^k$. The set of value functions $\mathcal{U}_\mathcal{D}$ compatible with the pairwise comparisons of all DMs is defined with $E^{\mathcal{U}_\mathcal{D}} = E^{\mathcal{U}} \cup E^k$, $k = 1, \ldots, s$. Note that $\mathcal{U}_\mathcal{D}$ corresponds to the intersection of sets of compatible value functions for all DMs in $\mathcal{D}$.

If $\varepsilon^* = max\ \varepsilon$, s.t. $E^{\mathcal{U}_k}$ $(E^{\mathcal{U}_\mathcal{D}})$, is greater than 0 and $E^{\mathcal{U}_k}$ $(E^{\mathcal{U}_\mathcal{D}})$ is feasible, the set of compatible value functions $\mathcal{U}_k$ $(\mathcal{U}_\mathcal{D})$ is non-empty. Otherwise, the provided preference information is inconsistent with the assumed preference model, which means that there is no value function that would reproduce the pairwise comparisons provided by DM_k (if $\mathcal{U}_k = \emptyset$) or all DMs (if $\mathcal{U}_\mathcal{D} = \emptyset$).

Representative Value Function. There is usually more than one compatible value function. The issue of selecting a single representative function has been discussed in detail in [6]. In this paper, we will use the most discriminant value function U_k^R $(U_\mathcal{D}^R)$, which is obtained by maximizing ε, subject to $E^{\mathcal{U}_k}$ $(E^{\mathcal{U}_\mathcal{D}})$. It discriminates comprehensive values of reference solutions related to the preference in the DM's (DMs') partial ranking.

Dealing with Incompatibility of Preference Information. In case of incompatibility, there is no value function compatible with the preference information provided by all DMs. Treating this problem, we will maximize a minimal number of pairwise comparisons of any DM which are consistent, being representable by a single additive value function [7]. It can be achieved by solving the following Mixed Integer Linear Programming (MILP) problem:

$$Maximize\ v, \text{ s.t.} \qquad (5)$$

$$\left. \begin{array}{l} \text{for } k = 1, \ldots, s, \text{ for } t = 1, \ldots, |B_k| : \\ [1 - v_t^k(a^*, b^*)] + U(a^*) \geq U(b^*) + \varepsilon, \\ v_t^k(a^*, b^*) \in \{0, 1\}, \\ \text{for } k = 1, \ldots, s : \\ v \leq \sum_{t=1}^{|B_k|} v_t^k(a^*, b^*), \\ E^{\mathcal{U}}. \end{array} \right\} E'^{\mathcal{U}_\mathcal{D}}$$

Apart from providing the minimal number of non-contradictory pairwise comparisons of all DMs (v^*), the solution of the above problem indicates which pairwise comparisons can be reproduced together by an additive value function (they are distinguished with $v_t^{k,*} = 1$). If for all DMs the numbers thereof are imbalanced, we arbitrarily choose the last v^* non-contradictory pairwise comparisons provided by each DM, so that none of them is favored. Then, we determine a representative (most discriminant) value function compatible with thus selected subset of holistic judgments.

3   Reminder on NSGA-II and NEMO

The role of genetic algorithms is to estimate meta-heuristically the Pareto fronts in MOO problems. In particular, NSGA-II [3] incorporates a fast non-dominated sorting algorithm to identify Pareto optimal solutions, and a diversity preservation mechanism for maintaining a well-spread Pareto front. It starts with the initialization of a random parent population P_0 of size N. Then, the offspring Q_0 of the same size is created using the usual selection, recombination and mutation operators. Further, the parents and their offspring ($R_t = P_t \cup Q_t$) are combined to obtain a population of size $2N$.

The new population (P_{t+1}) is filled with the best Pareto fronts from R_t (first $\mathcal{F}_1$ (i.e., non-dominated solutions), then $\mathcal{F}_2$ (i.e., solutions dominated only by some solutions from $\mathcal{F}_1$, etc.), until the size of the next front ($\mathcal{F}_l$) is larger than the number of free slots in P_{t+1}. To have exactly N members in the new population and to maintain diversity, the front $\mathcal{F}_l$ is ordered using the crowded distance comparison operator ($\succ_n$). The total crowding distance of a solution is the sum of its individual objectives' distances which are computed as the absolute normalized differences between the solution and its closest neighbors. Then, the $N - |P_{t+1}|$ solutions with the greatest crowding distance are added to P_{t+1}. The process is iterated until a stopping criterion is met.

NEMO [2] is an interactive evolutionary hybrid which combines NSGA-II with IMO approaches based on the principle of ordinal regression. Alike NSGA-II, NEMO uses the Pareto fronts as a primary criterion to rank individuals. The major innovation consists in asking the DM at regular intervals to compare a single pair of solutions (note that a set A is composed of solutions from the current population). The accumulated preference information is used to select a representative additive general value function U^R [6]. Then, the solutions within each Pareto front are ranked using a representative value comparison operator $\succ_{U^R}$ (the greater $U^R(a)$, the better the solution a). Algorithm 1 describes the use of NEMO for the t-th generation.

4   Using Ordinal Regression for Interactive Evolutionary Multiple Objective Optimization Group Decision

In this section, we propose a few approaches for interactive evolutionary multiple objective optimization incorporating preference information of several DMs.

Algorithm 1. A single NEMO iteration for constructing the t-th generation

$R_t = P_t \cup Q_t$
if Time to ask the DM {not conducted in NSGA-II} **then**
 Elicit DM's preferences {present to the DM a pair of non-dominated solutions and ask for a preference comparison}
 Determine the representative value function U^R
end if
$\mathcal{F} = \texttt{fast-non-dominated-sort}(\mathtt{R_t})$
Within each non-dominance rank, sort individuals according to representative value function {U^R replaces the crowding distance in NSGA-II}
$P_{t+1} = \emptyset$ and $i = 1$
while $|P_{t+1}| + |\mathcal{F}_i| \leq N$ **do**
 $\texttt{representative-value}(\mathcal{F}_\mathtt{i})$ {instead of $\texttt{crowding-distance-assignment}(\mathcal{F}_\mathtt{i})$}
 $P_{t+1} = P_{t+1} \cup \mathcal{F}_i$
 $i = i + 1$
end while
$\text{Sort}(\mathcal{F}_i, \succ_{U^R})$ {instead of $\text{Sort}(\mathcal{F}_i, \succ_n)$ in NSGA-II}
$P_{t+1} = P_{t+1} \cup \mathcal{F}_i[1 : (N - |P_{t+1}|)]$
$Q_{t+1} = \texttt{make-new-pop}(\mathtt{P_{t+1}})$

$t = t + 1$

Each of these approaches extends NEMO, originally designed for dealing with preferences of just a single DM. The scheme of this extension is common for all proposed variants with each DM being asked at regular intervals to compare a pair of randomly drawn solutions, and using the Pareto ranking as a primary criterion to rank individuals. The major differences concern a secondary criterion, and treating the population as a whole or evolving its parts individually for each DM. In particular, we propose the following variants of the **NEMO-GROUP** method:

- **NEMO-G1** determines a representative value function U_k^R for each $DM_k \in \mathcal{D}$ based on her/his pairwise comparisons only, and then ranks subsets of Pareto fronts according to their comprehensive values $U_{G1}^R(a) = \sum_{k=1}^{s} U_k^R(a)$;
- **NEMO-G2** determines U_k^R for each $DM_k \in \mathcal{D}$, and ranks subsets of Pareto fronts using $U_{G2}^R(a) = min_{k=1,\ldots,s} U_k^R(a)$;
- **NEMO-G3** maximizes the minimal number of pairwise comparisons of any $DM_k \in \mathcal{D}$ that can be represented together by an additive value function, determines the representative value function $U_{G3}^R(a)$ compatible with the consistent pairwise comparisons of all DMs (equal number of comparisons provided by each DM), and uses it to rank subsets of Pareto fronts;
- **NEMO-G4** divides a population into $1/s$ equal sub-populations, one for each $DM_k \in \mathcal{D}$, and evolves them separately using a representative value function $U_k^R(a)$ of each $DM_k \in \mathcal{D}$; a final population is obtained by combining together sub-populations of all DMs.

5 Experimental Results

To study the performance of different variants of NEMO-GROUP, we use ZDT1 and DTLZ2 with two (2D) and five (5D) objectives, respectively. We use artificial DMs who apply a pre-defined individual value functions for comparing pairs of solution whenever preference elicitation is conducted. Precisely, we use the linear functions, so the goal of each $DM_k \in \mathcal{D}$ is to minimize $U_k^{TRUE-LIN}(a) = \sum_{j=1}^{n} w_j^k g_j(a)$, where w_j^k, $j = 1, \ldots, n$, are weights of the n cost-type objectives. Thus, the whole group aims at minimizing $U_\mathcal{D}(a) = 1/s \sum_{k=1}^{s} U_k^{TRUE-LIN}(a)$. All these individual functions are unknown to the NEMO-GROUP algorithms, which instead use an additive value function defined in Sect. 2 as an internal preference model.

In our tests, we use a real-valued representation. We generate offspring by simulated binary crossover with probability of 0.9 and $\kappa_c = 1$, whereas mating selection is performed by tournament selection. We also apply Gaussian mutation with probability of 1/30. The population size is set to 60, and all methods are run for 500 generations.

5.1 Illustrative Examples

In this subsection, we use ZDT1-2D for an initial graphical comparison of the proposed approaches. We assume that preference elicitation is performed every 10 generation, and that DMs' value functions are parameterized with the following weights (w_1^k, w_2^k) for $k = 1, 2, 3, 4$: $DM_1 - (0.7, 0.3)$, $DM_2 - (0.4, 0.6)$, $DM_3 - (0.5, 0.5)$, and $DM_4 - (0.3, 0.7)$. Intuitively, the greater the weight, the more important it is to minimize the respective objective.

Figure 1 shows the results for different variants of NEMO-GROUP for three DMs ($DM_1 - DM_3$). To demonstrate the convergence to the Pareto front, for NEMO-G1, NEMO-G4, and NSGA-II, we depict populations obtained after 100, 300, and 500 generations. For clarity, for NEMO-G2 and NEMO-G3, we provide results only after 500 generations.

As can be seen, NSGA-II approximates the whole Pareto front, whereas all variants of NEMO-GROUP are focused on the solutions preferred to the DMs. For algorithms using an aggregated group value function as a secondary criterion (i.e., NEMO-G1, NEMO-G2, and NEMO-G3), the final population is narrowed to a single small part of the Pareto front composed of solutions which can be seen as the best compromise for all DMs. However, each of these approaches convergences to a slightly different region of the Pareto front (see the top-right part of Fig. 1). Typically, the populations constructed by NEMO-G1 and NEMO-G2 are closer to each other when compared with the population constructed by NEMO-G3. Indeed, NEMO-G1 and NEMO-G2 employ the group value functions aggregating the same DMs' individual representative value functions, though in a slightly different way, whereas NEMO-G3 constructs a single representative value function which is common for all DMs. Finally, since NEMO-G4 evolves a separate sub-population for each DM, the final population in our illustrative study

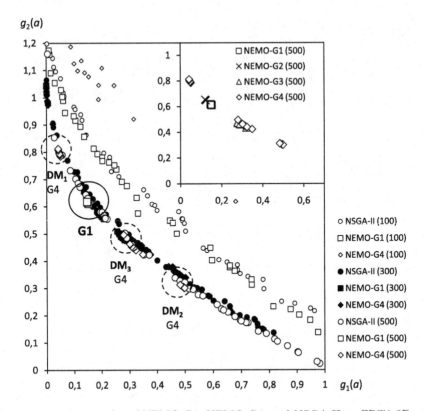

Fig. 1. Exemplary results of NEMO-G1, NEMO-G4, and NSGA-II on ZDT1-2D with three decision makers after 100, 300, and 500 generations. The results for NEMO-G2 and NEMO-G3 after 500 generation are provided in the top-right corner.

is composed of three clearly disjoint sets of solutions. These sub-populations contain solutions which are most preferred to a particular DM.

Such characteristic performance of different variants of NEMO-GROUP is confirmed by analogous results in the context of four DMs ($DM_1 - DM_4$; see Fig. 2). Obviously, NEMO-G1, NEMO-G2, and NEMO-G3 converge to different regions of the Pareto front than in case of three DMs, whereas NEMO-G4 approximates fourth sub-population with solutions more oriented to minimization of the second objective (as indicated by the weights for DM_4 - $(0.3, 0.7)$).

5.2 Convergence in Terms of a Utilitarian Value of the Solutions

In this subsection, we study the evolution of a utilitarian value for the best-of-population and average-in-population solutions in successive generations. These convergence factors permit to assess the performance of different variants of NEMO-GROUP from the point of view of a whole group of DMs. On the one hand, the best solution in the returned population may be perceived as a default outcome of the method that is most likely to be accepted by the DMs. On the

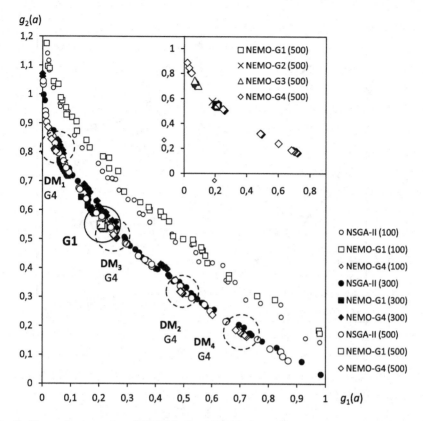

Fig. 2. Exemplary results of NEMO-G1, NEMO-G4, and NSGA-II on ZDT1-2D with four decision makers after 100, 300, and 500 generations. The results for NEMO-G2 and NEMO-G3 after 500 generations are provided in the top-right corner.

other hand, an average quality of the individuals contained in the population reveals if the search has been appropriately focused on the group consensus solutions [2].

All results presented in this section have been averaged over 100 independent runs, each for different weight vectors for the DMs' value functions. Figures 3 and 4 present the convergence plots for, respectively, value of the best solution and average value of all solutions in the population for ZDT1-2D with three DMs. The top-right parts of the figures depict the convergence plots for NEMO-G1 and NEMO-G4 starting from the first generation, whereas the main parts of the figures demonstrate the convergence between 100 and 500 generation for all considered algorithms. In this way, we can better illustrate when different approaches start to converge towards the Pareto front, what is their convergence speed measured in terms of a change of a utilitarian value, and what is value of the solution(s) at which their performance stabilizes.

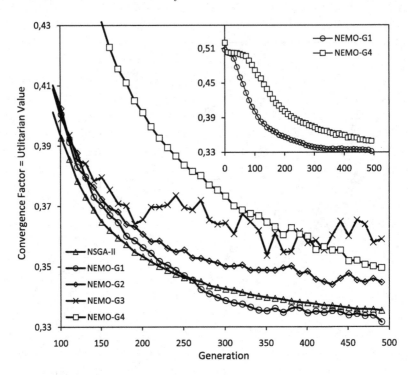

Fig. 3. Utilitarian value of the best-of-population solution in successive generations of NEMO-G1, NEMO-G2, NEMO-G3, and NEMO-G4 for three decision makers, and NSGA-II applied to ZDT1-2D (results averaged over 100 runs).

To compare the performance of proposed approaches for various benchmark problems and different numbers of DMs, we focus on the precise measures derived from the convergence plots. In this section, when presenting the experimental results in a tabular form, the text in bold and italics indicates the best performing algorithm across all 100 optimization runs. Additionally, we indicate in bold these approaches whose distance from the best performer proved to be statistically insignificant according to a Mann-Whitney-U test with 5 % significance level.

First, we refer to the minimal values obtained throughout the 500 generations (see Table 1). When it comes to the best utilitarian solution obtained during the optimization run (i.e., the minimal best-of-population value) for ZDT1-2D, NEMO-G1 and NEMO-G2 are most advantageous, while NEMO-G3 and NEMO-G4 are significantly worse than other approaches. Surprisingly, the value of the best solution discovered by NSGA-II is only slightly worse than that of the best performing variants of NEMO-GROUP. For DTLZ2-5D, NEMO-G4 and NEMO-G1 outperform other algorithms (except for the case with 2 DMs where the differences are statistically insignificant). Moreover, when moving to five dimensions, the best solution discovered by NSGA-II is much worse than in case of all NEMO-GROUP variants.

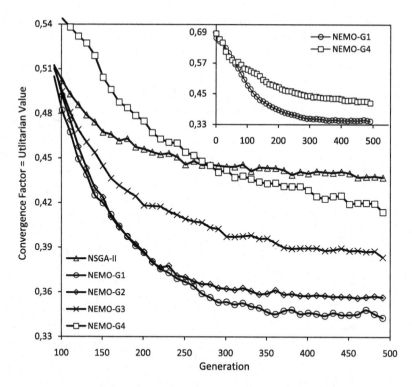

Fig. 4. Utilitarian value of the average-in-population solution in successive generations of NEMO-G1, NEMO-G2, NEMO-G3, and NEMO-G4 for three decision makers, and NSGA-II applied to ZDT1-2D (results averaged over 100 runs).

When comprehensively judging the returned population of solutions being most favorable from the point of view of the whole group (i.e., the best average-in-population value), NEMO-G1 performs the best for all considered problems and numbers of DMs except DTLZ2-5D and 3 DMs. However, when compared with NEMO-G2 or NEMO-G3, its advantage is statistically insignificant for some tested configurations. Furthermore, among all variants of NEMO-GROUP, NEMO-G4 proves to be the worst for all considered settings except DTLZ2-5D and 5 DMs. Finally, NSGA-II is significantly worse than NEMO-G1, NEMO-G2, and NEMO-G3, which construct only solutions which are relevant from the point of view of the whole group. This difference is particularly visible for a higher dimensional DTLZ2-5D.

As the other set of measures derived from the convergence plots we consider the average group utilitarian values observed throughout 500 generations. In this way, we are able to judge the overall performance of the algorithms from the point of view of either the best solution or a complete population returned after each generation. Since the value to which the algorithms converge highly affects the overall performance, the conclusions about the best and worst performing

Table 1. Minimal (best) value throughout 500 generations (results averaged over 100 runs; SD = standard deviation).

Approach	2 DMs $U_\mathcal{D}$	SD	3 DMs $U_\mathcal{D}$	SD	4 DMs $U_\mathcal{D}$	SD	5 DMs $U_\mathcal{D}$	SD
			Best-of-population value for ZDT1-2D					
NSGA-II	0.3009	0.0978	0.3352	0.0662	0.3574	0.0491	0.3616	0.0435
NEMO-G1	*0.2971*	0.1006	*0.3324*	0.0703	*0.3572*	0.0569	*0.3582*	0.0455
NEMO-G2	**0.2994**	0.1008	**0.3451**	0.0742	0.3661	0.0559	0.3691	0.0484
NEMO-G3	0.3121	0.1126	0.3567	0.0888	0.3892	0.0691	0.3991	0.0701
NEMO-G4	0.3052	0.1059	0.3467	0.0753	0.3816	0.0600	0.3979	0.0574
			Average-in-population value for ZDT1-2D					
NSGA-II	0.4372	0.0189	0.4372	0.0154	0.4392	0.0107	0.4370	0.0109
NEMO-G1	*0.3130*	0.1025	*0.3410*	0.0709	*0.3666*	0.0543	*0.3686*	0.0456
NEMO-G2	**0.3148**	0.0950	0.3551	0.0752	0.3802	0.0527	0.3816	0.0519
NEMO-G3	0.3321	0.1118	0.3838	0.1044	0.4096	0.0788	0.4232	0.0782
NEMO-G4	0.3528	0.1124	0.4150	0.0842	0.4490	0.0579	0.4659	0.0500
			Best-of-population value for DTLZ2-5D					
NSGA-II	0.1392	0.0531	0.1547	0.0552	0.1619	0.0451	0.1702	0.0463
NEMO-G1	**0.0943**	0.0528	**0.1097**	0.0520	**0.1279**	0.0614	**0.1239**	0.0442
NEMO-G2	**0.0919**	0.0608	0.1199	0.0731	0.1338	0.0664	0.1349	0.0543
NEMO-G3	**0.0954**	0.0600	0.1163	0.0546	0.1282	0.0538	0.1516	0.0693
NEMO-G4	*0.0887*	0.0533	*0.1045*	0.0458	*0.1102*	0.0366	*0.1194*	0.0388
			Average-in-population value for DTLZ2-5D					
NSGA-II	0.3932	0.0375	0.3906	0.0268	0.3915	0.0258	0.3872	0.0288
NEMO-G1	*0.1259*	0.0742	**0.1603**	0.0769	*0.1646*	0.0734	*0.1607*	0.0539
NEMO-G2	*0.1259*	0.0754	**0.1548**	0.0803	**0.1723**	0.0748	**0.1749**	0.0699
NEMO-G3	0.1346	0.0743	*0.1509*	0.0657	**0.1721**	0.0699	0.1951	0.0747
NEMO-G4	0.1411	0.0648	0.1683	0.0543	0.1769	0.0411	0.1863	0.0423

algorithms are analogous to the case of considering only the best results. The important differences are the following:

- NEMO-G4 performs poorly for ZDT1-2D, because it starts to converge later than other algorithms.
- NSGA-II and NEMO-G3 are even less advantageous for DTLZ2-5D than in case of considering the best results only, because their convergence curves are more erratic than the others, deteriorating several times in the phase when performance of other algorithms stabilizes or still slightly improves.
- NEMO-G2 shows clear advantages with respect to NEMO-G1 in terms of average-in-population value, because in the initial generations the performance of both algorithms is very similar and NEMO-G1 derives its comprehensive superiority from exploiting more favorable search directions only after 250 generation (Table 2).

Table 2. Average value throughout 500 generations (results averaged over 100 runs; SD = standard deviation).

	2 DMs		3 DMs		4 DMs		5 DMs	
	$U_{\mathcal{D}}$	SD	$U_{\mathcal{D}}$	SD	$U_{\mathcal{D}}$	SD	$U_{\mathcal{D}}$	SD
Approach			Best-of-population value for ZDT1-2D					
NSGA-II	0.3370	0.0997	*0.3693*	0.0706	*0.3910*	0.0530	*0.3951*	0.0479
NEMO-G1	*0.3367*	0.1032	**0.3701**	0.0733	**0.3913**	0.0558	**0.3959**	0.0495
NEMO-G2	**0.3368**	0.1029	**0.3772**	0.0765	**0.3991**	0.0578	**0.4027**	0.0513
NEMO-G3	0.3459	0.1081	0.3858	0.0799	0.4108	0.0513	0.4189	0.0453
NEMO-G4	0.3608	0.1079	0.4069	0.0840	0.4399	0.0677	0.4566	0.0675
			Average-in-population value for ZDT1-2D					
NSGA-II	0.4780	0.0320	0.4779	0.0239	0.4783	0.0182	0.4785	0.0171
NEMO-G1	0.3882	0.0897	*0.4142*	0.0634	*0.4323*	0.0501	*0.4359*	0.0436
NEMO-G2	*0.3846*	0.0866	**0.4207**	0.0696	**0.4394**	0.0576	**0.4409**	0.0521
NEMO-G3	0.3995	0.0973	0.4455	0.0815	0.4649	0.0528	0.4773	0.0426
NEMO-G4	0.4315	0.1016	0.4832	0.0778	0.5097	0.0570	0.5233	0.0551
			Best-of-population value for DTLZ2-5D					
NSGA-II	0.1357	0.0499	0.1551	0.0418	0.1625	0.0372	0.1695	0.0390
NEMO-G1	*0.0966*	0.0477	*0.1128*	0.0444	**0.1263**	0.0416	*0.1271*	0.0368
NEMO-G2	**0.0974**	0.0511	0.1173	0.0448	**0.1269**	0.0405	0.1325	0.0375
NEMO-G3	**0.0984**	0.0488	0.1235	0.0413	0.1343	0.0397	0.1468	0.0432
NEMO-G4	**0.0995**	0.0503	**0.1140**	0.0416	*0.1227*	0.0335	0.1331	0.0365
			Average-in-population value for DTLZ2-5D					
NSGA-II	0.3935	0.0177	0.3940	0.0135	0.3940	0.0128	0.3944	0.0118
NEMO-G1	*0.1449*	0.0503	*0.1640*	0.0484	*0.1755*	0.0406	*0.1741*	0.0343
NEMO-G2	**0.1460**	0.0517	**0.1682**	0.0444	**0.1799**	0.0407	0.1828	0.0389
NEMO-G3	0.1573	0.0527	0.1883	0.0477	0.2050	0.0482	0.2195	0.0504
NEMO-G4	0.1681	0.0580	0.1818	0.0480	0.1938	0.0399	0.2016	0.0415

6 Conclusions and Future Research

In this paper, we presented an interactive evolutionary multiple objective optimization method incorporating preference information of several decision makers into the evolutionary search. After recalling NEMO designed for interaction with a single decision maker, we extended it in different ways to group decision. In all proposed variants of NEMO-GROUP, the user interaction is based on ordinal regression. The main differences between these variants concern construction of an elitist function indicating values that solutions represent to the group members.

The experimental results confirm that the proposed approaches are able to focus the search on the group-preferred solutions. Nevertheless, they indicate that the proposed approaches differ with respect to both part of the Pareto front to which they converge as well as the convergence speed measured in terms of a change of a utilitarian value of the returned solutions.

We envisage the following developments:

- employing different procedures for selecting a single representative value function within ordinal regression (e.g., optimizing the misranking error ε in case of incompatibility) and using methods derived from Robust Ordinal Regression [5] for preserving elitism;
- verifying how different values of elicitation interval influence the convergence speed of the algorithms;
- testing the proposed approaches more thoroughly on a set of benchmark problems with different dimensions, various forms of the assumed DMs' true value functions, and different group value functions (e.g., egalitarian or elitist ones).

Acknowledgments. The authors acknowledge financial support from the Poznan University of Technology, grant DS-MLODA KADRA (2015).

References

1. Poles, S., Vassileva, M., Sasaki, D.: Multiobjective Optimization Software. In: Branke, J., Deb, K., Miettinen, K., Słowiński, R. (eds.) Multiobjective Optimization. LNCS, vol. 5252, pp. 329–348. Springer, Heidelberg (2008)
2. Branke, J., Greco, S., Słowiński, R., Zielniewicz, P.: Learning value functions in interactive evolutionary multiobjective optimization. IEEE Trans. Evol. Comput. **19**(1), 88–102 (2015)
3. Deb, K., Agrawal, S., Pratap, A., Meyarivan, T.: A fast and elitist multi-objective genetic algorithm: NSGA-II. IEEE Trans. Evol. Comput. **6**(2), 182–197 (2002)
4. Jacquet-Lagrèze, E., Siskos, Y.: Preference disaggregation: 20 years of MCDA experience. Eur. J. Oper. Res. **130**(2), 233–245 (2001)
5. Corrente, S., Greco, S., Kadziński, M., Słowiński, R.: Robust ordinal regression in preference learning and ranking. Mach. Learn. **93**(2–3), 381–422 (2013)
6. Kadziński, M., Greco, S., Słowiński, R.: Selection of a representative value function in robust multiple criteria ranking and choice. Eur. J. Oper. Res. **217**(3), 541–553 (2012)
7. Kadziński, M., Greco, S., Słowiński, R.: Selection of a representative value function for robust ordinal regression in group decision making. Group Decis. Negot. **22**(3), 429–462 (2013)

Fiscal-Monetary Game Analyzed with Use of a Dynamic Macroeconomic Model

Lech Kruś[✉] and Irena Woroniecka-Leciejewicz

System Research Institute, Polish Academy of Sciences,
Newelska 6, 01-447 Warsaw, Poland
{Lech.Krus,Irena.Woroniecka}@ibspan.waw.pl

Abstract. The paper deals with the fiscal-monetary game. In the game the fiscal and the monetary authorities take decisions on the choice of the optimum strategy from the point of view of realization of their respective economic objectives. A macroeconomic model has been constructed and used to represent the interrelations between, on the one hand, the instruments of fiscal policy and of the monetary policy, and, on the other hand – the economic effects resulting from their application. The best response strategies of the authorities and the Nash equilibrium state are analyzed. The simulation results obtained indicate that in a general case the Nash equilibrium is not Pareto optimal. It means that the policies should be coordinated and that respective negotiations leading to a Pareto-optimal consensus are needed.

Keywords: Macroeconomic modeling · Fiscal-monetary game · Nash equilibrium · Bargaining problem · MCDM · Negotiations

1 Introduction

The present paper concerns the problem of choice of fiscal and monetary policies in the context of mutual decision conditioning between the fiscal authority (the government) and the monetary authority (the central bank). Each policy is characterized by the definite degrees of restrictiveness and expansiveness. A noncooperative game is formulated in which the fiscal and monetary authorities play roles of players. Each authority tries to obtain his respective economic objective: a desired value of the GDP dynamics in the case of the fiscal authority, and a desired value of the inflation in the case of the monetary authority. Instruments of the policies, considered as strategies in the game include: the budget deficit in relation to the GDP value, in the case of the fiscal authority, and the real interest rate, in the case of the monetary authority.

A macroeconomic model is proposed and used in the game to express influences of the policies' instruments on the game outcomes. The model describes the business cycles mechanism and allows to analyze the state of the economy in time. Using the model the GDP growth and inflation can be derived as dependent on the instruments of the policies.

The fiscal-monetary game is analyzed under assumption of the independence of the fiscal and the monetary authorities. In many situations the Nash equilibrium in the game is not Pareto optimal. Therefore the multicriteria optimization is applied to derive Pareto optimal strategies with respect to the objectives of the authorities. Such

© Springer International Publishing Switzerland 2015
B. Kamiński et al. (Eds.): GDN 2015, LNBIP 218, pp. 199–208, 2015.
DOI: 10.1007/978-3-319-19515-5_16

strategies can be applied and resulting outcomes obtained under respective coordination of the policies.

The paper is constituted as follows. The macroeconomic model is presented in
Sect. 2 including descriptions of the product and the money markets as well as
interrelations between them. Section 3 includes formulation of the fiscal monetary
game and analysis of the game properties. The analysis is illustrated by selected results
of simulations. Conclusions and references finish (close) the paper.

The references include papers dealing with: analysis of monetary and fiscal policy
interactions, independence and coordination of the policies, policy games [1–3, 14, 20,
21], macroeconomic modeling [4–7, 16], methods of decision support in multicriteria
bargaining [8, 9], the reference point method of multicriteria optimization [17–19].
This paper continues the discussion presented by Nordhaus [14] related to independence versus coordination of the fiscal and monetary policies.

2   The Macroeconomic Model

A macroeconomic dynamic model is used in this research. It describes a mechanism of
the business cycles and allows for analysis of economic situation in time. On the other
hand it enables analysis of the monetary and fiscal policies, namely instruments of the
policies: the real interest rate, the deficit of the government budget in relations to GDP,
as well as their influence on the economy, especially on the GDP growth and on
inflation. The model and initial results of simulations have been published by
Woroniecka-Leciejewicz [22].

The model includes two modules describing: a product market and a money
market. The economy is assumed to be in the initial equilibrium state on the product
market as well as on the money market. The product market is described on the basis of
the multiplier model [7] and multiplier- accelerator model of business cycles [6, 16].
Undertaken assumptions and model relations are briefly presented below.

2.1   The Product Market

The production in the period t is calculated on the demand-side as a sum of the
consumption, investments and government expenditures:

$$Y(t) = C(t) + I(t) + G(t), \tag{1}$$

where: Y is the real production, C - the real consumption, I – the real investments,
G - government expenditures (spending on goods and services), t – time period.

It is measured as the real GDP value. The foreign exchange is not included in the
model.

The real consumption C in the period t:

$$C(t) = \beta(r)(CA + c(1 - t_n)Y(t - 1)), \tag{2}$$

where: CA denotes the autonomous consumption, c – a marginal propensity to consume, t_n – the net tax rate, r – the real interest rate, $r*$ - the neutral interest rate, α, λ are parameters, $\beta(r)$ is a coefficient expressing influence of the real interest rate on the consumption:

$$\beta(r) = 1 - \lambda\alpha(r - r*), 0 < \lambda < 1, \alpha > 0. \tag{3}$$

The real investments I in the period t:

$$I(t) = \alpha(r)(IA + k\Delta Y(t - 1)), \tag{4}$$

where:

$$a(r) = 1 - \alpha(r - r*), \alpha > 0, \tag{5}$$

IA denotes the autonomous investments, k – a capital-output ratio (usually, k is assumed to be in $(0,1)$ interval), $\Delta Y(t)$- the increase of the production in the period t, $\Delta Y(t) = Y(t) - Y(t - 1)$, (the increase of values for other model variables is denoted in an analogic way), $a(r)$ – a coefficient expressing influence of the interest rate on the investments.

The net tax revenue of the government budget (real) T_n in the period t:

$$T_n(t) = t_n Y(t), \tag{6}$$

where $t_n(t)$ is a net tax rate.

The government budget balance (real) BS in the period t:

$$BS(t) = T_n(t) - G(t), \tag{7}$$

where the net taxes value $T_n(t)$ equals revenues from the taxes minus the transfer payments, G are the expenditures including the government spending on goods and services without the transfer payments. In an analogical way the fiscal revenues take only into account the net tax revenues like in Eq. (6).

2.2 The Money Market

The equilibrium on the money market means that the money supply M is in a balance with the transactional demand for the money. The demand is determined by the level of prices and by the real production but depends also on velocity of money circulation.

$$M(t) = \frac{P(t)Y(t)}{v(t)}, \tag{8}$$

where: M denotes the money supply, P – an index of prices, v – velocity of money circulation.

The inflation p in the period t:

$$p(t) = \frac{\Delta P(t)}{P(t-1)} = \frac{\Delta M(t)}{M(t-1)} - \frac{\Delta Y(t)}{Y(t-1)} + \frac{\Delta v(t)}{v(t-1)}. \tag{9}$$

The inflation has been described in the model according to the monetarist theory by Milton Friedman [4, 5]. It is determined by an excessive growth rate of the money supply and by an increase of velocity of money.

It is assumed that the rate of change of velocity of money is in proportion to the rate in which the production is changed:

$$\frac{\Delta v(t)}{v(t-1)} = \rho \frac{\Delta Y(t)}{Y(t-1)}, \tag{10}$$

where ρ is a parameter.

The growth rate of the money supply in the period t:

$$\frac{\Delta M(t)}{M(t-1)} = m_e(t) + m_r(t) + m_b(t), \tag{11}$$

where: M is the money supply, m_e – the expected money supply growth (on the basis of expected production growth and inflation), $m_e(t)$ – the money growth being the result of the monetary policy, $m_b(t)$ – the money growth being the result of the fiscal policy.

The growth rate of the money supply as dependent on the policy of the interest rate in the period t has been described by a convex decreasing nonlinear function of the form:

$$m_r = \frac{\Delta M_r(t)}{M(t-1)} = \mu_0(r - \mu_1)^{\mu_2} + \mu_3, \quad \mu_2 < 0, \tag{12}$$

where: $\mu_0, \mu_1, \mu_2, \mu_3$ are parameters, μ_2 - the elasticity of the money growth with respect to the interest rate, $\mu_2 < 0$.

The growth rate of the money supply caused by the excess budget deficit in the period t has been described by a convex increasing nonlinear function:

$$m_b = \frac{\Delta M_b(t)}{M(t-1)} = \chi_0(b - b^*)^{\chi_1}, \quad \chi_1 > 0, \tag{13}$$

where: b^* - a level of the budget deficit in relations to GDP which does not require to increase the money supply, χ_0, χ_1 - parameters, χ_1 - elasticity of the money growth with respect to the excess of the budget deficit, $\chi_1 > 0$.

It is assumed that in the initial state the money market as well as the product market is in the equilibrium state. Cyclic changes in the economy are introduced by impulses in the form of changes of the private investments and changes of the monetary and the fiscal policies, i.e. by changes of the interest rate value and the budget expenditures. The cyclic changes tend in time to a new equilibrium state. Besides the cyclic changes the long-term trade expressing the exogenous technical progress is taken into account in the model. Effects of the applied monetary and fiscal policies are measured by the inflation in the new equilibrium state and by the average annual GDP growth rate in the considered interval of time.

The model has been tested using hypothetic data. Some number of simulations has been done. The model parameters, initial values of the model variables, as well as initial simulation results are presented in [10, 22].

3   The Fiscal-Monetary Game

3.1   Formulation

Relations between the fiscal authority and the monetary authority can be described by a noncooperative game. The game is defined in the strategic form as follows:

(i)   There are two players $i = 1$, 2: the fiscal authority (the government) and the monetary authority (the central bank).

(ii) For each player a set Ω^i of pure strategies is defined. The strategies of the fiscal authority are those of the budgetary policy – from the extremely restrictive to the extremely expansive. The measure, denoted by b, of the degree of restrictiveness/expansiveness of the fiscal policy is constituted here by the level of budget deficit in relation to GDP. The strategies of the monetary authority range from the extremely restrictive one to the extremely expansive. The degree of restrictiveness/expansiveness is equivalent simply to the value of the real interest rate and denoted by r. Let Ω denote the Cartesian product of the sets of the strategies $\Omega = \Omega^1 \times \Omega^2$.

(iii) For each player $i = 1$, 2, a function $h^i : \Omega \to R$ is given defining outcome of the player i for given strategies undertaken by the both players. The outcome of the fiscal authority is measured by the GDP growth rate, denoted by y, where $y = h^1(b, r)$. In the case of the monetary authority it is the inflation value, denoted by p, where $p = h^2(b, r)$. The functions h^i, $i = 1$, 2, are defined by the model relations.

(iv) For each player $i = 1$, 2, a preference relation is given in the set of the attainable outcomes. It is assumed here that each authority tries to achieve a given goal: the fiscal authority – a desired value of GDP growth, the monetary authority – a desired value of inflation.

Outcomes of the game in the discrete form are presented in Table 1 as in [20]. The strategies of the fiscal authority are shown in the first row and the strategies of the monetary authority - in the first column. The outcomes: GDP growth rate and inflation are denoted by y_{ij} and p_{ij}, respectively for assumed F_j (budget deficit b_j) and M_i (interest rate r_i) strategies of the authorities.

3.2   Analysis of the Game

Strategies, outcomes and payoffs of the game were analyzed using the macroeconomic model presented in Sect. 2. Relations (1–13) allow to derive the output quantities describing state of the economy, especially GDP growth rate and inflation, as

Table 1. The fiscal-monetary game – the table of outcomes

| | | **Central bank – the monetary policy** | | | |
| | | ← restrictive | | | expansive → |
		Monetary strategy M_1 (interest rate r_1)	Monetary strategy M_2 (interest rate r_2)	...	Monetary strategy M_n (interest rate r_n)
Government – fiscal policy ←expansive restrictive ↑	Fiscal strategy F_1 (budgetary deficit b_1)	p_{11} y_{11}	p_{12} y_{12}	... ...	p_{1n} y_{1n}
	Fiscal strategy F_2 (budgetary deficit b_2)	p_{21} y_{21}	p_{22} y_{22}	... ...	p_{2n} y_{2n}
	...			...	
	Fiscal strategy F_m (budgetary deficit b_m)	p_{m1} y_{m1}	p_{m2} y_{m2}	... ...	p_{mn} y_{mn}

dependent on strategies of the fiscal and monetary authorities. It is assumed that initially the economy is in an equilibrium state. Given fiscal and monetary policies i.e. given level of the budget deficit in relation to GDP and given value of the real interest rate create an impulse changing this state and generating oscillations of the output quantities. It has been observed that amplitude of the oscillations decrease in time and a new equilibrium state is achieved. The outcomes: GDP growth rate and inflation in this new equilibrium state are taken as the payoffs in the game.

Computer simulations have been made for wide range of values of the budget deficit in relation to GDP and values of the real interest rate. Selected results are presented and discussed below.

Figures 1 and 2 present outcomes of the authorities, as dependent on assumed strategies.

Inflation (Fig. 1) can be obtained on a low level when a restrictive monetary policy and a restrictive fiscal policy are applied. Expansive monetary and fiscal policies lead to enormous nonlinear increase of inflation. On the other hand restrictive monetary and restrictive fiscal policies lead to decrease of the economic growth (Fig. 2).

Let us assume given goals of the fiscal and the monetary authorities. Let the fiscal authority try to achieve the GDP growth rate on the level y^g, and let the monetary authority assume the inflation goal on the level p^g. Let Ω denotes the set of admissible pairs (b, r) of strategies. The respective best response strategies can be obtained as solutions of the optimization problems: Min $|h^1(b, r) - y^g|$ with respect to $b \in \Omega^1$ solved for all $r \in \Omega^2$, in the case of the fiscal authority and Min $|h^2(b, r) - p^g|$ with respect to $r \in \Omega^2$, solved for all $b \in \Omega^1$, in the case of the monetary authority, as solutions of the problem.

The best response strategies (marked in the figure by triangles) of the fiscal authority when the GDP growth $y^g = 2.2$ % can be achieved as well as the best response

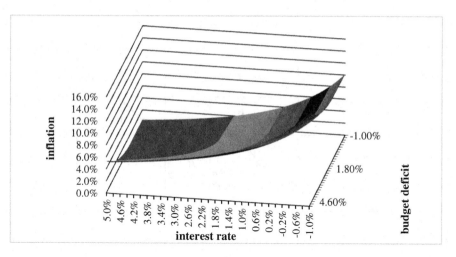

Fig. 1. Outcomes of the monetary authority

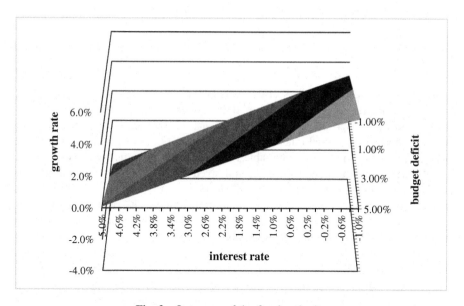

Fig. 2. Outcomes of the fiscal authority

strategies (marked by rhombs) of the monetary authority when the inflation $p^g = 2.5\ \%$ can be obtained, are presented in Fig. 3. Let us see that the lines presenting the strategies have not any joint point. The Nash equilibrium [12] exists for the combination of the most restrictive policy of the monetary authority and the most expansive policy of the fiscal authority.

Let us consider possible coordination of the policies. We assume that each player tries to minimize a distance to his goal, i.e. $d^y = |h^1(b, r)\text{-}y^g|$ and $d^p = |h^2(b, r)\text{-}p^g|$. In this case d^y and d^p can be treated as the criteria that should be minimized jointly. Let d^Ω

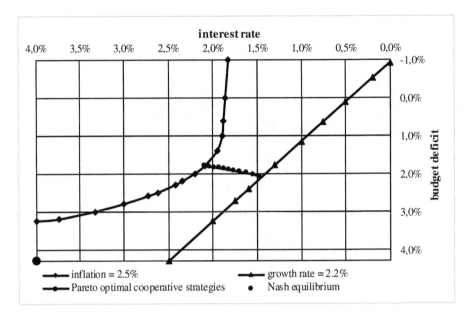

Fig. 3. The best response strategies of the authorities and possible cooperative – coordinated strategies

denote the set of attainable values of the pairs (d^y, d^p) for $(b, r) \in \Omega$. We say that the pair $(\hat{d}^y, \hat{d}^p)$ is Pareto optimal in the set d^Ω if there does not exists any pair $(d^y, d^p) \in d^\Omega$, $(d^y, d^p) \neq (\hat{d}^y, \hat{d}^p)$, such that $d^y \le \hat{d}^y$ and $d^p \le \hat{d}^p$. A representation of the strategies leading to the Pareto optimal outcomes is presented in Fig. 3 using circle marks. The Nash equilibrium is far from the possible Pareto optimal outcomes of the authorities.

The Pareto optimal strategies have been derived solving the following multicriteria optimization problem: **VMin** $(|h^1(b, r) - y^g|, |h^2(b, r) - p^g|)$ with respect to $(b, r) \in \Omega$, where **VMin** means that the criteria are minimized jointly. The representation of the Pareto optimal strategies and the respective Pareto optimal outcomes have been obtained applying the reference point method [17, 19].

Figure 4, presents the set of the Pareto optimal outcomes which can be obtained in the case of coordination of the policies. The point defined by the outcomes corresponding to the Nash equilibrium is also presented as well as the utopia point. The utopia point represents the outcomes at which the both goals of the authorities are achieved. The utopia point is not attainable in the considered case.

Let us see that looking for cooperative strategies we deal with the bargaining problem formulated by Nash [11] and studied by many scientists. Some ideas of computer based decision support methods in bargaining problems can be found in [8, 9]. The Pareto optimal strategies and the respective Pareto optimal outcomes derived and presented in Figs. 3 and 4, define frames for negotiations in which the fiscal and monetary authority may correct their goals and coordinate policies looking for a consensus. In further research a mediation process will be considered, in which mediation proposals can be derived with use of solution concepts to the bargaining problem. The solution concepts formulated by Nash [13] and Raiffa [15] could be used in this case.

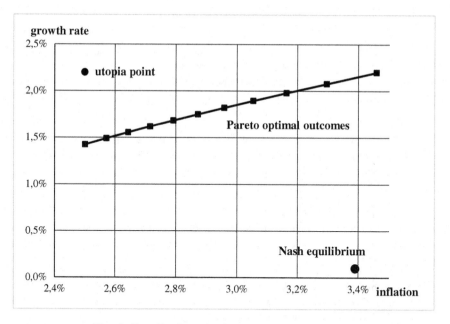

Fig. 4. Payoffs of cooperative – coordinated strategies

4 Conclusions

The paper presents selected results obtained within the research dealing with analysis of interactions of the fiscal and monetary policies with application of the game theory and the multicriteria optimization tools. Within the research the dynamic macroeconomic model has been constructed, the fiscal monetary game has been formulated and analyzed. The model describes a mechanism of the business cycles and allows for analysis of economic situation in time. On the other hand it enables analysis of the monetary and fiscal policies, namely instruments of the policies: the real interest rate, the deficit of the government budget in relations to GDP, as well as their influence on the economy, especially on GDP growth and on inflation.

The model relations have been implemented in the form of the computational algorithm. The algorithm is a part of a computer-based system used for simulations and analysis of the game in an interactive way. The simulations were done for some number of variants including different model parameters and initial values of model variables describing different states of economy.

The computational results presented show a typical case when the Nash equilibrium in the game is not Pareto optimal. In this case questions arise: how to support the players, when they are playing a non-cooperative game, in looking for a Pareto optimal consensus, how to reach the consensus taking into account specific, in generally conflicting objectives and preferences of the players. These questions indicate directions of further research. In this case a bargaining problem can be formulated and analyzed with use of the multicriteria optimization tools. Formulation and analysis of the bargaining problem can be basis for designing negotiations leading to a Pareto optimal consensus.

References

1. Beetsma, R., Jensen, H.: Monetary and fiscal policy interactions in a micro-founded model of a monetary union. J. Int. Econ. **67**(2), 320–352 (2005)
2. Bennett, N., Loayza, H.: Policy biases when the monetary and fiscal authorities have different objectives. Central Bank of Chile Working Papers, No. 66, pp. 299–330 (2001)
3. Blinder, A.S.: Issues in the coordination of monetary and fiscal policy. In: Monetary Policy in the 1980s, Federal Reserve Bank of Kansas City, pp. 3–34 (1983)
4. Friedman, M.: The role of monetary policy. Am. Econ. Rev. **LVIII**(1), 1–17 (1968)
5. Friedman, M.: Monetary theory and policy. In: Ball, R.J., Boyle, P. (eds.) Inflation. Penguin Modern Economics, Harmondsworth (1958)
6. Hicks, J.: A Contribution to the Theory of the Trade Cycle. Oxford, Clarendon (1950)
7. Keynes, J.M.: The General Theory of Employment, Interest and Money. Palgrave Macmillan, London (1936)
8. Kruś, L.: Multicriteria cooperative decisions, methods of computer-based support (in Polish: Wielokryterialne decyzje kooperacyjne, metody wspomagania komputerowego). Instytut BadańSystemowych PAN, Seria: Badania systemowe, Tom 70, Warszawa (2011)
9. Kruś, L.: Computer based support in multicritria bargaining with use of the generalised Nash solution consepts. In: Atanassov, K., et al. (eds.) Modern Appraches in Fuzzy Sets, Intuitionistic Fuzzy Sets, Generalized Networks and Related Topics: Applications, vol. II, pp. 43–60. SRI PAS, Warsaw (2014)
10. Kruś L., Woroniecka-Leciejewicz, I.: Analysis of policy-mix with use if the game theory and multicriteria optimization methods (in Polish: Analiza policy-mix z wykorzystaniem teorii gier i optymalizacji wielokryterialnej). Research Report, IBS PAN, Warszawa (2014)
11. Nash, J.F.: The bargaining problem. Econometrica **18**, 155–162 (1950)
12. Nash, J.: Non-cooperative games. Ann. Math. **54**(2), 286–295 (1951)
13. Nash, F.: Two-person cooperative games. Econometrica **21**, 129–140 (1953)
14. Nordhaus, W.D.: Policy games: coordination and independence in monetary and fiscal policies. Brookings Papers on Economic Activity, No. 2, pp. 139–215 (1994)
15. Raiffa, H.: Arbitration schemes for generalized two-person games. Ann. Math. Stud. **28**, 361–387 (1953)
16. Samuelson, P.A.: Interaction between multiplier analysis and the principle of acceleration. Rev. Econ. Stat. **21**, 75–78 (1939)
17. Wierzbicki, A.P.: On the completeness and constructiveness of parametric characterizations to vector optimization problems. OR Spectrum **8**, 73–87 (1986)
18. Wierzbicki, A.P., Kruś, L., Makowski, M.: The role of multi-objective optimization in negotiation and mediation support. Theor. Decis. **34**(2), 01–214 (1993)
19. Wierzbicki, A.P., Makowski, M., Wessels, J.: Model-based Decision Support Methodology with Environmental Applications. Kluwer Academic Press, Dordrecht, Boston (2000)
20. Woroniecka-Leciejewicz, I.: Decision interactions of monetary and fiscal authorities in the choice of policy mix. J. Organ. Transform. Soc. Change **7**(2), 189–210 (2010)
21. Woroniecka-Leciejewicz, I.: Equilibrium in the fiscal-monetary game and the priorities of the central bank and the government (in Polish: Równowaga w grze fiskalno-monetarnej a priorytety banku centralnego i rządu). In: Trzaskalik, T. (ed.) Modelowanie Preferencji a Ryzyko 2010, AE im. K. Adamieckiego, Katowice, pp.327–343 (2010)
22. Woroniecka-Leciejewicz, I.: An influence of policy-mix instruments on the economy – modeling approach (in Polish: Wpływ instrumentów policy-mix na gospodarkę – ujęcie modelowe). Zeszyty Naukowe WSISiZ, "Współczesne Problemy Zarządzania", pp. 7–33 (2015)

Voting and Collective Decision-Making

Warsaw School of Economics

A Framework for Aiding the Choice of a Voting Procedure in a Business Decision Context

Adiel Teixeira de Almeida[1(✉)] and Hannu Nurmi[2]

[1] Federal University of Pernambuco, Cx. Postal 7462, Recife 50630-970, Brazil
almeidaatd@gmail.com
[2] University of Turku, Turku, Finland
hnurmi@utu.fi

Abstract. In business organizations a group decision process, which takes aggregation of DMs' final choices into account, usually uses a voting procedure (VP). Therefore, a relevant part of the decision process consists of choosing the VP. Since all VPs have some drawbacks that may occasionally lead to undesirable outcomes it is important to characterize decision settings that make certain performance criteria particularly pertinent for choosing a VP. In this paper, it is assumed that this decision should be made by the DMs while the analyst will give some methodological and technical aid, and that a specific decision model will be used. Therefore, this paper presents a framework for aiding the choice of a VP in a business organization decision context, based on an MCDM model.

Keywords: MCDM voting choice · Choosing voting procedures · Preference analysis · Business organization decision context

1 Introduction

First ideas for proposing an MCDM model for aiding the choice of a voting procedure (VP) for a business organization decision problem have been considered with preliminary results [1]. Therefore, a framework is proposed in order to deal with the decision process related to this choice.

It is worth pointing out that, although one might think that VPs have been designed for political elections rather than for a business decision in a group context, these procedures are quite appropriate for a range of business decision problems.

The range of such problems analyzed in this paper has a few characteristics, including that of tackling the multiple objectives by each decision maker (DM) into account. In other word, each DM needs to undertake a multi-criteria decision making (MCDM) process.

In business organizations the MCDM group decision process, except in situations related to a negotiation process, may be of two kinds: (a) aggregation of DMs' initial preferences; (b) aggregation of DMs' individual choices [2]. In the former the DMs share the same objectives and the criteria are aggregated in an integrated way. In the latter, the DMs may have different objectives and criteria and integrating these is conducted over the alternative rankings given by each DM separately [2–4].

© Springer International Publishing Switzerland 2015
B. Kamiński et al. (Eds.): GDN 2015, LNBIP 218, pp. 211–225, 2015.
DOI: 10.1007/978-3-319-19515-5_17

The latter is the focus of this paper. For group decision processes that take the aggregation of DMs' final choice into consideration, a VP is a natural approach. However, another decision problem comes up in this situation, which is related to choosing the VP. Usually, this decision is based on technical issues associated with the characteristics and formal properties of the VPs. Although, this decision is not directly related to the actual decision-making faced by DMs and appears merely to be one of the technical decisions to be made during the process, we argue that this decision should be made by the DMs provided that they have some methodological and technical support [1].

In this paper, a framework is proposed in order to deal with the decision process of choosing a VP for a business organization decision context, and draws attention to include this in an MCDM model that will aid this choice.

The purpose of this paper is to propose a general framework. For this reason, specific details, such as which criteria should be used, how standard voting procedures actually score in these criteria, and particularly which weights should be assigned to the criteria are not straightforward given, because it should not be given in such way. For this reason, as it should be in a framework, all of this is presented as being dependent on the actual decision problem to be solved, so in the framework these questions are addressed as guidance for the decision process.

The next section discusses a few relevant aspects of business decision context. Section 3 points out a few issues for establishing the family of criteria in order to evaluate the VPs. Section 4 presents the framework for aiding the choice of the VP. Section 5 draws some conclusion and makes a suggestion for further research.

2   The Business Decision Process and the Modeling Process

The type of decision problem may have a great influence on choosing the procedure. However, the main distinction the type of decision may not be related to the two kinds of decision context mentioned; that is: a business decision and a political election. The main issue which makes a difference is related to either: choosing a person or choosing a policy. Both problems may be faced in business organizations, although the latter may be more common and usually may be referred to as a choosing an alternative course of action; for instance, choosing one project from several proposals. In business organizations, the choice of a person may be more related to recruiting a new employee (or a selecting a member of staff who will exercise some kind of function, for which specific skill are required) than choosing a representative of other people.

To choose such a procedure, two situations have to be considered. The first is related to choosing a procedure to be applied in every decision making process. This is a typical process of group decision on the Board of any business organization. Normally, the norms and formal procedures of the organization have to state which VP should be applied. The second situation is associated with choosing a procedure to be applied in a specific business decision problem in the organization. In this kind of situation, each decision problem requires differentiated considerations which may lead to a particularly suitable procedure.

The focus of this article is on the second situation, although an MCDM model for aiding the choice of a VP should also be applied for the first kind of situation.

Many examples may be given of this kind of decision problem. For instance, selecting suppliers is one of the most common problems in business organizations, in which many different group decision processes may be conducted. Some kinds of problems may include the customer in the decision process. For instance, the business of supplying movies by web is one of the situations, in which a group of DMs may be the users, whose opinions are taken into consideration in order to decide on what movies they should choose.

In the second situation, in most cases, the decision makers (DMs) may have already made their own ranking of alternatives, before an aggregation procedure starts to be considered.

The whole decision process may be divided into two specific decision processes:

- The decision process for choosing a voting procedure (DPVP), by means of aid from an MCDM model;
- The decision process for the business organization (DPBO), analyzed by means of a VP, which is directed to a specific decision problem.

The DPVP is the first decision process and is the main focus of this paper. The DPVP is related to a modeling step in the modeling process, in which an MCDM method is applied. The DPBO is the second decision process and is the main concern of the business organization, in which a VP is necessary in order to solve a group decision problem.

In general the DPBO is approached with little, if any consideration given to the DPVP. Usually, an analyst chooses a VP, according to some convenience in the modeling process, without following a structured process for making such a choice, although considering many technical concerns regarding Social Choice Theory.

2.1 Basic Elements for the DPVP

Some basic elements should be pointed out for the DPVP with regard to the modeling approach for building a MCDM model. First of all a family of criteria [5] has to be accounted for. The set of alternatives consists of the VPs to be evaluated for that particular DPBO. Since the set of alternatives is a discrete set, a consequence matrix may be considered as shown in Table 1.

Table 1. Consequence matrix.

VPs	Criteria 1	Criteria 2	Criteria 3	....	Criteria j..	...	Criteria n
VP1	C11	C12	C13	...		...	C1n
VP2	C21	C22	C23	...		...	C2n
...	...	...	...	...	...	...	...
VPi				...	Cij	...	...
...	...	...	...	...	...	...	...
VPm	Cm1	Cm2	Cm3	...	...	...	Cmn

The consequence C_{ij} consists of the outcome related to criterion j, that VP_i will give to the problem faced in the DPBO. Therefore, although a criterion may be related to some property, its consequence regarding a particular VP should be considered in the context of the particular decision problem, taking into account the business context of the DPBO. With regard to evaluating these properties in the context of a specific decision problem, some scholars may agree, whereas others do not for this topic. The main issue here is that this topic should have an occasion of being evaluated in the process, avoiding to be ignored, when it might be relevant in some contexts.

This consequence matrix should be translated into a decision matrix, as shown in Table 2, in which the consequences (or outcomes) are evaluated by a value function V (C_{ij}), which gives the value of consequence for VP_i according to its performance regarding criterion j in the context of the decision problem as faced in the DPBO. The scale of evaluation for this value function is an important factor when determining the MCDM method. For many methods a cardinal scale is required, such as an interval or ratio scale. However, in some circumstances only an ordinal scale may be obtained. If a consequence represents how some property (criterion) performs in a particular VP, considering the context of the DPBO, then the DM may obtain the value function by subjective evaluation. A typical scale that could be applied in this case is a scale of five levels (1 to 5). Level 5 indicates the most preferable level of performance that could be obtained with that property (criterion) for the particular DPBO and level 1 indicates the least desirable level of performance.

Table 2. Decision matrix.

VPs	Criteria 1	Criteria 2	Criteria 3	....	Criteria j..	...	Criteria n
VP1	V(C11)	V(C12)	V(C13)	...		...	V(C1n)
VP2	V(C21)	V(C22)	V(C23)	...		...	V(C2n)
...	...	...	...	...	...	...	...
VPi				...	V(Cij)	...	...
...	... .	...	...	...	...	...	...
VPm	V(Cm1)	V(Cm2)	V(Cm3)	...	...	...	V(Cmn)

It should be observed that the drawbacks a VP has may occasionally lead to undesirable outcomes, and therefore, it is important to characterize decision settings that make certain performance criteria particularly pertinent for choosing a VP. In other words, a drawback a VP has will not always happen for all input data set; that is, for all decision problems. To illustrate this, consider the property related to the independence of irrelevant alternatives. Let this property appear in the column of Tables 1 and 2, as property j. Then, consider three different VPs, such that, i = x, y, z. Let us assume that for i = x, there is complete independence of irrelevant alternatives, then $V(C_{xj}) = 5$. Let us assume that for i = y, the independence of irrelevant alternatives does not hold at all, then $V(C_{yj}) = 1$. Now, let us assume that for i = z, the independence of irrelevant alternatives does hold in 50% of cases, then $V(C_{yj}) = 3$. For another VP, in which the independence of irrelevant alternatives holds for less than in 50 % of cases, then $V(C_{yj}) = 2$. By the value function, the DM could assigns evaluations to these consequences, in accordance with their impact on the DPBO. Still, one may argue that independence of irrelevant alternatives is a characteristic that a VP have or do not have, regardless of the DPBO; and that is right. However, the question pointed out here is how often it appears and how this frequency affects that particular DPBO.

The decision matrix illustrated in Table 2 is the input for the MCDM method that is applied in order to evaluate each VP and the most adequate one is assigned for the group aggregation in the DPBO.

2.2   The Choice of a Voting Procedure

For the DPVP, it is assumed that such an aiding process includes the participation of an analyst or facilitator, whose role is to support all DMs in the group decision process.

One of the steps of the procedure requires the analyst to explain to the DMs what main VPs are available and what their main characteristics are, as well as their behavior regarding the paradoxes and main properties related to such procedures.

The analyst may adopt two different sequences for the decision process, as follows:

(a) DMs choose the VP before they rank the alternatives, related to the DPBO.
(b) DMs choose the VP after they rank the alternatives, related to the DPBO.

The second process is fine, if the DMs do not know each other's rankings, in the DPBO. The first process may lead to manipulation, by means of adopting strategic choices for the rankings, related to the alternatives in the DPBO. In the second process, part of the DPBO is conducted before the DPVP and finalized afterwards. In the first process, the DPBO is conducted completely only after the DPVP has been concluded.

It is expected that if one evaluates VP only after knowing to which problem they will be applied (including knowledge of alternatives and their evaluations - Table 2), this might introduce a bias in the selection of a VP. The DMs may consider an incentive to favor not the VP that is suited best for the problem, but the VP that will deliver the alternative they would like to see as winner. This issue will be always present and cannot be avoided, so the analyst has to deal with it.

3   Choice of Criteria for Selecting a Voting Procedure

Two kinds of criteria may be considered for this problem of the DPVP, in accordance with two main objectives. The first objective is directly related to the DPBO, in which the context of the business decision problem is considered, thus revealing the first kind of criteria. The second objective is related to the VPs themselves and their characteristics and how they affect the DPBO, thereby generating criteria associated with the properties and other characteristics, such as paradoxes that may be relevant for consideration when analyzing a VP.

There are quite a few studies in the literature related to the analysis and discussion of properties and paradoxes related to the use of the VP proposed. These properties and features of the VP have been relevant when choosing them for a particular problem. Selecting a set of criteria consisting of the most relevant properties for the VPs may be considered [6–8]. Also, the matrix which evaluates these properties for the main VPs can be built. This evaluation depends on the context of the decision problem and the scores this gives have to be consistent with the MCDM method.

3.1   Comparing Voting Procedures

The best-known results of modern Social Choice Theory pertain to the compatibility of various choice desiderata. Typically they aim to show that from a set of intuitively plausible principles of choice, only one proper subset can be adhered to by any given rule under all circumstances. Those circumstances are important for the incompatibility captured by Arrow's Impossibility Theorem [9]. This has been pointed out by many authors, e.g. Black [10] introduced the notion of single-peakedness to give a sufficient condition for the avoidance of the incompatibility. This notion turned out to be but one of several possible restrictions on the domain of preference profiles that would guarantee the satisfaction of Arrow's other conditions.

Most of us most do our daily shopping by simply revealing our true preferences (given the budget restrictions) when selecting goods for our basket. It would seem that this also holds for our responses to most opinion surveys. Some voters (perhaps the vast majority of them) also reveal their true opinions in political elections. This is called expressive voting.

Although a variety of criteria for comparing voting systems has been introduced over the past decades, it would seem that two of them are of particular importance since they can be related to rationality. The first is known as the participation condition and it can be viewed as an individual rationality criterion since a failure on the participation condition would conceivably confront an individual with a contingency where his/her vote would be harmful to his own interest in the sense that the outcome that would follow from his/her abstaining would be better for him/her.

The second similarly compelling and rationality-related criterion applicable in these circumstances is Pareto optimality. This can be viewed as a collective rationality criterion since it states that if each participant strictly prefers alternative x to alternative y, then y is not chosen. Clearly, a failure on Pareto optimality would be collectively irrational.

Of a somewhat more controversial nature are criteria connected with the name of Condorcet: the winner and loser criteria. The former dictates the choice of an alternative that would defeat all others in pairwise round-robin contests by a majority of votes. The latter, in turn, requires that an alternative that would lose against every other alternative in pairwise comparisons may not be elected.

Of these two Condorcet criteria especially the former has been very commonly advocated as a plausible desideratum for social choice rules. Those rules that satisfy it do, however, not satisfy another plausible condition, viz. positional domination [11]. An alternative x positionally dominates alternative y, if for each of ranks $j = 2, \ldots, k$, the number of voters assigning x to rank j or higher is larger than the number of voters ranking y to rank j or higher. The positional dominance criterion dictates that those alternatives that are positionally dominated by some other alternative may not be chosen.

In settings where the voters are primarily interested in the outcomes rather than expressing their opinions, the opinions expressed in balloting may deviate from the opinions held by the voters. Since the idea of taking a vote is to elicit the voters' opinions as accurately as possible, it would make sense to resort to systems where it is difficult to improve upon outcomes by misrepresenting one's opinions. But, how can this difficulty be defined in an objective way?

Successful preference misrepresentation requires information about the preferences (more precisely, the expressed preferences) of other voters. One way of measuring the difficulty of misrepresentation is to ask how detailed knowledge of the overall profile one needs to succeed in misrepresentation. E.g. in plurality voting one typically needs only information about the distribution of votes over all the alternatives that are the ranked first.

At the other extreme of difficulty is the single transferable vote (see [12]). Similarly, Nanson's and Kemeny's rules [7] would seem difficult to manipulate.

3.2   How to Deal with Voting Paradoxes

A brief discussion of some paradoxes, such as intransitivity and incompleteness may illustrate how they may be considered in such process.

It is not difficult to envision a setting where not only collective majority preferences but even individual ones could be intransitive (see, e.g. [13]). Consider for example an individual who has been given the task of ranking three universities. In his opinion, three criteria of equal importance should determine the ranking: research excellence, quality of education and external impact. Suppose that in terms of the first criterion the ranking is ABC, in terms of the second BCA and in terms of the third CAB. Using pairwise comparisons and majority rule in determining the pairwise winners, one ends up with an intransitive ranking: ABCA...

The occasional plausibility of intransitive individual preferences suggests that social choice rules could be based on pairwise comparison matrices representing individual opinions, i.e k-by-k matrices with entry $(i; j)$ equaling 1 if i^{th} alternative is viewed preferable to the jth one, equalling 0 if j^{th} alternative is preferred to the ith one. This approach has, in fact, a long history starting from Zermelo's seminal work [14].

Social choice rules can easily be defined using various tournament solution concepts: Pareto set, uncovered set, Copeland winners, the Banks set, etc.

Incomplete preferences can also be dealt with using the tournament apparatus. If an individual is unable to express a preference between two alternatives i and j, the tournament matrix can accommodate this by inserting 0 into both the position $(i; j)$ and $(j; i)$. Incomplete tournaments have been the focus of some scholarly attention for a long time. E.g. Zermelo [14] discussed chess tournaments with an unequal number of contests between various pairs of players. The methodology devised for these settings is immediately applicable also in voting settings.

Many paradoxical observations of voting systems turn out to be aggregation paradoxes. Some of them, e.g. inconsistencies of choice, can be avoided by resorting to consistent procedures. Usually, however, avoiding one paradoxical contingency leads to another type of paradox. So, there are trade-offs to be made in dealing with paradoxes (see e.g. [6, 8] for a summary).

3.3   Criteria Related to the Context of the Business Problem

As previously stated there are kinds of criteria and objectives, which is directly related to the DPBO, in which the context of the business decision problem is considered.

Of course these kinds of criteria have an impact on the VPs, since their consequences have to be explicitly indicated in Table 1. However, such criteria are not related to properties or paradoxes of the aggregation process of the VP itself, as is the case for the other kinds of criteria.

The main criterion of this kind may be the kind of input required by a VP. The input is related to the way in which the DMs will give their information about the alternatives in the DPBO. For instance, the DM may be required to rank the alternatives. In other methods the DM may be required to make pairwise comparisons of alternatives. In other situations, the DM may be required to rate the alternatives by means of evaluating them with some scale, such as a five-level (1 to 5) scale.

In a business organization, the DPBO in general consists of another MCDM model, which usually is much more complex than the MCDM model for the DPVP. In such situations, the criterion related to the kind of input may not be relevant, since an MCDM model will output this information.

On the other hand, for the problem previously mentioned related to a group of DMs choosing movies, this criterion is rather relevant and may have a great weight, compared with other criteria.

3.4   Influence of the Analyst on the Choice of Criteria

It should be emphasized that the set of criteria for choosing the VP should not be related to the analyst's interest in having some kind of influence on the decision process. The criteria should only be related to the DM's objectives related to the final business decision problem.

Therefore, the analyst should avoid exerting any kind of influence on choosing the criteria. An example is how best to reduce the computational complexity of the VP. This criterion should not be considered, unless it has some impact in the DPBO, which is the final purpose of the DPVP.

4   A Framework for Choosing a Voting Procedure

As mentioned, the DPVP is the first part of the whole decision process and is planned so as to choose a VP by means of an MCDM decision model. This section proposes a framework for building such a MCDM decision model, which is based on a procedure for building MCDM decision models [2].

This procedure has three phases:

- A preliminary phase, in which the basic elements of the decision model are considered
- A phase for preference modeling and choosing the MCDM method
- A finalization phase, in which a recommendation is given.

The preliminary phase has 5 steps which are concerned with identifying the DM and other actors in the decision process, and the establishing other elements of the decision model such as: the objectives, the criteria, and the set of alternatives.

The second phase has three steps, related to preference modeling, in which the MCDM method is chosen. Finally, in the last phase the alternatives are evaluated, a sensitivity analysis is made and a recommendation is proposed.

4.1 Framework for the Modeling Process and Choosing the Voting Procedure

Based on this procedure, Fig. 1 shows a framework for modeling the process for choosing the VP, which includes the steps that the analyst should follow and his/her interactions with the DM or a group of DMs.

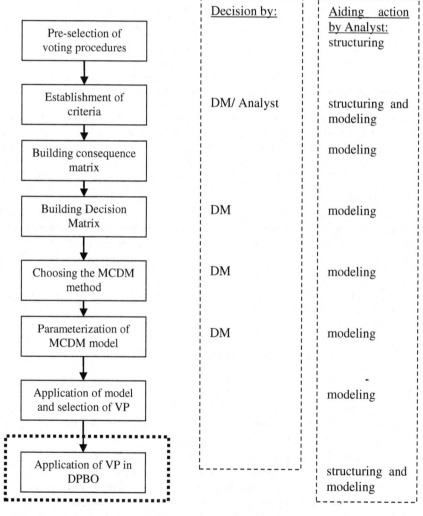

Fig. 1. Framework for modeling the process for choosing the voting procedure.

Figure 1 also shows the actions, at each step, to be taken by at least two actors: the DM (or a group of DMs) and the analyst. Actions by the DM consist of decision-making to be made, mainly regarding the inclusion of preference information within the model. Actions by the analyst may be of two types structuring and modeling.

The first step consists of a pre-selection of the set of VPs, to be included in the set of alternatives. The establishment of criteria is preceded by setting the objectives and involves decisions to be made by the DM. There are several structuring methods for conducting this step and depending on the complexity of the DPBO and the organizational context an appropriate approach should be applied [2, 5, 15]. The following steps consist of building the consequence matrix and the Decision Matrix.

As to choosing the MCDM method, the DM's preference structure should be considered. This step is aided by the analyst who models the evaluation of the DM's rationality, in accordance with what is discussed in Sect. 4.2. Next step regard the inter-criteria evaluation, involving mainly the establishing the criteria weights.

Then, it follows the application of the model, including implementing the model, conducting the sensitivity analysis and interpreting the results [2].

4.2   Choosing the MCDM Method for Comparing Voting Procedures

Choosing the MCDM method is one of the most important issues for building MCDM decision models [2, 16]. There are many MCDM methods, which may have a few different classifications. For the purpose of this study, some of these classifications may be useful. First, an MCDM method may be classified according to the action space, which can be either discrete of continuous. The former is of interest for the kind of problem analyzed.

Another classification considers the form of compensation, if any, for aggregating the criteria. Two situations may be considered: compensatory and non-compensatory methods [17, 18]. A number of methods may be included in the first type, for instance: MAUT (Multi-Attribute Utility Theory) and deterministic additive methods, such as AHP, SMARTS, MACBETH, among many others [18, 19].

The additive aggregation procedure is the most applied of these methods and can be represented by the following Eq. (1).

$$v(VP) = \sum_{i=1}^{n} k_i V_i(C_{ij}) \tag{1}$$

where:

$v(VP)$ is the global value of a particular VP

k_i is the parameter related to the inter-criteria evaluation of criterion i; this parameter is called either as a "weight" or a "scale constant" of criterion i.

$V_i(C_{ij})$ is the value of consequence for criterion i, as shown in Table 2.

C_{ij} is the consequence or outcome of VP j for criterion i, as show in Table 1.

Amongst the non-compensatory methods are included lexicographical and outranking methods, such as: ELECTRE, PROMETHEE, among many others [5, 17, 19].

A preference relation P is non-compensatory if the preference between two VPs x and y only depends on the subset of criteria in favor of x and y [20]. Let $P(x,y) = \{i: x_i P_i y_i\}$, then:

$$\left.\begin{cases} p(x,y) = p(z,w) \\ p(y,x) = P(w,z) \end{cases}\right\} \Rightarrow [xPy \Leftrightarrow zPw]. \tag{2}$$

In this case, it does not matter how high is the level of the performance of x or y in each criterion. That is, how high is the level of the performance $(V_i(C_{ij}))$, in Table 2, of a VP because a particular criterion is not taken into account. It is enough to know if the level of performance $(V_i(C_{ij}))$ of a VP is greater or less than another VP. That is, the only information needed is if $(V_i(C_{iz})) > (V_i(C_{iy}))$. This would mean that the performance of VP_z is greater than the performance of VP_y and VP_z is preferred to VP_y. This is the only information required in (2).

An example of this kind of evaluation may be illustrated by the PROMETHEE family of methods [17, 18, 21], in which the outranking degree $\pi(a, b)$ of each ordered pair of VPs (a, b) is obtained by (3):

$$\pi(a, b) = \sum w_j p(a, b). \tag{3}$$

where:
w_i is the weight of criterion i; the weights are normalized, such that: $\Sigma w_i = 1$.
$\pi(a, b)$ expresses to which degree a is preferred to b over all the criteria.
$P_j(a, b)$ is a preference function, established for each criterion by the DM; for the most simple function $P_j(a, b)$ is set to be 1, if the value of a is greater than the value of b in Table 2, and set to be 0, otherwise.

The weights are given by the DM, and for this kind of method, this straightaway represents the relative degree of importance of the criterion. For instance, the property related to the independence of irrelevant alternatives could have a minor impact over a particular DPBO, and in this case would receive a low weight from the DM. Also, this DM could consider that the property related to the Condorcet winner criterion would be twice as important as the former property, with a weight of 0.3. Therefore, the criterion related to the independence of irrelevant alternatives would have a weight equal to 0.15.

For this non-compensatory approach, (3) indicates that the outranking degree $\pi(a, b)$, comparing VPs a and b is the summation of the weights for those criteria, in which a has a better performance $(V_i(C_{ij}))$ than b, without considering how much better it is.

On the other hand, for the additive method the level of the performance $(V_i(C_{ij}))$ of a VP for a particular criterion is taken into account in the MCDM evaluation process, so this level is compensated for in the model given in (1). In this sense it is a compensatory method.

An issue to be taken into account is related to the scale of the value function $V_i(C_{ij})$. Of course for the additive model in (1) this has to be on a cardinal scale. In general, the interval scale is assumed for most elicitation procedures for the weights k_i. These procedures have to be applied in an appropriate way, since the weights in the additive model do not straightaway mean that these are the degrees of importance of a criterion.

On the other hand, for the non-compensatory methods an ordinal scale for the value function $V_i(C_{ij})$, in Table 2 may be applied, which can clearly seen analyzing (2). This ordinal scale is also applicable for most outranking methods, including the ELECTRE and PROMETHEE [5, 17, 18, 21]. However, there may be some exceptions in these methods. The computation of discordance indices may use difference in attribute values in ELECTRE method, if the analyst so designs the model. Alternatively, the analyst may consider discordance sets, instead of discordance index, so that qualitative criteria may be applied [17]. With regard to the PROMETHEE method, the ordinal scale may also be applied when using the "usual criterion" for $P_j(a, b)$ in (3), as explained above, which is consistent with the formulation in (2). This process of the method has no relation with the concepts of preference flow, which produces a cardinal scale for scores of the alternatives, since they are based on the summation of the criteria weights. However, when using thresholds in these methods one may consider a need for an interval scale, instead of an ordinal scale, since thresholds are applied in order to add or subtract to the value function $V_i(C_{ij})$. On the other hand, ordinal scale may be applied as an approximation, given the meaning of thresholds itself, contextualized for small amounts, with a high degree of approximation. Thus, using ordinal scales in this case may be seen as a good approximation, as is expected in the modeling process and it is conducted in some methods. For instance, some VPs use ordinal information (rank position) in order to produce scores for alternatives, such as the Borda procedure.

When choosing the MCDM method for the DPVP, this issue of compensatory or non-compensatory rationality [2] should be considered. Which of them is the most appropriate to the DM? This evaluation should be conducted. This is not a simple choice that the analyst can make. It should be based on the DM's preference structure. For instance, it should be checked if on comparing the performance of the VPs regarding the desiderata and other possible criteria, the DM is willing to make compensation.

An important consideration may be taken at this point with regard to the kind of rationality which would be more appropriate for the decision problem considered. That is, should the set of VPs be analyzed by a compensatory or non-compensatory approach?

Some reflections may be provided regarding which rationality would be more appropriate for the DMs in this particular decision problem.

One may find it hard to consider the possibility of a DM analyzing any two properties or characteristics of a VP, by making compensation between them. In this case, the value function of Table 2 is applied only to check which of the two VPs has a greater performance and there is no need to know by how much its performance is greater than the other.

On the other hand, one may think that it seems reasonable to consider that a DM may analyze two properties or characteristics of a VP, by comparing them in terms of which would be more acceptable. Conversely, a DM could consider which of them would be a more unsuitable presence in a VP.

If that is the case, it seems to be reasonable to assume that it would be more appropriate to apply a non-compensatory method in the DPVP. That is, the choice of the VP may be based on a formulation such as that in (3).

Alternatively, the DM may require that making comparisons between VPs require consideration to be given to the extent to which the performance on those properties (criteria) is analyzed in a model such that (1) should be applied. In this case, the scale of the value function, represented in Table 2, should be on a cardinal scale.

4.3   Additional Remarks for the Modeling Process

The main issue in the framework presented in Fig. 1 is that related to choosing the MCDM method in the DPVP and the possibility of a group decision process in the DPVP, which is the main issue following discussed. A few other issues are briefly emphasized as follows.

A major concern with this process is the requirement of the DM, mentioned in Fig. 1. In some DPBO there is a presence of a supra-DM, who has, in general, a hierarchical position in the organization's structure, above the other DMs. This is similar to the concept of the 'benevolent dictator problem' [22]. In this case this DM makes all decisions related to the DPVP. Otherwise, if the context is based on a 'participatory group problem' [20], then, all DMs acts jointly in the DPVP. That is, the DPVP turns out to be a group decision process. In this case, one may wonder that this may lead to a danger of an infinite regress. That is, a method would be needed to solve that second order group decision, for that another model would be needed, and so on. In this situation, the role of the facilitator is considerable relevant, in order to break a supposed circle of numerous regresses.

Since the second kind of criteria, as explained in Sect. 3, is a set of desiderata, would it be helpful to think that the DM's goal is to find a VP that satisfies all the desiderata? Of course that is not possible, since each VP satisfies a particular set of desiderata. Therefore, the aim of the DPVP is to find the VP which satisfies a set of desiderata that is comparatively more desirable to the DM, given a particular DPBO that this DM is facing. Besides, this is the purpose of the MCDM model in the DPVP.

An important modeling issue is related to establishing the DM's objectives and the criteria, and therefore to assigning consequences for building Table 1. Using a 5-level scale has been recommended for the subjective evaluations regarding the value function for Table 2. However, in some cases, it may happen that only two levels (0 or 1) can be applied. In this case, level 0 indicates the least desirable, meaning that a particular property (criterion) is undesired by the DM. Level 1, on the other hand, indicates the most desirable, meaning that the property in question is either desired by the DM or does not affect the decision process, at all.

It is noticeable that for this two-level scale, the two approaches (compensatory and non-compensatory) are equivalent. This can be easily verified by observing (1) that for all criteria i to which $V_i(C_{ij}) = 1$, the weights k_i of those criteria are added to the global value of the VP. This is similar to the effect of the preference function $(P_j(a, b))$ in (3). Therefore, (1) acts in the same way as (3). Thus, in such a cases as this, it would be easier to follow the non-compensatory approach, given the simplicity of the step for establishing the parameters of the model.

Finally it should be highlighted that while, the process illustrated in Fig. 1 is related to the DPVP, all its steps should be focused on the DPBO, which is the main purpose of the whole process. In this case the analyst should ensure that he/she considers the DM's preference throughout the process.

5 Conclusions

This paper presents a framework for aiding the choice of a VP for a business decision problem. The decision model has considered the following main issues: the non-compensatory rationality for the DM; the sequence of the decision process; the set of relevant criteria; and the evaluation matrix of properties by VPs.

The sequence of the decision process and the assumption of a non-compensatory approach for the MCDM method are justified based on the characteristics and typical context of this kind of decision.

The set of relevant criteria and the evaluation matrix of properties by VPs have already been suggested with several considerations to be included in the model [6–8].

The framework may facilitate the aiding process that the analyst will conduct by an analyst, although a few issues may be considered in future studies, such as: considerations related to a decision process with partial information.

Acknowledgments. This work had partial support from CNPq (Brazilian Research Sponsor) and from the University of Turku (financial support to the first author work there as a visiting professor in November of 2013).

References

1. de Almeida, A.T., Nurmi, H.: Aiding the choice of a voting procedure for a business decision problem. In: Proceedings of the Joint International Conference of the INFORMS GDN Section and the EURO Working Group on DSS, Toulouse University, pp. 269–276 (2014)
2. de Almeida, A.T., Cavalcante, C.A.V., Alencar, M.H., Ferreira, R.J.P., Almeida-Filho, A.T., Garcez, T.V.: Multicriteria and Multiobjective Models for Risk, Reliability and Maintenance Decision Analysis. International Series in Operations Research and Management Science. Springer, New York (2015)
3. Kim, S.H., Ahn, B.S.: Interactive group decision making procedure under incomplete information. Eur. J. Oper. Res. **116**, 498–507 (1999)
4. Dias, L.C., Climaco, J.N.: Dealing with imprecise information in group multicriteria decisions: a methodology and a GDSS architecture. Eur. J. Oper. Res. **160**(2), 291–307 (2005)
5. Roy, B.: Multicriteria Methodology For Decision Aiding. Kluwer Academic Publishers, Dordrecht (1996)
6. Nurmi, H.: Voting procedures: a summary analysis. Brit. J. Polit. Sci. **13**(2), 181–208 (1983)
7. Nurmi, H.: Comparing Voting Systems. D. Reidel Publishing Company, Dordrecht (1987)
8. Nurmi, H.: Voting Procedures under Uncertainty. Springer, Heidelberg (2002)
9. Arrow, K.: Social Choice and Individual Values, 2nd edn. Wiley, New York (1963)
10. Black, D.: Theory of Committees and Elections. Cambridge University Press, Cambridge (1958)

11. Fishburn, P.C.: Monotonicity paradoxes in the theory of elections. Discrete Appl. Math. **4**, 119–134 (1982)
12. Bartholdi, J.J., Orlin, J.B.: Single transferable vote resists strategic voting. Soc. Choice Welfare **8**, 341–354 (1991)
13. May, K.O.: Intransitivity, utility and aggregation of preference patterns. Econometrica **22**, 1–13 (1954)
14. Zermelo, E.: Die berechnung der turnier-ergebnisse als ein maximumproblem der wahrscheinlichkeitsrechnung. Math. Z. **29**, 436–460 (1929)
15. Keeney, R.L.: Value-Focused Thinking: a Path to Creative Decision Making. Harvard College, Cambridge (1992)
16. Roy, B., Słowinki, R.R.: Questions guiding the choice of a multicriteria decision aiding method. EURO. J. Decis. Process. **1**, 69–97 (2013)
17. Vincke, P.: Multicriteria Decision-Aid. Wiley, New York (1992)
18. Figueira, J., Greco, S., Ehrgott, M. (ed.): Multiple Criteria Decision Analysis: State of the Art Surveys. Springer, Heidelberg (2005)
19. Keeney, R.L., Raiffa, H.: Decision with Multiple Objectives: Preferences and Value Trade-offs. Wiley, New York (1976)
20. Fishburn, P.C.: Noncompensatory preferences. Synthese **33**, 393–403 (1976)
21. Brans, J.P., Vincke, Ph.: A preference ranking organization method: the promethee method for multiple criteria decision making. Manage. Sci. **31**, 647–656 (1985)
22. Keeney, R.L.: A group preference axiomatization with cardinal utility. Manage. Sci. **23**(2), 140–145 (1976)

Vote Swapping in Representative Democracy

Hayrullah Dindar[1], Gilbert Laffond[2], and Jean Lainé[2,3]([✉])

[1] Istanbul Bilgi University, Istanbul, Turkey
[2] Conservatoire National des Arts et Métiers,
Paris, France
[3] Murat Sertel Center for Advanced Economic Studies,
Istanbul Bilgi University, Istanbul, Turkey
jeanlaine@cnam.fr

Abstract. We investigate group manipulation by vote exchange in two-tiers elections, where voters are first distributed into districts, each with one delegate. Delegates' preferences result from aggregating voters' preferences district-wise by means of some aggregation rule. Final outcomes are sets of alternatives obtained by applying a social choice function to delegate profiles. An aggregation rule together with a social choice function define a constitution. Voters' preferences over alternatives are extended to partial orders over sets by means of either the Kelly or the Fishburn extension rule. A constitution is Kelly (resp. Fishburn) swapping-proof if no group of voters can get by exchanging their preferences a jointly preferred outcome according to the Kelly (resp. Fishburn) extension. We establish sufficient conditions for swapping-proofness. We characterize Kelly and Fishburn swapping-proofness for Condorcet constitutions, where both the aggregation rule and the social choice function are based on simple majority voting. JEL Class D71, C70.

Keywords: Representative democracy · Vote swapping · Vote exchange · Group manipulation

1 Introduction

A society has to choose a subset of a finite set of alternatives through an indirect voting procedure. Voters are first divided into a fixed number of districts, where each voter submits a preference described by a linear order over alternatives. Preferences are aggregated district-wise into some district preference. The final outcome is a subset of alternatives chosen from the profile of district preferences by means of a social choice function (SCF). It is well-known that the outcome of indirect voting may be sensitive to the distribution of voters across districts, even when the prevailing SCF is anonymous. This sensitivity may prevail even when one of two alternatives has to be chosen. Indeed, if district preferences are defined by means of simple majority voting while the final outcome is the majority winner

© Springer International Publishing Switzerland 2015
B. Kamiński et al. (Eds.): GDN 2015, LNBIP 218, pp. 227–239, 2015.
DOI: 10.1007/978-3-319-19515-5_18

among districts, it is fairly easy to find voters' preferences such that two different "district maps" lead to two different choices.[1]

The critical role played by the district map is often related to gerrymandering. Gerrymandering means a deliberate design of districts to influence the outcome of the elections. It may operate either by concentrating as many voters of one type into a single district to reduce their influence in other districts, or by spreading out voters of a particular type among many districts in order to prevent a sufficiently large voting block in any particular district. This may explain why the design of districts is always controversial.[2]

We adopt in this paper a different approach by considering how, for some distribution of voters among a given number of districts, voters can manipulate the procedure by exchanging votes in order to get a mutually preferred outcome. Instead of considering potential manipulation of the voter distribution by the incumbent party, we focus on the ability of voters to "re-shape" the district map by swapping votes. Hence, while vote swapping basically consists in re-distributing voters across districts, its scope does not address the same incentives as for gerrymandering. It can even be conceived as a strategic response to the potential manipulation by governing parties to design districts at their advantage. Vote swapping actually happens in practice, and has been favored by the existence of websites especially designed to solve the associated communication and bargaining problems. A good example is provided by the website www.votepair.ca/.[3] Other websites operated in the United States during the 2000 elections campaign, and launched intense debates about their legal status.[4] The final statement was made in 2007, confirming the legality of vote swapping.[5]

---

[1] Consider for instance the case of 15 voters divided into 5 districts with size 3 each, and where voters 1 to 9 prefer the alternative a to alternative b. Then place voters 1,2,3 in the first district, voters 4,5,6 in the second district, and dispatch each of the three remaining supporters of a in one of the last 3 districts. It follows that b wins. If now voter 3 exchange her location with any of the b supporters, a wins. This example relates to the referendum paradox (see [23,28]), which holds when indirect majority voting is inconsistent with direct majority voting (i.e. when the choice is made from the voters' preferences). Example 1 below provides another illustration of gerrymandering.

[2] One may consult the website: http://pjmedia.com/zombie/2010/11/11/the-top-ten-most-gerrymandered-congressionaldistricts-in-the-united-states/, where real districts exhibiting weird shapes are shown.

[3] On the welcome page of this website, the following text appears: "You should pair vote if either: You want to keep a political party from winning, You don't feel that there is any point in voting for who you want, as the candidate or party has no chance of getting elected, You are tired of your vote not being represented in Parliament".

[4] Once threatened by the California Secretary of States, the websites voteswap2000.com and votexchange2000.com immediately shut their virtual doors.

[5] On 8-6-2007, the 9th U.S. Circuit Court of Appeals ruled that "the websites' vote-swapping mechanisms as well as the communication and vote swaps they enabled were constitutionally protected. At their core, they amounted to efforts by politically engaged people to support their preferred candidates and to avoid election results that they feared would contravene the preferences of a majority of voters in closely contested states. Whether or not one agrees with these voters' tactics, such efforts, when conducted honestly and without money changing hands, are at the heart of the liberty safeguarded by the First Amendment."

Vote swapping describes a specific type of group-manipulation. Linking group strategy-proofness for SCFs to the immunity of voting procedures to vote swapping, or in short swapping-proofness, raises two issues. The first issue is that manipulation of an SCF is conditional to the way individuals compare sets, or put differently, to the way preferences over alternatives are extended to preferences over sets.[6] We consider two types of preferences over sets. The most conservative extension of preferences from alternatives to sets is the Kelly extension ([22]): given a linear order p over alternatives, a subset X of alternatives is preferred to another subset Y if p ranks all elements of X above all those in Y. Clearly, the Kelly extension allows to compare only disjoints sets. In contrast, the Fishburn extension ([20]) allows to compare some pairs of intersecting sets. It states that X is preferred to Y if p ranks all elements of $X \backslash Y$ above all those of $X \cap Y$, and all elements of $X \cap Y$ above all those of $Y \backslash X$. These two extensions respectively naturally define respectively the notions of Kelly swapping-proofness and Fishburn swapping-proofness. Clearly, if X is preferred to Y under the Kelly extension, it is so under the Fishburn extension. Therefore, Fishburn swapping-proofness implies Kelly swapping-proofness. The second issue when trying to use existing results on group strategy-proof SCFs is that the final outcome does not directly depend on individual preferences but instead, on district preferences. Hence misrepresenting individual preferences can alter the outcome only indirectly, by affecting aggregated preferences in one or several districts.[7]

The key notion we use in order to encompass this second issue is the notion of constitution. A constitution is a pair $\{\theta, F\}$ where θ is an aggregation rule, which maps every profile of voters' linear order into a complete binary relation over alternatives (district preferences), while F is an SCF that maps any profile of such binary relations into a subset of alternatives. Special attention is paid to two specific classes of constitutions. In a Condorcet constitution, θ is defined as the majority tournament built from voters' preferences while F is a tournament solution (which chooses from every tournament a subset of alternatives that uniquely reduces to the Condorcet winner whenever it exists). In a positional constitution, district preferences are built by using a score vector, and the choice from district preferences are computed by also using a (maybe different) score vector. Our results can be summarized as follows. First, we establish a set of sufficient conditions for Kelly (resp. Fishburn) swapping-proofness, and we prove that this set characterizes Kelly (resp. Fishburn) swapping-proofness for Condorcet constitutions. Second, we show that several constitutions based on well-known tournament solutions are Kelly swapping-proof, while very few are Fishburn swapping-proof. Finally, we prove that no positional constitution is Kelly swapping-proof (hence Fishburn swapping-proof).

---

[6] Strategy-proof SCFs (for different extended preferences over sets) are studied in particular in [3–7, 9, 11–13, 16, 19, 21, 22], and [29, 30].

[7] Note that "direct" manipulation by vote swapping cannot occur in the case of anonymous SCFs.

The paper is organized as follows. Sect. 2 is devoted to a simple model of indirect voting, where the notion of swapping-proof constitution is formally defined. Condorcet and positional constitutions are introduced in Sect. 3. Results appear in Sect. 4, and are further discussed in Sect. 5. Proofs are omitted and are available upon request.

2   Vote-Swapping in Representative Democracy

2.1   A Model of Representative Democracy

We consider a finite electorate $I_n = \{1, ..., i, ..., n\}$ involving a variable number $n \geq 3$ of voters confronting a finite set $A_m = \{a_1, ..., a_m\}$ of m alternatives where $m \geq 3$ is also variable. Voters' preferences are represented by linear orders on A_m. A *profile* is an n-tuple $p = (p_i)_{i \in I_n} \in \mathcal{L}(A_m)^n$, where $\mathcal{L}(A_m)$ stands for the set of all linear orders over A_m, and where p_i is voter i's preference. We denote by $\Pi(A_m)$ the set of all profiles over A_m, and we define $\Pi = \cup_{m \geq 3} \Pi(A_m)$. Given $p \in \Pi(A_m)$ together with $a, b \in A_m$, we define $p^{i:a,b}$ as the set of all preference profiles in $\Pi(A_m)$ such that $p' \in p^{i:a,b}$ if and only if $p'_j = p_j$ for all $j \neq i$, and $(a, b) \in p_i \Rightarrow (b, a) \in p'_i$. The electorate is divided into at least two mutually disjoint subsets, or districts, each sending one *delegate* to the upper electoral body. Formally, an *apportionment* is a partition $D = \{D_1, ..., D_K\}$ of I_n into $K \geq 2$ districts. The delegate of district D_k is denoted by k. Given a profile $p = (p_i)_{i \in I_n}$ together with an apportionment D, the profile of district k is the restriction $p \mid_{D_k}$ of p to D_k. We assume that each delegate aggregate voters' preferences in her district into a complete binary relation over A_m by means of an aggregation rule. Denoting by $\mathcal{Q}(A_m)$ the set of all complete binary relations over A_m, an aggregation rule is defined as a function $\theta : \Pi(A) \to \cup_{m \geq 3} \mathcal{Q}(A_m)$ such that $\forall m \geq 3$, $\forall p \in \Pi(A_m)$, $\theta(p) \in \mathcal{Q}(A_m)$. We assume throughout the paper that all delegates use the same aggregation rule. An aggregation rule is called *strict* if it generates an anti-symmetric district preference at any voters' profile, that is a tournament. We denote by $\mathcal{T}(A_m)$ the set of all tournaments over A_m . The aggregation rule θ is *strict* if $\forall m, n \geq 3$, $\forall p \in \mathcal{L}(A_m)^n$, $\theta(p) \in \mathcal{T}(A_m)$. Moreover, θ satisfies *representation-monotonicity* (hereafter *Rep-Mon*) if $\forall p \in \mathcal{L}(A_m)$, $\forall i \in I_n$, $\forall a \in A_m$, $\forall b \in A \backslash \{a\}$ with $a \varphi(p) b$, we have $(a, b) \in \varphi(p')$ for all $p' \in p^{i:a,b}$. In words, weakly strengthening an alternative in one voter's preference without altering other preferences cannot reverse its comparison with initially less preferred alternatives.

Given a voters' profile p together with an apportionment $D = \{D_1, ..., D_K\}$, delegate k's preferences are described by $\theta(p \mid_{D_k})$. In order to make notations as simple as possible, we write $P_k = \theta(p \mid_{D_k})$ for all $k = 1, ..., K$. Gathering all delegate preferences defines the *delegate profile* $p_{D,\theta} = (P_1, ..., P_K) \in \mathcal{Q}(A_m)^K$.

Final decisions are obtained by applying a voting rule to the delegate profile. Formally, a voting rule is a correspondence $F : \cup_{K \geq 2, m \geq 3} \mathcal{Q}(A_m)^K \to \cup_{m \geq 3} A_m$ such that $\forall K \geq 2$, $\forall n \geq 3$, $\forall Q \in \mathcal{Q}(A_m)^K$, $F(Q) \subseteq A_m$. Hence, if p and

$D = \{D_1, ..., D_K\}$ and are the prevailing voters' profile and apportionment, the final outcome is given by $F(p_{D,\theta}) = F(P_1, ..., P_K)$.

A *constitution* is a pair $\{\theta, F\}$ involving an aggregation rule (describing how delegate preferences emanate from citizens' preferences) and a voting rule (describing how one or several alternatives are chosen from delegate preferences). A constitution $\{\theta, F\}$ is strict if θ is strict. A *voting situation* for $\mathcal{C}$ is a 5−tuple $(I_n, A_m, p, D, \mathcal{C})$ involving a set I_n of n voters, a set A_m of m alternatives, a profile $p \in \mathcal{L}(A_m)^n$, an apportionment D describing how I_n is divided into districts, and a constitution $\{\theta, F\}$.

2.2 Swapping-Proofness

The well-known sensitivity of indirect elections outcomes to apportionment is usually associated to *gerrymandering*, defined as a strategic attempt by a running office party to establish a political advantage by manipulating district boundaries. It is already known that no reasonable constitution is immune to gerrymandering in the case where voters cast ballots naming only one alternative and where only one alternative is finally chosen ([10,14]). Moreover, every constitution based on simple majority voting is exposed to gerrymandering ([24]). This is illustrated by the following example.

Example 1. We consider a constitution such that whenever the number of districts is odd and each district contains an odd number of voters, (1) delegate preferences are built pairwise majority comparisons (delegate k ranks alternative a_j above another alternative a_h if more than half of voters in district D_k do so), and (2) the final outcome is the set of all alternatives which defeat the highest number of alternatives in the delegate profile according to the majority rule.[8] Consider the voting situation with 9 voters, 5 alternatives, 3 districts each with size 3, where voters' preferences are shown below

$$p = \begin{pmatrix} P_1, P_2, P_3 \ P_4 \ P_5 \ P_6 \ P_7, P_8, P_9 \\ \hline a_4 & a_5 \ a_5 \ a_2 & a_3 \\ a_5 & a_2 \ a_4 \ a_3 & a_1 \\ a_1 & a_3 \ a_2 \ a_1 & a_2 \\ a_2 & a_1 \ a_3 \ a_4 & a_4 \\ a_3 & a_4 \ a_1 \ a_5 & a_5 \end{pmatrix}$$

and where the apportionment is $D = (D_1, D_2, D_3) = (\{1, 2, 3\}, \{4, 5, 6\}, \{7, 8, 9\})$.

---

[8] Given a profile of district preferences $p_{D,\theta} = (\theta(p \mid_{D_1}), ..., \theta(p \mid_{D_K}))$, the majority tournament among districts $T(p_{D,\theta})$ is defined by $\forall a, b \in A_m$, $a\, T(p_{D,\theta})\, b$ iff $|\{k \in \{1, ..., k\} : a\theta(p \mid_{D_k})b\}| > \frac{K}{2}$. The final outcome we consider in this example is known as the Copeland set of $T(p_{D,\theta})$.

3   Condorcet and Positional Constitutions

Special attention is paid below to two specific classes of constitutions: Condorcet constitutions, where delegate preferences and the voting rule are based on simple majority voting, and positional constitutions, where simple majority is replaced by a scoring rule.

3.1   Condorcet Constitutions

A constitution is *Condorcet* if both delegate preferences and final choices are based on pairwise majority comparisons of alternatives. Given any integer L and any L-tuple of tournaments $\pi = (T_1, ..., T_L) \in \mathcal{T}(A_m)^L$ where L is odd, the tournament $T(\pi)$ over A_m is defined by: for all $h, j \in \{1, ..., m\}$, $a_h \, T(\pi) \, a_j$ if and only if $|\{i \in \{1, ..., L\} : a_h T_i a_j| > \frac{L}{2}$. We call $T(\pi)$ the majority tournament from π. Hence an alternative a_h defeats another alternative a_j in tournament $T(\pi)$ if a_h defeats a_j in more than half of the tournaments in π . Furthermore, the *Condorcet winner* of tournament T the (necessarily unique) alternative cw_T such that $cw_T \, T \, b$ for all $b \in A_m \backslash \{cw_T\}$. We call *admissible* a voting situation where each district preference is a tournament, and where the majority tournament across district is also well-defined. Since indifference is not allowed in voters' preferences, any voting situation involving tan odd number of districts, each involving an odd number of voters, is admissible. We define Condorcet constitutions in restriction to admissible voting situations. The majoritarian aggregation rule θ_{maj} is defined by: $\forall m \geq 3$, $\forall n \geq 3$ with n odd, $\forall p \in \mathcal{L}(A_m)^n$, $\theta_{maj}(p) = T(p)$. In Condorcet constitutions, each delegate compare alternatives according to the majority will in her district. Given a profile P together with an apportionment $D = (D_1, ..., D_K)$, delegate k's preferences are thus described by $T(p \, |_{D_k}) = T_k$, and the delegate profile is the K-tuple $p_{D,\theta_{maj}} = (T_1, ..., T_K)$. The tournament $T(p_{D,\theta_{maj}})$, called delegate tournament, is well-defined in any admissible voting situation. A voting rule F is *Condorcet consistent* if for any $m, n \geq 3$ with n odd and any $\pi = (T_1, ..., T_n) \in \mathcal{T}(A_m)^n$, $F(\pi) = \{cw_{T(\pi)}\}$ whenever $T(\pi)$ has a Condorcet winner.

Definition 1. *A constitution $\mathcal{C} = \{\theta, F\}$ is Condorcet (for admissible voting situations) if $\theta = \theta_{maj}$ and F is Condorcet consistent.*

In Condorcet constitutions, final choices $F(T(p_{D,\theta_{maj}}))$ are made from the delegate profiles by means of a voting rule that uniquely selects an alternative preferred to any other alternatives in a majority of districts when it exists. An interesting class of Condorcet constitutions involves *tournament solutions* as voting rules. A tournament solution is a Condorcet consistent correspondence $S : \cup_{m \geq 3} \mathcal{T}(A_m) \to \cup_{m \geq 3} A_m$ such that $S(T) \subseteq A_m$ for all $m \geq 3$ and all T in $\mathcal{T}(A_m)$, $S(T) \subseteq A_m$. Example 1 above involves a Condorcet constitution based on the tournament solution called the Copeland Set.

We consider several well-known tournament solutions that we briefly define below. An extensive survey of tournament solutions can be found in [25]. Given a tournament T over A_m,

If θ is built by using the Borda score vector for 5 alternatives, we get the following delegate profile

$$p_{D,\theta} = \begin{pmatrix} \dfrac{\theta(p\mid_{D_1}) \quad \theta(p\mid_{D_2})}{} \\ \begin{matrix} a_3 & a_2, a_5 \\ a_5 & a_3 \\ a_1 & a_4 \\ a_2 & a_1 \\ a_4 & \end{matrix} \end{pmatrix}$$

It follows that $I(p_{D,\theta}) = 5$, and thus the voting rule F is based on the same score vector as the one for θ. It is easily checked that $F(p_{D,\theta}) = \{a_3, a_5\}$.

4   Results

4.1   Kelly Swapping-Proofness

We first establish a sufficient condition for Kelly swapping-proofness.

Theorem 1. *A constitution $\{\theta, F\}$ is Kelly swapping-proof if θ is strict and satisfies (Rep-Mon) and F satisfies (SWIUA).*

An SCF satisfies the property of *independence of unchosen alternatives (IUA)* if its outcome is non-sensitive to changes in the delegate profile with respect to unchosen alternatives. Formally, F satisfies *(IUA)* if $F(P) = F(P')$ for all $P, P' \in \mathcal{Q}(A_m)^K$ such that $P_k\mid_{\{a,b\}} = P'_k\mid_{\{a,b\}}$ for all $(a,b) \in F(P) \times A$ and all $k \in \{1, ..., K\}$. It is easy to prove that $IUA)$ implies *(SWIUA)*. Hence, a corollary of Theorem 1 is

Corollary 1. *A constitution $\{\theta, F\}$ is Kelly swapping-proof if θ is strict and satisfies (Rep-Mon) and F satisfies (IUA).*

As a consequence of Corollary 1, we get the existence of Kelly swapping-proof Condorcet constitutions. Indeed, BP satisfies *(IUA)*. Moreover, θ_{maj} is strict at any admissible voting situation and clearly satisfies *(Set-Mon)*. Finally, it directly follows from the Kelly principle that if $\{\theta, G\}$ is Kelly swapping-proof and $F(p_{D,\theta}) \supseteq G(p_{D,\theta})$ at any voting situation, then $\{\theta, F\}$ is Kelly swapping-proof. As a consequence, one gets

Corollary 2. *If F is a superset of BP, then $\{\theta_{maj}, F\}$ is Kelly swapping-proof at any admissible voting situation.*

Since TC, UC, UC^∞ and MC are supersets of BP, a constitution based on each of those tournament solutions is Kelly-swapping-proof. Furthermore, if $F \in \{COP, SL\}$, then $\{\theta_{maj}, F\}$ is not Kelly swapping-proof. Indeed, Example 1 shows that $\{\theta_{maj}, COP\}$. Moreover, it is easily checked that in Example 1, when $S(\sigma) = \{3,6\}$, $SL(T(p_{D,\theta_{maj}})) = COP(T(p_{D,\theta_{maj}})) = \{a_5\}$ while

$SL(T(p_{D^\sigma, \theta_{maj}}) = COP(T(p_{D^\sigma, \theta_{maj}})) = \{a_4\}$. Since $a_4 p_i a_5$ for $i \in \{3, 6\}$, we conclude that $\{\theta_{maj}, SL\}$ is not Kelly swapping-proof.

Another consequence of Corollary 1 is that Kelly swapping-proofness holds under a property of the voting rule weaker than the property of *set monotonicity* introduced in [11], where group strategy-proofness is considered in direct voting (i.e. for one-district elections) and non-transitive asymmetric and complete preferences. Interpreting districts as voters, an SCF F satisfies *set-monotonicity (Set-Mon)* if for all $m, K \in \mathbb{N}$, for all $p \in \mathcal{Q}(A_m)^K$, for all $k \in \{1, ..., K\}$, for all $a \in A_m$ and for all $b \in A_m \backslash F(p)$, we have $F(p) = F(p^{k:a,b})$. Set-monotonicity means that the outcome is non-sensitive the weakening of some unchosen alternative. It is proved in [11] that a voting rule is Kelly group strategy-proof if and only if it satisfies set monotonicity.[9] It is easy to show that *(Set-Mon)* implies *(IUA)*. However, this does not imply Theorem 1, since in our setting the voting rule is not applied directly to voters' preferences.

It is obviously checked that θ_{maj} satisfies *(Rep-Mon)*. Our next result is that *(SWIUA)* actually characterizes Kelly swapping-proof Condorcet constitutions.

Theorem 2. *A Condorcet constitution $\{\theta_{maj}, F\}$ is Kelly swapping-proof if and only if F satisfies (SWIUA).*

In contrast with Condorcet constitutions, all positional constitutions are exposed to manipulation by vote swapping:

Theorem 3. *No positional constitution is Kelly swapping-proof.*

We actually prove the following stronger result: every positional constitution is Kelly manipulable at some voting situation where the outcome is a singleton and where all districts have the same size.

4.2 Fishburn Swapping-Proofness

Observe first that, since Fishburn swapping-proofness clearly implies Kelly swapping-proofness, Theorem 3 implies that no positional constitution is Fishburn swapping-proof.

We establish below a sufficient condition for Fishburn swapping-proofness. Beforehand we introduce three properties for SCFs.

An SCF F satisfies the property of *symmetric independence of unchosen alternatives (SIUA)* if $F(P) = F(P')$ for all $P, P' \in \mathcal{Z}(A)^n$, if $P \backslash P' \subseteq (A \backslash F(P) \times A \backslash F(P)) \cup (A \backslash F(P') \times A \backslash F(P'))$. In words, any two delegate profiles that differ only on alternatives unchosen for either of them lead to the same choice set. Clearly, *(SIUA)* is stronger than *(IUA)*. Moreover, F satisfies the property of *independence of jointly chosen alternatives (IJCA)* if $F(P) = F(P')$ for all $P, P' \in \mathcal{Z}(A)^n$, if $P \backslash P' \subseteq [(F(P) \cap F(P')) \times (F(P) \cap F(P'))]$. In words, any two delegate profiles that differ only on jointly chosen alternatives lead to the same choice set. Furthermore, F satisfies the property

---

[9] *(Set-Mon)* is actually sufficient for Kelly group strategy-proofness when preferences are transitive, as proved in [12].

of *strong independence of jointly chosen alternatives (SIJCA)* if it satisfies (IJCA) and if, in addition, $F(P) = F(P')$ for all $P, P' \in \mathcal{Z}(A)^n$, if $P \backslash P' \subseteq [(F(P) \cap F(P')) \times (A \backslash F(P) \cup F(P'))] \cup [(A \backslash F(P) \cup F(P')) \times (F(P) \cap F(P'))]$.

Theorem 4. *A constitution $\{\theta, F\}$ is Fishburn swapping-proof if θ is strict and satisfies (Rep-Mon) and F satisfies (SIUA) and (SIJCA).*

Another interesting sufficient condition can be obtained by completing *(Set-Mon)* with the property of *exclusive independence of chosen alternatives (EICA)* introduced in [13]. A voting rule satisfies *(EICA)* if unchosen alternatives remain unchosen when delegate preferences are modified only between chosen alternatives. Formally, F satisfies *(EICA)* if for all $m, K \in \mathbb{N}$ where $K > 2$, for all $P, P' \in \mathcal{Q}(A_m)^K$ such that $P_k |_{\{a,b\}} = P'_k |_{\{a,b\}}$ for all $k \in \{1, ..., K\}$ and for all $a, b \in A_m$ with $b \in A_m \backslash F(P)$, we have $F(P') \subseteq F(P)$. It is shown in [13] that in direct elections with complete asymmetric and non-transitive preferences, *(Set-Mon)* and *(EICA)* together are sufficient for Fishburn group strategy-proofness. While this result cannot be directly applied in our framework, we can nonetheless show the following

Theorem 5. *A constitution $\{\theta, F\}$ is Fishburn swapping-proof if θ is strict and satisfies (Rep-Mon) and F satisfies (Set-Mon) and (EICA).*

Since the Top-Cycle TC satisfies (Set-Mon) and (EICA), we get

Corollary 3. *The Condorcet constitution $\{\theta_{maj}, TC\}$ is Fishburn swapping-proof.*

Since the Copeland set COP, the Banks set B, the Slater set SL, the Uncovered set UC and its refinement UC^∞ violate (IUA) (see [25] for proofs), we get

Corollary 4. *If $F \in \{COP, SL, B, UC, UC^\infty\}$, no Condorcet constitution $\{\theta_{maj}, F\}$ is Fishburn swapping-proof.*

While (IUA) necessarily holds in a Fishburn swapping-proof Condorcet constitution, it is not sufficient. Indeed, the Bipartisan set BP and the Minimal Covering set MC both satisfy (IUA) but are not Fishburn swapping-proof.

Proposition 1. *The constitutions $\{\theta_{maj}, MC\}$ and $\{\theta_{maj}, BP\}$ are not Fishburn swapping-proof.*

Our last result in this section is that (SIUA) and (SIJCA) characterizes Fishburn swapping-proof Condorcet constitutions.

Theorem 6. *A Condorcet constitution $\{\theta_{maj}, F\}$ is Fishburn swapping-proof if and only if F satisfies (SIUA) and (SIJCA).*

5   Discussion

(1) In order to analyze the possibility of strategic manipulation by vote swapping, we model indirect elections as a two-stage procedure where in the first stage, individual votes (linear orders over alternatives) in each district are aggregated into a delegate preference, and where in the second stage a subset of alternatives is chosen from the profile of all delegate preferences. This two-stage procedure defines a constitution, that is a pair $\{\theta, F\}$, where θ is an aggregation rule, and where F is a multi-valued SCF. We show that if θ satisfies a monotonicity property and if F is satisfies the symmetric weak independence of unchosen alternatives (resp. satisfies symmetric independence of unchosen alternatives and strong independence of jointly chosen alternatives), $\{\theta, F\}$ is vote swapping-proof if voters compare sets according to the Kelly principle (resp. the Fishburn principle). Moreover, we prove that these conditions respectively Kelly and Fishburn swapping-proof Condorcet constitutions.

(2) To the best of our knowledge, this paper provides the first analysis of swapping-proofness. Previous studies of two-tiers elections focus on gerrymandering. In particular, every Condorcet constitution is exposed to gerrymandering ([24]). Moreover, the same hold for every constitution satisfying a mild separability property ([17]). Alternative frameworks are proposed in [1,2,10] and [14]. In [10] and [14,15], constitutions are defined such that in each district voters vote for a single alternative in A, but differ on the way the final outcome is chosen from district votes. If a "winners take all" assumption holds (each delegate is given a weight equal to the size of her district), unanimity, anonymity and immunity to the referendum paradox together characterize the very narrow class of partial priority rules ([14,15]). If instead district votes are built by means of some (maybe district-specific) voting function while the final vote is computed from district votes by using another voting function, and if the constant constitution (that gives the same outcome regardless the vote profile)) is banned, gerrymander-proofness combined with mild additional assumptions implies that each voter is pivotal, in the sense that she can change the winner by changing her vote at some unanimous profile ([10]). In [1,2], voters in each district do not vote directly on alternatives, but instead select a candidate defined as a linear order over alternatives. Voters rank candidates according to the Kemeny distance to their own ranking. If simple majority voting operates both for candidate selection and for choosing the set of winning candidates, then direct voting and indirect voting may fail to coincide even if voters' preferences are such that majority voting leads to a clear winner. In contrast with these approaches, a key feature of ours is that voters vote neither for one alternative nor for a candidate. Each voter report preferences over alternatives, and these preferences are aggregated district-wise into a district preference. Moreover, in contrast with [1,2], we do not model how delegates are chosen: each district is assigned one delegate whose preferences are emanating from voters' preferences. In particular, no choice is made at the district level. Furthermore, in contrast with [10] and [14,15], we do not restrict the agenda at the district level (all alternatives remain potentially

chosen from district preferences), and the final outcome is defined as a subset of alternatives rather than a single one.

(3) Manipulation by vote swapping extends to constitutions where district actually select alternatives, and by doing so restrict the agenda for the final outcome provide additional assumptions hold for the prevailing SCF. Given an aggregation rule θ and an SCF F, the sequential constitution $Seq\{\theta, F\}$ is formally defined as follows. Consider an apportionment $D = (D_1, ..., D_K)$ of the voter set I_n, and a voter profile p over A_m. Each district D_k chooses the subset $F(p \mid_{D_k})$ and rank all alternatives according to the district preference $\theta(p \mid_{D_k})$. The agenda at the upper level of choice is the set of all alternatives chosen in at least one district, that is $\cup_{1 \leq k \leq K} F(p \mid_{D_k})$, and the profile of district preferences over the agenda is thus $p_{D,\theta}|_{\cup_{1 \leq k \leq K} F(p|_{D_k})} = (\theta(p \mid_{D_1})|_{\cup_{1 \leq k \leq K} F(p|_{D_k})}, ..., \theta(p \mid_{D_K})|_{\cup_{1 \leq k \leq K} F(p|_{D_k})})$. Then $Seq\{\theta, F\}$ is Kelly swapping-proof if there is no $n, m \in \mathbb{N}$, no apportionment D, no voters' profile p, and no permutation σ of I_n such that $F(p_{D^\sigma, \theta})|_{\cup_{1 \leq k \leq K} F(p|_{D_k^\sigma})} \; p_i^K \; F(p_{D,\theta})|_{\cup_{1 \leq k \leq K} F(p|_{D_k})}$ for all $i \in S(\sigma)$. Fishburn swapping-proofness is similarly defined. Then we have the following

Proposition 2. *Let F be an SCF satisfying the Aizerman property (resp. the strong superset property). If a constitution $\mathcal{C} = \{\theta, F\}$ is not Kelly (resp. Fishburn) swapping-proof, then any sequential constitution $\{\theta, F, F_1, ..., F_k, ...\}$ is not swapping-proof.*[10]

Since MC and BP satisfies the strong superset property, Propositions 1 and 2 together imply that neither $Seq\{\theta_{maj}, MC\}$ nor $Seq\{\theta_{maj}, BP\}$ is Fishburn swapping-proof. Moreover, since scoring rules satisfy the Aizerman property, combining Theorem 3 and Proposition 2 shows that no positional sequential constitution is Kelly swapping-proof.

(4) While we mainly focus on Condorcet and positional constitutions, we can also make several statements about other constitutions. For instance, it follows from Theorem 2 that if θ satisfies (Rep-Mon) and F is either the Condorcet SCF (that uniquely selects the Condorcet winner whenever it exists, and all alternatives otherwise) or the Pareto SCF (that selects all alternatives not Pareto-dominated), then $\{\theta, F\}$ is Fishburn swapping-proof, hence Kelly swapping-proof.

References

1. Baldiga, K.: A failure of representative democracy, mimeo (2011)
2. Baldiga, K.: Representative democracy and the implementation of majority-preferred alternatives, mimeo (2012)
3. Bandyopadhyay, T.: Threats, counter-threats and strategic manipulation for non-binary group decision rules. Math. Soc. Sci. **2**, 145–155 (1982)

---

[10] F satisfies the Aizerman property if $\forall n, m \in \mathbb{N}$, $\forall a, b \in A_m$, $\forall P \in \mathcal{Q}(A_m)^n$ with $F(P) \subseteq B$, then $F(P/B) \subseteq F(P)$. Moreover, F satisfies the strong superset property if $F(P/B) = F(P)$.

4. Bandyopadhyay, T.: Multi-valued decision rules and coalitional non-manipulability. Econ. Lett. **13**, 37–44 (1983)
5. Barberà, S.: Manipulation of social decision functions. J. Econ. Theor. **15**, 266–278 (1977)
6. Barberà, S.: The manipulation of social choice mechanisms that do not leave "too much" to chance. Econometrica **45**, 1573–1588 (1977)
7. Barberà, S.: Strategy-proof social choice. In: Arrow, K., Sen, A.K., Suzumura, K. (eds.) Handbook of Social Choice and Welfare, vol. 2, pp. 731–832. Elsevier, New York (2010)
8. Barberà, S., Bossert, W., Pattanaik, P.K.: Ranking sets of objects. In: Barberà, S., Hammond, P., Seidl, C. (eds.) Handbook of Utility Theory, vol. 2, pp. 893–977. Springer, Heidelberg (2004)
9. Barberà, S., Dutta, B., Sen, A.: Strategy-proof social choice correspondences. J. Econ. Theor. **101**, 374–394 (2001)
10. Bervoets, S., Merlin, V.: Gerrymander-proof representative democracies. Int. J. Game Theory **41**, 473–488 (2012)
11. Brandt, F.: Group strategy-proof irresolute social choice functions, Technical report, Technische Universität München (2011)
12. Brandt, F.: Set-monotonicity implies Kelly-strategyproofness, Technical report, Technische Universität München (2014)
13. Brandt, F., Brill, M.: Necessary and sufficient conditions for the strategyproofness of irresolute social choice functions, Technical report, Technische Universität München (2011)
14. Chambers, C.P.: Consistent representative democracy. Games Econ. Behav. **62**, 348–363 (2008)
15. Chambers, C.P.: An axiomatic theory of political representation. J. Econ. Theor. **144**, 375–389 (2009)
16. Ching, S., Zhou, L.: Multi-valued strategy-proof social choice rules. Soc. Choice Welfare **19**, 569–580 (2002)
17. Dindar, H., Laffond, G., Lainé, J.: The Strong referendum paradox, Murat Sertel Center for Advanced Economic Studies, Working Paper 2013–3 (2013)
18. Duggan, J., Schwartz, T.: Strategic manipulability without resoluteness or shared beliefs: Gibbard-Satterthwaite generalized. Soc. Choice Welfare **17**, 85–93 (2000)
19. Feldman, A.: Manipulation and the Pareto rule. J. Econ. Theor. **21**, 473–482 (1979)
20. Fishburn, P.C.: Even-chance lotteries in social choice theory. Theor. Decis. **3**, 18–40 (1972)
21. Gardenförs, P.: Manipulation of social choice functions. J. Econ. Theor. **13**, 217–228 (1976)
22. Kelly, J.S.: Strategy-proofness and social choice functions without single-valuedness. Econometrica **45**, 439–446 (1977)
23. Lacy, D., Niou, E.M.S.: A Problem with referendums. J. Theor. Polit. **12**, 5–31 (2000)
24. Laffond, G., Lainé, J.: Representation in majority tournaments. Math. Soc. Sci. **39**, 35–53 (2000)
25. Laslier, J.F.: Tournament Solutions and Majority Voting. Springer Verlag, Heidelberg (1997)
26. Garvey, Mac: D.: A Theorem on the construction of voting paradoxes. Econometrica **21**, 608–610 (1953)
27. MacIntyre, I., Pattanaik, P.K.: Strategic voting under minimally binary group decision functions. J. Econ. Theor. **25**, 338–352 (1981)

28. Nurmi, H.: Voting Paradoxes and How to Deal with them. Springer Verlag, Heidelberg (1999)
29. Sato, S.: On strategy-proof social choice correspondences. Soc. Choice Welfare **31**, 331–343 (2008)
30. Umezawa, M.: Coalitionally strategy-proof social choice correspondences and the Pareto rule. Soc. Choice Welfare **33**, 151–158 (2009)

The Choice of Voting Rules Based on Preferences over Criteria

Hannu Nurmi[(⊠)]

Department of Political Science and Contemporary History,
University of Turku, Turku, Finland
hnurmi@utu.fi
http://users.utu.fi/hnurmi/homepage

Abstract. The community of voting system experts is largely divided on the issue of the best voting rule. Some – perhaps a majority – of the community stresses the performance related to Condorcet's intuition, while others take a more "positional" view of the voting rules. This paper approaches the choice of the rule from the viewpoint of the individuals that will subsequently be applying the chosen rule in solving opinion aggregation problems. Our first starting point is that each individual has a preference ranking over the criteria. This starting point reduces the rule selection into the classic social choice problem. Using the Borda count one is able to construct a vector of weights that reflects the importance that the individuals assign to various criteria. Using the analytic results on the compatibility of various rules and criteria we can then associate each rule with a value that reflects the aggregated opinion of the importance criteria. Hence, the choice of the rules gets its justification from the views that the individuals have on the significance of the criteria. Our second starting point is based on weights that individuals associate with the criteria. The collective weights are then determined as in range voting. Again a justification of the chosen rules can be expressed in terms of the importance that individual assign to criteria.

Keywords: Voting procedures · Kemeny's rule · Borda count · PROMETHEE

1 Introduction

Some years ago a group of voting theorists and electoral experts got together for a symposium in Normandy, France. The proceedings of the symposium were later on edited by Felsenthal and Machover [7]. At the end of the symposium an impromptu discussion was held among the participants about the best voting system to be used in a hypothetical situation involving the election of the director of a municipality. In other words, the system should be applicable in electing a single winner. In the discussion various procedures were proposed and the session was concluded with a vote. The alternatives – altogether 18 in number – were

© Springer International Publishing Switzerland 2015
B. Kamiński et al. (Eds.): GDN 2015, LNBIP 218, pp. 241–252, 2015.
DOI: 10.1007/978-3-319-19515-5_19

the voting systems proposed in the discussion and the ballot aggregation method was the approval voting.[1] The results are reproduced in Table 1.

Table 1. The number of approved procedures. *Source:* [10].

Number of approvals	0	1	2	3	4	5	6	7	8	9	10	> 10	total
Number of ballots		0	2	7	3	5	2	1	1	0	0	1	0	22

The table shows a fairly wide variation in the number of approved systems. Yet, a vast majority of voters approved $2-4$ systems. The procedures are listed in Table 2 which also indicates the number of approvals given to each one of them as well as the percentage of voters approving of each system. The reader unfamiliar with the procedures is referred to Laslier's article [10] which also provides a comprehensive analysis of the voting data.

A couple of observations about Table 2 are in order. Firstly, no procedure was approved of by all participants. Secondly, some proposed systems received no approval votes at all. Thirdly (and related to the preceding point), the most common voting system – the plurality or one-person-one-vote procedure – was voted for by no participant. Fourthly, the winner – the approval voting – was approved of by more than two thirds of the voters.

The first point provides the main motivation for this article. It shows that the expert community is not unanimous about the best voting procedure. Glancing at the statements that several voters give to support their ballots one immediately notices that the participants seem to emphasize somewhat different criteria when choosing their favorite systems. It is plausible to think that the very existence of many voting procedures can similarly be explained by the emphasis placed on different criteria of performance of procedures. The next section provides a theoretical reconstruction of some of the best-known systems in terms of this reasoning. Thereafter, we present the main contribution of this article, viz. a method for choosing a voting procedure on the basis of the participants' priorities regarding the performance criteria.

2   A Reconstruction of the Emergence of Some Voting Procedures

The literature on choosing the rule to make choices has a long history. Indeed, all constitutional thinking based on popular sovereignty touches upon the issue

---

[1] The impromptu nature of the proceedings is reflected by the somewhat light-hearted brainstorming debate preceding the vote as well as by the fact that the voters were not asked to reveal anything else but their approved systems. Several weeks after the meeting the participants were asked to disclose their reasons for voting the way they did, but at this time many didn't recall the systems they approved of, much less the reasons for doing so. Thus, we do not know how much the election outcome depends on the aggregation system adopted [10].

Table 2. The procedures and the distribution of approvals. *Source*: [10]

Voting rule	Approvals	Approving %
approval	15	68.18
alternative	10	45.45
Copeland	9	40.91
Kemeny	8	36.36
runoff	6	27.27
Coombs	6	27.27
Simpson	5	22.73
m. judgment	5	22.73
Borda	4	18.18
Black	3	13.64
range	2	9.09
Nanson	2	9.09
leximin	1	4.54
top cycle	1	4.54
uncovered	1	4.54
Fishburn	0	0
untrapped	0	0
plurality	0	0

of the principles of how the collectively binding choices are to be made and how the changes in those principles can legitimately be brought about. Three major works published about half a century ago set the stage and basic parameters for the recent contributions: Buchanan and Tullock's treatise, Rae's seminal article on majority rule and Niemi and Weisberg's edited volume [3,14,19].[2] All these focused on the rule that from an individual's point of view would be optimal either in minimizing the expected external and decision making costs (Buchanan and Tullock) or minimizing the expected probability of being on the losing side (Rae) or maximizing the long-term expected utility under various assumptions regarding other people's support probabilities (Niemi and Weisberg). More recently, the focus has shifted to the stability (or self-stability) of voting rules. The rule is defined as stable whenever it – once in use – will not be defeated by any other rule in pairwise contest when the said rule is being applied [1].

In the present paper these works are of limited applicability. To wit, they focus on optimal majority thresholds to be applied in dichotomous choice situations. Hence, they are not directly useful in settings involving more than two

---

[2] The problem of 'optimal' decision rule has, of course, a much longer history. See e.g. [22].

alternatives. Secondly, they assume individual utilities over outcomes (present or future) ensuing from the application of the chosen rules. As far as utilities over future outcomes are concerned, these are inherently ambiguous (see [18]): not only are the outcomes uncertain, but so are the utility functions of the participants. For our purposes a more useful approach is the one adopted by Diss and Merlin [5]. These authors start from individual preferences over procedures or procedural preferences. A given procedure is self-selecting in a profile (over procedures) if it results in itself being chosen in this profile. A set of procedures, on the other hand, is stable if in any profile over procedures at least one of its elements is self-selective. Our approach differs from that of Diss and Merlin in not assuming individual preferences over procedures. Instead we assume preferences over social choice desiderata, i.e. properties that choice rules may or may not be endowed with. In other words, we start from procedural preferences, but in a roundabout manner. Instead of assuming voter preference ranking over rules, we assume their preference rankings over values or norms or performance criteria. In this assumption we are closer to Dietrich's view [4] which is based on a property called procedural autonomy (see also [21]). This amounts to the requirement that the procedural judgments within the group should determine the social choice rule to be adopted. Now, procedural judgments can, of course, be simply rankings over procedures, but – at least conceptually – they can be properties of procedures as well.

Perhaps the most common of all voting procedures is the plurality rule: each voter has one vote at his/her disposal and the candidate or policy alternative receiving more votes than any of its contestants wins. The rationale of this rule is obvious: no other candidate gets as many votes as the winning one. However, it may happen that the plurality winning candidate gets less than 50 % of the votes. Hence it may not always get the support of the majority. To rectify this eventuality the plurality runoff system has been devised. It works precisely as the plurality procedure, but in case the plurality winner receives at most 50 % of the votes, a runoff is arranged between the two largest vote-getters. Whichever gets more votes than the other on this second round of voting is the winner. Thus, the winner can always claim to be supported by more than half of the electorate. More importantly, the runoff system guarantees that an eventual Condorcet loser is not elected. This simply follows from the fact that the plurality winner has to defeat by a majority of votes at least one other alternative, viz. its competitor on the second round of voting. If there is a winner already on the first round, i.e. there is a candidate ranked first in the opinion of a majority of voters, then of course the winner would defeat all the others in pairwise contests.

Another way of avoiding Condorcet losers being elected was discovered about two hundred years ago by Jean-Charles de Borda and is today known as the Borda count.[3] Thus, we have two solutions to the problem of electing a Condorcet

---

[3] The method was invented already in the 15'th century by Nicholas of Cusa, but arguably he did not emphasize the particular problem related to the plurality voting, viz. that it may result in the election of a candidate that would lose the pairwise contests against any other candidate (see [9,12]).

loser: the plurality runoff and the Borda count. Yet, the former is accompanied with a new problem, not faced with by the latter and, more importantly, by the plurality procedure, viz. nonmonotonicity. Table 3 illustrates this problem.

Table 3. Nonmononicity of the plurality runoff

6 voters	5 voters	4 voters	2 voters
A	C	B	**B**
B	A	C	**A**
C	B	A	C

Supposing that the voters vote according to their preferences listed in Table 3, there will not be a first-round winner, but a runoff that takes place between A and B. In this runoff the winner is A since it is preferred to B by those 5 voters whose favorite didn't make it to the runoff. Suppose now that the two voters on the right with ranking BAC would change their opinion with regard to A and B (marked by bold letters in the table) so that the winner A would be preferred to B by these two voters. In this new profile – which differs from the Table 3 one so that the winner (A) gets more support than originally – a runoff is still needed, but this time one between A and C. This runoff is won by C. This shows that additional support may, indeed, turn winners into non-winners under the plurality runoff system. Hence the procedure is nonmonotonic.

Similarly as plurality runoff can be seen as an attempt to improve upon the plurality system, Nanson's method can be seen – and was in fact seen by its inventor E. J. Nanson – as a way to rectify an apparent flaw in another system, viz. the Borda count (see [12]). For more than two centuries it has been known that the Borda count does not always end up with the Condorcet winner. Nanson set out to devise a system that would be as similar to the Borda count as possible, but still guarantee the choice of an eventual Condorcet winner. The system is based on the observation concerning the relationship between Condorcet and Borda winners. While it is known that the former winners are not necessarily ones with the highest Borda scores, it is still the case that they never have very low Borda scores. More specifically, an eventual Condorcet winner always has a higher than average Borda score. Nanson's method is based on this observation: it proceeds in rounds whereby the alternatives with an average or lower Borda score are eliminated and new scores are computed for the remaining alternatives until the winner is found. The criterion used in elimination guarantees that an eventual Condorcet winner is not eliminated.

So, the system invented by Nanson was, indeed, capable of solving a specific shortcoming of the Borda count. However, as was the case with plurality and plurality runoff systems, the solution procedure (here Nanson's method) has a flaw that the "flawed" system (here the Borda count) is not associated with. This is nonmonotonicity: while the Borda count is monotonic, Nanson's method isn't [8]. This illustrates the nature of many social choice results: they demonstrate incompatibilities between properties of choice functions. In short, procedures with all desirable properties do not exist. Trade-offs have to be made between desiderata.

This well-known state of affairs suggests a new angle to the problem of choosing a procedure of choice. Instead of fixing specific flaws in the systems that are being used – and thereby conceivably coming up with systems with flaws that the already used ones do not have – one could start from the criteria that one regards of primary importance. Different people may put different value on various criteria. This was clearly exemplified in the introduction of this article. Hence, it would make sense to take into account and make use of the information regarding differences in valuation by different voters when choosing a system to be resorted to in collective decisions. In the next section we outline several ways of going about this.

3   From Criterion Preferences to Voting Systems

The most straight-forward way to proceed is consider the problem as any preference aggregation problem, i.e. to use criterion preferences as inputs and, using some social choice rule, aggregate them into a collective preference ranking. Consider the following set of criteria (Table 4).[4]

Table 4. A set of choice criteria

a	the Condorcet winner criterion
b	the Condorcet loser criterion
c	the strong Condorcet criterion
d	monotonicity
e	Pareto
f	consistency
g	Chernoff property
h	independence of irrelevant alternatives
i	invulnerability to the no-show paradox

Table 4 exhibits but a relatively small subset of criteria discussed in the literature, but arguably some of the most important criteria are included in the list. More extensive sets are introduced and analyzed e.g. in [6] and in [20]. To work out a collective preference ranking over these 9 criteria, some aggregation rule has to be used. To do this, one would have to assume what one is aiming at, viz. a suitable choice rule. When due attention is given to their metric representations (see [13,15], two rules, however stand out: Kemeny's rule and the Borda count. The former chooses the collective ranking that is closest to

---
[4] For explanation of the criteria, see e.g. [16]).

the reported individual rankings in terms of binary reversals (inversion metric), while the latter counts for each alternative (choice rule) the number of binary preference reversals that are needed to make this alternative unanimously first ranked. Thus, both rules resort to the same metric, but different end state. In the present context Kemeny's rule would perhaps seem more appropriate since the choice procedure is to be chosen using the following performance table (Table 5).

Table 5. A comparison of voting procedures

Voting system	Criterion								
	a	b	c	d	e	f	g	h	i
Amendment	1	1	1	1	0	0	0	0	0
Copeland	1	1	1	1	1	0	0	0	0
Dodgson	1	0	1	0	1	0	0	0	0
Maximin	1	0	1	1	1	0	0	0	0
Kemeny	1	1	1	1	1	0	0	0	0
Plurality	0	0	1	1	1	1	0	0	1
Borda	0	1	0	1	1	1	0	0	1
Approval	0	0	0	1	0	1	1	0	1
Black	1	1	1	1	1	0	0	0	0
Pl. runoff	0	1	1	0	1	0	0	0	0
Nanson	1	1	1	0	1	0	0	0	0
Hare	0	1	1	0	1	0	0	0	0

The table indicates whether a procedure represented by a row satisfies (denoted by 1) or does not satisfy (denoted by 0) the criterion represented by the column $(a, \ldots, i)$. Again, the procedures are just a sample of the procedures discussed in the literature.

Suppose now that the collective ranking obtained by applying Kemeny's rule to the profile of reported rankings over criteria has criterion l ranked first. One then looks for all procedures that have a unity in the column l. If several procedures satisfy l, one then picks the criterion ranked second in the collective (Kemeny) ranking. Let this criterion be m. One then looks for procedures that satisfy both l and m. Again there may be several procedures, but continuing in this (lexicographic) manner one eventually ends up in a situation where all remaining procedures satisfy all top-most criteria in the collective preference ranking down to a point after which none of them satisfies the next one in the collective ranking. Those remaining procedures then constitute the choice set of procedures. To take an example, suppose that the Kemeny ranking is $d \succ e \succ b \succ f \ldots$. Then the outcome is a 3-way tie $\{Copeland, Kemeny, Black\}$ since all these satisfy monotonicity (d), Pareto (e) and Condorcet loser (b) criteria, but none of them is consistent (f).

Obvious objections can be presented against this system, perhaps the most important being its reliance of lexicographic ordering of criteria. A poor performance on the first ranked criteria cannot be 'bought' by good performance on criteria ranked lower in the collective ordering. This can be illustrated by a setting where the collective ranking puts consistency on the first place. It then follows that just three systems are left after the first criterion is considered. If the collective ranking puts the Condorcet loser criterion in the second place, the Borda count emerges as the chosen system. In other words, the other criteria have no role whatsoever in determining the chosen system.

In view of these considerations another set of procedures is suggested. The input is either the set of individual preference rankings over criteria or the distribution of utility values in a fixed interval, say $[0, 10]$, that each voter assigns to each criterion. We illustrate one version of the procedure by using Borda points given by each voter to each criterion. Suppose that there are three individuals and their preference ranking over the 9 criteria are as follows:

individual	1	abcdefghi
individual	2	dcbafeihg
individual	3	ihgfedcba

Criterion a, thus, gets 8 Borda points from 1, 5 points from 2 and 0 points from 3. It would then make sense to argue that procedures satisfying a, get 13 points from these three individuals, while the other procedures get no points. Similarly b gets 7 points from 1, 6 from 2 and 1 from 3. And so on. Those procedures that do not satisfy the criterion considered do not get any points from voters on that criterion. In effect, then, for each column of the table the entries are obtained by multiplying the points given by voters to the criterion represented by the column by the corresponding entry of Table 5. The results are seen in Table 6.

On the basis of criterion preferences and using the Borda count in the point assignment, the winning procedure is Kemeny's rule followed by a tie between Copeland's and Black's procedures.

In the preceding we have resorted to Kemeny's and Borda's rules in deriving collective rankings over choice desiderata. It is reasonable to ask the reason for adopting these two systems. Are we thereby not *assuming* something that we should find out, viz. a plausible way to choose rules? Yes and no. In the present context our aim has been to illustrate how aggregating views on properties can give us collective rankings over procedures in a systematic and objective manner. Other procedures could be used instead of these two. However, a more important justification for Kemeny's and Borda's rules can be given. To wit, they can both be given a metric representation that it is particularly plausible in the rule choice context. Given a profile of rankings over k alternatives, the Kemeny rule generates all $k!$ rankings and measures the distance of the observed profile from each of the generated rankings. The measurement uses inversion metric:

Table 6. The assignment of points to procedures on the basis of criterion preferences

Voting procedure	Criteria									
	A	B	C	D	E	F	G	H	I	sum
Amendment	13	14	15	16	0	0	0	0	0	58
Copeland	13	14	15	16	11	0	0	0	0	69
Dodgson	13	0	15	0	11	0	0	0	0	39
Maximin	13	0	15	16	11	0	0	0	0	55
Kemeny	13	14	15	16	11	12	0	0	0	81
Plurality	0	0	15	16	11	12	0	0	10	64
Borda	0	14	0	16	11	12	0	0	10	63
Approval	0	0	0	16	0	12	8	0	10	46
Black	13	14	15	16	11	0	0	0	0	69
Pl. runoff	0	14	15	0	11	0	0	0	0	40
Nanson	13	14	15	0	11	0	0	0	0	53
Hare	0	14	15	0	11	0	0	0	0	40

the distance between two rankings is the number of binary preference inversions that are needed (at the minimum) to make one of them identical to the other. The distance between a generated ranking and the observed profile is simply the sum of distances between each ranking of the profile and the generated ranking. Of the generated rankings the Kemeny rule singles out the one with the smallest distance to the observed profile. The singles out ranking – the Kemeny ranking – can be viewed as the nearest to consensus one. In the present context the Kemeny solution is a ranking over criteria that is closest to the views of the voters.

The Borda rule has a similar justification: the Borda ranking also measures the distance between the observed profile and a generated one. Each generated profile, however, resembles the observed one differing from the latter only in ranking one alternative first in every individual's opinion. In the Borda count one effectively tallies the number of binary switches needed to make any given alternative unanimously first-ranked. The alternative that needs the minimum number of such switches is the Borda winner. So, both Kemeny and Borda rules are based on the same metric, but different 'goal states', the former aiming at a consensus that applies to each position of the collective ranking, while the latter looks at distances from profiles having the same alternative ranked first. It would seem that the metric representation speaks in favour of these two systems.

We are here dealing with a set of procedures rather than a single method of aggregating opinions into a collective choice of a system. Instead of Borda points one could use Copeland scores (the number of criteria defeated by the criterion under scrutiny) or utility values. For reasons stated above (distance rationalizability), the Borda count and Kemeny's rule are, however, more appealing.

Anyway, the point is to outline a reasonable method for choosing the method of choice using systematically the information regarding voter 'values' or preferences over criteria.

This set of procedures is similar to the PROMETHEE methods [2]. There are essential differences as well. E.g. in the technique introduced above the ranking of a procedure is determined by the ranking assigned by various voters to those properties that the procedure satisfies. In PROMETHEE, on the other hand, the collective ranking is determined by pairwise comparisons of alternatives on various criteria of performance. Still, both sets of methods determine the collective ranking on the basis of weighted sum scores. In the next section we shall sketch how a variation of PROMETHEE II might look like in the present context.

4   The PROMETHEE II Approach: A Sketch

PROMETHEE methods start from the evaluation table k rows and m columns, each row representing an alternative, i.e. a voting rule and each column representing a criterion. Entry i, j thus gives the value of the i'th rule on j'th criterion.[5] This is essentially Table 5 above. In contrast to many MCDM problems, this is a $0 - 1$ matrix consisting of objective values. The distance of two rules, say, amendment and Borda, on a criterion, say Condorcet winning denoted by a in Table 5, is the difference of their values on the criterion under scrutiny. For example

$$d_a(amendent, Borda) = g_a(amendment) - g_a(Borda) = 1 - 0 = 1.$$

The degree of preference of a rule i over another m with respect to criterion j, denoted by $P_j(i, m)$, is a monotonically increasing function of the absolute value of distance which is $|d_j(g_j(i) - g_j(k))|$. The overall preference of rule i over rule m is defined as

$$\pi(i, m) = \sum_j P_j(i, m) w_j$$

Here w_j, where $w_j \geq 0$ and $\sum_j w_j = 1$, denotes the weight assigned to criterion j by the decision maker. Hence, even though the evaluation table consists of objectively determined entries, the preferences that different decision makers entertain over rules may be different due to the different weights assigned to criteria.

The overall preferences enable us to compute, for any rule i, the positive ($\phi^+(i)$) and negative ($\phi^-(i)$) outranking flows as follows:

$$\phi^+(i) = \frac{1}{k-1} \sum_x \pi(i, x)$$

$$\phi^-(i) = \frac{1}{k-1} \sum_x \pi(x, i)$$

---

[5] A few adaptations notwithstanding, the PROMETHEE description given here essentially follows that of [2,11].

The PROMETHEE II ranking is obtained as the net flow associated with each alternative i, that is, as the value $\phi(i) = \phi^+(i) - \phi^-(i)$. Alternatives with identical net flows are considered indifferent, while the ranking of net flows determines the asymmetric part of the ranking. The ranking of individual r thus computed is denoted by ϕ^r. Let now the weight of each individual r in a group N be W_r, with $W_r \geq 0$ for each r and $\sum_r W_r = 1$. Obviously with equal weight assigned to each individual we have $W_r = 1/|N|$. The PROMETHEE II group ranking of alternatives now is obtained as the ranking of the values:

$$\phi^G(i) = \sum_r \phi^r(i) W_r$$

In other words, the ranking of alternatives is determined by the summed net flows of alternatives weighted by the individual weights.

It can be seen that the methods outlined in the preceding section are similar in spirit to the PROMETHEE methods, but not identical with them. The main difference is that the latter approach the rule selection problem through individual preferences over rules, while the former start from the views that the individuals have on the properties of rules.

5   Conclusion

Given the plethora of voting systems currently in use in various contexts it is arguable that the designers have different desiderata in mind when devising those systems. Focusing on a single desideratum only is bound to cause problems because typically a good performance on one criterion is accompanied with bad performance on some others. Hence we suggest that the opinions regarding the desiderata ought to be made explicit in the choice of the system to be used. We have outlined a couple of ways of using voter opinions regarding criterion preferences in a systematic way in the choice of a voting procedure. These ways presuppose full rankings by voters over criteria of performance. Incomplete preference rankings constitute undoubtedly a major challenge to the methods outlined above. The present author has discussed them in the context of the Borda count in another paper [17].

Acknowledgements. The author is grateful to the referees for perceptive comments on an earlier version.

References

1. Barbera, S., Jackson, M.: Choosing how to choose: self-stable majority rules and constitutions. Q. J. Econ. **119**, 1011–1048 (2004)
2. Brans, J.-P., Mareschal, B.: Promethee methods. In: Figueira, J., Greco, S., Ehrgott, M. (eds.) Multiple Criteria Decision Analysis: State of the Art Surveys, pp. 163–196. Springer, Berlin (2005)

3. Buchanan, J., Tullock, G.: The Calculus of Consent. Logical Foundations of Constitutional Democracy. The University of Michigan Press, Ann Arbor (1962)
4. Dietrich, F.: How to reach legitimate decisions when the procedure is controversial. Soc. Choice Welfare **24**, 363–393 (2005)
5. Diss, M., Merlin, V.: On the stability of a triplet of scoring rules. Theor. Decis. **69**, 289–316 (2010)
6. Felsenthal, D.: Review of paradoxes afflicting procedures for electing a single candidate. In: Felsenthal, D., Machover, M. (eds.) Electoral Systems: Paradoxes, Assumptions, and Procedures, pp. 19–91. Springer, Heidelberg (2012)
7. Felsenthal, D., Machover, M. (eds.): Electoral Systems: Paradoxes, Assumptions and Procedures. Springer, Heidelberg (2012)
8. Fishburn, P.: Condorcet social choice functions. SIAM J. Appl. Math. **33**, 469–489 (1977)
9. Hägele, G., Pukelsheim, F.: The electoral systems of nicholas of cusa in the catholic concordance and beyond. In: Christianson, G., Izbicki, T., Bellitto, C. (eds.) The Church, the Councils & Reform: The Legacy of the Fifteenth Century, pp. 229–249. The Catholic University of America Press, Washington, DC (2008)
10. Laslier, J.-F.: And the loser is..plurality voting. In: Felsenthal, D., Machover, M. (eds.) Electoral Systems: Paradoxes, Assumptions and Procedures, pp. 327–351. Springer, Heidelberg (2012)
11. Macharis, C., Brans, J.-P., Mareschal, B.: The GDSS PROMETHEE procedure. J. Decis. Syst. **7**, 283–307 (1998)
12. McLean, I., Urken, A. (eds.): Classics of Social Choice. The University of Michigan Press, Ann Arbor (1995)
13. Meskanen, T., Nurmi, H.: Distance from consensus: a theme and variations. Mathematics and Democracy: Recent Advances in Voting Systems and Collective Choice, pp. 117–132. Springer, Heidelberg (2006)
14. Niemi, R., Weisberg, H. (eds.): Probability Models of Collective Decision Making. Charles E. Merrill, Columbus (1972)
15. Nitzan, S.: Some measures of closeness to unanimity and their implications. Theor. Decis. **13**, 129–138 (1981)
16. Nurmi, H.: Voting Systems under Uncertainty. Springer, Heidelberg (2002)
17. Nurmi, H.: Assessing Borda's rule and its modifications. In: Emerson, P. (ed.) Designing an All-Inclusive Democracy: Consensual Voting Procedures for Use in Parliaments, Councils and Committees, pp. 109–119. Springer, Heidelberg (2007)
18. Plott, C.: Individual choice of politico-economic process. In: Niemi, R., Weisberg, H. (eds.) Probability Models of Collective Decision Making, pp. 83–97. Charles E. Merrill, Columbus (1972)
19. Rae, D.: Decision rules and individual values in constitutional choice. Am. Polit. Sci. Rev. **63**, 40–56 (1969)
20. Richelson, J.: A comparative analysis of social choice functions. IV, Behav. Sci. **26**, 346–353 (1981)
21. Suzuki, T., Horita, M.: How to order the alternatives, rules and the rules to choose rules: when the endogenous procedural choice regresses. In: Kamiński, B., Kersten, G.E., Shakun, M.F., Szapiro, T. (eds.) GDN2015. LNBIP, vol. 218, pp. 47–59. Springer, Heidelberg (2015)
22. Tangian, A.: Mathematical Theory of Democracy. Springer, Heidelberg (2014)

Conflict Resolution in Energy and Environmental Management

Warsaw School of Economics

Controversy Over the International Upper Great Lakes Study Recommendations: Pathways Towards Cooperation

Monika Karnis[1(✉)], Michele Bristow[2], and Liping Fang[1]

[1] Department of Mechanical and Industrial Engineering,
Ryerson University, 350 Victoria Street, Toronto, ON M5B 2K3, Canada
{monika.karnis,lfang}@ryerson.ca
[2] Department of Systems Design Engineering, University of Waterloo,
200 University Avenue West, Waterloo, ON N2L 3G1, Canada
michele.bristow@uwaterloo.ca

Abstract. Unprecedented low water levels and a perception of inaction after a five-year study of the International Upper Great Lakes led activists to stir up controversy. This paper analyzes this conflict just prior to the release of the International Joint Commission's report on April 15, 2013 and proposes resolutions towards cooperation and improved public perception.

Keywords: Graph model for conflict resolution · Pareto optimal solution · Status quo analysis · Water levels regulation

1 Introduction

The water level on Lakes Michigan and Huron and the adjacent Georgian Bay in North America hit a record low in 2013 at 175.57 m referenced to the International Great Lakes Datum of 1985 (IGLD85) [1]. This occurred coincidentally after the completion of a five-year study by the International Upper Great Lakes Study (IUGLS) Board and two final reports on the impacts of the St. Clair River on the Upper Great Lakes water levels [2] and Lake Superior regulation [3]. The results of the IUGLS led to recommendations released by the International Joint Commission (IJC) on April 15, 2013 on Lake Superior regulation, multi-lake regulation, restoration of Lake Michigan-Huron water levels, and adaptive management and the Great Lakes-St. Lawrence Water Levels Advisory Board [4]. However, the adaptive management plan is criticized by Sierra Club Canada for being insufficient to remedy the persistent low water levels [5] or the recent experience of unusually high water level in Lakes Michigan and Huron [6]. This paper examines the contextual information surrounding the IJC's report released in April 2013 and how the preferences of the decision makers played a key role in determining the most stable scenario for this situation.

© Springer International Publishing Switzerland 2015
B. Kamiński et al. (Eds.): GDN 2015, LNBIP 218, pp. 255–267, 2015.
DOI: 10.1007/978-3-319-19515-5_20

2   Water Levels of the Laurentian Great Lakes

2.1   Economic Context

The Laurentian Great Lakes consist of large and complex ecosystems that provide water and means for economic activities to millions of people in the United States and Canada. There are several different groups of people that rely on the ebb and flow of the system's water for their own uses, including those involved in shipping industries, commercial fishing, power generation, manufacturing, recreation and tourism. In 2010, the Great Lakes-St. Lawrence Seaway system directly employed 92,923 people, who received a total of $4.5 billion dollars (CAD) in wages and salary [7]. The firms that provide vessel services, cargo handling, and inland transportation services earned $34.6 billion (CAD) in business revenue, which was split almost evenly between the United States and Canada [7]. These pursuits are spread out over two Canadian provinces and eight American states that border the system. Moreover, there are about 73 million tourist visits to the Great Lakes each year and over 100,000 cottage owners on the shoreline [8].

However, each of these interested parties can have a strong opinion about the water's use and distribution, which may lead to controversy over the desired water levels. The president of the Canadian Ship Owners Association, Robert Lewis-Manning, confirmed that "we are seeing lower cargo volumes just out of constraint from draft – the amount of water available to a ship – in certain sections of the system" which means that ships cannot carry as much cargo as before, but fortunately "it is nowhere near the point where the bottom line is being seriously impacted" [9]. Hence, controlling water levels may not be of benefit to the shipping industry if it incurs a high cost. On the other hand, Mary Muter, chair of the Sierra Club Ontario's Great Lakes section, advocates for a gradual increase of 25 cm to Lakes Huron and Michigan water levels to compensate for losses over the last 40 years [10]. Such intervention to restore water levels is supported by observed environmental impacts such as the destruction of wetlands due to lowered water levels. As a vital and finite resource, it seems obvious to environmental groups that water is something worth saving at any cost.

2.2   Environmental Context

The controversy is worsened by the considerable fluctuation of the water levels in the Great Lakes in recent history as shown in Fig. 1. The Great Lakes have receded to 60 cm below the historic average over the last 80 years [10]. These fluctuations are caused by anthropogenic factors as well as long-term climate trends. When the water levels reach the extremes of the ranges shown in Fig. 1, there can be impacts such as loss of beaches due to high water levels and destruction of wetlands due to low water levels [11]. High water levels can cause flood, soil erosion, loss of wetlands, greater susceptibility to storm damage and economic loss due to flooding of recreational lands. There is also a risk of high channel flows that can impede navigation for industrial shipping [11]. On the other hand, low water levels can lead to increased dredging to maintain shipping lanes, loss of marina services, shipping delays caused by ships needing to lighten their load, and exposure of mudflats. There is also an increased risk of near-shore water quality issues, loss of hydropower generation and risks to water supply infrastructure [11].

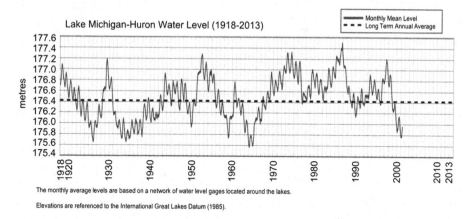

The monthly average levels are based on a network of water level gages located around the lakes.

Elevations are referenced to the International Great Lakes Datum (1985).

Fig. 1. Water level on Lakes Michigan and Huron over the historical record [12].

2.3 Historical Context

In an effort to regulate water levels, the physical systems have been altered by constructing diversions, dams, and locks, by dredging the connecting rivers and by hardening the shore line. There are two main facilities to regulate the water flow throughout the Great Lakes system: the Soo locks located in the St Mary's river connecting Lake Superior to Lake Huron near St. Mary's Falls Canal, and the Welland Canal connecting Lake Erie to Lake Ontario and traversing the Niagara Peninsula [13]. The St. Clair River has been dredged to make it deeper and wider so that commercial shipping may have easier access to the city of Detroit through Lake St. Clair and to the Atlantic Ocean via the St. Lawrence River [10]. The locks and canal are controlled water drainage points, whereas the St. Clair River is largely uncontrolled. These locations can be seen in Fig. 2.

Fluctuating water levels is an issue that has been studied many times by the IJC since the early 1960's. The IJC is mandated by the Canadian and United States governments to resolve disputes between the two countries under the Boundary Waters Treaty of 1909 and regulates projects affecting boundary waters [4]. The most notable studies in this time are listed in Table 1. The most recent study, titled the "International Upper Great Lakes Study", has created considerable controversy among interest groups about what is to be done about fluctuating water levels.

2.4 Cooperative Context

Although the interest groups in this case have different priorities, there is a meaningful reciprocal arrangement between them that should not be overlooked. The

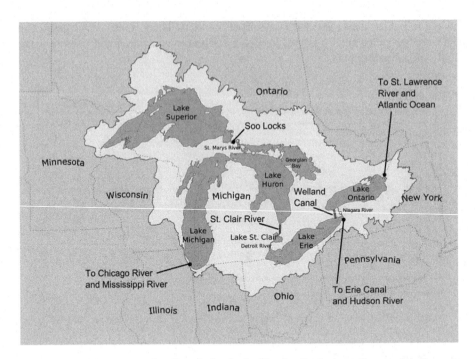

Fig. 2. A map of the Laurentian Great Lakes basin. The Soo Locks is used to regulate the water level in Lake Superior and the Welland Canal controls water flow from Lakes Erie to Ontario. Water flow through the St. Clair River is uncontrolled.

Table 1. Studies by the IJC concerning water levels [11].

Year	Study
1964–1973	Regulation of Great Lakes Water Levels Reference Study
1977–1981	Great Lakes Diversions and Consumptive Uses Reference Study
1977–1983	Limited Regulation of Lake Erie Study
1987–1993	Water Levels Reference Study
1999–2000	Report on the Protection of Waters of the Great Lakes
2001–2006	Lake Ontario – St. Lawrence River Study
2007–2012	International Upper Great Lakes Study

environmental groups rely on the local populace for support and the local populace relies on the shipping industry to provide economic stimulus to the area. These groups are consulted by the IJC, as well as the IJC's commissioned scientists [4]. However, these multiple interest groups each work to achieve their own objectives, which can result in a lack of cooperation that seriously hinders dialogue between groups. This lack of voluntary cooperation as well as enforceable cooperation between groups means that the conflict must be modeled as a non-cooperative game with an appropriate method like the graph model for conflict resolution (GMCR) methodology [14, 15].

In the next sections, the controversy over the recommendations is investigated by modelling and analyzing the conflict with the graph model for conflict resolution in order to gain insights on why the public reacted negatively to the IJC report and whether there are pathways to improve public perception by influencing the conflict towards win-win resolutions for all interested parties.

3   Modelling and Analysis of the International Upper Great Lakes Water Level Conflict

On April 15th 2013, the IJC released a report to the Canadian and United States governments containing advice on the recommendations of the International Upper Great Lakes Study [4]. The report was based on the findings of a technical study performed by the International Great Lakes Study Board which determined possible impacts of climate change and water variability on water levels in the Great Lakes. The study concluded that an active management plan and team was the only necessary solution to the water level problem, which would consist of active monitoring of water levels and the option to revisit the issue in the future [11].

There were public outcries from several stakeholders before and after the report was released. Many who lived, worked, and played on the Great Lakes felt that the recommendation from the IJC did not agree with their values, the objectives of which are mainly to protect the lakes environmentally and to keep the water levels steady so that business is not interrupted [16]. Shlozberg et al. [17] estimate a cost of $18.82 billion (CAD) by 2050 due to low water levels in the Great Lakes. In particular, the greatest economic impact will be to commercial fishing and recreational boating at $12.86 billion (CAD) [17]. The vice-chair of Restore Our Water International also agrees that the environmental effects are significant; for example, the increased drainage from the St. Clair River and other sources causes less dilution of fertilizer and other pollutants resulting in large algae blooms [5].

Given the significance of the IJC's role in making recommendations to both Canada's and the United States governments, the publishing of its advice on the recommendations of the International Upper Great Lakes Study [4] is a key event in this conflict's pathway to resolution. Hence, the point in time chosen to analyze the conflict among different parties in how to manage the water level of Lakes Michigan and Huron and Georgian Bay is just prior to the release of the IJC Report on April 15th 2013.

3.1   Modelling: Decision Makers, Options, and Preferences

To model a conflict between multiple participants, an appropriate systemic structure designed to encapsulate the characteristics of a conflict, known as a conflict model, must be chosen [14]. This structure is employed so that the conflicts between participants can be analyzed to reveal conflict resolutions or equilibria. The GMCR method can be visualized as a graph, where the conflict moves from state to state (the vertices of a graph) through transitions (the arcs of a graph) controlled by the participants or decision makers (DMs), who can influence the outcome of the conflict [14]. If a DM is not motivated to move from a particular state, the state is considered stable for that DM. If a state is stable for all of the DMs, the state constitutes an equilibrium [14]. Figure 3 is an example of the visual representation of this method. The GMCR method was chosen for this analysis because it best represents a situation where the DMs have different objectives [14]. The GMCR method also has the advantage of being able to incorporate irreversible moves, where a DM can move to a state but not from the same state, and can describe common moves, where multiple DMs can cause the conflict to move from one state to another [14].

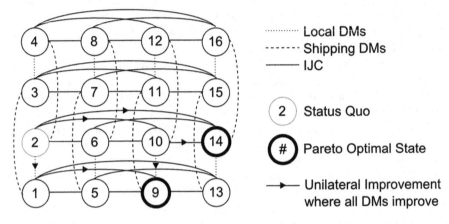

Fig. 3. Pareto optimal analysis using the graph representation of the conflict.

It is crucial to the integrity of the analysis to properly identify the decision makers. For this case, the groups most directly affected by the changing water levels would be the house or cottage owners and businesses living in and around the Great Lakes. These people would have a strong preference for keeping the water levels at their historical average so that their housing values, industry, fishing and general recreation are undisturbed. Property owners advocate their position through groups that require property ownership in the Great Lakes area, like the Georgian Bay Association and the Federation of Ontario Cottagers' Associations. Conservationists are concerned about the effect that water levels would have on the environment and have formed many environmental groups advocating for water levels consistent with the historical average, such as Great Lakes United and Lake Huron Center for Costal Conservation [8]. Another interested group consists of the shipping industries that use the Great Lakes-St. Lawrence Seaway to transport oil and other goods [8]. This group has an economic interest in keeping the water levels consistent, but has no interest in returning them to historical levels. Finally, the International Joint Commission is a neutral body comprised of members from the USA and Canada that is mandated to protect the shared waterways and to regulate shared water issues in an objective manner. The main DMs involved in the conflict before the IJC report was released can be categorized into three main groups: Local DMs, Shipping DMs and the IJC, as shown in Table 2.

The Local DMs would like there to be some intervention by the governments to return water levels to the historical average and stabilize them at that desired level, the Shipping DMs would like there to be no costly intervention because there is little for them to gain from one. Finally, the IJC is interested in coming to objective and scientifically sound recommendations for the governments of Canada and the United States to consider. In this case, the Local DMs consist of municipalities, local industries, and cottagers. The Shipping DMs in this conflict are the oil and shipping industries, which have an economic preference for the current water levels so that they will not need to alter shipping routes or equipment and ports [8]. The IJC is by mandate neutral between the Canadian and American governments and is also mandated to solution.

Table 2. Decision makers and their options.

Decision maker	Option	Description
Local DMs	1. Protest publicly	If the Local DMs feel that their needs are not being represented, they can protest publicly in the form of demonstrations and through the media
Shipping DMs	2. Reduce shipping via Great Lakes	If the Shipping DMs feel that the cost burden is too great to transit through the Great Lakes, they can substitute marine shipping with another transportation mode
IJC	3. Recommend a permanent solution	A permanent solution would be a series of locks, a canal, or speed bumps at one of the major uncontrolled drainage points like the St. Clair River. Requires a large investment from both governments in terms of time, money and other resources
	4. Recommend a temporary solution	A temporary solution needs a much smaller investment than a permanent solution, but still requires the approval of both nations because it affects them both. It would provide a short-term solution to fluctuating water levels. A temporary solution would be the use of large floating buoys to displace water and temporary structures in the St. Clair River
	5. Recommend active management	An active management solution would be to wait, watch and determine if there is enough evidence to support the implementation of a permanent or temporary solution

There are 16 feasible states as shown in Table 3. State 2 is known as the status quo state, or the state of the decision makers' positions before the game is started. In this state, the Local DMs are protesting publicly as a way to influence change in the management of the water levels while the Shipping DMs have not reduced shipping on the Great Lakes and the IJC has not made any recommendations yet.

The preferences for each DM are described in Table 4. This table uses the option numbers from Table 3 and a series of prioritized logic statements to specify the nature

Table 3. Feasible states.

	States															
	1	2	3	4	5	6	7	8	9	10	11	12	13	14	15	16
Local DMs																
1. Protest	N	Y	N	Y	N	Y	N	Y	N	Y	N	Y	N	Y	N	Y
Shipping DMs																
2. Reduce	N	N	Y	Y	N	N	Y	Y	N	N	Y	Y	N	N	Y	Y
IJC																
3. Permanent	N	N	N	N	Y	Y	Y	Y	N	N	N	N	N	N	N	N
4. Temporary	N	N	N	N	N	N	N	N	Y	Y	Y	Y	N	N	N	N
5. Active management	N	N	N	N	N	N	N	N	N	N	N	N	Y	Y	Y	Y

of the preference. The resulting ordinal ranking of states according to each DM's preferences is listed in Table 5.

3.2   Analysis: Static Stability, Status Quo Analysis, and Pareto Optimal States

A state is considered an equilibrium if it is a state where none of the DMs are motivated to move to another state based on a defined solution concept. Using GMCR II [19, 20] only states 14 and 16 are found to be equilibria. In both equilibria, the public protests while the IJC recommends active management. They are only differed by the reaction of the Shipping DMs, whereas in state 14 the Shipping DMs maintain the status quo, in state 16 the Shipping DMs decide to reduce shipping through the Great Lakes. State 14 is considered a strong equilibrium based on Nash stability, general metarationality (GMR), symmetric metarationality (SMR), sequential stability, and nonmyopic stability. On the other hand, State 16 is a weak equilibrium as it is found to be an equilibium only by GMR and SMR solution concepts. Relative to DMs' preferences, states 14 and 16 are highly preferred states for Shipping DMs and the IJC and in the lower end of the spectrum for the Local DMs.

Table 6 shows a status quo analysis, where it can be seen that state 14 is reachable from the status quo of state 2 simply by IJC making a unilateral improvement from state 2 to state 14. On the other hand, state 16 is not reachable from the status quo as it would require the Shipping DMs to disimprove its position unilaterally either from state 14 to move to state 16 or from state 2 to state 4. As a result, state 14 is the most likely resolution of the modelled conflict.

Pareto optimal states are states in which it is not possible for a DM to make an improvement without making another DM worse off. Figure 3 shows the graph representation of the conflict where the vertices denote states and the edges represent allowable state transitions. In this model, state transitions are reversible, hence a DM can move in both directions of an edge. An arrow denotes a unilateral improvement for the DM that controls the transition as well as an improvement for other DMs. State transitions that would make at least one DM worse-off do not have an arrow head.

Table 4. Prioritized preference statements for all decision makers.

Decision maker	Preference statement	Explanation
Local DMs	3	From this perspective, a permanent solution would return water levels to the historical average
	1 IFF 5	Local DMs will protest if active management is recommended
	4	A temporary solution would also be preferred over no solution as it also provides some stability (albeit in a short-term way)
	−5	An active management solution has been deemed unacceptable to many activist groups [5]
	−2	The city and local industries do not want to impede the shipping industry
Shipping DMs	5	The current set up of docks, shipping lanes and canals is profitable for the oil and shipping industry. Therefore, they would prefer the solution that does not disrupt shipping and is the least expensive: active management
	2 IFF 3	If shipping in the Great Lakes were to become too expensive, i.e. more than $3.6 billion CAD in transportation costs a year [18], the goods could be shipped over land.
	4	However, a temporary solution costs less than a permanent one, which means less taxes and toll increases
	−3	Shipping DMs would prefer to not have a permanent solution due to shipping disruptions and increased tolls
	−1	They would like to remain on friendly terms with the cities to prevent business disruptions
IJC	5	The IJC has a preference for an active management solution because the IUGLS found that the water levels are adequate and consistent enough to not require immediate action [3]
	4	A temporary solution is preferred over a permanent one because its study shows that the water levels do not require drastic action [3]
	−1	The IJC is tasked with finding an objective solution to the problem without thought to the opinion of other stakeholders, but as a government-sponsored commission, it must have a slight preference for the welfare of citizens over that of the commercial industry
	−2	
	3 IFF 1 & 2	A permanent solution with the support of the public and industry is slightly preferred over its singular opinions due to the collaborative mandate of the IJC

Starting from the status quo state, unilateral improvements where all DMs improve are marked. From the graph, the Pareto optimal states that are reachable from the status quo state are vertices that have at least one arrow directed into the vertice and no arrows leading out. Hence, states 9 and 14 are Pareto optimal states.

Table 5. Ordering of states based on DMs' preferences from most preferred to least preferreed states (left to right).

	←Most preferred state															
Local DMs	5	7	6	8	9	11	1	3	14	16	10	12	2	4	13	15
Shipping DMs	13	14	15	16	9	10	1	2	7	8	11	12	3	4	5	6
IJC	13	15	14	16	9	11	10	12	1	5	3	7	2	6	8	4

Table 6. Status quo analysis.

	Status Quo 2	✓	14	✗	16		Status Quo 2	✗	4	✓	16
Local DMs											
1. Protest	Y		Y		Y		Y		Y		Y
Shipping DMs											
2. Reduce	N		N	→	Y		N	→	Y		Y
IJC											
3. Permanent	N		N		N		N		N		N
4. Temporary	N		N		N		N		N		N
5. Active management	N	→	Y		Y		N		N	→	Y

3.3 Discussion

The stability, status quo and Pareto optimal analyses show that the most likely resolution to the conflict is state 14, where the IJC recommends that the governments of Canada and the United States implement an active management solution to manage the water levels in the Upper Great Lakes, the Shipping DMs do not reduce shipping traffic in the Great Lakes-St. Lawrence Seaway, and the Local DMs protest publicly. Indeed, local DMs continued to protest publicly through statements in newspaper articles that appeared shortly after the IJC report was released. Headlines such as "Activists denounce inaction on Great Lakes' low water 'crisis'" appeared on April 18, 2013 [5] and "It will take all our efforts to restore the Great Lakes" appeared on July 3, 2013 [21]. If the IJC had hoped to reduce protests by Local DMs which play a role in shaping public perception, the IJC would need to assuage fears that the water levels will continue to decrease. This could perhaps be achieved by publishing counter-argument articles in newspapers and other media on the reasons why an active management solution is best and backed by sound science to support its conclusions. Its findings will have to be presented in a format that is more accessible and intuitive than the documents outlining its recommendations to the Canadian and American governments. The status quo analysis further confirmed that state 14 is the likely resolution to the conflict since the only other equilibrium is not reachable via unilateral improvements from the status quo.

The search for Pareto optimal states was undertaken to find pathways towards possible win-win resolutions. As it turns out, state 14 is a Pareto optimal state. From the perspective of the Local DMs, state 14 is generally not considered a win. However, the analysis also revealed state 9, where IJC recommends a temporary solution and this recommendation is essentially supported by both Local and Shipping DMs. While this state is less preferred than state 14 for both IJC and Shipping DMs, it is a compromise which is much more satisfying for Local DMs than the status quo and compared to active management. As an evolutionary resolution, a temporary solution could be argued as a reasonable and reachable compromise, however, not a win-win. Consequently, it should be noted that state 9 may not appeal to Local DMs with extremist perspectives and who are not willing to compromise.

4 Conclusions

This paper has examined the circumstances and participants in a situation that led to a controversial recommendation being submitted to the Canadian and American governments. It was found that there were three main groups of DMs: the Local DMs, who advocated for a permanent solution to restore water levels to the historical average, the Shipping DMs who generally do not wish to face business disruptions, and the IJC, which was tasked with recommending an objective solution to the involved governments. The model was found to produce results that reflected the real-world conflict. The most likely resolution to the modelled conflict is an active management solution recommended to the governments where Local DMs continue to protest and Shipping DMs maintain shipping traffic. Further discussion determined that this state was both reachable through status quo analysis and further strengthened by Pareto optimality. However, Pareto optimal analysis also revealed an alternative future in which a temporary solution is recommended, representing a compromise rather than a win-win resolution unfortunately. Arguably however, it may be in the best interests of all parties and help to nurture collaboration by taking action towards preserving the environmental integrity of the Great Lakes without causing undue difficulties to any participant. This model and analysis indicates that this goal is not only possible, it is attainable. An opportunity for further study would be an analysis of the conflict after the IJC published its report, thereby changing to role of the IJC from a DM role to a support role, and how the Canadian and American governments enter the case as decision makers.

References

1. US Army Corps of Engineers: FINAL 2013 and Long-Terrm (1918–2013) Mean, Max & Min Monthly Mean Water Levels (Based on Gage Networks), 14 February, 2014. http://www.lre.usace.army.mil/Portals/69/docs/GreatLakesInfo/docs/WaterLevels/LTA-GLWL-Metric_2013.pdf. Accessed 31 December, 2014
2. IUGLS: Impacts on Upper Great Lakes Water Levels: St. Clair River. Final report to the International Joint Commission. Ottawa, Canada-Washington, D.C., USA (2009)

3. IUGLS: Lake Superior Regulation: Addressing Uncertainty in Upper Great Lakes Water Levels. Final report to the International Joint Commission. Ottawa, Canada-Washington, D.C., USA (2012)

4. IJC: International Joint Commission's Advice to Governments on the Recommendations of the International Upper Great Lakes Study. A Report to the Governments of Canada and the United States. Ottawa, Canada-Washington, D.C., USA (2013)

5. Desjardins, L.: Activists denounce inaction on Great Lakes' low water 'crisis'. Radio Canada International, CBC, Montreal, Quebec, Canada, 18 April, 2013. http://www.rcinet. ca/en/2013/04/18/activists-denounce-inaction-on-great-lakes-low-water-crisis/. Accessed 31 December, 2014

6. Van Brenk, D.: Lakes Huron, Michigan Unusually High. The London Free Press, London (2014). http://www.lfpress.com/2014/12/19/lakes-huron-michigan-unusually-high. Accessed 31 December, 2014

7. Martin Associates: The Economic Impacts of the Great Lakes-St. Lawrence Seaway System. Lancaster, PA: The St. Lawrence Seaway Management Corporation and The St. Lawrence Seaway Development Corporation (2011). http://www.greatlakes-seaway.com/en/pdf/eco_ impact_full.pdf. Accessed 31 December, 2014

8. Zebryk, N.: Water Levels and Recreation in the Great Lakes Basin. Policy Brief, Great Lakes Policy Research Network (2013). http://www.greatlakespolicyresearch.org/wp-content/uploads/2013/09/Water-Levels-Policy-Brief.pdf. Accessed 31 December, 2014

9. Brennan, R.J.: Great Lakes water levels mean some boaters 'playing Russian roulette'. The Star, Toronto, Ontario, Canada, 18 September, 2012. http://www.thestar.com/news/canada/ 2012/09/18/great_lakes_water_levels_mean_some_boaters_playing_russian_roulette.html. Accessed 31 December, 2014

10. Mackarel, K.: Why the Lakes are slowly getting less Great. The Globe and Mail, Toronto, Ontario, Canada, 25 August, 2012. http://www.theglobeandmail.com/news/national/why-the-lakes-are-slowly-getting-less-great/article4499074/. Accessed 31 December, 2014

11. International Great Lakes-St. Lawrence River Adaptive Management Task Team: Building Collaboration Across the Great Lakes-St. Lawerence River System, An Adaptive Management Plan for Addressing Extreme Water Levels, Breakdown of Roles, Responsibilities and Proposed Tasks. Final Report to the International Joint Commission. Ottawa, Canada-Washington, D.C., USA (2013). http://ijc.org/files/tinymce/uploaded/ documents/reportsAndPublications/FinalReport_Adaptive%20Management%20Plan_ 20130530.pdf. Accessed 31 December, 2014

12. US Army Corps of Engineers: Great Lakes Water Levels (1918–2013). http://www.lre. usace.army.mil/Portals/69/docs/GreatLakesInfo/docs/WaterLevels/LTA-GLWL-Graph_ 2013.pdf. Accessed 31 December, 2014

13. Chamber of Marine Commerce: Great Lakes Lock Infrastructure (2014). http://www. marinedelivers.com/great-lakes-lock-infrastructure. Accessed 31 December, 2014

14. Fang, L., Hipel, K.W., Kilgour, D.M.: Interactive Decision Making: The Graph Model for Conflict Resolution. Wiley, New York (1993)

15. Kameda, H., Altman, E., Touati, C., Legrand, A.: Nash equilibrium based fairness. Math. Meth. Oper. Res. **76**(1), 43–65 (2012)

16. Porter, C.: Georgian Bay loses water while International Joint Commission does nothing. The Star, Toronto, Ontario, 19 September, 2012. http://www.thestar.com/news/ontario/2012/ 09/19/georgian_bay_loses_water_while_international_joint_commission_does_nothing. html. Accessed 31 December, 2014

17. Schlozberg, R., Dorling, R., Spiro, P.: Low Water Blues: An Economic Impact Assessment of Future Low Water Levels in the Great Lakes and St. Lawrence River. Mowat Centre and Council of the Great Lakes Region, Toronto, Ontario, Canada (2014)

18. St. Lawrence Seaway Management Corporation: Delivering Economic Value: Annual Report 2011–2012, Cornwall, Ontario, Canada (2012). http://www.greatlakes-seaway.com/en/pdf/slsmc_ar2012_nar_en.pdf. Accessed 15 October, 2014
19. Fang, L., Hipel, K.W., Kilgour, D.M., Peng, X.: A decision support system for interactive decision making, part 1: model formulation. IEEE Trans. Syst. Man Cybern. Part C **33**(1), 42–55 (2003)
20. Fang, L., Hipel, K.W., Kilgour, D.M., Peng, X.: A decision support system for interactive decision making, part 2: analysis and output interpretation. IEEE Trans. Syst. Man Cybern. Part C **33**(1), 56–66 (2003)
21. Nies, J.: It will take all our efforts to restore the Great Lakes. Manitoulin Expositor, Manitoulin Island, Ontario, Canada, 3 July, 2013. http://www.manitoulin.ca/2013/07/03/it-will-take-all-of-our-efforts-to-restore-the-great-lakes/. Accessed 1 January, 2015

Option Prioritization for Three-Level Preference in the Graph Model for Conflict Resolution

Yuhang Hou[1], Yangzi Jiang[2], and Haiyan Xu[1(✉)]

[1] College of Economics and Management, Nanjing University of Aeronautics and Astronautics, Nanjing, Jiangsu, China
596811629@qq.com, xuhaiyan@nuaa.edu.cn
[2] Department of Management Science, University of Waterloo, Waterloo, ON, Canada
y2jiang@uwaterloo.ca

Abstract. A three-level preference (or called strength of preference) ranking structure based on option prioritization is developed within the paradigm of the Graph Model for Conflict Resolution. In a strategic conflict, a decision maker usually controls various courses of actions which are referred to as options. An option-based preference structure could efficiently model preferences under a complex conflict situation. There are three preference representations in a graph model for simple preference (or two-level preference), including Option Weighting, Direct Ranking, and Option Prioritizing in which the Option Prioritizing approach is the most effective. Therefore, the Option Prioritizing approach is extended to three-level preference from the two levels of preference in this paper. This proposed approach is more effective and convenient for modeling preference and is easy to implement into a decision support system. A specific case study is provided to show how three-level preference is calculated using the proposed approach.

Keywords: Option-based preference · Option prioritizing · Three-level preference · Graph model for conflict resolution · Decision makers

1 Introduction

The graph model for conflict resolution (GMCR) proposed by Kilgour et al. [1] contains modeling module and analysis module. The key ingredients in any conflict model are the decision makers (DMs), states or scenarios that could take place, and the preferences of each DM. Preference plays an important role in strategic conflicts. Different types of preference structures are developed and integrated into GMCR, which include two-level preference (or simple preference) [2], unknown preference [3], multiple-level preference [4–6], and hybrid preference [7]. In 2004, Hamouda et al. [4] proposed a new preference framework called "strength of preference" that includes two new binary relations, "≫ greatly preferred", and, "> mildly preferred", to express DM i's strong and mild preferences for one state over another, respectively, as well an equal relation. This is referred to as a 3-level preference structure in this paper. If one does

© Springer International Publishing Switzerland 2015
B. Kamiński et al. (Eds.): GDN 2015, LNBIP 218, pp. 269–280, 2015.
DOI: 10.1007/978-3-319-19515-5_21

not consider strength in preferences, the three levels of preference will reduce to the two-level structure defined by Fang et al. [2].

Three methods are available to generate preference for a DM: direct ranking for preference over states; and two implicit ranking methods based on options including option weighting and option prioritizing [8]. Option weighting process contains weights assigned to each option choice and total weights are used to determine an ordering of states. For an option prioritizing approach, it is based upon a set of lexicographic statements about options. Let m and h denote the number of states and the number of options, respectively. Then, $m = 2^h$ that means the number of options is much less than the number of states. For small models, direct ranking technique is the most convenient method of ranking. However, for a complex conflict, it is more efficient to use option weighting or option prioritizing method. Among the three approaches, the option prioritizing approach is more flexible, effective and convenient for modeling preference with regard to nearly all sizes of models owing to that it is easier for a DM to provide an ordered set of preference statements that the DM likes to see about the available options.

Until now, option prioritizing is available and integrated into GMCR II [8, 9] for two-level preference, but it cannot be used for complex situations with three-level preference or unknown preference. Another decision support system (DSS) based on matrix representation for stabilities in GMCR has very strong functions to analyze stabilities for three levels of preference [10], unknown preference [11], and hybrid preference [12]. However, the preference modeling of this DSS contains direct ranking only, so it is hard to be used in practice with complicated situations.

In this research, option prioritizing approach for two-level preference is extended to model three levels of preference. The option form of preference representation is especially useful for practical applications because it can easily handle conflicts having any finite numbers of DMs, each of whom controls a finite number of option or courses of action. Consequently, as is done throughout this research, often option form is employed for writing down a conflict as part of the GMCR methodology. Because the number of states is typically much larger than the number of options in a conflict, when option form is employed in practice, the user only has to provide a relatively short list of options, for which it is easy to expand the option prioritizing to handle general and flexible preference structures, and easy to implement into a DSS. In addition, the concept of preference trees for two-level preference [13] is extended to present for three-level preference framework in this research.

The remainder of this research is organized as follows. Section 2 provides some important definitions related to options in GMCR and introduces option prioritizing for two-level preference. Then option prioritizing is extended to strength of preference in Sect. 3. Section 4 consists of a case study of a model of the Gisborne Lake conflict (Newfoundland, Canada) that demonstrates how the proposed method can be employed in practice with three-level preference. Finally, some conclusions and ideas for future work are presented in Sect. 5.

2 Option Prioritizing for Two-Level Preference

2.1 Game in Option Form

In a strategic conflict, a DM usually controls various courses of actions which are referred to as options. Let O_i denote the option set of DM i, where o_{ij} is DM i's j^{th} option. Then, the set of all options in a conflict model is $O = \bigcup_{i \in N} O_i$ in which N is the DMs' set and i indicates which DM controls the options. Let $n = |N|$ be the number of DMs. Let h_i stand for the number of options for DM i, then $h = \sum_{i=1}^{n} h_i$ is the total number of options available to the DMs. When a given DM decides which of his or her options to select or do not select a specific strategy is formed.

Definition 1 (Strategy in Option Form). *Let O_i denote the option set of DM i for $i \in N$ for which $o_{ij} \in O_i$. A strategy for DM i is a mapping $g: O_i \rightarrow \{0,1\}$, such that*

$$g(o_{ij}) = \begin{cases} 1 & \text{if DM } i \text{ selects option } o_{ij}, \\ 0 & \text{otherwise.} \end{cases}$$

where o_{ij} is DM i's j^{th} option.

One can assign $g(o_{ij})$ a value of 1 to indicate that DM i will select option o_{ij}. Similarly, $g(o_{ij}) = 0$ means that DM i will not choose this option. A state is formed when each DM has selected a specific strategy. In other words, for each option the DM controlling the option has decided whether or not he or she will choose it. The formal definition for a state is as follows.

Definition 2 (State in Option Form). *Let $O = \bigcup_{i \in N} O_i$ be the set of all options in a conflict for $o_{ij} \in O_i$, $i = 1, 2, \cdots, n$. A state is a mapping $f: O_i \rightarrow \{0,1\}$, such that $f(O) = (f(O_1), f(O_2), \cdots, f(O_n))$ in which $O_i = (o_{i1}, \cdots, o_{ih_i})$ and $f(O_i) = g(O_i) = (g(o_{i1}), \cdots, g(o_{ih_i}))$ for $i = 1, 2, \cdots, n$*

Therefore, a state can be treated as an h-dimensional column vector having an element of 0 or 1. One often uses f_s to express the h-dimensional column vector to denote state s. A concise way to represent the set of all possible states in a conflict is to use the concept of a power set written as $\{0,1\}^O$, where O is the set of all options, each of which can be not chosen or selected as indicated by 0 or 1, respectively. Therefore, the set of all mathematically possible states in a conflict model is $\{0,1\}^O$.

The option form of a game is formally defined as follows.

Definition 3 (Game in Option Form). *A game G in option form is usually written as $G = \langle N, \{O_i\}_{i \in N}, S, \{\succ_i, \sim_i\}_{i \in N} \rangle$, Where*

- *$N = \{1, 2, \cdots, n\}$ is a non-empty set of DMs;*
- *for each DM $i \in N$, O_i is the non-empty option set of DM i;*
- *$S = \{s_1, s_2, \cdots, s_m\}$ is a non-empty set of feasible states;*
- *for each DM $i \in N$, $\{\succ_i, \sim_i\}$ represents i's preference where $s_k \succ_i s_t$ means that DM i prefers state s_k to state s_t while $s_k \sim_i s_t$ indicates that DM i has equal preference for these two states or is indifferent between them.*

Note that $\{\succ_i, \sim_i\}$ is called two-level preference. Specifically, $s \succ_i q$, indicates that DM i strictly prefers state s to state q, and $s \sim_i q$ means that DM i is indifferent between states s and q (or equally prefers s and q). $\succeq_i$ means an either strictly preferred or equally preferred relation, i.e., $s \succeq_i q$ indicates that DM i may strictly prefer state s to state q, may equally prefer s and q. It is assumed that the preference relations of each DM $i \in N$ have the following properties:

(i) $\succ_i$ is asymmetric;
(ii) $\sim_i$ is reflexive and symmetric; and
(iii) $\{\succ_i, \sim_i\}$ is strongly complete.

In addition to the above three properties, note that the strict preference relation, $\succ$, and the equal preference relation, $\sim$, arse transitive. The above preference representation is over states, which is a complicated process to input preference information into the DSS GMCR II for a complicated case. The procedure of option prioritizing is presented for two-level preference as follows.

2.2   Preference Representation Based on Option Prioritizing

The option prioritizing approach in GMCR II constitutes a generalization of the "preference tree" method originally suggested by Fraser et al. [13]. In option prioritizing, the user is asked to provide an ordered set of preference statements for each decision maker. Preference statements consist of options and logical connectives. Each preference statement takes a truth value, either True (T) or False (F), at a particular state. The relative importance of preference statements is reflected by its position in the list: a statement that occupies a higher place in the list is more important in determining the decision maker's preferences.

Preference between any two states is determined using the statements $\Omega_1, \Omega_2, \cdots,$ Ω_k in the order of priority. State $s \in S$ is preferred to state $q \in S$ $(s \neq q)$ for a DM if and only if there exists j, $1 \leq j \leq k$, such that

$$
\begin{aligned}
\Omega_1(s) &= \Omega_1(q) \\
\Omega_2(s) &= \Omega_2(q) \\
&\vdots \quad \vdots \quad \vdots \\
\Omega_{j-1}(s) &= \Omega_{j-1}(q) \\
\Omega_j(s) &= T \text{ and } \Omega_j(q) = F
\end{aligned}
\tag{1}
$$

In GMCR II, preference statements are expressed using options and logical connectives as shown in Table 1 in which "$-$", "&", and "|" stand for nonconditional logical relations "not", "and", and "or", respectively, as well as conditional relationships between two nonconditional statements, "IF" and "IFF" [14].

A scheme that can rank states is to assign a "score" $\Psi(s)$ to each state s according to its truth values when the statements are employed. Assume k is the total number of statements that have been provided, and $\Psi_j(s)$ is defined by

Table 1. True-value for simple preference connectives

A	B	−A	A & B	A | B	B IF A	B IFF A
T	T	F	T	T	T	T
T	F	F	F	T	F	F
F	T	T	F	T	T	F
F	F	T	F	F	T	T

$$\Psi_j(s) = \begin{cases} 2^{k-j} & \textit{if } \Omega_j(s) = T, \\ 0 & \textit{otherwise.} \end{cases} \tag{2}$$

And $\Psi(s) = \sum_{j=1}^{k} \Psi_j(s)$. This idea for determining two-level preference is extended to three levels of preference as follows.

3 Preference Representation Based on Option Prioritizing for Three-Level Preference

Each DM has preferences among the possible states that can take place. The ordinal preferences (ranking of states from most to least preferred, with ties allowed) and the cardinal preferences (the value of preference function for each state represented by a real number) are often required by some models. The graph model requires only the relative preference information for each DM. The proposed approach to generate the three-level preference based on option prioritizing is carried out in this section. The framework of the three levels of preference is introduced as follows.

3.1 The Three-Level Preference Structure

A triplet relation on S that expresses strength of preference according to indifferent, mild, or strong preference, was developed by Hamouda et al. [4, 5]. For states $s, q \in S$, the preference relation $s \sim_i q$ indicates that DM i is indifferent between states s and q, the relation $s >_i q$ means that DM i mildly prefers s to q, and $s \gg_i q$ denotes that DM i strongly prefers s to q. Similar to the properties for simple preference, the characteristics of the preference structure, $\{\sim_i, >_i, \gg_i\}$, containing three kinds of preference for each DM $i \in N$ are as follows:

(i) $\sim_i$ is reflexive and symmetric;
(ii) $>_i$ and $\gg_i$ are asymmetric; and
(iii) $\{\sim_i, >_i, \gg_i\}$ is strongly complete.

Notably, the three binary relations, "$\gg$ greatly preferred", "$>$ mildly preferred", and "$\sim$ equally preferred", are transitive. With regard to the transitivity, the three-level preference has a vital property that DM i mildly prefer s_1 to s_2 and strongly prefers s_2 to s_3 signifies the DM strongly prefers s_1 to s_3, that is $s_1 \gg_i s_3$ in the event of $s_1 >_i s_2$ and

$s_2 \gg_i s_3$. Likewise, $s_1 \gg_i s_2$ and $s_2 >_i s_3$ implies $s_1 \gg_i s_3$. The preference type "$\gg_i$" has similar properties to "$>_i$". The notation is introduced above to present DM i's preference between two states. The preference representation is often employed for the direct ranking approach. These preferences are presented based on states. It will be a complicated process for a large conflict model. The ranking approach based on "option" called Option Prioritizing is introduced as follows.

3.2   Option Prioritizing for Strength of Preference

If a DM is strongly preferred a statement Ω_t, then the notation "Ω_t^{+}" is applied to express the DM's strong preference over state s. The analysis process is presented as follows. Assume k is the total number of statements that have been provided. The weight is firstly defined by $W_j = 2^{k-j}$. Taking "Ω_t^{+}" into account, the weight is redefined as

$$W_j^* = \begin{cases} 2^{k-j} + 2^k & \text{if } 1 \leq j \leq t \\ 2^{k-j} & \text{if } t < j \leq k \end{cases} \tag{3}$$

Then, a scheme that can rank states is to assign a "score" $\Psi(s)$ to each state s according to its truth values when the statements contain information with strength of preference. Specifically, based on the definition for W_j^*, $\Psi_j(s)$ is defined by

$$\Psi_j(s) = \begin{cases} W_j^* & \text{if } \Omega_j(s) = T, \\ 0 & \text{otherwise.} \end{cases} \tag{4}$$

Equation 4 is employed if some DM strongly prefers the statement Ω_t, denoted $(\Omega_t)^{+}$. Otherwise, Eq. 2 is used.

Based on the Eqs. 3 and 4, it is easy to get that if a DM is strongly preferred the statements $\Omega_{t_1}, \Omega_{t_2}, \cdots, \Omega_{t_g}, 1 \leq t_1 < t_2 < \cdots < t_g \leq k$, then the weight in Eq. 3 turn into the W_j^{**} in consideration of $(\Omega_{t_1})^{+}, (\Omega_{t_2})^{+}, \cdots, (\Omega_{t_g})^{+}$ in the Eq. 5.

$$W_j^{**} = \begin{cases} 2^{k-j} + g \cdot 2^k & \text{if } 1 \leq j \leq t_1 \\ 2^{k-j} + (g-1) \cdot 2^k & \text{if } t_1 < j \leq t_2 \\ 2^{k-j} + (g-2) \cdot 2^k & \text{if } t_2 < j \leq t_3 \\ \quad \vdots & \quad \vdots \\ 2^{k-j} + 2^k & \text{if } t_{g-1} < j \leq t_g \\ 2^{k-j} & \text{if } t_g < j \leq k \end{cases} \tag{5}$$

Accordingly, based on the definition for W_j^{**}, the Eq. 4 translates into the Eq. 6.

$$\Psi_j(s) = \begin{cases} W_j^{**} & \text{if } \Omega_j(s) = T, \\ 0 & \text{otherwise.} \end{cases} \tag{6}$$

"Preference tree" [13] can be extended to rank each states for a conflict with three levels of preference. Assume that there are statements $\Omega_1, (\Omega_2)^+, \cdots, (\Omega_{k-1})^+, \Omega_k,$ which contains strong preferred statements shown in the left of Fig. 1. All states are ranked from the most preferred to the least preferred for some DM as shown in Fig. 1 according to Eqs. 5 and 6. For example, state s_{l_1} is combined by some DM who selects statement Ω_1 true "T", is strongly preferred statement $(\Omega_2)^+$ with "T". Similarly, the DM is strongly preferred statement $(\Omega_{k-1})^+$ with "T" and mildly preferred statement Ω_k with "T". Therefore, $\Psi_1(s_{l_1}) = 2^{k-1} + (k-2) \cdot 2^k$, $\Psi_2(s_{l_1}) = 2^{k-2} + (k-2) \cdot 2^k$, $\cdots$, $\Psi_{k-1}(s_{l_1}) = 2^1 + 2^k$, $\Psi_k(s_{l_1}) = 2^0$. The process is shown in the first column in Fig. 1. If the score of state s_{l_1} is $\Psi(s_{l_1}) = \sum_{j=1}^{k} \Psi_j(s_{l_1}) = W$, then $\Psi(s_{l_2}) = W - 1$ since the only difference between s_{l_1} and s_{l_2} is Ω_k with true "T" and false "F", respectively. For state s_{l_3}, the DM does not select $(\Omega_{k-1})^+$ true, so $\Psi_{k-1}(s_{l_3}) = 0$ rather than $\Psi_{k-1}(s_{l_3}) = 2^1 + 2^k$ as state s_{l_1}. Hence, the score of state s_{l_3} is $\Psi(s_{l_3}) = W - 2 - 2^k$. The scores of the other states can be calculated, similarly. According to the difference of scores between two states, the strength of preference over state can be defined as follows.

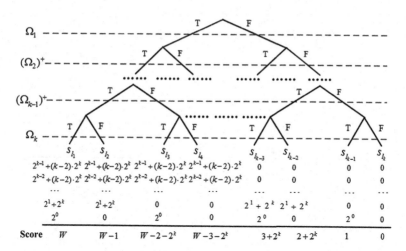

Fig. 1. Preference tree for a conflict with strong preferred statements

Definition 4. *Let $s_1, s_2 \in S$. If the difference of scores of s_1 and s_2 is the same, then the preferences of the two states are indifferent denoted $s_1 \sim s_2$; if the difference of two scores is more than "0" and less than "2^k", then the preferences of s_1 and s_2 are with "mildly preferred" relation denoted $s_1 > s_2$ or $s_2 > s_1$; if the difference of two scores is greater than or equal to "2^k", then the preferences of s_1 and s_2 are with "strongly preferred" relation denoted $s_1 \gg s_2$ or $s_2 \gg s_1$.*

Therefore, the states in Fig. 1 can be ranked by sequences

$$s_{l_1} > s_{l_2} \gg s_{l_3} > s_{l_4} > \cdots > s_{l_{k-3}} > s_{l_{k-2}} \gg s_{l_{k-1}} > s_{l_k}$$

The tree-level preference or strength of preference over state is generated based on option prioritization.

The following will introduce why the "score" $\Psi(s)$ to each state s should be computed according to Formula 3 instead of any other formula if a DM is strongly preferred a statement Ω_t.

Assume that there are statements $\Omega_1, \Omega_2, \cdots, \Omega_{k-1}, \Omega_k$, which contains simple preferred statements shown on the left of Fig. 2. All states are ranked from the most preferred to the least preferred for some DM as shown in Fig. 2 according to Eq. 2.

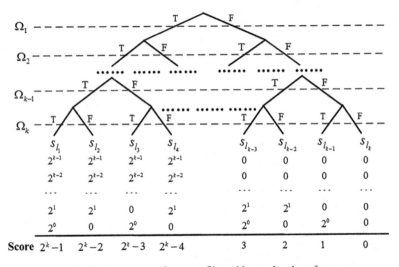

Fig. 2. Preference tree for a conflict with two-level preference

As can be seen from Fig. 2, the two-level preference or simple preference over state is $s_{l_1} \succ s_{l_2} \succ \cdots \succ s_{l_{k-1}} \succ s_{l_k}$. The scores of s_{l_1}, $s_{l_{k-1}}$, s_{l_k} are $\Psi(s_{l_1}) = 2^k - 1$, $\Psi(s_{l_{k-1}}) = 1$, $\Psi(s_{l_k}) = 0$, respectively, then the difference of s_{l_1} and $s_{l_{k-1}}$ is $2^k - 2$, $s_{l_{k-1}}$ and s_{l_k} is 1. If the DM strongly prefers $s_{l_{k-1}}$ to s_{l_k}, nevertheless, and mildly prefers s_{l_1} to $s_{l_{k-1}}$, the difference of s_{l_1} and $s_{l_{k-1}}$ should be greater than the difference of $s_{l_{k-1}}$ and s_{l_k}. Therefore, we should make an adjustment to the original scores of s_{l_1} and $s_{l_{k-1}}$, specifically, adding 2^k to the original scores of s_{l_1} and $s_{l_{k-1}}$ to reflect the preference with strength. If a DM is strongly preferred a statement Ω_{k-1} as shown in Fig. 2, then $s_{l_1} > s_{l_2} \gg s_{l_3} > s_{l_4}, \cdots, s_{l_{k-3}} > s_{l_{k-2}} \gg s_{l_{k-1}} > s_{l_k}$. In order to reflect the strength of preferences, the weight of Ω_{k-1} should add 2^k. As $\Omega_1, \Omega_2, \cdots, \Omega_{k-2}$ have higher priority in despite of the statement Ω_{k-1} strongly preferred by the DM, 2^k should also be added to the weights of $\Omega_1, \Omega_2, \cdots, \Omega_{k-2}$. Therefore, the "score" $\Psi(s)$ should be computed according to Formula 3. It is easy to reach the Definition 4 through the above analysis.

The procedure to implement the calculation of three-level preference using option representation is carried out using a real application next.

4   Application on the Lake Gisborne Conflict

In this section, the proposed option prioritizing is applied to a practical problem to show its processes. Lake Gisborne is located near the south coast of a Canadian Atlantic province of Newfoundland and Labrador. In June 1995, a local division of the McCurdy Group of Companies, Canada Wet Incorporated, proposed a project to export bulk water from Lake Gisborne to foreign market. On December 5, 1996, the government of Newfoundland and Labrador approved this project because of the potential economic benefits from this project. However, this proposal immediately aroused considerable opposition from a wide variety of lobby groups who cited the unpredictable harmful impacts on local environment. The Federal Government of Canada supported the opposing groups and introduced a policy to forbid bulk water export from major drainage basins in Canada. Because of the great pressure, in 1999, the government of Newfoundland and Labrador introduced a new bill to ban bulk water export from Newfoundland and Labrador. Therefore, Canada Wet had to abandon the Gisborne Water Export project. (See details in [3]).

Since several groups support the project, the provincial government might restart the project at an appropriate time in the future for the urgent need for cash. This case is economics-oriented. However, the provincial government might oppose this project because of the devastating consequences to the environment. This is environment-oriented. The economics-oriented provincial government and the environment-oriented provincial government result in uncertainty in preferences for the Gisborne conflict model. The details can be found in [3]. This conflict is modeled using three DMs: DM 1, **Federal (Fe)**; DM 2, **Provincial (Pr)**; and DM 3, **Support (Su)**; and a total of three options, which are presented in Table 2. The following is a summary of the three DMs and their options [3]:

Table 2. Feasible states for the Lake Gisborne model [3].

Federal								
1. Continue	N	Y	N	Y	N	Y	N	Y
Provincial								
2. Lift	N	N	Y	Y	N	N	Y	Y
Support								
3. Appeal	N	N	N	N	Y	Y	Y	Y
State number	1	2	3	4	5	6	7	8

- Federal government of Canada (**Federal**): its option is to continue a Canada wide accord on the prohibition of bulk water export (**Continue**),
- Provincial government of Newfoundland and Labrador (**Provincial**): its option is to lift the ban on bulk water export (**Lift**), and
- Support groups (**Support**): its option is to appeal for continuing the Gisborne project (**Appeal**).

In the Lake Gisborne model, the three options are combined to form 8 feasible states listed in Table 2, where a "Y" indicates that an option is selected by the DM controlling it and a "N" means that the option is not chosen. The graph model of the Lake Gisborne conflict is shown in Fig. 3. The labels on the arcs of the graph indicate the DM who can make the move.

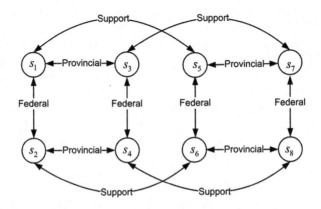

Fig. 3. Graph model for the Gisborne conflict.

The procedures to determine preference for the Lake Gisborne model with different situations are provided as follows. According to the preference statements of Federal's preference which are "1", "−2", "−3" analyzed from the case, means that "Federal choose option 1", "Provincial does not select option 2", and "Support does not select option 3", respectively, the Federal's preference over states in simple preference is $s_2 \succ s_6 \succ s_4 \succ s_8 \succ s_1 \succ s_5 \succ s_3 \succ s_7$ using the Formula 1 presented in Sect. 2.

In the same way, Support's preference in simple preference can be calculated. The preference statements of Support's preference are "2", "−3", "−1", indicating that "Provincial select option 2", and "Support does not choose option 3", "Federal does not choose option 1", respectively. Therefore, the two-level preference for Support is $s_3 \succ s_4 \succ s_7 \succ s_8 \succ s_5 \succ s_6 \succ s_1 \succ s_2$ using the Formula 1.

When the economics-oriented Provincial Government strongly prefers option "lift", "2$^+$" is added to preference tree presented in Fig. 4. Table 3 illustrates the preference statements of Provincial's three-level preference. Provincial Government's "scores" for states s_3, s_7, s_4, s_8 are "15, 14, 13, and 12", respectively, using Formula 5 and 6 in Sect. 3. The difference of "scores" between any two adjacent states is less than "2^3". Therefore the preference for DM 2 over the four states is $s_3 > s_7 > s_4 > s_8$. Similarly, the preference for DM 2 over states s_1, s_5, s_2, s_6 is $s_1 > s_5 > s_2 > s_6$. It is clear to see that the difference of "scores" between states s_8 and s_1 is "9", greater than "2^3", so Provincial Government's preference with strength is $s_3 > s_7 > s_4 > s_8 \gg s_1 > s_5 > s_2 > s_6$.

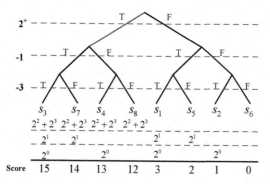

Fig. 4. Preference tree for the Gisborne conflict with the three-level preference for DM 2

Table 3. Provincial's three-level preference statements

Statements	Descriptions
2^+	Provincial strongly lift the ban on bulk water export
-1	Federal does not continues to prohibit bulk water export
-3	Support does not appeal for continuing the Gisborne project

According to the case background, the preference statements of environment-oriented Provincial's three-level preference are "$(-2)^+$", "1", "−3" while the preference statements are "2^+", "−1", "−3" for the economics-oriented Provincial Government. In the same way, the preference with strength for environment-oriented Provincial Government is $s_2 > s_6 > s_1 > s_5 \gg s_4 > s_8 > s_3 > s_7$ using Formulas 5 and 6 in Sect. 3.

From the new option prioritization approach, one can get strategic insights about why and how Provincial's preference information is three-level. Since the target of this research is to develop an efficient method to model three-level preference of DMs involved in strategic conflicts and is not to analysis the stability of a conflict, the stability calculations for the case with three-level preference are not preformed here.

5   Conclusion and Future Work

In this paper, a flexible method, option prioritization, is extended and improved from simple preference to preference with strength. The approach is convenient to present a DM's preference for a complicated case. The case of the Lake Gisborne model demonstrates how the new approach can be applied to generate the three levels of preference. In the near future, the proposed approach will be extended to include unknown preference and hybrid preference, and is incorporated into a new decision support system.

Acknowledgments. The authors appreciate financial support from the National Natural Science Foundation of China (71471087) National Social Science Foundation of China (12AZD102) and the Foundation of Universities in Jiangsu Province for Philosophy and Social Sciences (2012ZDIXM014).

References

1. Kilgour, D.M., Hipel, K.W., Fang, L.: The graph model for conflicts. Automatica **23**, 41–55 (1987)
2. Fang, L., Hipel, K.W., Kilgour, D.M.: Interactive Decision Making: The Graph Model for Conflict Resolution. Wiley, New York (1993)
3. Li, K.W., Hipel, K.W., Kilgour, D.M., Fang, L.: Preference uncertainty in the graph model for conflict resolution. IEEE Trans. Syst. Man Cybern. Part A: Syst. Hum. **34**, 507–520 (2004)
4. Hamouda, L., Kilgour, D.M., Hipel, K.W.: Strength of preference in the graph model for conflict resolution. Group Decis. Negot. **13**, 449–462 (2004)
5. Hamouda, L., Kilgour, D.M., Hipel, K.W.: Strength of preference in graph models for multiple-decision-maker conflicts. Appl. Math. Comput. **179**, 314–327 (2006)
6. Xu, H., Hipel, K.W., Kilgour, D.M.: Multiple levels of preference in interactive strategic decisions. Discrete Appl. Math. **157**, 3300–3313 (2009)
7. Xu, H., Hipel, K.W., Kilgour, D.M., Chen, Y.: Combining strength and uncertainty for preferences in the graph model for conflict resolution with multiple decision makers. Theory Decis. **69**(4), 497–521 (2010)
8. Fang, L., Hipel, K.W., Kilgour, D.M., Peng, X.: A decision support system for interactive decision making. Part 1: Model formulation. IEEE Trans. Syst. Man Cybern. Part C Appl. Rev. **33**(1), 42–55 (2003)
9. Fang, L., Hipel, K.W., Kilgour, D.M., Peng, X.: A decision support system for interactive decision making. Part 2: Analysis and output interpretation. IEEE Trans. Syst. Man Cybern. Part C Appl. Rev. **33**(1), 56–66 (2003)
10. Xu, H., Kilgour, D.M., Hipel, K.W.: An integrated algebraic approach to conflict resolution with three-level preference. Appl. Math. Comput. **216**, 693–707 (2010)
11. Xu, H., Kilgour, D.M., Hipel, K.W.: Matrix representation of conflict resolution in multiple-decision-maker graph models with preference uncertainty. Group Decis. Negot. **20**(6), 755–779 (2011)
12. Xu, H., Kilgour, D.M., Hipel, K.W., McBean, E.A.: Theory and application of conflict resolution with hybrid preference in colored graphs. Appl. Math. Model. **37**, 989–1003 (2013)
13. Fraser, N.M., Hipel, K.W.: Decision support systems for conflict analysis. In: Singh, M.G., Salassa, D., Hindi, K.S. (eds.) Managerial Decision Support Systems, pp. 13–21. North-Holland, Amsterdam (1988)
14. Rubin, J.E.: Mathematical Logic: Applications and Theory. Saunders, Philadelphia (1990)

Negotiation Support Systems and Studies

Warsaw School of Economics

The Role of Communication Support for Electronic Negotiations

Mareike Schoop[(✉)]

Information Systems I, University of Hohenheim, 70593 Stuttgart, Germany
m.schoop@uni-hohenheim.de

Abstract. Each (electronic) negotiation consists of communication and decision making. We will discuss relevant theories for a strong communication support, in particular for electronic negotiations. To this end, we will looks at the research area of communication modelling that has provided the Language-Action Perspective (LAP) with its underlying theories. We will show how LAP can be operationalised for e-negotiations using the negotiation support system Negoisst as the one example having implemented these concepts. In general, we will argue for the vital role of communication support in e-negotiation processes.

Keywords: Negotiation support systems · Communication support · Theory of communicative action · Speech act theory · Negoisst

1 Introduction

Electronic negotiations have been defined as having additional potential compared to traditional face-to-face negotiations due to the usage of information and communication technology (cf. [1]). The potential can be exploited for various types of support such as communication support, decision support, document management, and conflict management [2, 3].

Whilst decision support has been the core of most negotiation support systems and indeed was the historic basis for such systems [4], communication support has long been neglected [5].

The few dedicated approaches to communication support for e-negotiations range from structuring different types of communication in a negotiation [6] to complete support on all semiotic levels [7, 8].

Outside of negotiation research, the research area of communication modelling deals with supporting organisational communication. The so-called Language-Action Perspective (LAP) argues that language does not only have a descriptive but also a performative role, hence its name indicating that language can be action [9]. In particular, the Speech Act Theory of John Searle [10] has served as the theoretical foundation for LAP approaches. Later approaches have also used the Theory of Communicative Action by Jürgen Habermas [11] as their underlying theory.

We will revisit the key elements of both theories and apply them to electronic negotiations in the remainder of the paper concluding that dedicated communication support is a key element of success in electronic negotiations.

© Springer International Publishing Switzerland 2015
B. Kamiński et al. (Eds.): GDN 2015, LNBIP 218, pp. 283–287, 2015.
DOI: 10.1007/978-3-319-19515-5_22

2   Theoretical Foundations of Communication Support

There are many communication theories that can be relevant to supporting complex interactive communication processes. The two theories that we will discuss for e-negotiations stem from philosophical-political linguistics. The goal of these theories is to answer the question as to what constitutes understanding. In general, the goal of each communicative act is to create mutual understanding. This overall goal is relevant for negotiations as well. However, negotiations also consider individual goals represented by the desired negotiation agreement.

2.1   Speech Act Theory

In his Speech Act Theory [10], John Searle argues that understanding is achieved if and only if the communication partners understand what the utterance is about and understand the way the utterance is meant. This leads to a distinction between the *propositional content* (i.e. the content of the utterance) and the *illocutionary point* (i.e. the mode of communication). Taken together, they must be understood in order to achieve understanding for each utterance which is called a speech act.

The illocutionary point serves as the basis for a classification of speech acts. The *assertive illocutionary point* (present e.g. in statements or reports) represents facts of the real world. The *commissive illocutionary point* commits the author to the action described in the propositional content, e.g. used in promises. Using the *directive illocutionary point* (e.g. in requests or questions), the author tries to get the recipient to perform the action represented in the propositional content. The *expressive illocutionary point* represents the author's psychological states or feelings as in, for example, apologies, anger, or praise. Finally, the *declarative illocutionary point* is the archetype of a performative speech act as its mere utterance leads to factual changes. Prominent examples include the declaration of marriage by a registrar or a priest, the proclamation of guilt of an accused, or the final acceptance of an offer in a negotiation process. All of these latter examples clearly show that declaratives (short for speech acts with a declarative illocutionary point) are also uttered against an existing normative background regulating the author's professional role and the context of the exchange. For example, only a judge can pronounce the guilt of a person accused of a crime.

2.2   Theory of Communicative Action

Jürgen Habermas has published his Theory of Communicative Action more than a decade later than Speech Act Theory [11]. Whilst he agrees on the distinction between the propositional content and the illocutionary point, he does not agree that understanding both leads to mutual understanding between the communication partners. Rather he argues that even if content and mode are understood, understanding might not be achieved since the recipient might not agree with certain claims by the author. Habermas argues that the recipient must say "yes" to the so-called validity claims that the author implicitly or explicitly raises with each utterance.

The validity claim of comprehensibility means that the recipient must understand the speaker and no terminological or language problems exist. Furthermore, the author's intention as represented by the illocutionary point must also be understood. If truth is fulfilled, then the recipient agrees with the author in the *truth* of the statement and can thus share the speaker's knowledge or experiences. If the claim of truthfulness is fulfilled, then the speaker believes in the sincerity of the expressed feelings or psychological attitudes. The forth claim of appropriateness is closely related to underlying norms and values. If the recipient agrees on this claim, (s)he acknowledges that the author has the relevant role to make such a statement. If any of the claims is not fulfilled, the author must initiate reparative actions to overcome these disagreements as they represent communication problems which prevent mutual understanding.

Habermas' theory consists of additional elements which we will not discuss for the present context.

3   Speech Acts and Communicative Action in Electronic Negotiations

Since negotiation consists of communication and the main goal of the current paper is to show how dedicated communication support for electronic negotiations can work and which positive effects is has, we will now apply the theoretical constructs to the domain of electronic negotiations. We focus on electronic negotiations since the threat of misunderstandings and miscommunication is much more severe without the additional help of gestures, mimics, tone of voice, signs etc. We will illustrate the implementation of the theory using the negotiation support system Negoisst which is one of the few systems offering complex communication support and the only system using such support on all semiotic levels (i.e. syntactic, semantic, and pragmatic level) [2, 7, 8].

To enable complete understanding, the distinction into propositional content (represented by the message content) and illocutionary point (represented by the message type) is vital.

The illocutionary point is relevant for both the syntactic and the pragmatic level of support. The intention of the author is represented by the message type which shows the context of the message and is equivalent to the illocutionary point. The recipient can directly interpret the mode and thus the pragmatic aspect of the message is conveyed. This method is called *pragmatic enrichment* of the message in Negoisst. The illocutionary point is also the basis for the negotiation protocol which regulates message exchange, communication roles, order of messages etc. This *syntactic enrichment* is implemented in Negoisst.

The propositional content of an electronic negotiation utterance is equivalent to the message content, i.e. what the negotiation partners write in their messages. In Negoisst, messages are written in natural language to enable the richest form of expression. In order to avoid the disadvantage of natural language which is its ambiguity and missing structure, the rich language content is linked to the structured negotiation agenda representing the issues under negotiation. This means that particular words are tagged with a clearly defined semantics to avoid misunderstandings about the content of an offer or request. Thus, *semantic enrichment* is performed in Negoisst. This structure that is added to the natural language message also enables automated document extraction, i.e. each message leads to a contract version that is automatically created from the messages preventing later editing and enabling transparency and traceability.

The validity claims as introduced by Habermas are also operationalised in Negoisst. Comprehensibility is represented by the fact that a message can be of message type "question" or "clarification"; both of which enable comprehensibility problems to be addressed. Discussions about facts (i.e. validity claim truth) or about sincerity of the negotiation partner (i.e. validity claim truthfulness) are enabled likewise. Finally, the validity claim of appropriateness is dealt with by defining clear roles for all participants in a negotiation process and by choosing the right negotiation protocol that only allows appropriate message exchange. If the content of a message is deemed to be inappropriate, the discussion function can help to solve this issue.

4 Discussion

This paper argues for a strong communication support that is as strong as the decision support present in most negotiation support systems. Whilst decision support is quantitative and thus highly structured, communication support deals with the rich content of natural language that can be extended ad infinitum. Therefore, a strong theoretical basis is even more vital. The role of communication support for electronic negotiations is the prime one. If communication does not go smoothly, it will affect the decision making, ultimately leading to sub-optimal agreements [12].

Nevertheless, communication support, decision support, and document management are all interwoven and need to be supported as a whole [2].

The negotiation support system Negoisst has been in use for trainings and international negotiation experiments for the past 15 years. It provides a holistic support as described above and has provided a rich database of over a thousand negotiations. This data is the basis for our research into communication quality, decision support for incomplete preferences, document-centred negotiations, conflict management, and blended learning approaches to e-negotiations to name but a few.

An electronic negotiation that fulfils the relevant validity claims, that creates understanding on the content as well as on the mode and that supports the right decisions has the ultimate potential to lead to a successful agreement.

References

1. Ströbel, M., Weinhardt, C.: The Montreal taxonomy. Group Decis. Negot. **12**, 143–164 (2003)
2. Schoop, M.: Support of complex electronic negotiations. In: Kilgour, D.M., Eden, C. (eds.) Handbook of Group Decision and Negotiation, pp. 409–423. Springer, Heidelberg (2010)
3. Dannenmann, A., Schoop, M.: Conflict resolution support in electronic negotiations. In: Bernstein, A., Schwabe, G. (eds.) International Conference Wirtschaftsinformatik, paper 31, Zurich (2011)

4. Jarke, M., Jelassi, M.T., Shakun, M.F.: Mediator: towards a negotiation support system. Eur. J. Oper. Res. **31**, 314–334 (1987)
5. Weigand, H., de Schoop, M., Moor, A., Dignum, F.: B2B negotiation support – the need for a communication perspective. Group Decis. Negot. **12**, 3–29 (2003)
6. Yuan, Y., Rose, J.B., Archer, N.P.: A web-based negotiation support system. Electron. Markets **8**, 13–17 (1998)
7. Schoop, M., Jertila, A., List, T.: Negoisst – a negotiation support system for electronic business-to-business negotiation in e-commerce. Data Knowl. Eng. **45**, 371–401 (2003)
8. Schoop, M., Reiser, A., Duckek, K., Dannenmann, A.: Negoisst – sophisticated support for electronic negotiations. In: Proceedings of Group Decision and Negotiation (2010)
9. Winograd, T., Flores, F.: Understanding Computers and Cognition – A New Foundation for Design. Addison Wesley, Boston (1987)
10. Searle, J.R.: Speech Acts – An Essay in the Philosophy of Language. Cambridge University Press, Cambridge (1969)
11. Habermas, J.: Theorie des Kommunikativen Handelns, vol. 1. Suhrkamp, Frankfurt am Main (1981)
12. Schoop, M., van Amelsvoort, M., Gettinger, J., Koerner, M., Koeszegi, S.T., van der Wijst, P.: The interplay of communication and decision in electronic negotiations – communicative decision or decisive communication? Group Decis. Negot. **23**, 167–192 (2014)

More Than Words: The Effect of Emoticons in Electronic Negotiations

Johannes Gettinger[1]([⊠]) and Sabine T. Koeszegi[2]

[1] Institute of Interorganisational Management and Performance,
University of Hohenheim, 70593 Stuttgart, Germany
Johannes.Gettinger@wil.uni-hohenheim.de
[2] Institute of Management Science,
Vienna University of Technology, 1040 Vienna, Austria
Sabine.Koeszegi@tuwien.ac.at

Abstract. While affect plays a similar fundamental role in both, electronic and face-to-face negotiations, the expression of emotions in computer-mediated communication differs considerably from face-to-face settings. The aim of this experimental study is to analyze how the systematic use of emoticons – facilitated with software – affects negotiation behavior in alternative computer-mediated negotiation settings. With a 2×2 design comparing system-induced emoticon use with a text-only condition in synchronous chat or asynchronous e-mail mode we isolate effects of emoticons in these different communication settings. Results show that emoticons are used in different functions, i.e. mainly to supplement and support text messages and less often to mitigate its content. Furthermore, emoticon support increases the communication of positive affect in asynchronous negotiations while it decreases communication of negative affect and distributive negotiation behavior in synchronous negotiations. These findings propose that advancing communication quality via contextualization of affective information in negotiation support systems is promising.

Keywords: Emoticons · Contextualization of information · Emotions · Negotiations

1 Introduction

New information- and communication technologies provide a variety of different media for synchronous and asynchronous communication within and between individuals and private and public entities. E-mail, social networks, chat systems, electronic diaries, and similar technologies are used to coordinate day-to-day business, to interact with colleagues and customers at different locations, and also to a great extent to conduct negotiations. There has been a historic dispute on whether or not computer-mediated communication (CMC) has similar socio-emotional qualities compared to face-to-face communication (F2F). Researchers representing a cues-filtered-out perspective [1, 2] propose that with reduced media bandwidth, social presence is reduced: Due to missing non-verbal and para-verbal cues in CMC contextual information and emotions are filtered out resulting in less friendly, less emotional and less personal communication.

© Springer International Publishing Switzerland 2015
B. Kamiński et al. (Eds.): GDN 2015, LNBIP 218, pp. 289–305, 2015.
DOI: 10.1007/978-3-319-19515-5_23

This has been opposed particularly by the social information processing (SIP) model [3, 4] which advances that communicators in CMC may need more time to compensate the missing non-verbal and para-verbal cues in order to bring relational effects to similar levels compared to F2F. A recent review of empirical evidence concludes, that *"there is no indication that CMC is a less emotional or less personally involving medium than F2F"* [5, p. 766]. Nevertheless, communication technologies affect how we communicate and what we communicate [6]. While more traditional approaches like the Media Richness theory postulate that communicators should base their media choice on task characteristics [7, 8], more recent theories advance that communicators should adapt their communication strategy according to task characteristics and the communication process, including the communication goal and strategy, the medium and message form, and communication complexity [6]. For instance, in case communicators need to resolve a conflict with less rich media, e.g. e-mail, they have to contextualize task-oriented communication with relational or emotional information to compensate for the lean medium. Information- and communication systems should be designed to support these processes [9, 10]. Therefore, a particular focus is now laid on emotion encoding as well as on contextualization in CMC since these aspects shape interpersonal communication aiming for mutual understanding and relationship management [6].

One possibility to contextualize and encode emotions in CMC is the use of emoticons (standing for *emotion* and *icon*), which are referred to as relational icons, visual cues or pictographs serving as surrogates for non-verbal communication to express emotion and adding a para-linguistic component to a message [11]. Within the last decade, there has been some empirical research analyzing the effect of emoticons in CMC in general (for an analysis see e.g. [12]) which shows that emoticons - beyond serving as paralanguage - are morpheme-like structural markers with illocutionary force [12].

Yet, in electronic negotiation research, the analysis of the influence of emoticons on negotiation behavior, processes and outcomes is lagging. Since the significance of emotions in negotiations is undisputed, the aim of this experimental study is to analyze how the systematic use of emoticons affects negotiation behavior in computer-mediated negotiation settings. Referring to Media Richness Theory and its three dimensions [7, 8], i.e. (1) multiple information cues, (2) personal address, and (3) feedback immediacy, we are particularly interested in how feedback immediacy and emoticons interrelate in negotiations. With a 2 × 2 design comparing system-induced emoticon with a text-only condition in both modes, synchronous chat and asynchronous e-mail mode, we can isolate effects of emoticons in different communication settings. A more profound understanding of the effects of emoticons on synchronous and asynchronous electronic negotiation processes and outcomes will contribute to the further development of communication support for negotiations.

The remainder of this paper is organized as follows: the next section reviews the literature on computer-mediated communication, system design and the role of affect in communication. Based on the state of the art, we formulate hypotheses for the experimental study. In section three the experiment and methods of data analysis are introduced. Results are presented in section four and in section five we discuss the findings and limitations of this study and provide suggestions for future research.

2   Theoretical Background

2.1   Cognitive-Affective Model of Organizational Communication

The so-called "Cognitive-Affective Model of Organizational Communication" is based on the notion that the main objective of interpersonal communication is to reach mutual understanding between the communicators and to manage the relationship [6]. The degree to which communicators are able to reach these objectives depends on communication inputs, the communication process, and communication complexity. While the communication inputs describe antecedents of the process such as task characteristics, in the communication process the communicators' goals define their strategies, which in turn interact with the use and the characteristics of the medium and the specific message form. These interactions are shaped by the communication complexity surrounding the entire construct (see Fig. 1). Communication complexity consists of three elements acting as barriers to mutual understanding and relationship management: (1) cognitive complexity as a function of information intensity, multiplicity of views and incompatibility between information representation and use, (2) dynamic complexity referring to feedback processes and time constraints and (3) finally, affective complexity which refers to the sensitivity and change of attitudes towards the communication partner or the subject matter. Affective complexity considers relational aspects in the communication process including affective behavior and trust building. Low levels of trust and a lack of an appropriate normative context in which the communication is embedded impede reaching mutual understanding and the establishment of a good relationship.

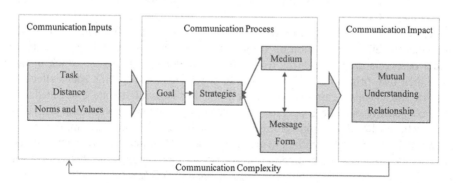

Fig. 1. Cognitive-affective model of organizational communication (adapted from [6], p. 256)

Negotiations are described as a process of conflict resolution between parties who interact with each other shaping their outcomes and their relationship [13]. Mutual understanding and a good relationship are important prerequisites to reach integrative solutions satisfying the needs and interests of the involved parties [14]. Negotiations are tasks in which all three dimensions of complexity are reflected. Therefore, handling the affective complexity is closely related to providing the basis for integrative negotiations. Communication directed at the management of affective complexity requires the inclusion of explicit affective statements typically requiring higher channel capacity [6]. Alternatively, negotiators can externalize the provision of additional affective information to the communication system [10].

2.2 Emotion as Social Information in (Electronic) Negotiations

In either way, emotion management in negotiations is a very promising avenue. The impact of emotions on negotiation processes between inherently social and emotional human beings is well established. Emotions can induce both, competitive or cooperative behaviors: negative emotions tend to increase competitiveness, while positive emotions tend to increase cooperation (e.g. [15, 16]). Nevertheless, positive emotions may also negatively affect negotiations. They may induce irrational behavior [17] and lead to biased judgments [18] or expectations [19, 20]. Similarly negative affect can also have positive effects: Empirical studies show that negative affect may motivate or induce negotiators to provide information [20, 21], help to overcome a crisis if power differences between negotiators exist [21, 22] or induce cooperative behavior [23].

The analysis of emotions has also found some attention in electronic negotiations research. It has been shown that in the restricted environment of CMC, emotions provide important contextual meaning for the interpretation of messages [24]. Furthermore, affect is contagious and tends to be reciprocated also in CMC [22]. However, the presence of emotions is not always salient and their particular influence may differ from F2F-negotiations. Earlier research has suggested that emotions contribute to a form of "hyperpersonal" communication [25] or to extreme behaviors like flaming [1]. Therefore, researchers have pointed out that more work is needed in order to develop a more comprehensive understanding of how emotions work and evolve in virtual environments [26].

Van Kleef et al. [27, 28] proposed the Emotion as Social Information (EASI) Model to predict the effect of emotion on communication processes: according to the EASI model observers draw inferences from others' emotional expressions, which in turn guide their own behavior and/or response. In negotiations, the specific social effects of positive and negative emotions depend on whether negotiations are embedded in a more competitive or more cooperative context. In a competitive context the display of anger reflects that the negotiator expressing anger has reached her limits, e.g. her aspiration level, and is not willing to give in anymore. Therefore, negative emotions are more likely to induce the counterpart to react with more cooperative behavior. Conversely, in a cooperative environment the expression of negative emotions is more likely to reduce cooperative behavior as it conveys adverse signals and elicits similar emotions in the counterpart.

Empirical evidence on the EASI model proves that both the verbal expression [15, 29] as well as the nonverbal expression [30] of emotions can be used strategically to influence others in face-to-face settings. In CMC, emotions are (implicitly) conveyed in text messages by the wording and phrasing of utterances [31]. Darics [32] provide a comprehensive linguistic discussion of threats and promises in negotiations demonstrating how semantic and syntactic features of the speech acts influence their perception and effectiveness. They show that specifically language intensity and language immediacy reveal the strength and the direction of a negotiator's affect. While language intensity is determined by the lexical choice, language immediacy is not expressed explicitly but is an unconscious process in which a speaker's affective states influence his or her lexical and syntactic choices [32, p. 162]. Additionally, communicators can use contextualization cues such as non-standard spelling, letter and punctuation mark repetition (e.g. ???) or lexical surrogates (e.g. hmmm) and the like as linguistic form to express affect. These cues contribute to the signaling of "contextual presuppositions" that allow for inferences about the meanings people intend to convey in a specific situation (Gumperz, cited in [33], p. 142). Particular forms of contextualization cues are emoticons. They are considered to be the socio-emotional suppliers in CMC (e.g. [33]). Compared to verbal description of emotions or implicit conveyance through specific lexical choices, emoticons provide a direct, precise and convenient way of emotional expression in CMC. Because of this they have received a worldwide acceptance [5] and several CMC software programs, such as for example MSN, ICQ, Hotmail, etc. have embedded an emoticon function (converting ASCII based character strings into pictograms). They are mostly used to express emotions, to strengthen a message or to express humor [34]. Their use also affects impression formation in relationships: when communicators use positively valenced emoticons their personality is perceived to be more extrovert [35], friendlier and warm [36].

In principle, emoticons can be classified in pictorial emoticons, e.g. Santa Claus * <\:-) or emotional-attitudinal emoticons [37] that represent[1]:

(a) *facial expressions:* happy :-), laughing :-D, sad :-(, angry :-{, wink;-)...
(b) *action:* kiss :-*, yawn 1-O, screaming :-@, big hug ((H))... or
(c) *appearance:* big nose :o), wearing glasses B-)...

The use of emoticons in CMC differs from medium to medium and shows even a huge variety within different types of media: Rezabek & Cochenour [38] analyzed emails on listservs and found that they were used between 1 to 25 % of the emails depending on the listserv. Tossel et al. [39] found in a longitudinal analysis of text messaging from smartphones that only 4 % of messages contained at least one emoticon. Garrison et al. [40] investigated emoticons in instant messaging discourses and found that out of the eighteen different emoticons which were used at all, just three of them – all with a positive valence – were used in 75 % of all cases. Furthermore, emoticons tend to be used more often in synchronous than in asynchronous communication [34] and – in accordance with Hall's theory of contextuality of cultures - are used more by users coming from high-context cultures [41]. There also seems to be a gender difference in emoticon use [12].

2.3   Hypotheses

It has been proposed that verbal and nonverbal cues are equally important for text interpretation in CMC [42]. Furthermore, the interpretation of non- and para-verbal cues is analogue to F2F encounters context-sensitive, see e.g. [33, 42, 43]. Emoticons and

---

[1] There are hundreds of emoticons which have been catalogued in dictionaries and on Websites (e.g. Wikipedia).

other CMC cues interact with other verbal and nonverbal cues and have – depending on the context of communication – mainly the following three different functions [11, 12, 37]:

i. *Supplement:* When emoticons are added as paralinguistic in order to convey an important aspect of a linguistic utterance, they help to clarify the meaning of a text and to eliminate misunderstandings: e.g. "When I returned home, she was already there :-(". The textual message could be understood both, in a positive and negative way. With the added emoticon the meaning is clarified unequivocally. Here, the emoticon is used as an illocutionary force [12] that augments the meaning of textual message by substituting non-verbal cues.

ii. *Support:* Emoticons are also used to support a text message: e.g. *"I am happy :-)"*. In this case, there is a denotative correspondence (congruence) between the text and the emoticon.

iii. *Antiphrasis:* When emoticons are used to contradict or annul the verbally expressed meaning, they produce ambiguity. Used in this way, they are used to express sarcasm: *"I am happy :-("*, irony: *"This has to be taken very seriously ;-)"*, or to mitigate disagreement: *"Sorry, but I do not agree :-)"*

These functions have been analyzed in empirical studies which show that emoticons indeed can strengthen the impact of a verbal message [11, 34, 43], help to emphasize a meaning during message creation and interpretation [38, 43], and to clarify textual messages [11]. Furthermore, emoticon use is significantly correlated to perceived information richness and perceived usefulness [44]. Huang et al. [44] also shows that emoticon use increases enjoyment and personal interaction by eliminating difficulties in expressing feelings in words. In this light, emoticons fulfill similar functions as non-verbal displays in F2F communication [5, 38]. We therefore expect that negotiation processes that are facilitated with systems enforcing users to complement their textual messages with emoticons will differ in several aspects to negotiation processes without emoticon support.

First of all, we expect that negotiators provided with emoticon support write less text, i.e. exchange less thought units. The emoticons serve as supplements to written text:

H1 : Emoticons serve as supplement to text messages in electronic negotiations.

Most research concerning the influence of positive and negative emoticons on the message interpretations supports the idea that positive emoticons make the message more positive and negative more negative correspondingly [11, 45]. Consequently, we hypothesize that positive emoticons will support cooperative behavior and negative emoticons will support competitive behavior. We therefore expect a positive correlation between use of positive emoticons (happy, laughing) text messages containing agreement or concession and a positive correlation between the use of negative emoticons (sad, angry) with text messages containing rejection or disagreement, respectively.

H2 : Emoticons are used to support textual messages (congruence)

Although it can be assumed that emoticons are more often used to supplement or support text messages, we expect that we also find communication in which emoticons are used as antithesis. Vandergriff [42] found in her analysis that emoticons and other CMC cues are sometimes used to mitigate or aggravate disagreement. Emoticons can be used to create ambiguity and to express sarcasm varying the valence of the emoticon and the valence of the message [46]. When used as antiphrasis, positive emoticons (happy, laughing) are used to mitigate textual disagreement (e.g. "*I am sorry, but this offer does not meet my expectations :-)*". Negative emoticons (sad, angry) used with agreement in text messages are used to express sarcasm (e.g. this is really a wonderful offer :-(, 'winking' and 'eat my shorts' emoticons used with positive/negative agreement are used to express irony/sarcasm (e.g. "*Your competitive offers make me very happy;-)*".

H3 : Emoticons are used to mitigate disagreement or to express sarcasm (antiphrasis)

The use of the smile emoticon has been found to be a positive politeness strategy that creates a collaborative work environment [33]. Vandergriff [42] also finds in her analysis of chat negotiations, that emoticons and CMC cues were used mainly in the service of politeness. However, we assume that this effect is mediated by communication mode as this impacts feedback immediacy. Communication media with low feedback immediacy (asynchronous media) are less adequate to handle high affective communication complexity [6]. Furthermore, in the context of e-negotiation, negative emotions can be better controlled in asynchronous communication modes [47]. Thus, we hypothesize a positive effect of emoticon support particularly in asynchronous negotiations. Since asynchronous negotiations tend to be less emotional, the threat of escalating emotional statements is limited and emoticons can unfold their supportive function. In synchronous negotiations, on the contrary, we expect that the use of emoticons boosts (negative) emotions. We assume that this eventually results in more competitive behavior.

H4 : Emoticon support facilitates integrative negotiation behavior in asynchronous but not in synchronous negotiations.

Referring to F2F negotiations, researchers use emotional states as predictors of negotiation outcomes [48]. For instance, Kopelman et al. [48] find that negotiators were more likely to accept a proposal from a negotiator displaying positive emotion than from a negotiator displaying negative emotion. Due to the expected increase in integrative behavior – in particular in asynchronous negotiations – and the increased mutual understanding combined with a better handling of the relationship, should also influence the likelihood to reach an agreement.

H5 : **Emoticon support leads to more agreements in asynchronous but not in synchronous negotiations.**

3 Methodology

To test our research questions, we collected data in a 2×2 laboratory experiment at two European Universities (within one country). The 2×2 design varies the communication mode (synchronous/asynchronous) and the availability of emoticon support (available/ not available) (see Table 1). The used negotiation case described a situation in which the subjects had to decide in bilateral negotiations on a Friday evening program considering three alternatives following predefined and conflicting preferences.

Table 1. Treatments (n = number of participants)

	Emoticons	No Emoticons
Synchronous	n = 24	n = 32
Asynchronous	n = 26	n = 16

The treatments of the design were implemented in a text-based electronic communication support system N-SWAN, which facilitated participants to exchange, store and retrieve messages. The implemented emoticon feature forced users in the "emoticon treatment" to tag each text message with an emoticon; otherwise users could not send the text message. Negotiators could choose among six different types of emoticons: 'happy' & 'laughing' are emoticons with a positive valence whereby the latter one expresses stronger arousal than the former, 'sad' & 'angry'' are emoticons with a negative valence whereby the latter expresses a stronger arousal than the former, and finally two emoticons 'winking' & "eat my shorts'. Winking has a positive valence, while "eat my shorts" has a strong negative valence. We decided to use pictograms instead of ASCII-based character strings, because empirical evidence points to a somewhat stronger impact of pictograms on communication [35]. In the experiment, negotiations in the synchronous communication mode could last up to one hour whereas asynchronous negotiations were allowed to take up to a maximum period of three days. Demographic data was collected at the beginning of the experiments via questionnaires, while exchanged messages and emoticons were recorded by the system.

We applied content analysis as means to transform the 2,972 exchanged messages into quantitative data [49]. The text messages of all 49 negotiation transcripts were unitized into 3,686 thought units and coded into the pre-tested category scheme of [50] (see Table 2) by two independent coders (Guetzkow's U of 0.054, Cohen's kappa 0.848). Additionally, emoticon functions (supplement, support, antiphrasis) were also coded in reference to the text messages by two independent raters (Cohen's kappa 0.855).

Table 2. Category scheme

	Integrative	Distributive
Action	(1) Agree, accept, concede	(4) Reject or disagree
	(2) Show positive emotions	(5) Show negative emotions
	(3) Make a new offer	(6) Use tactics or threats
Information	(7) Provide information	(9) Use persuasive arguments
	(8) Request information	
Off-task categories	**(10) System-related communication, (11) Off-task communication, (12) Communication protocol**	

4   Results

In the 895 text messages of the emoticon treatment, emoticons are overwhelmingly used as a supplement (766 times, 84.3 %). Only 30 times (3.4 %) emoticons were coded as being redundant to the text messages and in 110 cases (12.3 %), the emoticon was used as antiphrasis to the text. While the use of the three functions differs significantly ($X^2(2) = 1059. 27$, $p < .001$) they are distributed similarly between synchronous (supplements 87.1 %, redundancy 1.7 %, antiphrasis 11.3 %) and asynchronous negotiations (supplements 80.2 %, redundancy 5.9 %, antiphrasis 13.8 %). These results provide support that emoticons serve different functions. In the following, we further elaborate the data in more detail.

H1 suggests that emoticons are used to supplement text messages and therefore substitutes written text in messages. Therefore, we also test for differences in number of thought units in the respective treatments (see Table 3). Factorial ANOVA shows a significant main effect of emoticon support $F(1,94) = 19.012$, $p < .001$, a non-significant main effect of the communication mode, $F(1,94) = 1.432$, $p = .234$, and a non-significant interaction effect, $F(1,94) = .464$, $p = .497$ on the amount of exchanged thought units. ANOVA post hoc tests verify that the differences between groups due to the used communication mode are significant, within the synchronous treatment, $p < .01$, as well as within the asynchronous treatment, $p < .05$, however, not between the two groups using emoticons, $p > .1$, and between the two groups not using emoticons, $p > .1$[2] Therefore, H1 is clearly supported.

Table 3. Number of absolute thought units across the four different treatments

Treatments		Mean	Std. Dev.
Emoticons	Synchronous	26.79	18.57
Emoticons	Asynchronous	24.04	13.46
No emoticons	Synchronous	53.72	28.64
No emoticons	Asynchronous	43.69	39.84

---

[2] All reported post-hoc tests in our manuscript are 2-tailed and based on bootstrapping, n = 1000, using either Bonferroni or Games-Howell comparisons in case that the Levene test indicates a violation of homogeneity of variances, $p < .05$.

H2 and H3 propose that emoticons are also used to support text messages (congruence between text message and emoticon) or to mitigate disagreement, to express sarcasm or irony (antiphrasis of emoticon to text message). Therefore, we analyzed in addition to the frequency analysis reported above, whether emoticons are related to specific behavioral patterns reflected in the communication categories. On average, negotiators used most often the "happy"-emoticon (M = 11.70 per negotiation, SD = 12.30). The emoticons "sad" (M = 2.28, SD = 4.11) and "winking" (M = 2.74, SD = 2.77) were similarly often used. Every second negotiator used one time the emoticon "angry" (M = 0.50, SD = 1.39) or the "laughing" emoticon (M = 0.52, SD = 1.18). Negotiation dyads referred least frequently to the "shorts" emoticon (M = .16, SD = 0.47). To test whether emoticons are linked with specific communication patterns we ran correlation analysis using Spearman's rho and performed bootstrapping (n = 1000) to increase reliability of our confidence intervals. Results show that frequencies of used emoticons correlate with the relative frequencies of integrative and distributive action categories but only to a minor degree with information categories: The "happy" emoticon correlates positively with "agreeing, accepting, conceding", $r_s = .297$, p < .1, and expressing positive emotions, $r_s = .325$, p < .05, and negatively with communication used to reject or disagree, $r_s = -.225$, p < .1, and expressions of negative emotions, $r_s = -.328$, p < .05. The use of the "sad" emoticon is negatively linked to "agreeing, accepting, conceding", $r_s = -.281$, p < .1, and expressing positive emotions, $r_s = -.257$, p < .1. Similarly, the "angry" emoticon is less used in combination with the communication categories "agree, accept, concede", $r_s = -.284$, p < .05, but rather with statements expressing negative emotions $r_s = .284$, p < .1. The emoticon "winking" is related to providing information to the counterpart, $r_s = .238$, p < .1. Last, the emoticon "shorts" is positively linked to expressions of rejection and disagreement, $r_s = .220$, p < .1, and expressions of negative emotions $r_s = .329$, p < .1. Consequently, we find empirical evidence for H2 proposing that emoticons are indeed used in congruence with text messages.

To evaluate how emoticons are used as antiphrasis (to mitigate the statement of the textual message or to express sarcasm or irony) we ran additional nonparametric group analyses. Of all 110 emoticons serving as antiphrasis, the happy emoticon is used 76 times, the sad emoticon 12 times, the winking emoticon 16 times, the laughing emoticon 5 times, and the shorts emoticon once, while the angry emoticon was used never as an antiphrasis. Considering this distribution, we only analyzed text messages with the happy or sad emoticon using the Holm-Bonferroni approach to control for type I errors. Results indicate that the sad emoticon is more often used than the happy emoticon in combination with categories "new offers", "tactics or threats" and "request information", (all p < .05). Furthermore, we find no differences in the used relative communication units when either the sad or the winking emoticon are used. Therefore, we find no clear pattern how emoticons are used as antiphrasis (H3).

In the next step we evaluate how emoticon support and communication mode affect communication behavior (H4). Investigating the effect of the use of emoticons in synchronous and asynchronous negotiations, we run a factorial MANOVA including all communication categories.[3] We find significant effects on the communication behavior depending on emoticon support, $F(12,83) = 2.271$, $p < .05$, communication mode, $F(12,83) = 1.944$, $p < .05$, and the interaction between both treatments, $F(12,83) = 3.103$, $p < .005$. Due to problems with assumptions of MANOVA,[4] we additionally run individual factorial ANOVAs for the action- and information-oriented communication categories and additionally perform multiple comparisons with bootstrapping ($n = 1000$) for the conditional main effects. The results indicate that the relative frequencies of integrative and distributive action- and information-oriented communication differs significantly in the treatments (see Table 4).

Table 4. Effect of emoticons and communication mode on communication behavior (p° results based on bootstrapped multiple comparisons, n = 1000)

Communication Category	Integrative Action			Communication Category	Distributive Action		
	Ind. Variable	F	p (p°)		Ind. Variable	F	p (p°)
Agree, Accept, Concede	Emoticon	1.522	.220 (.273)	Reject or Disagree	Emoticon	2.648	.107 (.148)
	Comm. Mode	0.041	.839 (.857)		Comm. Mode	0.122	.728 (.744)
	Interaction	7.061	.009		Interaction	1.203	.275
Positive Emotions	Emoticon	2.955	.089 (.085)	Negative Emotions	Emoticon	1.324	.253 (.194)
	Comm. Mode	0.263	.610 (.596)		Comm. Mode	9.129	.003 (.001)
	Interaction	5.987	.016		Interaction	6.458	.013
New Offer	Emoticon	0.639	.426 (.452)	Tactics or Threats	Emoticon	2.081	.152 (.170)
	Comm. Mode	3.405	.068 (.067)		Comm. Mode	0.124	.726 (.739)
	Interaction	0.369	.545		Interaction	0.001	.978
	Integrative Information				Distributive Information		
Provide Information	Emoticon	3.640	.059 (.070)	Persuasive Arguments	Emoticon	11.366	.001 (.004)
	Comm. Mode	0.040	.841 (.851)		Comm. Mode	0.016	.898 (.903)
	Interaction	0.001	.970		Interaction	4.114	.045
Request Information	Emoticon	1.033	.312 (.310)				
	Comm. Mode	0.362	.549 (.533)				
	Interaction	0.423	.517				

In more detail, integrative negotiation behavior differs in the treatments (see Fig. 2): In the synchronous treatment, post hoc tests show that negotiators in the emoticon support treatment express more often communication intended to "agree, accept or concede", $p < .01$. Furthermore, when negotiators have emoticon support, they use more communication for approval in the synchronous than in the asynchronous treatment, $p < .05$.

We also find a weak significant main effect of emoticon support ($p < .1$) and a significant interaction with communication mode ($p < .05$) on the expression of positive emotions in text messages. Within the asynchronous treatment, negotiators express

---

[3] Testing the assumption of normality, all relative subcategories show significant deviations from normality. To cope with the skewed data, we apply a Box Cox transformation anchoring all values at 1 and using a λ of −6.5. The transformation of the data is rendered necessary as initial calculations of F-values in MANOVAs are not supported by bootstrapping.

[4] Checking the assumptions of MANOVA, the Box's Test of equality of covariance matrices indicates a violation of the assumption of equality of covariance matrices ($p < .000$). Furthermore, the Levene test indicates that the assumption of equality for error variances is violated for several communication categories.

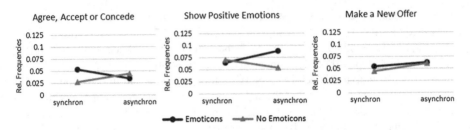

Fig. 2. Impact of emoticons and communication mode on integrative communication behavior

more positive emotions when they have emoticons support compared to no emoticons support p < .05. Furthermore, when negotiators have emoticons support, they tend to express more positive emotions when engaging in asynchronous compared to synchronous negotiations, p < .1.

Regarding the exchange of offers, bootstrapped group comparisons find negotiators in the synchronous treatment without emoticons support to use less new offers, than in the asynchronous treatment with emoticons support, p < .05, as well as without emoticons support, p < .1.

Furthermore, the expression of negative emotions is substantially influenced by the communication mode, p < .01, and the interaction between the communication mode and the emoticon support, p < .05. Negotiators in the synchronous treatment without emoticons support express most negative emotions (see Fig. 3). They express more negative emotions than negotiators in the synchronous treatment with emoticons support p < .05, and more than negotiators in both asynchronous treatments – with emoticons support, p < .01, and without emoticons support, p < .01. Furthermore, negotiators with emoticon support in the synchronous group express slightly more negative emotions than negotiators in the asynchronous group without emoticons support, p < .1.

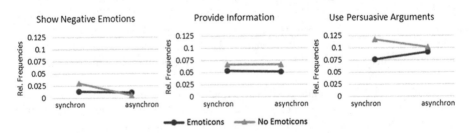

Fig. 3. Impact of emoticons and communication mode on distributive communication behavior

Regarding the use of exchanged information, we find again a tendency for emoticons to act as substitute to the written exchange of information, p < .1, accompanied by a tendency that in synchronous negotiations without emoticons support more information is exchanged than in asynchronous negotiation with emoticons support, p < .1. Finally, the use of persuasive arguments differs substantially between groups. Negotiators in synchronous treatment without emoticons support use more persuasive communication than negotiators in both treatments with emoticons support (both, p < .01) as well as negotiators without emoticon support in asynchronous negotiations (p < .1). Subjects in asynchronous treatment without emoticons support engage in more persuasive behavior than subjects negotiating synchronously with emoticons support, p < .1.

Analyzing the effect of emoticon support and communication mode on the likelihood of reaching an agreement (H5), we find a significant association between our treatments and the agreement-rate, $X(3) = 6.330$, p < .05. This finding, however, is caused by the asynchronous communication mode that increases the agreement-rate, 17 out of 21 dyads reached an agreement in asynchronous negotiations, while in synchronous negotiations only 13 out of 28 dyads reached an agreement, $X(1) = 6.025$, p < .05, but not by emoticon support, 17 out of 25 dyads reached an agreement with emoticon support, while 13 out of 24 dyads reached an agreement without emoticon support, $X(1) = .987$, p = .387.

5 Discussion, Conclusion and Outlook

The management of affective complexity is fundamental for negotiators to reach mutual understanding in communication and a positive relationship [6]. We have proposed that integrative negotiations can be supported with communication tools that facilitate the contextualization of communication by providing emoticons. Results of the 2 × 2-designed laboratory experiment support this claim: First of all, emoticon support makes communication more effective. Our results show that negotiators with emoticon support need less words/text to reach agreements compared to negotiators without emoticon support (emotions are used to supplement text messages). Secondly, negotiators exhaust the full range of functions of emoticons by additionally using emoticons to support text messages (emoticons are used in congruence), to mitigate the content of text messages (emoticons are used as antiphrasis). Finally, emoticon support significantly changes negotiation behavior by facilitating integrative negotiation behavior.

However, the impact of emoticon support differs in communication modes related to feedback immediacy. Integrative negotiations are typically characterized by a dual focus on the task and the relationship, e.g. [14]. The contextualization of information via emoticons seems to mitigate effects of feedback immediacy of the medium: In asynchronous negotiations which are typically described as task-oriented and "cool conversations" [47] because of low feedback immediacy, emoticon support induces negotiators to more often express positive affect. Apart from using predominantly the happy emoticon to tag the messages, they also express more often positive emotions in the written text. In contrast to that, synchronous negotiations have been referred to "hot debates" [47]: high feedback immediacy induces negative affective and inhibited behavior. In synchronous negotiations, we have witnessed that emoticon support – again apart from supplementing text messages with positive affect through the happy emoticons – reduces distributive behavior reflected in less persuasive behavior and less expressions of negative emotions in the text messages. Furthermore, we also observe more integrative negotiation behavior with emoticon support reflected in more positive affective behavior. To put it in other words, emoticon support heats up (too cool) asynchronous communication with positive emotions, while it helps to cool down (too hot) synchronous communication. These results support the importance of contextualizing information in social interactions via lean media [6].

Recently more and more research areas traditionally rooted in more analytic approaches have created awareness that the consideration of behavioral human factors increases the ability of model-based problem solving to help decision makers [51]. Yet, in negotiation support, most systems still focus on traditional analytic support rooted in economic considerations of their designers. However, our results support the notion that system designers should give more weight to communication aspects of negotiations. Currently, only the system Negoisst puts a clear focus on communication support on the syntactic, semantic and pragmatic level [9]. The enrichment of text messages via a contextualization of factual context of a message similarly to Negoisst's communication enrichment should make affective dimensions of messages more explicit. While the necessity to develop affective systems is undisputed, given the current state-of-the-art, the use of text-based (pro-active) affective systems is a major challenge [10]. However, affective systems can already be used for training purposes. In such an "emotional training" negotiators could be confronted with counterparts differing in their emotional reactions. Affective systems could use predefined sentences or entire text messages transmitting specific emotions, as already done for research purposes, see e.g. [15].

While our study delivers interesting insights, is not without limitations. Our results are based on one experiment using a single case and student subjects. The impact of the used emoticon support might also differ for varying degrees of conflict intensity. While the used case describes a realistic scenario for students rooted in the private life, business negotiations might require a different approach to contextualize affective information. Furthermore, in our analysis we have focused rather on integrative and distributive elements of the communication process and less on traditional economic elements of negotiations, like utility values used to e.g. compare characteristics of agreements. However, recent research directly comparing behavioral and economic support in negotiations postulates that effects of both support approaches are not limited to the respective support dimension, but actually show several spillovers [52]. Therefore, one promising avenue for future research is to untangle the relationship of contextualized information and its impact on economic dimensions of negotiations like concessions patterns and efficiency of agreements.

Acknowledgments. We thank Ronny Mitterhofer for the programming of the N-Swan system used in the experiment. Furthermore, we want to thank Evelyn Braumann, Olga Preveden, Sharjeel Saleem and Ying Xu for their contribution to data coding.

References

1. Rice, R.E., Love, G.: Electronic emotion: socioemotional content in a computer-mediated network. Commun. Res. **14**, 85–108 (1987)
2. Sproull, L., Kiesler, S.: Reducing social context cues: electronic mail in organizational communication. Manag. Sci. **32**, 1492–1512 (1986)

3. Walther, J.B.: Interpersonal effects in computer-mediated interaction: a relational perspective. Commun. Res. **29**, 52–90 (1992)
4. Walther, J.B.: Anticipated ongoing interaction versus channel effects on relational communication in computer-mediated interaction. Hum. Commun. Res. **20**, 473–501 (1994)
5. Derks, D., Fischer, A.H., Bos, A.E.R.: The role of emotion in computer-mediated communication: a review. Comput. Hum. Behav. **24**, 766–785 (2008)
6. Te'eni, D.: Review: a cognitive-affective model of organizational communication for designing IT. MIS Q. **25**, 251–352 (2001)
7. Daft, R.L., Lengel, R.H.: Organizational information requirements, media richness and structural design. Manag. Sci. **32**, 554–571 (1986)
8. Daft, R.L., Lengel, R.H., Trevino, L.K.: Message equivocality, media selection, and manager performance: implications for information systems. MIS Q. **11**, 355–368 (1987)
9. Schoop, M.: Support of complex electronic negotiation. In: Kilgour, D.M., Eden, C. (eds.) Handbook of Group Decision and Negotiation, vol. 4, pp. 409–423. Springer, The Netherlands (2010)
10. Broekens, J., Jonker, C.M., Meyer, J.-J.C.: Affective negotiation support systems. J. Ambient Intell. Smart Environ. **2**, 121–144 (2010)
11. Walther, J.B., D'Addario, K.P.: The impacts of emoticons on message interpretation in computer-mediated communication. Soc. Sci. Comput. Rev. **19**, 234–347 (2001)
12. Dresner, E., Herring, S.C.: Funtions of the nonverbal in CMC: emoticons and illocutionary force. Commun. Theory **20**, 249–268 (2010)
13. Lewicki, R.J., Saunders, D.M., Barry, B.: Negotiation. McGraw-Hill, Boston (2006)
14. Pruitt, D.: Negotiation Behavior. Academic Press, New York (1981)
15. Van Kleef, G.A., De Dreu, C.K.W., Manstead, A.S.R.: The interpersonal effects of anger and happiness on negotiation behaviour and outcomes. J. Pers. Soc. Psychol. **86**, 57–76 (2004)
16. Van Kleef, G.A., De Dreu, C.K.W., Manstead, A.S.R.: The interpersonal effects of emotions in negotiations: a motivated information processing approach. J. Pers. Soc. Psychol. **87**, 510–528 (2004)
17. Adler, R.S., Rosen, B., Silverstein, E.M.: Emotions in negotiation: how to manage fear and anger. Negot. J. **14**, 161–179 (1998)
18. Thompson, L.: Negotiation behavior and outcomes empirical evidence and theoretical issues. Psychol. Bull. **108**, 515–532 (1990)
19. Kramer, K.M., Newton, E., Pommerenke, P.L.: Self-enhancement biases and negotiator judgement: effects of self-esteem and mood. Organ. Behav. Hum. Decis. Process. **56**, 110–113 (1993)
20. Kumar, R.: The role of affect in negotiations: an integrative overview. J. Appl. Behav. Sci. **33**, 84–100 (1997)
21. Morris, M.W., Keltner, D.: How emotions work: the social functions of emotional expression in negotiations. In: Staw, B.M., Sutton, R.I. (eds.) Research in Organizational Behavior, vol. 22, pp. 1–50. JAI, Amsterdam (2000)
22. Friedman, R.A., Currall, S.C.: E-Mail Escalation: Dispute Exacerbating Elements of Electronic Communication. Vanderbilt University (2004)
23. Eisenberg, N., Fabes, R.A., Miller, P.A., Fultz, J., Shell, R., Mathy, R.M., Reno, R.R.: Relation of sympathy and personal distress to prosocial behavior: a multimethod study. J. Pers. Soc. Psychol. **57**, 55–66 (1989)
24. Brett, J.M., Olekalns, M., Friedman, R., Goates, N., Anderson, C., Lisco, C.C.: Sticks and stones: language, face, and online dispute resolution. Acad. Manag. J. **50**, 85–99 (2007)
25. Walther, J.B.: Computer-mediated communication: impersonal, interpersonal hyperpersonal interaction. Commun. Res. **23**, 3–43 (1996)

26. Martinovski, B.: Emotion in negotiation. In: Kilgour, D.M., Eden, C. (eds.) Handbook of Group Decision and Negotiation, vol. 4, pp. 65–86. Springer, The Netherlands (2010)

27. Van Kleef, G., De Dreu, C., Manstead, A.: An interpersonal approach to emotion in social decision making: the emotions as social information model. In: Zanna, M. (ed.) Advances in Experimental Social Psychology, vol. 42, pp. 45–96. Academic Press, Burlington (2010)

28. Van Kleef, G.A.: How emotions regulate social life: the emotions as social information (EASI) model. Curr. Dir. Psychol. Sci. 18, 184–188 (2009)

29. Clark, M.S., Taraban, C.B.: Reactions to and willingness to express emotion in two types of relationships. J. Exp. Soc. Psychol. 27, 324–336 (1991)

30. Pietroni, D., Van Kleef, G.A., De Dreu, C.K.W., Pagliaro, S.: Emotions as strategic information: effects of other's emotions on fixed-pie perception, demands and integrative behavior in negotiation. J. Exp. Soc. Psychol. 44, 1444–1454 (2008)

31. Griessmair, M., Koeszegi, S.T.: Exploring the cognitive-emotional fugue in electronic negotiations. Group Decis. Negot. 18, 213–234 (2009)

32. Gibbons, P., Bradac, J.J., Busch, J.D.: The role of language in negotiations: threats and promises. In: Putnam, L.L., Roloff, M.E. (eds.) Communication and Negotiation, pp. 156–175. Sage, Newbury Park (1992)

33. Darics, E.: Non-verbal signalling in digital discourse: the case of letter repetition. Discourse Context Media 2, 141–148 (2013)

34. Derks, D., Bos, A.E.R., von Grumbkow, J.: Emoticons in computer-mediated communication: social motives and social context. CyberPsychol. Behav. 11, 99–101 (2008)

35. Ganster, T., Eimler, S.C., Krämer, N.: Same same but different!? the differential influence of smilies and emoticons in person perception. Cyberpsychol. Behav. Soc. Netw. 15, 226–230 (2012)

36. Taesler, P., Janneck, M.: Emoticons und Personenwahrnehmung: Der Einfluss von Emoticons auf die Einschätzung unbekannter Kommunikationspartner in der online-Kommunikation. Gruppendynamik und Organisationsberatung, 41, 375–384 (2010)

37. Amaghlobeli, N.: Linguistic features of typographic emoticons in SMS discourse. Theory Pract. Lang. Stud. 2, 348–354 (2012)

38. Rezabek, L.L., Cochenour, J.J.: Visual cues in computer-mediated communication: supplementing text with emoticons. J. Vis. Literacy 18, 201–215 (1998)

39. Tossel, C.C., Kortum, P., Shepard, C., Barg-Walkow, L.H., Rahmati, A.: A longitudinal study of emoticon use in text messaging from smartphones. Comput. Hum. Behav. 28, 659–663 (2012)

40. Garrison, A., Remely, D., Thomas, P., Wierszewski, E.: Conventional faces: emoticons in instant messaging discourse. Comput. Compos. 28, 112–125 (2011)

41. Pflug, J.: Contextuality and computer-mediated communication: a cross cultural comparison. Comput. Hum. Behav. 27, 131–137 (2011)

42. Vandergriff, I.: Emotive communication online: a contextual analysis of computer-mediated communication (CMC) cues. J. Pragmatics 51, 1–12 (2013)

43. Riordan, M.A., Kreuz, R.J.: Emotion enconding and interpretation in computer-mediated communication: reasons for use. Comput. Hum. Behav. 26, 1667–1673 (2010)

44. Huang, A.H., Yen, D.C., Zhang, X.: Exploring the potential effects of emoticons. Inf. Manag. 45, 466–473 (2008)

45. Derks, D., Bos, A.E.R., von Grumbkow, J.: Emoticons and social interaction on the internet: the importance of social context. Comput. Hum. Behav. 23, 842–849 (2007)

46. Derks, D., Bos, A.E.R., von Grumbkow, J.: Emoticons and online message interpretation. Soc. Sci. Comp. Rev. 26, 379–388 (2008)

47. Pesendorfer, E.-M., Koeszegi, S.T.: Hot versus cool behavioural styles in electronic negotiations: the impact of communication mode. Group Decis. Negot. 15, 141–155 (2006)

48. Kopelman, S., Rosette, A.S., Thompson, L.: The three faces of eve: strategic displays of positive, negative, and neutral emotions in negotiations. Organ. Behav. Hum. Decis. Process. **99**, 81–101 (2006)

49. Srnka, K.J., Koeszegi, S.T.: From words to numbers - how to transform rich qualitative data into meaningful quantitative results: guidelines and exemplary study. Schmalenbach's Bus. Rev. **59**, 29–57 (2007)

50. Weingart, L.R., Brett, J.M., Olekalns, M., Smith, P.L.: Conflicting social motives in negotiating groups. J. Pers. Soc. Psychol. **93**, 994–1010 (2002)

51. Hämäläinen, R.P., Luoma, J., Saarinen, E.: On the importance of behavioral operational research: the case of understanding and communicating about dynamic systems. Eur. J. Oper. Res. **228**, 623–634 (2013)

52. Gettinger, J., et al.: Impact of and interaction between behavioral and economic decision support in electronic negotiations. In: Hernández, J.E., Zarate, P., Dargam, F., Delibašić, B., Liu, S., Ribeiro, R. (eds.) EWG-DSS 2011. LNBIP, vol. 121, pp. 151–165. Springer, Heidelberg (2012)

Online Collaboration and Competition

Warsaw School of Economics

A Longitudinal Case Study on Risk Factor in Trust Development of Facilitated Collaboration

Xusen Cheng, Shixuan Fu[(⊠)], and Yuxiang Peng

University of International Business and Economics, 100029 Beijing, China
fsx8888@163.com

Abstract. Computer-mediated collaboration is widely used in various organizations. Trust has proved to have an influence on online collaboration. This paper aims to conduct an in-depth investigation on an important trust factor during online collaboration, which is risk. The research samples were collected from Chinese part-time MBA students. They were invited to use the group support system (GSS) designed under the theory of facilitated collaboration with the thinkLets method to support the online collaboration. During this longitudinal research, questionnaires were collected at three stages, namely, at the beginning, during and at the end of the experiment, interviews were also conducted. Results show the level of trust was raised over time. Among all the trust factors, risk shows the most significant change, and the level of risk is decreased. Finally, the correlation analysis was conducted to detect the relationship between risk and trust in facilitated collaboration.

Keywords: Risk · Trust development · Facilitated collaboration · Trust factors

1 Introduction

Rigorous business competition drives people to take inter-organizational alliances, when an individual's ability is limited. Collaborations among team members turn out to be essential. Research shows that most fortune 100 companies collaborate frequently, but only 13 % of the team collaborations were considered to be effective [1]. This may result from a lack of trust among team members. Trust has its importance to collaboration teams [2, 3], and plays an important role in overcoming barriers, such as conflicts avoidance [4].

Increasingly advanced information technology has made online collaboration possible and popular. Teleconference, social network services, video conferences and discussion groups are adopted by a growing number of companies [5]. Moreover, as noted by Serçe et al. [6], compared with traditional collaboration, an online team is a set of geographically dispersed and functionally diverse organization, calling for a higher level of trust to improve work efficiency.

Researchers have attempted to decompose trust into different parts, namely trust factors that have an influence on trust. The overall level of trust can be figured out by evaluating each trust factor. Measuring trust over time could help assess the role of

B. Kamiński et al. (Eds.): GDN 2015, LNBIP 218, pp. 309–320, 2015.
DOI: 10.1007/978-3-319-19515-5_24

collaboration tools, but most studies on trust factors in computer mediated collaborations failed to take the change of trust factors into account [7, 8]. Cheng et al. [9] investigated trust development over time according to the six factors in the use of computer mediated collaboration tools, but has not systematically explored one certain factor in detail and found out the most significant one.

Among various trust factors, risk is the most frequently mentioned one which could be seen as the anticipated hazard of interpersonal relationships [10]. Minimizing risk makes team members willing to trust each other and contributes to effective collaboration [11]. Therefore we investigate the following questions related to the development of risk factor.

Research question 1: What is the change trend of risk factor in the context of facilitated collaboration? After collaboration over time, does the overall level of trust change?

Research question 2: What is the correlation between risk and the overall level of trust?

This paper is structured as followed. Section 2 begins by laying out the background of relevant studies. Research method and data collection will be given in Sect. 3. Section 4 is concerned with the investigation of research data through the mix use of qualitative and quantitative analysis. Then, in Sect. 5, we analyze the results and have a discussion, then give a brief summary and critique of the findings as well as our limitations.

2 Research Background

2.1 Facilitated Collaboration

As an approach that aims to design collaboration process, collaboration engineering has been developed into an emerging research field [12]. Collaboration engineering can be considered as a combination of facilitation, design and training approach that can be supported with group collaborating tools [13]. A facilitator is needed to decompose tasks and instruct processes. Facilitation is a participative leadership, and contributes to improve a group's communication and information flow [14]. Facilitated collaboration had been applied to various areas, such as education, military, business community and so on. By offering sustained collaboration support, collaboration process is designed and deployed for a recurring task.

ThinkLets is a core concept in facilitated collaboration which include generate, reduce, clarify, organize, evaluate and build consensus [15]. It is the smallest unit of intellectual capital for the creation of collaborative tools, and provides a transferable, reusable block for process design [15]. Based on various tasks, users could choose the most appropriate thinkLets methods to simplify the collaboration process [13].

Group support system (GSS) is a suite of software to support groups in their collaborative effort. The importance of the design of collaboration process is amplified when GSS is used [12].There are various kinds of group support systems, such as GroupSystems (developed at the University of Arizona), SAMM (from the University of Minnesota) and discussion platform which is developed with the agile method on the WAMP platform (Windows/Linux+ Apache+ Mysql+ Php) [16].

2.2 Trust Development

According to Holton [17], trust is a situation when individuals feel comfortable and open in sharing their insights and concerns. There is a volume of published studies describing trust. The topic of trust issues in online collaboration has been addressed by many scholars [3, 18], and trust is considered as a dynamic construct [19].

Costa et al. [20] examined the development of social trust in project teams, research data was collected at the beginning, middle and end of the project. On the basis of what Lewicki and Bunker [21] identified during three stages of trust development, namely calculus based trust, knowledge based trust and identification based trust. Recently in 2013, Bhati et al. [22] examined how trust developed between branch managers and loan officers in different phases over a period of time. In distributed teams, trust development is also investigated through longitudinal study [23, 24]. However, most of the research data on trust development are pure students who hardly have any work experience.

In an investigation into trust development of business online community, Nolan et al. deconstructed six factors in the perspective of individual trust which represents the conflicting priorities. Those factors are presented in their research as: risk, benefit, utility value, interest, effort and power [25]. The ideal state of those components is minimizing risk and effort, maximizing other parts [9].

2.3 Risk and Trust

Among six trust factors, risk is associated with providing information to unknown recipients and acting upon information received from them [25].Willingness to take risks has been suggested as one of the few characteristics common to all trust situations [26]. Risk is evaluated on every possible outcome of a particular action. Risk and trust are two facets of decision-making [10]. Besides, risk and perspective-taking were considered as two elements of trust in behavioral economics [27].

Under the condition of risk, the tendency to trust is relatively weaker [28]. Therefore, it is necessary to discover a way of minimizing risk in virtual collaboration. Scholars once analyzed risk management through repeatable distributed collaboration processes, showed the trend in risk management, and tried to identify possible risks in the early stage to control them [29]. Besides, in facilitated collaboration thinkLets were thought to reduce risk in online collaboration [30, 31]. We investigate risk that have an influence on trust in online collaboration with the help of collaboration engineering.

3 Research Method

3.1 Case Background

Various methods have been developed and introduced to measure trust development, in which a case study approach was used to conduct an in-depth, holistic investigation [32]. Considering the approach used in other similar studies [25, 33], we are going to use an exploratory case study approach.

In our research, 73 part-time MBA students with 38 males and 35 females are selected. As part-time MBA students, almost all of these participants have a minimum of three years work experience. We divide the students into 15 groups composed of four or five students randomly. The groups are all assigned the same project task to find out the problem in an E-business website and work out the solutions.

In the classroom, team members could directly exchange, share and discuss ideas. While after class, they can use QQ, Wechat, Skype, and other online communication software for collaboration. Besides, participants are encouraged to use discussion platform to facilitate their collaboration process. Discussion system is a self-developed online platform designed according to the process of thinkLets, and is instructed by the principle of collaboration engineering [16]. In general, with fixed class time as well as suggested instructions, the influence of irrelevant variables can be reduced efficiently.

3.2   Data Collection

In order to track the development of trust, we conduct survey three times during the project. That is to say, we divide the whole period into three equal stages, the initial stage, the middle stage and the final stage. At the start of each stage, the professor assigns the corresponding task. As an after class assignment, all the students are required to complete questionnaires designed by Cheng et al. [34]. In different stages, we have received 219 pieces in total. Gross error and redundant data were eliminated by statistical means. Finally, valid obtained data was 71 for each stage.

Especially, we have adopted a combination of semi-structured interviews to explore and analyze trust development in different stages during team collaboration. The design of the interview questions were based on the theoretical basis of former researchers [9]. In an attempt to make each interviewees feel as comfortable as possible, the pilot interviews were conducted informally by professionally trained interviewers, then we've modified the possible misunderstanding of the interview questions which may mislead our target participants. We have also investigated the backgrounds, group culture of the target participants, for the ease of improving interview questions and making the data of in-depth interviews more effective.

A total of 34 students are volunteered to be interviewed at the final stage. According to the transcripts of interviews, the interviewees include facilitators and ordinary group members during their team collaboration.

4   Data Analysis

4.1   Reliability and Validity Tests of the Questionnaires

We test the questionnaire's validity and applicability in order to measure targets' attitudes or behaviors accurately and comprehensively.

Cronbach's α is a statistic referring to the average of split-half reliability coefficient obtained from all the possible scale project division methods, which is the most commonly used method of reliability measurement. Different scholars hold different views on the boundary value of the reliability coefficients. Some believe that in general studies it should be at least 0.8 to be accepted, and at least 0.7 in exploration studies. In practice, it only need to be 0.6, while further revision is needed when the questionnaire has a Cronbach's α which is less than 0.6. We used Cronbach's α to analyze the reliability of the six trust factors and found all of them above 0.8, which explained a high reliability and research value.

4.2 Average Values of Six Trust Factors over Time

Research data were collected in three different stages, namely the initial stage, the middle stage, and the final stage. For each stage, we conducted a questionnaire survey for each student and calculated the arithmetic means of six trust factors of each group. We further calculated average values of six factors in each group according to three stages, see Table 1.

With effective communication and clear goals, six trust factors will gradually approach the ideal value [9]. Among them, the ideal values of risk factor and effort factor are both 1. The decrease of the two factors means an increase in trust. Meanwhile, the ideal values of benefit factor, utility value factor, interest factor, and power factor are 5. The rise of these four factors shows increase in the trust.

In order to exhibit the change trend of trust factors over time, we have adopted a spider diagram according to three different stages. Figure 1 shows that risk has a significant downward trend while the four factors of benefit, utility value, interest and effort display an upward trend, but the trend is less pronounced. Effort factor shows a downward trend after the first rise.

Table 1. Average value of six factors of part-time MBA students

	Risk	Benefit	Utility value	Interest	Effort	Power
Stage1	2.31	4.30	3.96	4.35	3.94	3.09
Stage2	2.08	4.30	3.99	4.35	4.13	3.28
Stage3	1.82	4.36	4.21	4.45	4.06	3.22

4.3 The Significant Change of Risk Factor

As for changes of six trust factors, we need to measure their rates of change in order to get a further understanding of the influence of their tendency towards the trust, so we introduced the calculation method of the year-on-year rate [35].

$$\text{year - on - year change ratio} = \frac{(\text{current value - base - period value})}{\text{base -perriod value}} \times 100\%$$

We defined the initial stage as the basic period and calculated the year-on-year rate of the middle stage and the final stage, as is showed in the Table 2.

From Table 2, it follows that risk shows the most obvious change of a downward trend, while the changes of benefit factor and interest factor are less obvious. By drawing a line chart of the change of the risk, we further validate the most obvious change of risk, which is in a decline trend Fig. 2.

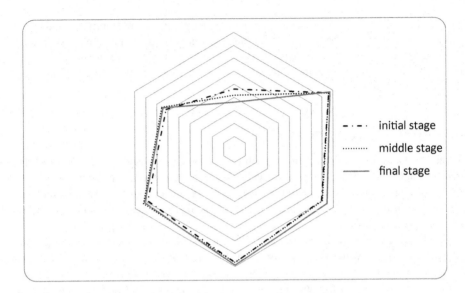

Fig. 1. The spider diagram of trust factors over time

Table 2. Year-on-year rate of change of the part-time MBA students

	Risk	Benefit	Utility value	Interest	Effort	Power factor
Stage1						
Stage2	-9.96 %	0.00 %	0.76 %	0.00 %	4.82 %	6.15 %
Stage3	-21.21 %	1.40 %	6.31 %	2.30 %	3.05 %	4.21 %

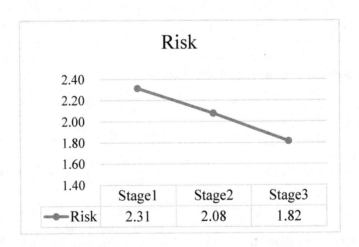

Fig. 2. The change of risk of the part-time MBA students

4.4   The Development of the Overall Level of Trust

We use the standardized residual between the trust model of the part-time MBA students and ideal trust model [9] to measure the index of the trust. The smaller the trust standardized residual is, the higher the trust will be.

$$\text{standard residual} = \frac{\sum_1^n (\text{observation} - \text{regression estimate})^2}{n-1}$$

$$\text{Trust Measure} = sr_{Risk} + sr_{Benefit} + sr_{UtilityValue} + sr_{Interest} + sr_{Effort} + sr_{Power}$$

Thus, we gathered the standardized residual of trust of each stage and the corresponding year-on-year rate of change

Table 3 shows the values of the trust of the sample rise as the experiment proceeds, which is in the decline trend of the standardized residual of trust. The year-on-year rate of change reaches 7.55 % at the final stage, which indicates that the use of thinkLets teamwork system helps improve the trust and the efficiency and effectiveness of teamwork. Meanwhile, the ratio of 7.55 % also shows the statistical validity of the method using the standardized residual to calculate the index of trust.

Table 3. Change of the trust standardized residual

	Trust standardized residual	Year-on-year rate of change
Stage1	23.12875	
Stage2	22.97709	0.66%
Stage3	21.38339	7.55%

4.5   The Correlation Between Risk and the Overall Level of Trust

Through the calculation of correlation index, we efficiently proved that the rise of the trust is attributed to the significant decrease of risk.

Correlation is a description of the uncertainty of the relationship between two or more variables. Correlation Analysis refers to the statistical analysis method or process on Correlation between variables. We use Pearson's Correlation Coefficient for correlation analysis on Trust Measure and six trust factors, which can be obtained from:

$$r = \frac{\sum X - \frac{\sum x \sum y}{N}}{\sqrt{\left(\sum X^2 - \frac{(\sum x)^2}{N}\right)\left(\sum Y^2 - \frac{(\sum y)^2}{N}\right)}}$$

Pearson's Correlation Coefficient r is used to determine if there is a correlation between the two data sets, X and Y. It varies between -1 and 1. When r>0, it shows a positive correlation; when r < 0, it shows a negative correlation. The absolute value of r indicates the degree of correlation between the variables. The closer the absolute value is to 1, the stronger correlation it shows. The results of correlation analysis on risk and trust standardized residual are shown in Table 4.

Table 4. Correlation analysis between risk value and the standardize residual

Stage	Risk	Trust standardized residual	Pearson correlation index	Correlation's year-on-year rate of change
Stage1	2.1623	23.12875	0.338	
Stage2	2.0722	22.97709	0.351	3.85 %
Stage3	1.8651	21.38339	0.481	42.31 %

As is shown in Table 4, the trust standardized residual shows a downward trend, that is, with the decrease of the risk, trust rises. The Pearson correlation coefficient between them at the initial stage is 0.338, up to 0.481 at the final stage, with an increase rate of 42.31 %. From the significant increase of the correlation, we conclude that: as for the sample of the part-time MBA students, by collecting data from the time series, risk presents the most obvious trend in the six trust factors. risk has a significantly negative correlation with the trust and the decline of observably result in the rise of trust, further leading to the rise of the efficiency of the team collaboration.

5 Discussion and Conclusion

Previous researches have recognized that trust is an important factor influencing the outcomes of online collaboration [18, 36]. In this paper, based on the previously proposed trust factors, we conducted a case study using facilitated collaboration tool, the discussion system.

5.1 Discussion of Research Findings

The overall level of trust is increased through computer-supported facilitated collaboration

We have calculated the standardized residual to evaluate the level of trust in the three stages respectively. The standard residual decreases by 7.55 % from stage2 to stage3, which shows the increase of the overall level of trust.

This finding are also supported by the interview comments, Someone has mentioned that I feel involved in my team because sometimes my opinion get the most votes, sometimes my vote is quite important for our team. I'm all satisfied with other teammates. The level of trust is obviously increased. While another participant holds that according to three months collaboration, we're familiar with each other. The trust level is indeed improved.

Among six trust factors, risk is the most significantly changed factor with a decreasing trend.

The spider diagram shows the obvious change compared with other factors. Besides, according to quantitative analysis of year-on-year change rate, from early stage to middle stage, the change range of benefit, utility value and interest is smaller than 1 %, while risk decreased by about 9.96 %. From middle stage to final stage, the result is more desirable that the change rate of risk is 21.21 %. However, the second significantly changed factor, the utility value, only increased by 6.31 %.

At the meanwhile, one participant said that the system has a simple but efficient function to break the emotional barriers. Especially in the later stage, we are familiar with each other, we don't hesitate to deliver our opinions. There is almost no risk. Another active participant told us that I felt that sending message anonymously helps me to share my opinions freely. With the collaboration going on, the level of risk is decreased.

Through facilitated collaboration, the level of risk decreases over time and trust increases accordingly.

Through simple calculation of mean value of trust over time, the level of trust decreases. Then, the correlation analysis of the level of risk and trust shows their negative correlation over time. Besides, the year-on-year change rate is increased significantly, from 3.85 % to 42.31 %, that means during the mid-to-late period of collaboration, the decrease of risk significantly increase the level of trust.

The correlation between risk and trust is also highlighted by the qualitative analysis. If someone holds that through long period of collaboration, then personal preference is no longer a private one, so risk is decreased. From strangers to acquaintances, the level of trust is indeed increased. Besides, a facilitator in another group told us that the platform is easy to use and makes our collaboration effective, I'm accustomed to this software, so risk is decreased, thus at least, trust toward the software is increased.

5.2 Theoretical and Practical Implications

Theoretically, through quantitative analysis, this study shed light on the investigation of trust factors. Risk is validated to exist in the initial level of online collaboration, and changes significantly through longitudinal research. According to facilitated process, the level of risk is decreased over time, which is consistent with the research findings of previous researches in different background [6, 7], similar experimental setting in different case context [33, 34], and towards the ideal states of six trust factors [9]. By investigating risk in facilitated collaboration, we fill the research gap of deep investigation of a certain factor through the use a thinkLets method, and find out an important role of facilitated collaboration, that is, to reduce risk in online collaboration.

From a practical viewpoint, the results show that facilitated collaboration contributes to reduce risk over time. It offers valuable reference to introduce facilitated process to real business online collaboration. Through the introduction of a facilitator, online business discussion may be more effective. Additionally, it also provides clues for software developers to design more useful tools.

5.3 Limitations and Future Research

We have conducted a case study using the part-time MBA students, however, this is a special context which has not been tested in other contexts and it may not be applied. Therefore, future research will be considered using various sources of research samples. An in-depth interview analysis of the reasons for our conclusions could also be considered. Moreover, the emphasis of our research is one of the trust factors, risk. In future research, we would like to make a correlation analysis of six trust factors and the overall level of trust, and compare all the trust factors in facilitated collaborations.

Acknowledgments. The authors would like to thank the National Natural Science Foundation of China (GrantNo.71101029) and the Program for Young Excellent Talents of UIBE (Grant No. 13YQ08) for providing funding for part of this research.

References

1. Daley, D.M.: Interdisciplinary problems and agency boundaries: exploring effective cross-agency collaboration. J. Public Adm. Res. Theory **19**(3), 477–493 (2009)
2. Brown, H.G., Poole, M.S., Rodgers, T.L.: Interpersonal traits, complementarity, and trust in virtual collaboration. J. Manag. Inf. Syst. **20**(4), 115–138 (2004)
3. Chua, R.Y., Morris, M.W., Mor, S.: Collaborating across cultures: cultural metacognition and affect-based trust in creative collaboration. Organ. Behav. Hum. Decis. Process. **118**(2), 116–131 (2012)
4. Lauring, J., Selmer, J.: Openness to diversity, trust and conflict in multicultural organizations. J. Manag. Organ. **18**(6), 795–806 (2012)
5. Ransbotham, S., Kane, G.C.: Membership turnover and collaboration success in online communities: explaining rises and falls from grace in Wikipedia. MIS Q. **35**(3), 613–620 (2011)
6. Serçe, F.C., Swigger, K., Alpaslan, F.N., Brazile, R., Dafoulas, G., Lopez, V.: Online collaboration: collaborative behavior patterns and factors affecting globally distributed team performance. Comput. Hum. Behav. **27**(1), 490–503 (2011)
7. Jarvenpaa, S.L., Knoll, K., Leidner, D.E.: Is anybody out there? Antecedents of trust in global virtual teams. J. Manag. Inf. Syst. **14**(4), 29–64 (1998)
8. Beldad, A., De Jong, M., Steehouder, M.: How shall I trust the faceless and the intangible? A literature review on the antecedents of online trust. Comput. Hum. Behav. **26**(5), 857–869 (2010)
9. Cheng, X., Macaulay, L., Zarifis, A.: Modeling individual trust development in computer mediated collaboration: a comparison of approaches. Comput. Hum. Behav. **29**(4), 1733–1741 (2013)
10. Jøsang, A., Presti, S.L.: Analysing the relationship between risk and trust. In: Jensen, C., Poslad, S., Dimitrakos, T. (eds.) iTrust 2004. LNCS, vol. 2995, pp. 135–145. Springer, Heidelberg (2004)
11. Thomalla, F., Downing, T., Spanger-Siegfried, E., Han, G., Rockström, J.: Reducing hazard vulnerability: towards a common approach between disaster risk reduction and climate adaptation. Disasters **30**(1), 39–48 (2006)
12. Kolfschoten, G.L., Vreede, G.J.: A design approach for collaboration processes: a multi-method design science study in collaboration engineering. J. Manag. Inf. Syst. **26**(1), 225–256 (2009)
13. Kolfschoten, G.L.: Theoretical Foundations for Collaboration Engineering. Delft University of Technology, The Netherlands (2007)
14. Griffith, T.L., Fuller, M.A., Northcraft, G.B.: Facilitator influence in group support systems: intended and unintended effects. Inf. Syst. Res. **9**(1), 20–36 (1998)

15. De Vreede, G.J., Kolfschoten, G.L., Briggs, R.O.: ThinkLets: a collaboration engineering pattern language. Int. J. Comput. Appl. Technol. **25**(2), 140–154 (2006)
16. Cheng, X., Li, Y., Sun, J., Zhu, X.: Easy collaboration process support system design for student collaborative group work: a case study. In: 47th Hawaii International Conference on System Sciences, pp. 453–462. IEEE Press, Hawaii (2014)
17. Holton, J.A.: Building trust and collaboration in a virtual team. Team Perform. Manag. **7**(3), 36–47 (2001)
18. Tariq, A., Aslam, H.D., Habib, M.B., Siddique, A., Khan, M.: Enhancing employees collaboration through trust in organizations:(an emerging challenge in human resource management). Mediterr. J. Soc. Sci. **3**(1), 559–565 (2012)
19. Lai, C.S., Chen, C.S., Chiu, C.J., Pai, D.C.: The impact of trust on the relationship between inter-organisational collaboration and product innovation performance. Technol. Anal. Strateg. Manag. **23**(1), 65–74 (2011)
20. Costa, A.C., Bijlsma-Frankema, K., de Jong, B.: The role of social capital on trust development and dynamics: implications for cooperation, monitoring and team performance. Soc. Sci. Inf. **48**(2), 199–228 (2009)
21. Lewicki, R.J., Bunker, B.B.: Developing and maintaining trust in work relationships. In: Kramer, R., Tyler, T. (eds.) Trust in Organisations: Frontiers of Theory and Research, pp. 114–139. Sage, Thousand Oaks (1996)
22. Bhati, S.S., De Zoysa, A.: Stages of trust development in banking relationship. Banks Bank Syst. **8**(1), 36–44 (2013)
23. Pavlova, E.: Trust development in distributed teams: a latent change score model. Doctoral dissertation, University of South Florida, Florida (2012)
24. Zolin, R., Hinds, P.J., Fruchter, R., Levitt, R.E.: Interpersonal trust in cross-functional, geographically distributed work: a longitudinal study. Inf. Organ. **14**(1), 1–26 (2004)
25. Nolan, T., Brizland, R., Macaulay, L.: Individual trust and development of online business communities. Inf. Technol. People **20**(1), 53–71 (2007)
26. Lapidot, Y., Kark, R., Shamir, B.: The impact of situational vulnerability on the development and erosion of followers' trust in their leader. Leadersh. Quart. **18**(1), 16–34 (2007)
27. Evans, A.M., Krueger, J.I.: Elements of trust: risk and perspective-taking. J. Exp. Soc. Psychol. **47**(1), 171–177 (2011)
28. Mukherjee, D., Renn, R.W., Kedia, B.L., Mukherjee, D.: Development of interorganizational trust in virtual organizations: an integrative framework. Eur. Bus. Rev. **24**(3), 255–271 (2012)
29. Van Grinsven, J., de Vreede, G.: Addressing productivity concerns in risk management through repeatable distributed collaboration processes. In: 36th Hawaii International Conference on System Sciences, pp. 10–15. IEEE Press, Hawaii(2003)
30. De Vreede, G.J., Briggs, R.O., Massey, A.P.: Collaboration engineering: foundations and opportunities: editorial to the special issue on the journal of the association of information system. J. Assoc. Inf. Syst. **10**(3), 7–9 (2009)
31. Briggs, R.O., De Vreede, G.J., Nunamaker Jr., J.: Collaboration engineering with ThinkLets to pursue sustained success with group support systems. J. Manag. Inf. Syst. **19**(4), 31–64 (2003)
32. Feagin, J., Orum, A., Sjoberg, G.: A Case for Case Study. University of North Carolina Press, Chapel Hill (1991)
33. Wilson, J.M., Straus, S.G., McEvily, B.: All in due time: The development of trust in computer-mediated and face-to-face teams. Organ. Behav. Hum. Decis. Process. **99**(1), 16–33 (2006)

34. Cheng, X., Nolan, T., Macaulay, L.: Don't give up the community: a viewpoint of trust development in online collaboration. Inf. Technol. People **26**(3), 298–318 (2013)
35. Gyani, A., Shafran, R., Layard, R., Clark, D.M.: Enhancing recovery rates: lessons from year one of IAPT. Behav. Res. Ther. **51**(9), 597–606 (2013)
36. Kim, D.J., Ferrin, D.L., Rao, H.R.: A trust-based consumer decision-making model in electronic commerce: the role of trust, perceived risk, and their antecedents. Decis. Support Syst. **44**(2), 544–564 (2008)

Intention to Repurchase Group Coupon Service: The Intertwined Effect of Service Quality of Vendor and Service Provider

Hsiangchu Lai[1]([⊠]) and Shu-Hwa Hsu[2]

[1] National Taiwan Normal University, Taipei, Taiwan
hclai@ntnu.edu.tw
[2] Foxconn Technology Group, Taipei, Taiwan

Abstract. The success of Groupon creates many followers who attract consumers by providing location-based electronic coupons with big discount. Despite of its great growth rate, there were many complaints from consumers. The purpose of this research is to learn how service quality of vendors (i.e. the websites selling the electronic coupons) and service providers (i.e. the stores where consumers can redeem coupons) interactively affect the intention to repurchase coupons from a vendor or a service provider. The results indicate that service quality of both the vendor and the service provider will affect the intention to repurchase coupons from vendor, but the service provider's service quality has higher effect. It means that if a service provider does not provide good quality, it will affect the sales of the coupon vendor. The service quality of service provider will affect the intention to buy service from the service provider at either the regular price or the coupon price. These findings indicate the coupon vendor should be careful in recruiting service providers.

Keywords: Service quality · SERVQUAL · Electronic coupon · Group purchase · Repurchase intention

1 Introduction

Over the past two decades, online advertising has been developing at quickly along with the e-commerce and has been undergoing dramatic changes in the use of different online advertising formats [23]. Recently, with the launch of Groupon in late 2008, a whole new advertising format based on "group coupon" has been introduced. The idea is that group coupon websites sell electronic coupons with significant discounts of 50 to 70 % and consumers can redeem the coupons from the service providers who issue the coupons. However, such promotion is valid only if a certain minimum number join the deal [10].

In contrast to traditional advertising formats, there are three participants: vendor (i.e. the websites selling the electronic coupons), service provider (i.e. the stores where consumers can redeem coupons) and consumers. Consumers pay the coupon price upfront and get discount vouchers from vendors; they can later redeem them from the service provider within a certain period of time. However, despite the fast development of group coupon advertising format, vendors have received many negative comments

© Springer International Publishing Switzerland 2015
B. Kamiński et al. (Eds.): GDN 2015, LNBIP 218, pp. 321–332, 2015.
DOI: 10.1007/978-3-319-19515-5_25

from industry experts, consumers, and service providers. Industry experts think it is hard to convert deal-users into repeat buyers loyal to the service provider. Group coupon users complain that it is difficult to make reservations due to various reasons or they are disappointed with the services they get from service providers.

These negative comments from group coupon users reflect gaps which arise from inconsistent perceptions of expectations and experiences of the participants in this advertising format. These gaps increase the difficulties with satisfying and retaining coupon users. It is a challenging issue for both the vendors and the service providers.

Although some studies focused on the effectiveness of group coupon promotion and its profitability [6, 10–13], little attention has been paid to how service quality in group coupon advertising affects the customers' repurchase intentions. It is likely that group coupon advertising would easily attract consumers with high price sensitivity and lower brand loyalty [3].

To better understand how service quality and customer price sensitivity influence the repurchase intention, this study intended to: (1) explore the source of service gaps; (2) understand the independent effect of service quality of both the vendor and the service provider on different kinds of repurchase intentions; (3) understand the moderating impact of price sensitivity on the impact of the service quality of service provider on the consumers' intention to repurchase service at the regular price from service providers.

2 Service Quality Model for Group Coupon Advertising

Perceived service quality is defined as "the difference between consumer expectations and perceptions, which in turn depends on the size and direction of gaps associated with the delivery of service quality on the marketer's side" by Zeithaml, et al. [27] (p. 36). They also conceptualize the service quality model known as SERVQUAL using five dimensions: tangibles, reliability, responsiveness, assurance, and empathy. Based on the group coupon advertising model and in particular SERVQUAL model, we present the possible service gaps of group coupon advertising in Fig. 1. We can see that the vendor and the service provider work together to provide service to consumers and therefore, both Gap 2 and Gap 3 in SERVQUAL proposed by Zeithaml, et al. [27] are divided into two parts. One gap may be caused by the vendor and the other may result from the service provider.

To conceptualize service quality of the vendor, we utilize the dimensions in E-S-QUAL developed by Parasuraman et al. [20] and modify some dimensions as suggested by literature and customers' redeeming experiences collected from the Internet. The E-S-QUAL scale measures four dimensions of service quality. They are efficiency, system availability, fulfillment, and privacy. For group coupon services, efficiency and system availability both have positive influence because whether group coupon website is friendly or not is important to consumers. Additionally, we use assurance dimension instead of the privacy dimension in E-S-QUAL because group coupon vendors and customers are concerned about the vendor guarantee policy regarding delivery and returns. Different from tradition, customers pay before getting the merchandise and service; vendors have to promise the customers that they not only can get what they

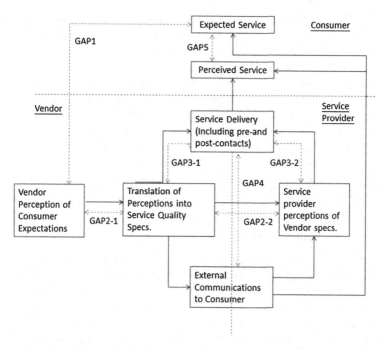

Fig. 1. Service gaps of group coupon advertising

expect from the service provider but also that they will get money back if they do not redeem the voucher successfully. Therefore, the Assurance dimension here involves trust and confidence for customers to deal with.

Information quality which defines "the quality of the content of the site" in [2] research, especially for the accurate information item, was viewed as an important issue in the WebQual instrument and in numerous studies [1, 7, 19, 22, 28]. In our research, most customers get information about merchandise/service from the vendor but redeem their vouchers and get what they want from the service provider. The website service quality also considers information about products, services, return policies, and guarantee policies [28] (p. 4). The reliability of the website can be reflected in the reliability of the information that is captured by information quality [28]. Most customers are new to service providers. Hence, the information quality of website plays an important role in this kind of advertisement format. Whether a vendor provides truthful and believable information to customers affects their perceptions of the service quality of the vendor.

Based on the above discussions, we conceptualize the service quality of vendor using four dimensions: efficiency, system availability, assurance, and information quality. The service quality of the service provider is measured with other four dimensions: interaction quality, physical environment quality, fulfillment and policy. Additionally, to better understand the importance of each role in group coupon service, we divide repurchase intention into three types: the intention to repurchase coupons from vendors, the intention to repurchase service from the service provider at the original price, and the intention to repurchase service from the service provider with a coupon.

Chang et al. [7] stated that e-service quality affects customers' satisfaction and then generates customers' loyalty which can be measured by the repurchase intention and WOM. On the other hand, for retailer's service quality, satisfied customers have intentions to revisit or return to the same service provider having experienced good service quality [8]. Service quality of vendors has positive and direct influence on consumers' intention to repurchase coupons from vendors because customers would go back to repurchase group coupons due to good service experiences. However, customers' intention to repurchase from vendor also is affected by the service delivered by service provider. Customers would also go back to buy group coupons if they have great experience of redeeming coupons from service providers. They may come back not only for the same type of coupons with different service providers but also for different type of group coupons. Therefore, we propose the following hypotheses:

Hypothesis 1a: In group coupon advertising service, service quality of the vendor has positive effect on consumers' intention to repurchase coupons from vendor.

Hypothesis 1b: In group coupon advertising service, service quality of the service provider has positive effect on consumers' intention to repurchase coupons from vendor.

Further, service quality of service providers has positive and direct effect on consumers' intention to repurchase service from service providers too. Based on consumers' comments on the Internet, customers would only come back if they have experienced great service from service providers even paying the original price. Therefore, the following hypothesis is proposed:

Hypothesis 2: In group coupon advertising service, service quality of service providers has positive effect on consumers' intention to repurchase service from service providers at the original price.

A group coupon advertisement with high price discount would attract customers who have high price sensitivity. Garretson et al. [15] demonstrated that price sensitivity is a major driving factor for coupon usage. Moreover, advertisement may increase customers' price sensitivity [16] (p. 210). Whether consumers can repurchase service is a major concern for service providers. It is important therefore to investigate whether those who have higher price sensitivity will go back to consume again at the original price. Hence we propose the following hypothesis:

Hypothesis 3: In group coupon advertising service, price sensitivity has a moderating effect on consumers' intention to repurchase service from service providers at the original price.

Great service of service providers would/should attract customers to go back to buy the merchandise or services. However, some customers' comments indicate that consumers would go back only if they can buy group coupons provided by the service provider again. Therefore, we propose that consumers would also have the intention to repurchase service from the service provider with a coupon if they have experienced good services provided by service provider.

Hypothesis 4: In group coupon advertising service, service quality of service provider has positive effect on consumers' intention to repurchase service from service provider with coupon.

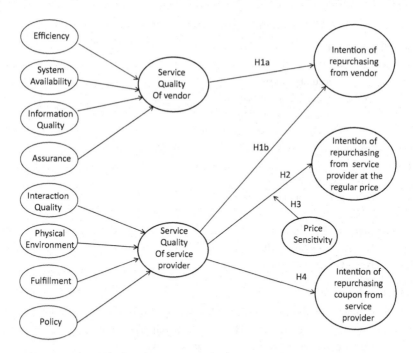

Fig. 2. Group coupon service model

All of the above proposed hypotheses are summarized as a group coupon service model shown in Fig. 2.

3   Research Methodology and Data Collection

3.1   Measurements

We adopted survey as a research methodology. The questionnaire was developed based on literature [2, 4, 19, 20, 24, 28] and users' coupon redeeming experiences shared on the Internet. For pretest, we asked 25 subjects who had group coupon redeeming experience to finish the questionnaire. Ambiguous responses were removed after the pretest. The survey questions are shown in Table 1. The items were measured on a seven-point Likert scale, anchored from "1" (strongly disagree) to "7" (strongly agree).

3.2   Data Collection

Online survey was adopted. Respondents were required to have at least one purchase and redeeming coupon experience. They were asked to answer questions about their most impressive group coupon redeeming experience to determine if they had sufficient experience with group coupon advertisement format. We collected 366 valid samples. 44.1 % of our respondents answered the questionnaire based on the experience of Groupon Taiwan as their group coupon vendor, following with Gomaji and Yahoo!

Table 1. Measurement items

Service quality of vendor	
Information quality	
INF1	Provides accurate information
INF2	Provides believable information
INF3	Provides relevant information
INF4	Provides easy to understand information
INF5	Provides information at the right level of detail
Efficiency	
EFF1	This site is well organized
EFF2	This site is simple to navigate
EFF3	Information at this site is well organized
System availability	
SYS1	This site does not crash
SYS2	This site launches and runs right away
SYS3	This site is always available for business
Assurance	
ASS1	It protects information about my Web-shopping behavior
ASS2	It does not share my personal information with other sites
ASS3	This site protects information about my credit card
ASS4	The online company is trustworthy
ASS5	I am happy with the vendor guarantee policy
Service quality of service provider	
Interaction quality	
IQ1	Employees readily respond to requests
IQ2	Employees understand my specific needs
IQ3	Employees are always willing to help
IQ4	Employees have the knowledge to answer questions
IQ5	Employees are consistently courteous
Physical environment	
PQ1	Physical facilities are visually appealing
PQ2	Modern looking equipment and fixtures
PQ3	Materials associated with the store are visually appealing
Fulfillment	
FUL1	Waiting time is predictable
FUL2	It sends out the items ordered
FUL3	It delivers orders when promised
Policy	
POL1	This store has operating hours convenient to all their customers
POL2	This store has good booking services
POL3	This store has good coupon redeeming services

(*Continued*)

Table 1. (*Continued*)

Service quality of vendor	
Repurchase intention – vendor	
	If I needed this product or service in the future...
RP-V1	I would likely buy it from this website
RP-V2	I would probably revisit this website
RP-V3	I would probably try this website
Repurchase intention – service provider with regular price	
	If I needed this product or service in the future...
RP-S1	I would likely buy it with regular prices from this service provider
RP-S2	I would probably revisit with regular prices from this service provider
RP-S3	I would probably try with regular prices from this service provider
Repurchase intention – service provider with coupon price	
	If I needed this product or service in the future...
RP-Q1	I would likely buy it with group coupons from this service provider
RP-Q2	I would probably revisit with group coupons from this service provider
RP-Q3	I would probably try with group coupons from this service provider
Price sensitivity	
PS1	I will change what I had planned to buy in order to take advantage of a lower price for what I want
PS2	I am sensitive to differences in prices
PS3	I would change what I had planned to buy because of significant price discount

Discount +; and 82 % of all participants answered the questionnaire based on the experience of redeeming restaurant/bar coupons.

4 Data Analysis and Results

4.1 Measurement Model

The adequacy of the measurement model was assessed by evaluating the reliability, convergent validity and discriminant validity [9]. Reliability testing was conducted on the data to examine the internal consistency between items expected to measure the same construct, and it was examined using the composite reliability values which should be greater than 0.7. As shown in Table 3, all factors demonstrated high internal consistency with α ranging from 0.86 – 0.99 for group coupon service model.

Regarding convergent validity, the average variance explained (AVE) by each construct must exceed 0.5, as suggested by Fornell and Larcker [14] and all indicator loadings to be significant should exceed 0.7 [9]. In our research, there's evidence of convergent validity of group coupon service model with the AVE for all factors exceeding 0.5, indicating that the majority of the variance was explained by the constructs [14]. Additionally, our loading value of each item for its reflective construct was greater than 0.7.

Regarding discriminant validity, we assessed it in two ways. First, the square root of the average variance extracted is greater than all corresponding correlations [14].

Table 2. Demographic information

Variables	Categories	Sample size (n = 366)
Gender	Male	159
	Female	207
Age	<20	43
	21–25	219
	26–30	70
	>30	34
	High school	22
	College or university	249
	Advanced degree	95
Average income level	< 5,001	80
	5001–10,000	171
	10,001–15000	62
	15,001–20,000	14
	> 20,000	39
Numbers of experience on group coupon	1–3	221
	4–6	104
	7–10	19
	> 11	22
Types of experience on group coupon	Restaurant/Bar	342
	Daily goods	178
	Salon & Spa	68
	Tourism-related services	37
	Auto services others	22

In our case, all of the diagonal values exceed the inter-construct correlations. Second, an examination of the theta matrix confirmed that no item loaded more highly on another construct than it did on its associated construct [25]. Based on these two tests, all constructs in group coupon service model exhibited satisfactory discriminant validity.

4.2 Path Analysis

The adequacy of the measurement model was assessed by evaluating the reliability, convergent validity and discriminant validity [9]. All constructs passed the tests. The final result of the structural model is shown in Fig. 3. It reports the beta coefficients and t-values for the group coupon service model along with the R-square scores for each endogenous construct which assesses the explanatory power of a structural model.

Table 3. Descriptive statistics of constructs

Constructs	Items	C.R.	AVE
Information quality (INF)	5	0.93	0.72
Efficiency (EFF)	3	0.92	0.80
System Availability (SYS)	3	0.92	0.79
Assurance (ASS)	5	0.94	0.77
Interaction Quality (IQ)	5	0.96	0.82
Physical Environment (PQ)	3	0.95	0.86
Fulfillment (FUL)	3	0.92	0.79
Policy (POL)	3	0.96	0.89
Price Sensitivity (SEN)	3	0.86	0.68
Repurchase Intention on Vendor (RPI_V)	3	0.97	0.92
Repurchase Intention on Store with original price (RPI_S)	3	0.98	0.95
Repurchase Intention on Store with coupon (RPI_Q)	3	0.99	0.97

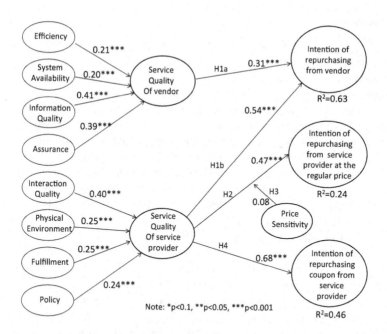

Fig. 3. Structural model

Path analysis with the latent variables was used to test the proposed research model [17]. The path coefficients for the full model are reported in Fig. 3. Firstly, as we can see from Fig. 3, all path coefficients are positive except for the price sensitivity to repurchase intention from the service provider at the regular price. For the service quality of the vendor in group coupon service model, information quality has the most significant impact ($\gamma = 0.41$), followed by assurance ($\gamma = 0.39$). For the service quality of service provider in the group coupon service model, interaction quality has the most significant impact ($\gamma = 0.40$). It is necessary to compare the paths leading to each repurchase intention to understand the importance of each independent variable in the group coupon advertising format.

The results show that the service quality of both the vendor and the service provider has impact on repurchase intention. First, regarding the intention of repurchasing from the vendor, the service quality of service provider ($\gamma = 0.54$) has higher impact than that of the vendor ($\gamma = 0.31$). Second, the service quality of service provider impacts not only the intention of repurchasing from the vendor but also the intention of repurchasing from the service provider itself ($\gamma = 0.68$ for repurchasing coupon; $\gamma = 0.47$ for repurchasing at the regular price). It indicates that customers are more willing to go back to purchase with coupon than at the regular price in group coupon advertising services. In addition, consistent with the earlier research, the service quality of service provider is the most important construct influencing customers to go back and repurchase at the regular price. Third, the hypothesis that price sensitivity has a moderating effect on the repurchase intention of service provider at the regular price is not supported.

4.3   Discussion

This research provided new insights into two important aspects of our understanding of consumer behavior in group coupon advertisement format. Our findings suggest that vendors and service providers should determine the most important dimension for them. Information quality is the most important dimension for vendors followed by the assurance quality. In order to earn customers' trust and narrow the service gap between customers and them, they should provide accurate information reflecting the real picture of the service provider. Incorrect information, exaggerated beautiful-looking pictures and the difference in price between the vendor and the service provider should be avoid. Additionally, vendors should also pay great attention to fulfill commitment to customers as they promised. To service provider, the most important dimension is the interaction quality. For example, customers with experiences of redeeming coupons (collected from the Internet) pointed out that they received poor service, attributed to employees' frustration with increased workload from group coupon shoppers. It is important for service providers to know how to balance their employees' workload and their handling of group coupon customers. Service providers want to get profit from group coupon promotion, however, there seem to be some problem with service delivery.

Second, though group coupon advertising format makes the service providers popular with new customers, it could also result in customers' higher price sensitivity [21]. Table 2, shows that group coupon promotions attract customers with high price sensitivity. Yoo, Donthu, and Lee [26] suggested that managers should avoid frequent price cuts or a consistent low-price strategy (e.g., everyday low price) because they lower the perceived quality and product image. Also, the optimal price level decreases when advertising attracts additional consumers who are more price sensitive than the existing customers [18]. Whether those customers who have high price sensitivity have higher repurchase intention is an important issue for service providers to consider.

Third, we discussed our findings for each repurchase intention separately. For repurchase intention of the vendor, the service quality of the service provider is even more important than the service provider of a vendor in a regular case. We suggested that the vendor should find service providers more purposefully. Based on our findings, though the service provider gives customers good services, the coupon shoppers prefer to go back to repurchase with coupons; the service quality also influences their willingness to repurchase at the regular prices. For the intention to repurchase from service provider at the regular price, the service quality of the service provider still plays an important role.

5 Conclusions

There are four major findings. Firstly, we found that information quality and assurance are the two important factors in the service quality of vendors, while the service quality of service provider is largely influenced by its interaction quality. Second, for consumers' intention to repurchase coupons from vendors, the service quality of service providers is more significant than that of vendors. It indicates that vendors should choose their cooperation partners carefully.

The third major finding is that service quality of service providers does not only impact consumers' intention to repurchase coupons from vendors but it also influences their intention to repurchase service from service providers with either coupon or at the regular price. It implies that service providers should deliver the service very carefully and try to follow whatever the coupons promise particularly since this is an advertising model. If the service quality is not good enough, it will result in a negative advertising effect. Finally, price sensitivity does not have significant moderating effect on the relation between the service quality of service provider and consumers' intention to repurchase service from the service provider at the regular price. However, we can draw conclusions from the statistic data that most customers have high price sensitivity which may lead to lower loyalty to the service provider.

Acknowledgments. This work is supported by research grant NSC 98-2410-H-110-010-MY3 from the Ministry of Science and Technology, Taiwan, Republic of China.

References

1. Barnes, S.J., Vidgen, R.:. WebQual: an exploration of web-site quality. In: Proceedings of the Eighth European Conference on Information Systems (ECIS), Vienna, Austria (2000)
2. Barnes, S.J., Vidgen, R.T.: An integrative approach to the assessment of e-commerce quality. J. Electron. Commer. Res. **3**, 114–127 (2002)
3. Bawa, K., Shoemaker, R.W.: The effects of a direct mail coupon on brand choice behavior. J. Mark. Res. **24**, 370–376 (1987)
4. Brady, M.K., Cronin, J.J.: Some new thoughts on conceptualizing perceived service quality: a hierarchical approach. J. Mark. **65**, 34–49 (2001)
5. Brown, S.W., Swartz, T.A.: A gap analysis of professional service quality. J. Mark. **53**, 92–98 (1989)

6. Byers, J. W., Mitzenmacher, M., Potamias, M., Zervas, G.. a month in the Life of Groupon. Arxiv preprint arXiv: 1105.0903 (2011)
7. Chang, H.H., Wang, Y.H., Yang, W.Y.: The impact of e-service quality, customer satisfaction and loyalty on e-marketing: moderating effect of perceived value. Total Qual. Manag. **20**, 423–443 (2009)
8. Chen, C.M., Chen, S.H., Lee, H.T.: Interrelationships between physical environment quality, personal interaction quality, satisfaction and behavioural intentions in relation to customer loyalty: the case of kinmen's bed and breakfast industry. Asia Pac. J. Tourism Res. **18**, 262–287 (2013)
9. Chiu, C.M., Chang, C.C., Cheng, H.L., Fang, Y.H.: Determinants of customer repurchase intention in online shopping. Online Inf. Rev. **33**, 761–784 (2009)
10. Dholakia, U.: How Effective are Groupon Promotions for Businesses? (2010)
11. Dholakia, U.: How Businesses Fare with Daily Deals: A Multi-Site Analysis of Groupon, Livingsocial, Opentable, Travelzoo, and Buywithme Promotions (2011)
12. Dholakia, U.M., Tsabar, G., Meals, G.P.: a startup's experience with running a Groupon promotion: Mimeo. Rice University (2011)
13. Edelman, B., Jaffe, S., Kominers, S.D., School, H.B.: To Groupon or not to Groupon: the profitability of deep discounts.: Harvard Business School (2010)
14. Fornell, C., Larcker, D.F.: Evaluating structural equation models with unobservable variables and measurement error. J Mark. Res. **18**(1), 39–50 (1981)
15. Garretson, J.A., Clow, K.E.: The influence of coupon face value on service quality expectations, risk perceptions and purchase intentions in the dental industry. J. Serv. Mark. **13**, 59–72 (1999)
16. Kalra, A., Goodstein, R.C.: The impact of advertising positioning strategies on consumer price sensitivity. J. Mark. Res. **35**, 210–224 (1998)
17. Kang, G.D., James, J.: Service quality dimensions: an examination of Grönroos's service quality model. Managing Serv. Qual. **14**, 266–277 (2004)
18. Kaul, A., Wittink, D.R.: Empirical generalizations about the impact of advertising on price sensitivity and price. Mark. Sci. **14**, 151–160 (1995)
19. Loiacono, E.T., Watson, R.T., Goodhue, D.L.: WebQual: an instrument for consumer evaluation of web sites. Int. J. Electron. Commer. **11**, 51–87 (2007)
20. Parasuraman, A., Zeithaml, V.A., Malhotra, A.: E-S-QUAL: a multiple-item scale for assessing electronic service quality. J. Serv. Res. **7**, 213–233 (2005)
21. Petrick, J.F.: Segmenting cruise passengers with price sensitivity. Tour. Manag. **26**, 753–762 (2005)
22. Swaid, S.I., Wigand, R.T.: Key dimensions of e-commerce service quality and its relationships to satisfaction and loyalty. In: 20th Bled eConference eMergence: Merging and Emerging Technologies, Processes, and Institutions, Bled, Slovenia, pp. 414–428 (2007)
23. Tutaj, K., van Reijmersdal, E.A.: Effects of online advertising format and persuasion knowledge on audience reactions. J. Mark. Commun. **18**, 5–18 (2012)
24. Wakefield, K.L., Inman, J.J.: Situational price sensitivity: the role of consumption occasion, social context and income. J. Retail. **79**, 199–212 (2003)
25. White, J.C., Varadarajan, P.R., Dacin, P.A.: Market situation interpretation and response: the role of cognitive style, organizational culture, and information use. J. Mark. **67**, 63–79 (2003)
26. Yoo, B., Donthu, N., Lee, S.: An examination of selected marketing mix elements and brand equity. J. Acad. Mark. Sci. **28**, 195–211 (2000)
27. Zeithaml, V.A., Berry, L.L., Parasuraman, A.: Communication and control processes in the delivery of service quality. J. Mark. **52**, 35–48 (1988)
28. Zhang, X.N., Prybutok, V.R.: A consumer perspective of e-service quality. IEEE Trans. Eng. Manage. **52**, 461–744 (2005)

Defining Human-Machine Micro-Task Workflows for Constitution Making

Nuno Luz[1,2(✉)], Marta Poblet[3], Nuno Silva[1], and Paulo Novais[2]

[1] GECAD (Knowledge Engineering and Decision Support Group),
Polytechnic of Porto, Porto, Portugal
{nmalu, nps}@isep.ipp.pt
[2] ALGORITMI Centre, Departamento de Informática,
Universidade do Minho, Braga, Portugal
pjon@di.uminho.pt,
[3] Graduate School of Business and Law,
RMIT University, Melbourne, Australia
marta.pobletbalcell@rmit.edu.au

Abstract. This paper presents a novel task-oriented approach to crowdsource the drafting of a constitution. By considering micro-tasking as a particular form of crowdsourcing, it defines a workflow-based approach based on Onto2Flow, an ontology that models the basic concepts and roles to represent workflow-definitions. The approach is then applied to a prototype platform for constitution-making where human workers are requested to contribute to a set of tasks. The paper concludes by discussing previous approaches to participatory constitution-making and identifying areas for future work.

Keywords: Micro-tasking · Micro-tasks · Workflows · Ontologies · Political crowdsourcing · Legal crowdsourcing · Constitution-making · Participation

1 Introduction

Constitution-making can be broadly defined as a set of activities intended to produce a constitution, the highest law of a state. To the UN, constitution-making "covers both the process of drafting and substance of a new constitution, or reforms of an existing constitution" [1]. Klein and Sajo have also defined it as a "decision-making process carried out by political actors, responsible for selecting, enforcing, implementing, and evaluating societal choices" [2]. Given that constitution-making may only happen once in a generation, it is often seen as a unique moment shaping both the present and the future of a country. As Elster has put it, "if there is one task for which 'wisdom' would seem highly desirable, it is that of writing a constitution" [3].

This paper reviews a few examples of how the wisdom of the crowd has been tapped in recent constitution-making processes across the world and proposes a new approach to write a constitution based on micro-tasking, a particular form of crowdsourcing.

Section 2 provides definitions of crowdsourcing and micro-tasking and additional background knowledge on recent examples of constitutional crowdsourcing. Section 3

© Springer International Publishing Switzerland 2015
B. Kamiński et al. (Eds.): GDN 2015, LNBIP 218, pp. 333–344, 2015.
DOI: 10.1007/978-3-319-19515-5_26

briefly reviews ontology-based micro-tasking workflows and presents Onto2Flow, an ontology designed to retrieve structured and semantically enriched data from micro-tasks. Section 4 applies this framework to a prototype platform that enables the micro-tasking of a constitutional text. Section 5 discusses both the potential and limitations of this approach. The conclusion, finally, suggests future work in this area.

2 Background

The word crowdsourcing was coined by Jeff Howe and Mark Robinson in 2006 to represent "the act of taking a job traditionally performed by a designated agent (usually an employee) and outsourcing it to an undefined, generally large group of people in the form of an open call" [4]. This broad conceptualization has been followed by a myriad of definitions of crowdsourcing drawn from different but connected approaches: collective intelligence (CI), human computation, social intelligence, and social computing. It also has been noted that "while human computation (HC) is a term that is mostly used by the scientific community, crowdsourcing (CS) is a term highly employed in the business world [5]. Despite the variety of perspectives, all approaches highlight three key elements in crowdsourcing: crowds, tasks, and mediating technologies.

Micro-task crowdsourcing, in particular, is a special kind of human computation where relatively complex tasks are divided into smaller and independent micro-tasks [5]. These micro-tasks are then modelled and published through a computational platform (e.g. Mechanical Turk and CrowdFlower), which distributes them through a crowd of workers.

Micro-tasks are often employed for solving large-scale problems that are often too complex for computers to solve on their own [6]. These problems usually require a degree of creativity (or just common sense), plus some background knowledge [7, 8]. In our view, the drafting of a constitution: (i) can be represented as a large-scale problem that can be divided into smaller tasks; (ii) these micro-tasks can be completed by a crowd of heterogeneous citizens with different degrees of legal expertise (from none to expert).

2.1 Crowdsourced Constitution-Making

In the political and legal domains, crowdsourcing methods and tools have been used as a means to collect input from citizens on a variety of areas, such as legal drafting, legal reform, legal education, policy-making and human rights advocacy [9–12]. Crowdsourced constitution-making, in particular, was famously displayed in Iceland in 2011 with the use of social media to collect peoples' views and opinions on the constitutional draft [13]. Similar initiatives were taking place almost simultaneously in Kenya (2010), Ghana (2010-2011), Somalia (2011), Egypt (2012), and Libya (2012), among other countries [14]. Likewise, Morocco announced a constitutional reform in early 2011 and, shortly after, a citizen-based initiative launched reforme.ma, a dedicated crowdsourcing platform fully integrated with Facebook and Twitter where citizens could like or dislike the proposed articles and comment on them [15].

In the effort to make constitution making as participatory as possible, these initiatives have all tapped into social media (and, in some cases, e-mail and text messages) to elicit comments from the public. In all cases, and regardless of the final number of participants, thousands of comments were posted and eventually collected. The analysis of how these contributions were classified a posteriori and their eventual impact on the final drafts would require a case-by-case approach. Yet, it seems clear that in all mentioned examples the public was invited to comment, answer questions, vote, or "like", but not to "write" the constitution itself. To date, crowdsourced constitution-making has heavily relied on online deliberation, but the impact of such deliberative processes on the final outcome is yet to be fully assessed. While deliberative processes are core to constitution-making, we aim at a complementary approach where the constitutional draft is also the product of coordinated micro-tasking via the participation of a large number of participants.

2.2   Ontologies in Description Logics

Our approach adds a new layer to constitution-making by considering a micro-task workflow-based approach to the drafting and refinement of the document. Drafting and refinement workflows are modelled using ontologies, which allow a formal, explicit and shared conceptual representation while maintaining machine interpretability. Ontologies are formal because they are supported by unambiguous formal logics; explicit since they make domain assumptions explicit for reasoning and understanding; and shared for its ability to provide consensus.

Ontologies "represent the best answer to the demand for intelligent systems that operate closer to the human conceptual level" [16]. Thus they are an appropriate representation mechanism for environments where both human and machine agents must interpret the data and perform a particular set of actions. Furthermore, the inherent extensibility of ontologies allows the growing set of domain ontologies in the Semantic Web to be re-used in the representation of workflows.

3   Ontology-Based Micro-Task Workflows

Micro-tasks (or simply "tasks" from now on) can be seen as atomic operations that produce a specific set of data. These atomic operations occur within a specific domain of operation involving certain domain knowledge. Given a task, its domain of operation is defined by its input and output specifications.

Onto2Flow is an approach to the representation, instantiation and execution of workflows that represents workflows of tasks as extensions of other domain ontologies. These extensions are called workflow-definition ontologies. Workflow-definition ontologies assemble two different data dimensions: (a) the static domain dimension (corresponding to the domain ontology) and (b) the dynamic task and workflow dimension (corresponding to the Onto2Flow ontology). In this perspective, task-definitions (or task representations) are extensions of the domain ontology, which add an operational dynamic dimension. Figure 1 illustrates these two dimensions and their assemblage.

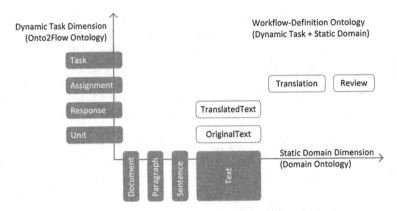

Fig. 1 Static and dynamic dimension in a workflow-definition ontology.

3.1 The Onto2Flow Process

Onto2Flow assumes that domain ontologies represent the structure and semantics of the data presented to (and retrieved from) workers. Accordingly, the approach considers two steps (outlined in Fig. 2): (i) task-definition and workflow-definition (the ontology of the workflow), and (ii) the instantiation and execution of the workflow on a particular input dataset.

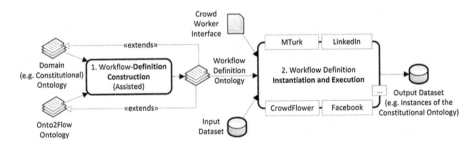

Fig. 2 Overview of the Onto2Flow approach.

At the stage of workflow-definition (1), the requester must clearly define the activities involved in the workflow through a semantic model of the input and output data and create a workflow-definition. For workflow-definitions containing task-definitions that human workers have to solve, crowd (user) interface templates must be supplied along with the workflow-definition ontology. The interface templates present the task data to the worker and retrieve the submitted response.

At stage 2 (instantiation and execution of the workflow-definition), workflow-definition ontologies can be instantiated multiple times and executed by any workflow engine that is able to interpret the Onto2Flow ontology and apply the ground rules established by the proposed method. Furthermore, Onto2Flow-based workflow engines may dispatch the execution of the tasks to external micro-task execution communities such as Mechanical Turk and CrowdFlower, or provide their own task resolution interfaces that may interact with external social networks.

3.2 The Onto2Flow Ontology: Concepts and Roles

The Onto2Flow ontology defines the basic concepts and roles required to represent workflow-definitions (see Fig. 3). It captures concepts and lessons learnt from workflow-definition languages and approaches such as the XPDL (XML Process Definition Language) and BPMN (Business Process Modelling Notation) [17]. Furthermore, it incorporates concepts that support the crowdsourcing, distribution, and delivery of tasks.

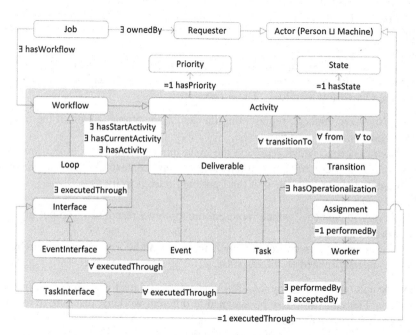

Fig. 3. Overview of the Onto2Flow ontology.

The concept *Job* represents a workflow execution environment created by a *Requester*, which may contain more than one *Workflow*.

Activities are the interconnected components that form a workflow. There are three main types of activities: the *Workflow*, the *Task* and the *Event*. Among these, *Deliverables*, which include *Task* and *Event*, represent a group of activities that require worker or external interaction through some kind of *Interface*.

Two main types of actors are considered by the Onto2Flow ontology: the *Requester* (the one requesting the execution of a workflow) and the *Worker* (the one solving the tasks of the workflow), which are either *Human* or *Machine*.

Each *Actor* may belong to several *ActorGroup*. Actor groups allow requesters to associate and filter groups of actors for participation in particular tasks. An *ActorGroup* may include a wide set of attributes, including social network analysis clustering measurements (e.g. clusterability), improving the control of the requester over the selection of workers. Inclusively, each *ActorGroupMembership* may feature a wide set of actor specific attributes and measurements (e.g. centrality and prestige).

Workflows are graphs of activities linked through transitions, which establish a process that delivers a specific result dataset given an input dataset.

The flow of activities in a workflow is established through *Transitions*. There are six types of transitions, depending on the set of (i) incoming activities (*BasicTransition*, *MergeTransition* or *SynchronizationTransition*), (ii) outgoing activities (*Parallel-Transition* or *DisjunctTransition*), (iii) whether there is one or more conditions to be fulfilled in order to continue its execution (*ConditionalTransition*).

An *Event* is an external occurrence that either triggers the continuation of a running workflow (*RunningEvent*) or triggers the execution of a new workflow (*InstantiationEvent*).

A *Task* is a set of assignments and operations on top of input data, which must be performed by workers. The representation of a task involves multiple concepts and roles in the Onto2Flow ontology. These concepts are:

- The *Assignment* concept, representing the actual operationalization of the task;
- Input concepts:
- The *Unit* concepts, which represent the input unit of work given to the worker;
- The *UnitContext* concepts, which represent relevant contextual input data that must be presented along with the unit (and possibly related to it);
- Output concepts:
- The *Response* concepts, which represent the top-level response or output given by the worker;
- The *ResponseContext* concepts, which represent additional output given by the worker, usually related to the response.

Each work unit (represented by the *Unit* concept) is assigned to a worker through an *Assignment*. The same unit may be assigned to different workers, resulting in different solutions to the same problem.

The execution of a workflow requires interaction with external actors and services during the execution of *Event* and *Task* activities. While an *Event* is typically listened for, and arrives through an *EventInterface*, a *Task* must be delivered to and retrieved from workers through a *TaskInterface*. Thus, interfaces represent logical and/or physical components through which the interaction with workers (machine or human) is performed (e.g. a Web service interface, a graphical user Web interface).

The ability to represent different types of interfaces enables the specification of distinct interfaces, commonly used on user-centric environments [18]:

- Simple, where a single medium or modality is used. For instance, tasks can be delivered to workers through a visual interface, a sound interface, or simply through a web interface (the common case for crowdsourcing applications);

- Multi-modal, i.e. capable of merging and coordinating multiple mediums and modalities as a single interface.

Accordingly, and of particular interest in the crowdsourcing scenario, different types of user interface implementations, such as a game interface or a mobile interface, can be used to distribute tasks through human workers.

4 Catalan Constitution-Making Scenario

The Catalan constitution-making scenario is a prototype of a micro-tasking platform to crowdsource the elaboration of a constitutional text. This scenario uses the Constitute project ontology as the static domain dimension. The Constitute project is a database of constitutional texts to search and compare constitutions across the world [19]. On top of the Constitute project ontology, a workflow-definition following the Onto2Flow method was built. The resulting workflow-definition, as shown in Fig. 4, aims to take the ontology-based representation of a proposed Catalan constitution and crowdsource its elaboration, stemming from a basic initial text [20].

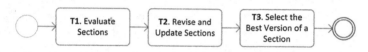

Fig. 4. Overview of the Catalan constitution-making workflow-definition.

The process contemplates the following tasks, all performed by human workers:

- T1 - evaluates sections of the current constitution document and is performed by any worker;
- T2 - revises and updates sections of the current constitution document marked in the previous task and is performed by expert workers;
- T3 - selects the best version of a section from the set of proposed sections in the previous task and is performed by any worker.

The Constitute project ontology represents the constitution document through sections. A partial illustration of the Constitute project ontology is presented in Fig. 5. An additional set of concepts was added to the static domain dimension in order to represent the opinion and the assessment of the constitution sections.

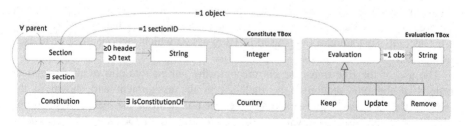

Fig. 5. Partial Constitute project ontology and additional assessment concepts.

4.1   The Workflow-Definition

The constitution-making workflow-definition was built using both a construction framework prototype implementation and the Protégé ontology editor. The Protégé ontology editor was used to establish some common axioms that are not yet featured by the construction framework, such as the union of input and output concepts, and inverse roles. A detailed illustration of the workflow-definition is presented in Fig. 6. Notice how each task-definition contains a complete representation of all the concepts and relationships involved. Also, this representation is directly mapped to the Constitute project ontology.

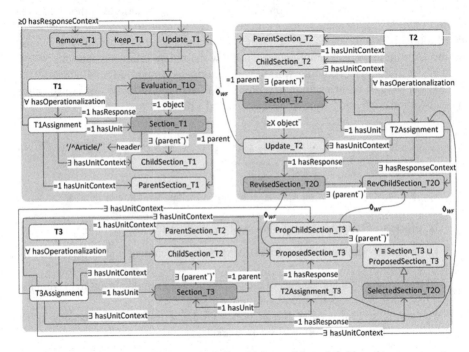

Fig. 6. Task-definitions in the constitution-making workflow-definition ($\Diamond_{WF}$ represents a dependency relationship, which can be reduced to a subsumption).

In T1, the amount of assignments per unit will correspond to the amount of evaluations given to each section. Thus, T1 must have an amount of assignments per unit greater or equal to X, where X is the amount of evaluations that request an update of the section. This amount (X) is used in T2 to assess which sections must be revised and updated.

The use of the role transitive closure onto the *parent* role allows all descendant sections of the unit section to be included in the assignment and shown to the worker. Also, regular expressions may be used to restrict the value of data-type roles. Such is the case of the value of the *header* role in T1 ($Section\_T1 \sqsubseteq header :$ "$/^\wedge Article/$").

4.2   The Task-Definition UI Templates

In the Catalan constitution-making scenario all tasks are solved by human workers (volunteers). Volunteers contribute by adopting two different profiles: non-experts or experts. Non-experts are the large majority of citizens who sign into the platform to complete tasks in T1; experts are those volunteers designated by the requesters with an editing role of the outputs produced by non-experts (classification, collation, amendments). In both cases, the workflow-definition includes an UI (User Interface) template. The UI template of T1 presents the unit section, its parent section, and all its descendant sections to the non-expert volunteer. The volunteer is then invited to evaluate the contents of the section (an article of the constitution) and assess whether it needs to be: (i) updated (rewritten), (ii) removed or (ii) accepted as it is. Volunteers can access the complete initial constitutional draft at any time to situate their assignment into the broader picture of the full text.

The UI template of T2 presents the unit section to expert workers in the same way as T1, including any modifications of the constitutional text by non-experts in T1. The expert volunteer is then asked to submit a new revised section with all outputs collected T1 classified and, if necessary, edited and collated.

Finally, the UI template of T3 presents each of the previously submitted sections (during T2), along with the original section. In T3, all volunteers are requested to select the best version. Figure 7 below offers an example of the UI template of T1 as presented to non-expert volunteers.

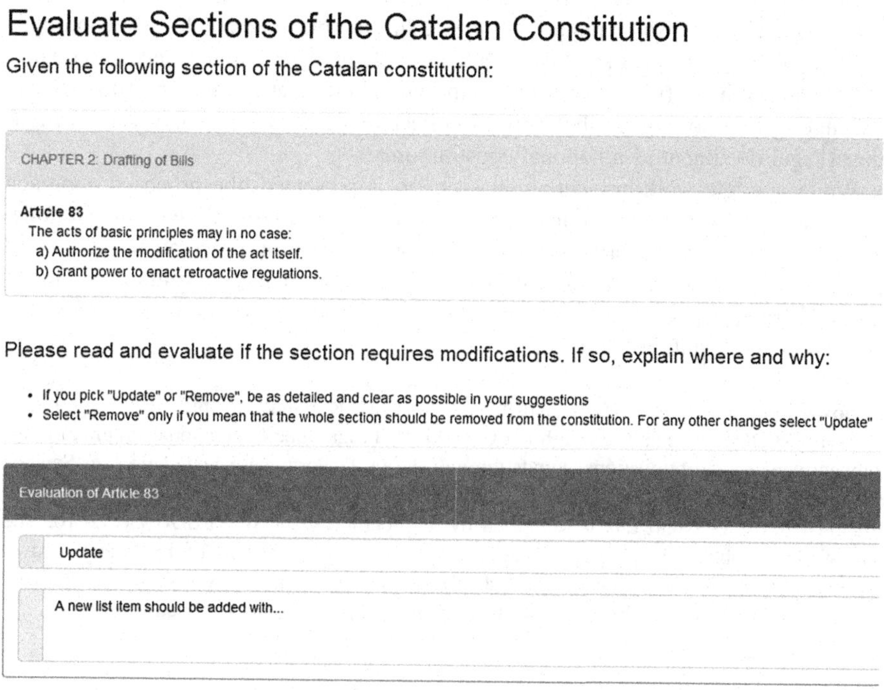

Evaluate Sections of the Catalan Constitution

Given the following section of the Catalan constitution:

CHAPTER 2: Drafting of Bills

Article 83
The acts of basic principles may in no case:
a) Authorize the modification of the act itself.
b) Grant power to enact retroactive regulations.

Please read and evaluate if the section requires modifications. If so, explain where and why:

- If you pick "Update" or "Remove", be as detailed and clear as possible in your suggestions
- Select "Remove" only if you mean that the whole section should be removed from the constitution. For any other changes select "Update"

Evaluation of Article 83

Update

A new list item should be added with...

Fig. 7. Example assignment with the UI template of T1.

5   Conclusions and Future Work

Crowd sourcing the writing of a constitution to a large number of citizens is a complex task that can be addressed by subdividing it to smaller units (micro-tasks). While there are a number of examples of participatory constitution-making that involve online deliberation, none of them offers a platform for citizens to edit the articles of the text. Rather, their focus on eliciting and collecting opinions from public deliberation, generally via social media, makes crowd sourcing initiatives accessory to the drafting process developed elsewhere (e.g. in constitutional commissions). Ultimately, this contingent aspect of crowd sourcing makes it difficult to assess the impact of online participation on both the drafting process and the final outcome.

In our approach, writing a constitution becomes the core task. We rely here on two well-researched conditions in the literature on the "wisdom of the crowd effect": (i) independence of judgment and (ii) heterogeneity of the crowd [21–23]. When these two conditions are met, the crowd can perform better than individual experts.

To date, the platform has been tested by a reduced group of 8 experts who have provided useful feedback. Future work involves expanding the testing to larger groups of volunteers and refine the following issues: (i) identification of sub-topics within an article and further division of micro-tasks; (ii) credentials and role of experts; (iii) aggregation mechanisms in T3 (e.g. ratings, rankings) to avoid inconsistencies, and (iv) generally, mechanisms to detect and resolve conflicts between different sections in a constitution.

Beyond addressing these different issues dealing with coordination mechanisms, further research will also be required to tackle substantive issues on how to coordinate the crowd itself: (i) motivation; (ii) incentives to participate; (ii) relevance and quality of the contributions; (iii) monitoring spam and sabotage attempts, etc. The ultimate challenge is how to engage the crowds' collective wisdom in drafting such a high-impact legal document as a national constitution.

While a single workflow-definition scenario is presented, the proposed approach allows the construction of multiple workflow-definitions from different constitutional ontologies. Furthermore, the workflow-definition construction process can be assisted and performed semi-automatically through the analysis of the constitutional domain ontology structure [24]. This is only possible by using the Onto2Flow ontology as the core workflow-definition ontology.

Acknowledgments. This work is part-funded by FEDER Funds, by the ERDF (European Regional Development Fund) through the COMPETE Programme (Operational Programme for Competitiveness) and by National Funds through the FCT (Portuguese Foundation for Science and Technology) within the project FCOMP-01-0124-FEDER-028980 (PTDC/EEI-SII/1386/ 2012). The work of Nuno Luz is supported by the doctoral grant SFRH/BD/70302/2010. The work of Marta Poblet draws from previous research within the framework of the project "Crowd sourcing: instrumentos semánticos para el desarrollo de la participación y la mediación online" (DER 2012-39492-C02-01) by the Spanish Ministry of Economy and Competitiveness.

References

1. United Nations Rule of Law - Guidance Note of the Secretary-General: United Nations Assistance to Constitution-making Processes (2009)
2. Klein, C., Sajo, A.: Constitution-making: process and substance. In: Rosenfeld, M., Sajé, A. (eds.) The Oxford Handbook of Comparative Constitutional Law, pp. 419–441. Oxford University Press, Oxford (2012)
3. Elster, J.: The optimal design of a constituent assembly. In: Landemore, H., Elster, J. (eds.) Collective Wisdom: Principles and Mechanisms, pp. 148–172. Cambridge University Press, New York (2012)
4. Howe, J.: The rise of crowdsourcing. Wired Mag. **14**, 1–4 (2006)
5. Luz, N., Silva, N., Novais, P.: A survey of task-oriented crowdsourcing. Artif. Intell. Rev. **43**, 1–27 (2014)
6. Von Ahn L.: Human computation. In: 46th ACM IEEE Design Automation Conference, pp. 418 – 419 (2009)
7. Chklovski T.: Learner: a system for acquiring commonsense knowledge by analogy. In: Proceedings of the 2nd ACM International Conference on Knowledge Capture, pp. 4–12, Sanibel Island, FL (2003)
8. Singh, P., Lin, T.: Open mind common sense: knowledge acquisition from the general public. In: Meersman, R., Tari, Z. (eds.) On the Move to Meaningful Internet Systems 2002: CoopIS, DOA, and ODBASE, pp. 1223–1237. Springer, Heidelberg (2002)
9. Orozco D.: Democratizing the law: legal crowdsourcing (lawsourcing) as a means to achieve legal, regulatory and policy objectives. Regul. Policy Objectives. **53**(1), 1–43 (2014)
10. Poblet, M.: Visualizing the law: crisis mapping as an open tool for legal practice. J. Open Access Law. **1**(1), 1–20 (2013)
11. Aitamurto, T.: Crowdsourcing for Democracy: New Era In Policy-Making. Committee for the Future Parliament of Finland, Helsinki (2012)
12. Casanovas, P.: Legal crowdsourcing and relational law: what the semantic web can do for legal education. J. Australas. Law Teach. Assoc. **5**, 159–176 (2012)
13. Landemore, H.: Inclusive constitution-making: the icelandic experiment. J. Polit. Philos. **23**(1), 1–26 (2014)
14. Gluck, J., Ballou, B.: New Technologies in Constitution Making. USIP, Washington (2014)
15. Deely S., Nesh-Nash T.: The future of democratic participation: an online constitution making platform, pp. 43–62 (2014)
16. Obrst, L., Liu, H., Wray, R.: Ontologies for corporate web applications. AI Mag. **24**, 49 (2003)
17. Hornung T., Koschmider A., Mendling J.: Integration of heterogeneous BPM Schemas: the case of XPDL and BPEL. In: CAiSE Forum, vol. 231 (2006)
18. Luz, N., Pereira, C., Silva, N., Novais, P., Teixeira, A., Oliveira e Silva, M.: An Ontology for human-machine computation workflow specification. In: Polycarpou, M., de Carvalho, A.C., Pan, J.-S., Woźniak, M., Quintian, H., Corchado, E. (eds.) HAIS 2014. LNCS, vol. 8480, pp. 49–60. Springer, Heidelberg (2014)
19. Constitute Project. https://www.constituteproject.org/
20. Constitucio de Catalunya. Reagrupament.cat (2010)
21. Davis-Stober, C.P., Budescu, D.V., Dana, J., Broomell, S.B.: When is a crowd wise? Decision **1**, 79 (2014)
22. Levine, S.S., Prietula, M.J.: The hazards of interaction: why isolation can benefit performance. In: Academy of Management Proceedings 2013, vol. 10736 (2013)

23. Ober, J.: Democracy's wisdom: an aristotelian middle way for collective judgment. Am. Polit. Sci. Rev. **107**, 104–122 (2013)
24. Luz, N., Silva, N., Novais, P.: Generating human-computer micro-task workflows from domain ontologies. In: Kurosu, M. (ed.) HCI 2014, Part I. LNCS, vol. 8510, pp. 98–109. Springer, Heidelberg (2014)

Creating Value Through Crowdsourcing:
The Antecedent Conditions

Michael Rowe[✉], Marta Poblet, and John Douglas Thomson

RMIT University Graduate School of Business and Law,
405 Russell Street, Melbourne, VIC 3000, Australia
{michael.rowe,marta.pobletbalcell,
doug.thomson}@rmit.edu.au

Abstract. The benefits of crowdsourcing are becoming more widely understood and there is a methodological move towards organisations using "participatory models" to engage stakeholder communities and align decision making more closely to the needs of stakeholders. Many tasks can now be distributed to "the crowd" for action. Our research aims to understand the antecedent conditions that inform management decisions to adopt crowdsourcing techniques as a means of value creation. Our preliminary findings suggest that to be successful, three antecedent criteria must be met – the task being crowdsourced must be modular in nature, a community of interest must be engaged, and there needs to be a structural capability within the organisation to be able to facilitate the engagement of the crowd and utilise the output from the crowd in a manner that creates value.

Keywords: Crowdsourcing · Strategy · Open-innovation · Online community · Social media

1 Introduction

Crowdsourcing has been defined as a "type of participative online activity in which an individual, an institution, a non-profit organization (...) proposes to a group of individuals (...) via a flexible open call, the voluntary undertaking of a task" [1]. This is usually through the use of social media technologies - "a group of Internet-based applications that...allow the creation and exchange of user generated content" [2]. A review of the literature on this topic demonstrates that challenges facing organisations seeking to utilize crowdsourcing include developing an operational perspective of how sustainable competitive advantage can be appropriated through meaningful e-engagement with stakeholders.

The aim of this research is to establish an understanding of those antecedent considerations that inform management decisions to adopt crowdsourcing as a means of creating value.

1.1 Crowdsourcing Profile

At the outset it must be recognised that crowdsourcing is not one single thing, rather it covers a variety of activities, behaviours and outcomes. Typologies have been

© Springer International Publishing Switzerland 2015
B. Kamiński et al. (Eds.): GDN 2015, LNBIP 218, pp. 345–355, 2015.
DOI: 10.1007/978-3-319-19515-5_27

proposed by a range of theoreticians including Schenk and Guittard [3] who define the nature of the process of crowdsourcing as either integrative (through using pooled and unedited data), or selective (by identifying and integrating only part of the full set of responses). They further categorize the type of task being offered to the crowd as routine, complex or creative [3] and in doing so provide an intuitive framework for identifying and classifying crowdsourcing activity.

It has been demonstrated that crowd-based inputs can enable better decisions, are typically less expensive, and more suitable to adaption than in-house equivalents [4–6]. As the diversity of application grows, crowdsourcing is transitioning from being the fundamental business model of purpose-built entities (for example, TripAdvisor providing crowdsourced guidance to travelers, and iStockphoto a platform for the sale of crowdsourced photographic images) to a management practice that can be selectively employed within parts of an enterprise to create value.

While the general awareness of crowdsourcing in the business community has increased as online modalities of value creation become more widespread, the utilization of the technique remains contingent on a belief in the minds of management that outcomes so obtained will be in some measure better, cheaper or favourably distinguished from outcomes realized through conventional outsourcing practice. The boundaries delineating the opportunity to crowdsource are currently ill-defined and management perspectives of the actual practice of crowdsourcing, and the operational constraints that may impact on the technique's ability to contribute to value creation, are not well understood.

1.2   Literature and Methodology

While a body of literature exploring the role of crowdsourcing across a range of applications is emerging, it is mostly focused on crowdsourcing itself – processes, taxonomies, performance and constraints – rather than seeking to understand the circumstances that may lead a decision-maker to the consideration of crowdsourcing as an appropriate technique for value creation. In a comprehensive survey of publications related to crowdsourcing, Zhao and Zhu [7] note that while 64 % of articles used empirical methods, almost all of these articles related to events and/or processes. In other words the literature is oriented towards classifying existing models rather than understanding the preconditions that enable those models to function in the first place.

Where recent research seeks to explore the decision to crowdsource, it draws from literature rather than interaction with those active in the field. For instance, Thuan, Antunes & Johnstone [8] utilised a structured literature review to derive a model that positioned the decision to crowdsource as mediated by four factors; environment, management, people and the particulars of the task. This model does not anticipate a broader set of drivers of behavior, nor necessarily preconditions whereby a crowdsourced solution may provide greater opportunities for value creation than conventional methods.

To begin addressing these issues, open-ended conversations with eight participants were undertaken in order to obtain the perspective from experienced decision-makers in this area. The open-ended conversation format adopted aimed at: (i) building rapport with participants; (ii) obtaining detailed and nuanced perceptions, and (iii) developing an accurate narrative that includes the meaning of the experiences from those involved in the situation (in the social constructionism tradition of Berger & Luckmann [9] and, more recently, Eriksson and Kovalainen [10]). In this perspective, we have elicited a narrative as a "way of knowing that is different but complementary to logical-scientific knowledge" [11]. Two initial outcomes arose from this approach: the first relates to the meaning to the respondent; the second informs the literature by identifying aspects not previously considered.

1.3   Routine, Complex or Creative

The organisations identified for involvement in this research typically address issues that are either inherently complex to the point of being "wicked" problems, or ones requiring novel or creative approaches with the potential to lead to truly innovative outcomes.

Crowdsourcing of purely process-based tasks - those that require little if any domain specific knowledge - can be undertaken through engagement of undifferentiated individuals without specialist insight or alignment with a community of interest. For example the citizen science site Galaxy Zoo [12] requires simply that the user identify features on satellite photographs of indistinct objects in space. The degree of expertise required is minimal, and lack of prior association with the subject matter will not yield less valuable results for the organisation.

When the nature of the task begins to require a greater depth of understanding, the harnessing of the thoughts of random individuals may provide results with a poor signal to noise ratio [13]. For this reason, where opinions or specialist insight is required to fulfil a task, the organisation may seek out communities of interest, or introduce moderating mechanism to filter usable information from that of less practical contributions [14].

As part of its 10 year plan, the City of Melbourne, Australia has developed a virtual budget simulator tool that enables ratepayers to provide their preferred apportionment of the City's overall budget across the five main categories of Deliver Community Services, Activate City, Advance Melbourne, Design, Build and Manage Assets, and Regulate, and numerous sub-categories [15]. The simulator shows current levels of expenditure in each category and provides controls for the user to propose variations to future spending according to their own individual preference. As the pre-dispositions of individuals participating may make their inputs inconsistent with the broad responsibilities of the City, the data is collated and referred to a panel of 43 residents for moderation. Membership of this panel reflects the demographic composition of the city. The panel then considers the respondent data and provides recommendations to the Council's budgeting process. This is an example of a *community* being engaged, with a moderation process refining crowd inputs. Membership of this community is implied by being a ratepayer of the municipality, and having the interest to participate [16].

Communities are not necessarily passive in nature. A prominent example of this is the Danish toy manufacturer, Lego. Lego practices a form of open innovation that formally places the user community at the centre of the product innovation effort [17].

"Adult Fans of Lego" (AFOLs) form Lego User Groups (LUGs) based around either geographic location or common interests. Lego puts in place relationship agreements to officially recognize these groups and this provides the basis of a formal and legally constituted means of interacting and soliciting ideas for new products and new strategic directions for the company. These user groups form what's known as LUGNET – or the Lego User Group Network. The Lego communities developed spontaneously, on forums that are operated independently from the company. Activity on these sites is driven by the needs of the members to associate and share their passion for the product [18]. As such these communities can be described as authentic and autonomous.

Contrast this with innovative camera developer Lytro [19] and Australian software developer MYOB [20]. Both of these companies operate moderated forums on their own company websites through which they engage customers and stakeholders in the product development process. These communities may be considered "captive" as all activity happens on a forum site owned and operated by the respective companies. It may be argued that authenticity is critical when engaging communities of interest but if the organisation is embedded or closely moderating the group a form of adverse selection may take place where the community feeds back to the company what they think the company wants to hear [21].

Stakeholder Engagement or Community Conversations. One alternative approach organisations can adopt is to side-step the stakeholder engagement process altogether and turn instead to the data contained in the community conversations [22]. This marks a transition from *asking* the community, to *watching* the community, then analysing and interpreting directly from the conversations taking place within that community. New cloud-based artificial intelligence algorithms coupled with semantic connectivity and topic modelling tools enable deep and coherent insights to be developed from text-based datasets. While still in its infancy, his represents a compelling and possibly controversial option for enterprises seeking to better understand the needs and priorities of their involved stakeholder groups [23].

To summarize these perspectives Fig. 1 depicts a typology of crowdsourcing that illustrates an empirical relationship between community type and crowdsourced task type. It demonstrates the potential for organizations to transition from engaging their communities interactively, to surveillance, data mining and subsequent semantic analysis of authentic and spontaneous discussion threads.

2   Issues

While the promise of crowdsourcing is attractive the reality may be more problematic. Tasks that can be crowdsourced are often (if not always) tasks that have previously been undertaken using "conventional" means – there are few if any crowdsourced outcomes that cannot be obtained some other way. If, for example a firm seeks to better understand the features its customer wants included in its next model release, a market research program would normally be undertaken. This prompts the question of what antecedent conditions need to be satisfied for a manager to utilize the crowd in place of a specialized resource, and how might the crowd's participation in the decision making

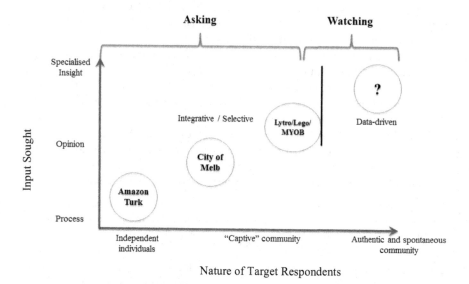

Fig. 1. Categorisation of crowdsourcing participation models by community

of the organisation provide management with greater value than alternative courses of action? Implicit in this is an understanding of the inflection point when the nature of the proposed task moves from the domain of mediated interaction with third party service provider to pure crowdsourcing. An antecedent set of criteria must be in place and satisfied for organizational decision-makers to select crowdsourcing as a viable alternative to more conventional forms of interaction, or indeed no interaction at all. This requires both an awareness and understanding of the role crowdsourcing might play on behalf of the manager, and a capability for the organisation to be able to undertake the crowdsourcing activity.

2.1 Decision Making Methodology

From an organizational perspective, crowdsourced tasks can be seen to satisfy two types of need: operational or strategic. Operational tasks are routine and integrative in nature, and are typical of the tasks that are performed through platforms such as Amazon Turk [24]. These are pure outsourced business processes and do not engage the collective intelligence of the crowd [25]. Contrast this to strategic tasks which move the locus of option generation effectively beyond the walls of the organisation and locates it amongst an undifferentiated but not disinterested crowd.

Dibbern [26] provides a useful survey of decision theory literature and methodological frameworks including Agency Theory, Transaction Cost Theory, and a number of other methodology approaches that are focused on perspectives such as the impact of politics within an organisation, the nature of the organization's relationship with external parties, and the resource base of the organisation. This assemblage of methodological foundations does not however reveal the steady emergence of a dominant model but rather summarizes the theoretician's struggle to account for the range of factors influencing management decision-making.

If the theory cannot agree then modelling the practice may provide a methodology for the reflection of reality. An early process model proposed four stages of decision making: intelligence, design, choice and implementation [27]. "Intelligence" denotes the identification of the issue to be addressed, "design" is the formulation of the range of potential methods to address the issue, "choice" is the selection of the desired solution, and "implementation" is the execution of that solution. As a generic decision making model this has value but it assumes a purely rational approach. Simon subsequently built on the work of Barnard [28] to propose two additional elements that influence the management decision making process: intuition and emotion.

> "The sources of these non-logical processes lie in physiological conditions or factors, or in the physical and social environment, mostly impressed upon us unconsciously or without conscious effort on our part. They also consist of the mass of facts, patterns, concepts, techniques, abstractions, and generally what we call formal knowledge or beliefs" [27].

Combining the rapidly changing nature of methodological tools that connect communities to organisations and the expectation of users embracing this technology, purely rational decision-making models fall short of capturing the effects of uncertainty in the process. Methodological models based on the inclusion of emotional attributes may be too ill-defined to offer predictive or interpretive value.

2.2   Sensemaking Attitudes

In the context of uncertain and rapidly changing environments issues of organizational sensemaking and knowledge creation become inextricably interwoven with the decision making process [29]. Sensemaking "constructs the shared meanings that define the organization's purpose and frames the perception of problems or opportunities that the organisation needs to work on" [29]. In this context it is an action concerned as much with looking forward as it is with constructing a narrative in retrospect. It is into this context that the participants in this study will fall.

Two attitudes were prevalent among the organisations observed. The first related to the potential for disadvantage through incomplete knowledge. When constructing a forward-facing view of the environment there was a clear sense that while the manager may not have understood the competitive advantages or limitations of the new technology, failure to include it in the planning process would represent a form of failure. In this case there was a perceived disconnect between those that had responsibility for making the decision to crowdsource, from those that had the technical ability to implement that decision.

The second attitude was the belief that this was a phenomenon driven by social forces and not business needs. There was a very clear indication of technology leading the development of strategy rather than serving it. In general the push to sensemake was seen as a net reducer of opportunity and a distraction to "business as usual". Accommodating it in a way that created value was perceived to be risky and in many cases to attract additional costs that could not easily be offset by strategic gains.

3   Three Antecedents for Crowdsourcing

A prime purpose of strategy formation is to align the activities of the organisation with the unmet needs of the stakeholder. This implies an antecedent condition requiring that the subject of the task in strategic crowdsourcing will be designed so as to enable change - the subject must have modular characteristics i.e. be able to change one aspect to optimize that characteristic without the necessity to change the entire subject. Modularity in product design has been held to improve the acceleration of innovation. [30, 31]. This modularity may extend to product features, policy settings or reconfiguration of core competencies. Products or services that are tightly bound to one form (because of regulatory, intellectual property, market share constraints or simply the inherent properties of the product or service) will derive little value from adopting crowdsourcing techniques.

A second antecedent that must be satisfied is the presence of an accessible and engaged community. This can be either fostered by the organisation (less authentic) or one that has spontaneously organized outside the organisation (more authentic) [32]. The degree of authenticity is perhaps correlated with the quality of commitment and thus the sincerity of response. It was observed that not all communities of interest are equal. Spontaneous communities that self-organize with neither the knowledge nor the guidance of the offeror were seen to provide better quality of input than that obtained by communities maintained on an organization's website and moderated by members of the organisation. This is consistent with prior research, particularly in respect of dedicated online brand communities (OBCs) [33]. However interacting with the communities that formed independently of the organisation was perceived to carry with it the potential for greater reputational damage as the entity was unable to moderate or influence discussions directly. Management's awareness of the need for community is a given. Management's understanding that better results come from uncontrolled and spontaneously formed communities is less clear.

A third antecedent is an organizational structure that respects and resources the process and provides forward budgeting that allows for the inherent uncertainty that goes with devolving the creation of new ideas and insights to external parties. In practice this was seen to be problematic. Crowdsourcing may consume more resources and be more difficult to manage than expected. This is broadly consistent with research on the diffusion of technological innovation throughout business [34]. The operating structure of organisations is shaped by the existing demands of customers and stakeholders. Management efforts to make processes more efficient reinforce existing practice and reduce opportunity for variation [35]. When a new category of business activity is identified, the understanding of both the operational overhead required to implement the technology and the nature of returns to be expected from the activity is frequently unrealistic. This happened historically with the introduction of desktop computing, development and integration of Internet sales channels, and the adoption of social media into strategic marketing plans. It is only when a dominant design emerges across a range of organisations and industry sectors that a degree of predictability emerges in the planning and execution of initiatives [36]. There is a need to appreciate that this is a dynamically developing and specialist area. A piecemeal approach and lack of dedicated resources will not necessarily lead to desired outcomes.

3.1 Decisions to Resource Crowdsourcing

A participant in this research noted that management decisions to resource programs of innovation and change are budgeted on the basis of cost, time to complete and anticipated contribution to the achievement of strategic aims of the organisation - the project is defined in advance of resources being committed. The observation was made that when crowdsourcing is employed to generate strategic direction the decision to resource must be made before the specific nature of the proposed activities is known. Most organisations manage resources well but inherently leave relatively little slack available to be flexibly deployed in the service of emergent ideas. Attempting to adopt crowdsourced outcomes within an organizational environment such as this will compromise outcomes and cause unnecessary stresses within the organisation.

Decision making without the power to apply those decisions is disabling not enabling. Adopting an organizational structure that doesn't merely include crowd responses as an input to the decision making process, but that embraces them (with some qualification) as the answer to the task, achieve better results than other approaches [37]. In all cases the crowd inputs from decision-making activities were filtered by the offeror prior to being accepted. This mediating role of the responsible manager provides the opportunity for qualitative assessments to be made to ensure congruence with the strategic aims of the organisation. Novel mindsets and "left-field" thinking is valuable but only when it doesn't conflict with defining organizational intangibles that are often built up over a considerable time period. Managers quoted the need for pragmatism, and the need to satisfy internal constraints and often complex policy prerogatives as reasons for this filtering process. The risk is that inputs that are judged to be inconsistent with existing management views are discarded thereby limiting the potential effectiveness of the crowdsourcing activity. Part of management thinking before embarking on crowdsourcing is that a "safety valve" is required and peace of mind is gained through management control over the degree of utilization of final inputs.

The presence of these three conditions enables a mode of market interaction which, rather than reproduce organisations as systems of control, configures operations as a "discursive contested place of encounter and exchange" [38].

4 Conclusions

The practice of blurring the boundaries between organisations and their constituent stakeholders has considerable merit when considered under the right circumstances. The awareness of crowdsourcing as a management option has perhaps never been higher. Misapplied, or applied in situations not naturally conducive to the inclusion of outside parties may lead to problematic outcomes. For this reason, studies of crowdsourcing practice as it is happening, and observing the limitations and basic criteria for successful implementation are an important step forward. As the model transitions out of specialist pure-plays and becomes a feature of everyday life so can incremental advantages be expected to accrue. When organisations no longer have to take best guesses at stakeholder requirements but can integrate the stakeholder's viewpoint in an empowered, authentic and immediate manner, outcomes for all may reasonably be expected to improve.

The research has found that in order for crowdsourcing to be successfully undertaken three criteria must be met – the subject of the task being crowdsourced must be modular in nature i.e. elements of the subject must be able to be changed without compromising the integrity of the whole. Secondly a community of interest must be engaged. With the widespread adoption of social media technologies identifying or creating these communities is often straightforward. Finally, there needs to be a structural capability within the organisation to be able to both engage the crowd and utilize the output from the crowd in a manner that creates value. The potential for using semantic connectivity methodology and cloud-based artificial intelligence algorithms to interrogate data collected from user discussion forums is apparent, but no examples of this have come to the researchers' attention.

Implications for management of crowdsourcing projects are that structural capabilities must be in place and resourced ahead of the commencement of a crowdsourcing program.

Acknowledgments. The work of Marta Poblet draws from previous research within the framework of the project "Crowdsourcing: instrumentos semánticos para el desarrollo de la participación y la mediación online" (DER 2012- 39492 -C02 -01) by the Spanish Ministry of Economy and Competitiveness.

References

1. Estellés-Arolas, E., González-Ladrón-de-Guevara, F.: Towards an integrated crowdsourcing definition. J. Inform. Sci. **38**(2), 189–200 (2012)
2. Kaplan, A.M., Haenlein, M.: Users of the world, unite! The challenges and opportunities of Social Media. Bus. Horiz. **53**(1), 59–68 (2010)
3. Schenk, E., Guittard, C.: Towards a characterization of crowdsourcing practices. J. Innov. Econ. Manage. **7**(1), 93–107 (2011)
4. Barbier, G., Zafarani, R., Gao, H., Fung, G., Liu, H.: Maximizing benefits from crowdsourced data. Comput. Math. Organ. Theory **18**(3), 257–279 (2012)
5. Ogawa, S., Piller, F.T.: Reducing the risks of new product development. MIT Sloan Manage. Rev. **47**(2), 65 (2006)
6. Von Hippel, E.: Democratizing Innovation. MIT Press, Cambridge (2005)
7. Zhao, Y., Zhu, Q.: Evaluation on crowdsourcing research: current status and future direction. Inform. Syst. Front. **16**(3), 417–434 (2014)
8. Thuan, N.H., Antunes, P., Johnstone, D.: Factors influencing the decision to crowdsource. In: Antunes, P., Gerosa, M.A., Sylvester, A., Vassileva, J., de Vreede, G.-J. (eds.) CRIWG 2013. LNCS, vol. 8224, pp. 110–125. Springer, Heidelberg (2013)
9. Berger, P., Luckmann, T.: The Social Construction of Reality: A Treatise on the Sociology of Knowledge (1967)
10. Eriksson, P., Kovalainen, A.: Qualitative Methods in Business Research. Sage, London (2008)
11. Bruner, J.S.: Actual Minds, Possible Worlds. Harvard University Press, Cambridge (2009)

12. Raddick, M.J., Bracey, G., Gay, P.L., Lintott, C.J., Murray, P., Schawinski, K., Vandenberg, J.: Galaxy zoo: exploring the motivations of citizen science volunteers. Astron. Educ. Rev. **9**(1), 010103 (2010)

13. Starbird, K., Muzny, G., Palen, L.: Learning from the crowd: collaborative filtering techniques for identifying on-the-ground Twitterers during mass disruptions. In: Proceedings of 9th International Conference on Information Systems for Crisis Response and Management, ISCRAM (2012)

14. Bojin, N., Shaw, C.D., Toner, M.: Designing and deploying a 'compact' crowdsourcing infrastructure: a case study. Bus. Inform. Rev. **28**(1), 41–48 (2011)

15. City of Melbourne, Budget Simulator Report, 10 Year Financial Plan Community Engagement (2014)

16. http://participate.melbourne.vic.gov.au/projects/10yearplan/peoples-panel-report-guide-10-year-financial-plan/

17. Antorini, Y.M., Muñiz, A.M.: Invited article: the benefits and challenges of collaborating with user communities. Res. Technol. Manage. **56**(3), 21–28 (2013)

18. Lakhani, K.R., Lifshitz-Assaf, H., Tushman, M.: Open innovation and organizational boundaries: task decomposition, knowledge distribution and the locus of innovation. In: Grandori, A. (ed.) Handbook of Economic Organization: Integrating Economic and Organizational Theory, pp. 355–382. Edward Elgar Publishing, Northampton (2013)

19. https://support.lytro.com/hc/communities/public/topics?locale=en-us

20. http://community.myob.com/t5/forums/recentpostspage/post-type/thread/interaction-style/idea

21. Chen, J., Xu, H., Whinston, A.B.: Moderated online communities and quality of user-generated content. J. Manage. Inform. Syst. **28**(2), 237–268 (2011)

22. Nandi, G., Das, A.: A survey on using data mining techniques for online social network analysis. Int. J. Comput. Sci. Issues (IJCSI) **10**(6), 162–167 (2013)

23. Feldstein, A.P.: Analyzing online communities: a narrative approach. MiT6: Stone and Papyrus, Storage and Transmission. International Conference (2009)

24. Acosta, M., Zaveri, A., Simperl, E., Kontokostas, D., Auer, S., Lehmann, J.: Crowdsourcing linked data quality assessment. In: Alani, H., et al. (eds.) ISWC 2013, Part II. LNCS, vol. 8219, pp. 260–276. Springer, Heidelberg (2013)

25. Brabham, D.C.: Crowdsourcing as a model for problem solving an introduction and cases. Convergence Int. J. Res. New Media Technol. **14**(1), 75–90 (2008)

26. Dibbern, J., Goles, T., Hirschheim, R., Jayatilaka, B.: Information systems outsourcing: a survey and analysis of the literature. ACM SIGMIS Database **35**(4), 6–102 (2004)

27. Simon, H.A.: The New Science of Management Decision. Harper, New York (1960)

28. Barnard, C.: The functions of the executive. Cambridge/Mass (1938)

29. Choo, C.W.: Sensemaking, knowledge creation, and decision making: organizational knowing as emergent strategy. In: Choo, C.W., Bontis, N. (eds.) The Strategic Management of Intellectual Capital and Organizational Knowledge, pp. 79–88. Oxford University Press, Oxford (2002)

30. Ulrich, K.: The role of product architecture in the manufacturing firm. Res. Policy **24**(3), 419–440 (1995)

31. Henderson, R.M., Clark, K.B.: Architectural innovation: the reconfiguration of existing product technologies and the failure of established firms. Adm. Sci. Q. **35**, 9–30 (1990)

32. Brogi, S., Calabrese, A., Campisi, D., Capece, G., Costa, R., Di Pillo, F.: Effects of online brand communities on brand equity in luxury fashion industry. Int. J. Eng. Bus. Manage. **5**(1), 1–9 (2013)

33. Lee, D., Kim, H.S., Kim, J.K.: The impact of online brand community type on consumer's community engagement behaviors: consumer-created vs. marketer-created online brand community in online social-networking web sites. Cyberpsychology, Behav. Soc. Network. **14**(1–2), 59–63 (2011)
34. Zhu, K., Kraemer, K.L., Xu, S.: The process of innovation assimilation by firms in different countries: a technology diffusion perspective on e-business. Manage. Sci. **52**(10), 1557–1576 (2006)
35. Lam, A.: Organizational innovation (2004)
36. Lee, J.R., O'Neal, D.E., Pruett, M.W., Thomas, H.: Planning for dominance: a strategic perspective on the emergence of a dominant design. R&D Manage. **25**(1), 3–15 (1995)
37. Ansari, S., Munir, K.: Letting users into our world: some organizational implications of user-generated content. Res. Sociol. Organ. **29**, 79–105 (2010)
38. Anderson, C.: Ephemeral architectures: towards a process architecture (2009)

On Integrating an IS Success Model and Multicriteria Preference Analysis into a System for Cloud-Computing Investment Decisions

Rangaraja P. Sundarraj and Sathyanarayanan Venkatraman[(✉)]

Department of Management Studies, IIT Madras, Sardar Patel Road,
Chennai 6000361, Tamil Nadu, India
rpsundarraj@iitm.ac.in, sathya.venkatraman68@gmail.com

Abstract. Investing in cloud computing technology is one of the latest trends in IT. This is a multi criteria investment decision involving many stakeholders and a sequence of coordinated assessment activities that are strategic, qualitative (technical) and quantitative (financial) in nature. This paper integrates an information system (IS) success model with preference elicitation techniques drawn from the multi-criteria decision-making literature. We also show how this decision model can be extended as a vendor negotiation tool. Finally, we describe a prototype Decision Support System (DSS) featuring this model.

Keywords: Cloud · Business value of IT · Delone & McLean IS success model · Decision support · Group decision · AHP · DEA · MCDM · Negotiations · IT risk · IT evaluation

1 Introduction

Cloud computing enables organizations to avoid huge IT capital expenses and instead adopt a pay-per-use model to gain access to resources such as: servers and storage through "Infrastructure as Service" (IaaS), development platforms through "Platform as Service" (PaaS) and software/applications through "Software as service" (SaaS) [1]. As per NIST (National Institute of Standards and Technology) there are also many cloud deployment models available such as Private, Public, Hybrid and Community Clouds [2]. Some of the distinct technological capabilities of cloud computing include controlled interface, location independence, sourcing independence, ubiquitous access, virtual business environments, addressability and traceability, and rapid elasticity [3].

Many of the current day businesses are transforming their business with cloud technology. A decision to invest in cloud computing to enable the business transformation involves a complex evaluation process and involvement from various organizational stakeholders [4], who have to view the investment from many perspectives such as business IT alignment, technical quality, risks and financial outcomes etc. Although there are research works considering the evaluation of certain key aspects of cloud computing (e.g. ROI model for cloud [5, 6]), the ranking of commercial cloud computing vendors based on technical features and quality of services [7], and the

© Springer International Publishing Switzerland 2015
B. Kamiński et al. (Eds.): GDN 2015, LNBIP 218, pp. 357–368, 2015.
DOI: 10.1007/978-3-319-19515-5_28

development of frameworks for information assurance/security on cloud computing e.g., [8], what we find as a research gap is the availability of a "cloud investment decision model" which integrates various investment evaluation aspects holistically.

In this paper, we first develop a framework for cloud investment decision-making, integrating factors such as business IT alignment (strategic fitness) analysis, technical (qualitative) analysis and financial (quantitative) outcomes evaluation. We then detail one element of this framework, by bringing in theoretical extensions of a well known IS success by DeLone and Mclean (D&M) [9] along with two multi criteria evaluation approaches [10, 11]. A DSS outlining the incorporation of this model is also proposed. This paper is written based on the ongoing research we have been undertaking on developing the DSS for cloud computing investment.

The paper is further organized as follows: Sect. 2 details the framework for cloud computing investment decision and reviews the literature. Section 3 presents the adoption of D&M model and enhances it by adding *Risk Mitigation* as an additional criterion; this is followed by the application of multi criteria approach. Section 4 details the application of AHP & DEA in our cloud evaluation with an example and Sect. 5 summarize & propose a future research work.

2   Framework for Cloud Computing Investment Decision

The IS evaluation literature refers to various types of evaluations such as strategic evaluations [12] and financial evaluations [13] of IT Investments. Berghout and Renkema detail four approaches viz. the financial approach, the multi-criteria approach, the ratio approach and the portfolio approach to IT evaluation [14]. Bacon carried out an empirical research on criteria used for allocation of IT resources and technology investments across 80 organizations [15]. The criteria identified fall into (i) management domain, which essentially deals with the strategic fitness of IT investments, (ii) financial domain, which deals with cost & returns of investment; and (iii) development domain, which deals with technical aspects of IT investment. Adapting from this and other follow-up to this, we break-up the process of making cloud-computing investment decisions into three stages namely strategic fitness assessment, criteria selection, assessment of technical features and financial returns before the negotiation, and final decision (see Fig. 1). We discuss these stages below.

2.1   Strategic Fitness

Many researchers in the past have created conceptual models for strategic alignment of IT and business (or strategic IT fitness). Examples of some well-known ones include: (i) Strategic Alignment Model (SAM) [16] (ii) A conceptual model extending the idea of SAM and depicting the relationship between IS strategy and business performance [17], (iii) IT business value model with a resource based view [18], and (iv) A model focused on exploring the relationship between IT-business plan alignment and performance [19].

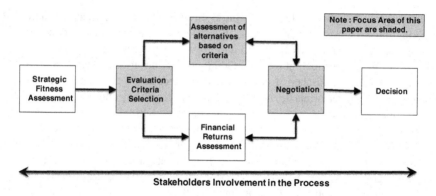

Fig. 1. Cloud computing investment decision framework

There are also models focusing on measurement of IT-business alignment. Examples of some well-known models include: (i) Multi Level Strategic Fit Model for measurement at overall organization level and also individual business units [20]. (ii) model based on dynamic capabilities theory factoring-in the time lags between IT implementation and business value creation [21] and extending further a practically usable framework that can help assess, monitor and achieve the alignment [22].

With cloud computing, an enterprise's product-centric and firm-based model for applications and systems can be transformed to a global, distributed, service-centric model [3]. Iyer and Henderson posit that to understand cloud computing and to make appropriate investment decisions, companies have to focus on strategy aspect of the cloud, not just technology [3]. Aligning IT capabilities (cloud capabilities) to the business and market needs, and in turn ensuring the strategic fitness of technology to the business are critical for value creation through technology [16].

As a part of the cloud strategy, organizations need to evaluate two aspects of business IT alignment viz. efficiency and effectiveness. Efficiency is about reduction of operational costs for a given quality of service & effectiveness is enabling flexibility & responsiveness to changing market needs [23]. Hence, while deciding on the cloud computing investments, the assessment of business IT alignment (strategic fitness of IT investment to business) becomes the logical first step before we move in to technology evaluation. The prima facie decision to go ahead with further evaluation of cloud computing investment would be based on the output of strategic fitment analysis. For example, for a banking business, there could be legal & compliance restrictions on hosting confidential data on public cloud applications, making a cloud computing prima facie unviable.

2.2 Evaluation Criteria and Assessment

An investment on cloud computing is meaningful only if it can create business benefits, either tangible or intangible. The types of value include financial (e.g., ROI), inter-mediate (e.g., process-related) or affective (e.g., perception-related) [24]. The value creation of investment depends on many factors like usage of IT [25, 26], process efficiency or the design of IT [27] etc., But at the time of making a decision on cloud computing, the business benefits are just assessments and estimations as compared to technical attributes, which can be tested and proven. This makes the qualitative evaluation or technical evaluation of the cloud computing more reliable and hence important. There are a number of parameters that needs to be evaluated viz. accountability, agility, cost, performance, assurance, security & privacy and usability and its is also important that vendors are evaluated based on a few KPIs (Key Process Indicators) such as service provider's response time, sustainability, suitability, accuracy, transparency, interoperability, availability, reliability, stability, cost, adaptability, elasticity, throughput & efficiency, scalability etc. [7]. If we look through this list of attributes cited, they are either system-based features, information based features, services based features or risk based features. This is in line with the well-known models like DeLone & McLean IS Success model which have been used in IS evaluation literature for long.

2.3   Stakeholders Involvement in Technology Decision Process

In the current times where the organizations are transforming through technology, the IT investments and decision-making processes have major impact on organizational success [28]. Typically the technology decision process consists of a sequence of actions that begins with the identification of an IS-related crisis, problem, or opportunity and culminates in the approval of an IT project [28]. Clearly, the decision to invest on information technology is a complex multi stage process involving a variety of stakeholders from many domains at different levels & different expectations within a firm [29]. Renkema highlights that there are 4 Ps (Product, Process, Participation and Politics) in information technology investments [4]. Since a variety of organizational actors influence the decision-making process, all the participants are accountable for the outcome of IT investments [30]. It is also critical to note that, the relationship between the CIO with the business executives also significantly impacts IT assimilation [31] and in turn determines the success of IT projects. With cloud computing there is a dependency on third party service providers and hence increased exposure to the organization including losing control on critical data. Hence, the involvement of many senior executives and their buy-in on the decision becomes more important for cloud computing. In summary, cloud computing technology evaluation & decision involves stakeholder's participation, group dynamics and it is beyond pure technical or financial evaluation. Next, we focus on the evaluation and assessment parts of the framework.

3   Integrating IS Success Model with Preference Analysis

We have highlighted the importance of selecting the right set of criteria to evaluate cloud computing. To identify a basic model for the selection of our evaluation criteria, we reviewed the previous research in the area technology acceptance models and the IS success models. The Technology Acceptance Model (TAM) [32] and its extensions have been used last couple of decades extensively and they do a good job in emphasizing the people aspects and acceptance by users as the measure of IS success.

TAM postulates that the attitude towards the usage of IT and the actual usage are dependent on the ease of use and perceived usefulness of IT. However, it does not explain the attributes (systems and the service management aspects) of IT that helps in the user adoption and perception of usefulness. Another model is Seddon's IS effectiveness matrix [33], which focuses on people issues related to technology, but does not tie back business performance impact of IT.

3.1   IS Success Model

DeLone & McLean (D&M) developed an IS Success Model [34], which is a simple and useful model explaining how system quality and information quality of the IT under consideration could help in user adoption of technologies and in turn create business benefits. It emphasizes more on the role of systems & information quality in creating benefits for the organization, and has formed the basis of a number of prominent works. However, one thing that D&M does not capture is the importance of IT services supporting the technology. In 2003, the D&M model was modified to address the said limitation and it now includes services quality as an additional parameter that creates business value. It also captures the people aspect of technology by including the intention to use and actual use of technology.

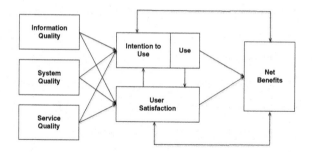

Fig. 2. DeLone & McLean IS success model

The D&M Model. We chose the updated D&M Model [9] as our base because it can sufficiently explain not just the aspects of technology & system, but can also the service attitude of cloud service provider towards the customer and the attitude of user towards the cloud computing. It has to be noted here that "the user" includes both the internal user and the customer. With D&M model as a base, we select the attributes or parameters pertaining to System Quality, Information Quality & Services Quality as evaluation criterions. In the context of cloud computing, the system quality deals with the attributes of information system itself, which produces the information such as performance, throughput, elasticity, stability, flexibility, availability, ease of use etc., The information quality deals with the information product from the system for desired characteristics such as accuracy, meaningfulness, and timeliness of access to information etc., The services quality deals with the quality of IT services offered by the vendor/internal IT team such as services efficiency, reliability, empathy, responsiveness, transparence, assurance, accountability etc.,

Extending the D&M Model. It has to be noted that even though D&M model captures the effect of systems, information and service qualities on the use & adoption of technology, it is still not sufficient from the context of cloud computing. A very important factor called "risks" arising out of moving into cloud should be given a due importance and hence, as a part of evaluation we also need to assess the vendors' risk mitigation features & capabilities (Fig. 2). Karami and Guo identify risk factors in cloud services and present an approach for risk-benefit assessment in sourcing cloud services [35]. There are many other articles which deal specifically with the information security risks of cloud computing [36–38]. Oh, Gallivan and Kim studied the market's perception of the transactional risks of information technology outsourcing using stock market data and conclude that the outsourcing does influence the investors' perception [39]. It is important that, when investing in (especially new/unproven) technology, organizations should ensure that, there is no bias on evaluation process of risky IT projects [40]. Cloud computing being a new trend (or unproven) and also typically involves outsourcing IT services (especially for public clouds), the risk evaluation gains an even more importance. We conclude that pertaining to evaluation of cloud technology investments, analyzing "Risk Mitigation" features/capabilities should be added to the capabilities pertaining to system quality, information quality and services quality (Fig. 3).

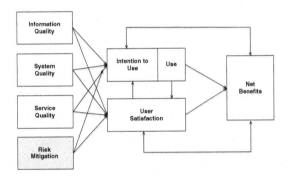

Fig. 3. DeLone & McLean IS success model with risk mitigation

3.2 Preference Elicitation and Analysis

Organizations first evaluate the strategic fitness of cloud computing investment to decide the prima facie worthiness of the investment. Once they find a very clear fitness, then they proceed with technical (qualitative) and financial (quantitative) analysis of the alternatives (vendor solutions) available for investment and choose the solution amongst the alternatives. Our goal is to use AHP-DEA integrated technique to ascertain the cloud computing solution vendor who offers the value for the investment and in this section we demonstrate the same with an example.

The evaluation of each solution alternative is based on various criteria and thus the whole problem could be thought as a multi-criteria decision-making (MCDM) problem.

We use the AHP-DEA integration method used by Ramanathan [41] and enhance the same with additional steps to include the elements of vendor negotiation during cloud computing evaluation (Fig. 4).

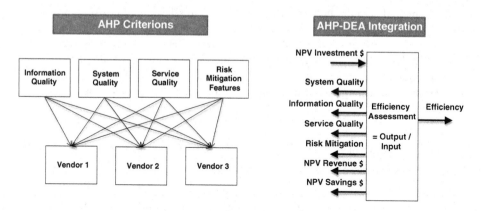

Fig. 4. AHP-DEA integration

Evaluation Criterions (Alternative Solutions from 3 vendors)

- **Technical (Qualitative) Criterions:** Based on our theoretical development (§3.1), we select four criteria viz. information quality, system quality, service quality and risk mitigation features.
- **Financial (Quantitative) Criterions:** Unlike the technical criteria, the financial criteria would change depending on different business needs of the organizations. In the example shown, we have chosen net present value (NPV) of the investment as an input and NPV revenue and NPV cost savings that the investment generates as outputs. The financial efficiency is determined by the ratio of output and input.

Procedure:

Table 1. Example of AHP Based evaluation of cloud computing alternatives

Technical evaluation (AHP)	Weighage	Vendor 1	Vendor 2	Vendor 3
System quality	41 %	58 %	31 %	11 %
Information quality	8 %	67 %	23 %	10 %
Services quality	35 %	39 %	41 %	20 %
Risk mitigation	16 %	67 %	23 %	10 %
Overall technical preference*		53 %	33 %	14 %

Based on the preference elicitation on the importance of evaluation criteria and using AHP pairwise comparison we initially determine weightages for the criteria. As a next step we evaluate the alternatives against each criterion to create criterion-wise weightages of the alternatives and the ranking. We see in our example (Table 1) that vendor 1 ranks highest in overall technical evaluation.

Now using DEA technique & the linear programming (with an objective function of maximum efficiency and constraints of max value of 1), we determine the weightage factor of each outputs in every alternative such that the determined value is the maximum possible efficiency for the given alternative. In our example, we take the results from AHP and fix them as output parameters in DEA. In other words, we apply DEA technique with 1 input (NPV Investment) and 6 outputs (weightage factors of system quality, information quality, service quality & risk mitigation and NPV Revenue & NPV Cost Savings) to ascertain the overall ranking of alternatives considering both the technical and financial parameters. In our example we see that considering both technical and financial criteria, vendor 2 comes on top as the preferred one.

Table 2. Example of AHP-DEA integrated evaluation of cloud computing alternatives

DEA parameters (2 years)	Vendor 1	Vendor 2	Vendor 3	Remarks
NPV investment $	40000	30000	32000	Input
NPV revenue $	45000	35000	40000	Outputs
NPV savings $	15000	20000	15000	
System quality $	58 %	31 %	11 %	
Information quality	67 %	23 %	10 %	
Service quality	39 %	41 %	20 %	
Risk mitigation	67 %	23 %	10 %	
Overall preference	72 %	100 %	74 %	Efficiency

3.3 Extension of AHP-DEA Technique for Negotiations

As an extension of the above, the AHP-DEA technique can also be used effectively for vendor negotiations. In real life scenarios of organizations investing in cloud computing technology there would be many rounds of discussions & negotiations with the vendors and typically vendors do not reveal their best price or technical options until many discussions are completed. In our example we determined vendor 1 to be having the highest preference for the technical quality (AHP output) and suppose this is a baseline quality we do not want to compromise, then using Linear Programming we can fix the constraint that "vendor 1 also needs to be highest in overall efficiency". We can now make the input cost as a variable to solve by LP & determine the input price point at which vendor 1 becomes the overall best. In our example we can show that at 28,333$ input price vendor 1 becomes the overall best. This is roughly 71 % of the original price of 40,000$. This input would be of great value for the decision maker who interacts with the vendors for negotiations. The AHP-DEA process can be carried out repeatedly for every change in technical quality enhancements or price reductions after every round of negotiation. This equips the negotiator with a useful information pre & post negotiations.

4   A Design Science Approach to DSS for Cloud Computing Investment

One of the objectives in this research is the creation of a DSS for cloud computing investment. We outline how this can be done using the core principles of design science approach [42] As an initiative toward that goal, we created a prototype MS Excel based tool (Fig. 5), which would be the part of envisaged DSS in future. Similar approaches to spreadsheet based DSS have been done in the past [43, 44] This tool automates the AHP-DEA computing. Some of the key features of the tool include provision for preference elicitation, pairwise comparison, criterion weighage calculation, integration of AHP and DEA, linear programming with Excel Solver, ranking of alternatives and simulation of what-if scenarios. The screen shot shown (Fig. 5) below is the worksheet for DEA calculations that we carried out in our example. The results Tables 1, 2 come from this tool.

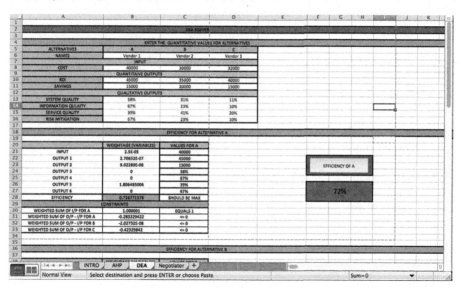

Fig. 5. Screenshot of the AHP-DEA tool

Henver described the design science as a build and evaluate process in order to produce artifacts [45] and this is precisely our goal to take to test this tool based on publicly available data on cloud features and pricing of the vendors and refine the artifacts. Peffedlxrs [42] in his DSRM highlights six stages and currently we are in Stage two, in the process of finalizing the objectives of the DSS.

5   Conclusions and Proposed Future Research

In this paper we created a decision framework for cloud computing technology investment by combining the strategic, qualitative (technical) and quantitative (financial) evaluation activities. We used D&M IS Success model for criteria determination and added the enhancements needed from the cloud computing context. As a part of the process we used AHP & DEA integrated technique and also included the provision for stakeholders preference elicitation and negotiation. However, the following are the proposed improvements of this paper, which are planned to be addressed in continued future research:

- The current decision model assumes that current cost and future outflow is known while using DEA. In complex IT projects involving cloud computing, this is not the typical case and the inputs costs & returns are dynamic. They also stagger & vary over the time and many a times we will not even have the perfect knowledge of the complete lifecycle of the project. To address this nature of cloud IT investment scenario, there is a scope of enhancement in our DSS with the integration of real options valuation (ROV) in the decision model.
- There is a more detailed work needed to enhance the cloud computing investment decision framework, covering all the elements. In this paper we have covered only the evaluation criteria and assessment part in detail.
- A detailed process for cloud computing strategic assessment technique needs to be created.
- As a part of the future research we plan to create a decision support system (tool/instrument) for cloud computing investment and this DSS would be designed based on "Design Science Theories".

Acknowledgements. We would like to acknowledge and thank the following internstudents for their contribution in developing the AHP-DEA excel based tool:

1. Abhishiekh Ramesh - abhishiekh1@gmail.com
2. R. Arvind Srivatsan - arvindsrivatsan@yahoo.co.in
3. Raghav R - rraghav94@gmail.com
4. Sudharshan Ashok - sudharshan93@gmail.com.

References

1. Weinhardt, C., Anandasivam, A., Blau, B., Stosser, J.: Business models in the service world. IT Prof. **11**(2), 28–33 (2009). IEEE Computer Society
2. Mell, P., Grance, T.: The NIST definition of cloud computing. Natl. Inst. Stand. Technol. **53** (6), 50 (2009). www.csrc.nist.gov/publications/nistpubs/800-145/SP800-145.pdf
3. Iyer, B., Henderson, J.C.: Preparing for the future: understanding the seven capabilities of cloud computing. MIS Q. Executive. **9**(2), 117–131 (2010)
4. Renkema, T.: The four P's revisited: business value assessment of the infrastructure impact of IT investments. J. Inf. Technol. **13**, 181–190 (1998)
5. Nanath, K., Pillai, R.: A model for cost-benefit analysis of cloud computing. J. Int. Inf. Manag. Assoc. **22**(3), 93–118 (2013)
6. Misra, S.C., Mondal, A.: Identification of a company's suitability for the adoption of cloud computing and modelling its corresponding return on investment. Math. Comput. Model. **53** (3–4), 504–521 (2011)

7. Garg, S.K., Versteeg, S., Buyya, R.: A framework for ranking of cloud computing services. Future Gener. Comput. Syst. **29**(4), 1012–1023 (2013)

8. Information Assurance Framework for Cloud Computing. The European Network and Information Security Agency (ENISA) (2009). http://www.enisa.europa.eu

9. Delone, W.H., McLean, E.R.: The DeLone and McLean model of information systems success: a ten-year update. J. Manag. Inf. Syst. **19**(4), 9–30 (2003)

10. Liu, F.F., Hai, H.L.: The voting analytic hierarchy process method for selecting supplier. Int. J. Prod. Econ. **97**, 308–317 (2005)

11. Ramanathan, R.: Supplier selection problem: integrating DEA with the approaches of total cost of ownership and AHP. Supply Chain Manag. Int. J. **12**(4), 258–261 (2007)

12. Sarkis, J., Sundarraj, R.P.: Factors for strategic evaluation of enterprise information technologies. Int. J. Phys. Distrib. Logist. Manag. **30**(3/4), 196–220 (2000)

13. Anandarajan, A., Wen, H.J.: Evaluation of information technology investment. Manag. Decis. **37**(4), 329–339 (1999)

14. Berghout, E., Renkema, T.: Methodologies for investment evaluation: a review and assessment. In: Grembergen, W.V. (ed.) Information Technology Evaluation: Methods & Management, pp. 78–97. Idea Group Publishing, Hershey (2001)

15. Bacon, C.J.: The use of decision criteria in selecting information systems/technology investments. MIS Q. **16**(3), 335–353 (1992)

16. Henderson, J.C., Venkatraman, N.: Leveraging information systems and transforming organizations. IBM Syst. J. **32**(1), 198–221 (1993)

17. Chan, Y.E., Huff, S.L., Barclay, D.W., Copeland, D.G.: Business strategic orientation, information systems strategic orientation and strategic alignment. Inf. Syst. Res. **8**(2), 125–150 (1997)

18. Melville, N., Kraemer, K., Gurbaxani, V.: Review: information technology and organizational performance: an integrative model of IT business value. MIS Q. **28**(2), 283–322 (2004)

19. Kearns, G., Lederer, A.: The effect of strategic alignment on the use of IS-based resources for competitive advantage. J. Strateg. Inf. Syst. **9**(4), 265–293 (2000)

20. McLaren, T.S., Head, M.M., Yuan, Y., Chan, Y.E.: A multilevel model for measuring fit between a firm's competitive strategies and information systems capabilities. MIS Q. **35**(4), 909–929 (2011)

21. Schwarz, A., Kalika, M., Keffi, H., Schwarz, C.: A dynamic capabilities approach to understanding the impact of IT-enabled businesses processes and IT-business alignment on the strategic and operational performance of the firm. Commun. Assoc. Inf. Syst. **26**(4), 57–84 (2010)

22. Avison, D., Jones, J., Powell, P., Wilson, D.: Using and validating the strategic alignment model. J. Strateg. Inf. Syst. **13**, 223–246 (2004)

23. Tallon, P.P., Kraemer, K.L., Gurbaxani, V.: Executives' perceptions of the business value of information technology: a process-oriented approach. J. Manag. Inf. Syst. **16**(4), 145–173 (2000)

24. Devaraj, S., Kohli, R.: Measuring information technology payoffs: a meta analysis of structural variables in firm-level empirical research. Inf. Syst. Res. **14**(2), 127–145 (2003)

25. Devaraj, S., Kohli, R.: Performance impacts of information technology: is actual usage the missing link? Manag. Sci. **49**(3), 273–289 (2003)

26. Soh, C., Markus, M.L.: How IT creates business value: a process synthesis. Proceedings of the Sixteenth International Conference on Information Systems, pp. 29–41. Association for Information Systems, Atlanta (1995)

27. Lucas, H.C.: The business value of information technology: a historical perspective and thoughts for future research. In: Banker, R.D., Kauffman, R.J., Mahmood, M.A. (eds.) Strategic Information Technology Management: Perspectives on Organizational Growth and Competitive Advantage, pp. 359–374. IGI Publishing, Hershey (1993)
28. Boonstra, A.: Structure and analysis of is decision-making process. J. Eur. Inf. Syst. 12(3), 195–209 (2003)
29. Bower, J.L.: Managing the Resource Allocation Process. Harvard Business School Classics, New York (1970)
30. Xue, Y., Liang, H., Boulton, W.R.: Information governance in information technology investment decision processes: the impact of investment characteristics, environment, and internal context. MIS Q. 32(1), 67–96 (2008)
31. Armstrong, C.P., Sambamurthy, V.: Information technology assimilation in firms: the influence of senior leadership and IT infrastructures. Inf. Syst. Res. 10(4), 304–327 (1999)
32. Venkatesh, V., Davis, F.D.: A theoretical extension of the technology acceptance model: Four longitudinal field studies. Manag. Sci. 46(2), 186–204 (2000)
33. Seddon, P. B., Staples, D. S., Patnayakuni, R., Bowtell, M. J.: The IS effectiveness matrix: the importance of stakeholder and system in measuring IS success. In: Proceedings of the International Conference on Information Systems, pp. 165–176. Association for Information Systems (1998)
34. Delone, W.H., McLean, E.R.: Information systems success: the quest for the dependent variable. Inf. Syst. Res. 3(1), 60–95 (1992)
35. Karami, A., Guo, Z.: A fuzzy logic multi-criteria decision framework for selecting IT service providers. In: Proceedings of the 45th Hawaii International Conference on System Sciences, pp. 1118–1127 (2012)
36. Nanavati, M., Colp, P., Aiello, B., Warfield, A.: Cloud security: a gathering storm. Commun. ACM 57(5), 70–79 (2014)
37. Demirkan, H., Goul, M.: Taking value-networks to the cloud services: security services, semantics and service level agreements. Inf. Syst. E-Bus. Manag. 11, 51–91 (2013)
38. Ronga, C., Nguyena, S.T., Jaatun, M.G.: Beyond lightning: a survey on security challenges in cloud computing. Comput. Electr. Eng. 39, 47–54 (2013)
39. Oh, W., Gallivan, M.J., Kim, J.W.: The market's perception of the transactional risks of information technology outsourcing announcements. J. Manag. Inf. Syst. Crossing Boundaries Inf. Syst. Res. 22(4), 271–303 (2006)
40. Rose, J.M., Rose, A.M., Norman, C.S.: Evaluation of risky information technology decisions. J. Inf. Syst. 18(1), 53–66 (2004)
41. Ramanathan, R.: Data envelopment analysis for weight derivation and aggregation in the analytic hierarchy process. Comput. Oper. Res. 33, 1289–1307 (2006)
42. Peffers, K., Tuunanen, T., Rothenberger, M.A., Chatterjee, S.: A design science research methodology for information systems research. J. Manag. Inf. Syst. 24(3), 45–77 (2008)
43. Sarkis, J., Sundarraj, R.P.: A decision model for strategic evaluation of information technologies Strategic Evaluation. Inf. Syst. Manag. 18(3), 62–72 (2001)
44. Olavson, T., Fry, C.: Spreadsheet decision-support tools: lessons learned at hewlett-packard. Interfaces- INFORMS 38(4), 300–310 (2008)
45. Hevner, A.R., March, S.T., Park, J., Ram, S.: Design science in information systems research. MIS Q. 28(1), 75–105 (2004)

Demand Management with Energy Generation and Storage in Collectives

Ronghuo Zheng[1]([✉]), Ying Xu[1], Nilanjan Chakraborty[2], Michael Lewis[3], and Katia Sycara[4]

[1] Tepper School of Business, Carnegie Mellon University, Pittsburgh, PA 15213, USA
{ronghuoz,yingx1}@andrew.cmu.edu
[2] Stony Brook University, Stony Brook, NY 11794, USA
nilanjan.chakraborty@stonybrook.edu
[3] School of Information Science, University of Pittsburgh, Pittsburgh, PA 15260, USA
ml@sis.pitt.edu
[4] Robotics Institute, Carnegie Mellon University, Pittsburgh, PA 15213, USA
katia@cs.cmu.edu

Abstract. In this paper, we focus on demand side management in consumer collectives with community owned renewable energy generation and storage facilities for effective integration of renewable energy with the existing fossil fuel-based power supply system. The collective buys energy as a group through a central coordinator who also decides about the storage and usage of renewable energy produced by the collective. Our objective is to design coordination algorithms to minimize the cost of electricity consumption of the consumer collective while allowing the consumers to make their own consumption decisions based on their private consumption constraints and preferences. Minimizing the cost is not only of interest to the consumers but is also socially desirable because it reduces the consumption at times of peak demand. We develop an iterative coordination algorithm in which the coordinator makes the storage decision and shapes the demands of the consumers by designing a virtual price signal for the agents. We prove that our algorithm converges, and it achieves the optimal solution under realistic conditions. We also present simulation results based on real world consumption data to quantify the performance of our algorithm.

1 Introduction

Facing the rapid depletion of fossil fuel reserves and the increasing carbon emission, one main objective in energy industry is to reduce the usage of fossil fuels for electricity generation. This can be achieved by (a) increasing the penetration of renewable energy sources in (e.g., wind energy, solar energy) in electricity supply and (b) reducing the usage of fossil fuel-powered generators that are usually used to meet peak demand by shifting energy use of consumers at peak demand. To enhance the penetration of renewable energy in electricity supply, there has been several recent initiatives at using *community-owned generation facilities*

© Springer International Publishing Switzerland 2015
B. Kamiński et al. (Eds.): GDN 2015, LNBIP 218, pp. 369–381, 2015.
DOI: 10.1007/978-3-319-19515-5_29

for supplying part of the electricity needs of a community [1]. To reduce the variation of supply of renewable energy due to uncontrolled factors like weather, it is usually recommended to install storage facilities. Therefore, in this paper, we focus on demand side management in consumer collectives with community owned renewable energy generation and storage facilities for effective integration of renewable energy with the existing fossil fuel-based power supply system.

Effective integration of renewable energy should be able to not only reduce aggregate demand for the energy generated by fossil fuel, but also to reduce peak demand from the traditional power plants. To reduce peak demand, electric utilities use differential pricing systems like time-of-use-pricing (where higher prices are charged at times of expected high load). Thus, we want to minimize the electricity consumption cost of the collective which reduces the consumption at times of peak demand and is also of interest to the consumers. In particular, we study a collective of consumers in the presence of a collectively owned renewable generation facility (e.g., solar panels) and a collectively owned storage facility (e.g., battery). The electricity demand of the group of consumers is fulfilled by renewable generation and the electricity market. A central coordinator purchases electricity from the market on behalf of the consumers and also makes decisions on the usage and storage of renewable energy. The consumers have their individual private constraints and preferences, which they may not want to share with other group members or the coordinator. The consumers make their consumption decisions based on their private constraints and preferences. Our goal is to *design coordination algorithms so that the coordinator (with information of the market price of electricity and a forecast of the renewable generation) minimizes the total cost of electricity procurement of the collective by (a) managing the operation of the storage and (b) shifting the consumers' demand while allowing them to make their own decisions about their electricity usage subject to their private constraints.*

Although our problem can be formulated as an optimization problem, solving the problem becomes challenging because of the lack of information of the agents about the whole problem. Each consumer only knows its own demand constraints but has no information about the constraints of other agents. The no information assumption has been shown to be a challenge in other group decision problems such as multi-agent negotiation (e.g., [2,3]). The coordinator also does not have any information about the agents' constraints. Thus there is no agent who has all the "data" about the problem to be solved. The only common information is about the forecast of available renewable energy (which we assume to be reasonably accurate) and the market price of electricity over the planning horizon. These assumptions are reasonable for *day ahead* planning. The coordinator is similar to a mediator in a negotiation among the consumer collective (e.g., [4,5]) who needs to search for the optimal mediation protocol but also needs to decide the storage policy in our problem. The coordinator is not a market maker or a traditional demand response aggregator, but is akin to a social planner (similar to [6,7]). But the task of our coordinator is more complicated: in addition to coordinating consumers' demand, the coordinator has to solve the *coupled problem* of designing a method for charging and discharging storage.

To solve the coupled demand and storage management problem, we design an iterative algorithm consisting of two primary steps. In one step, the coordinator assumes a profile for charging and discharging the battery and uses virtual price signals to coordinate the consumers to obtain an optimal demand profile. In the second step, the coordinator uses the demand profile given by the consumers to compute an optimal storage solution. In step one, through the use of the virtual price signal, we ensure that (a) when each agent minimizes its own energy consumption cost, the total cost is also minimized and (b) at the optimal solution, each agent's *virtual* energy cost calculated based on its virtual signal equals its real payment. Using (a) and (b) above, we prove that our iterative algorithm converges to the optimal demand profile for the agents that minimizes the total energy cost. Furthermore (b) above ensures that the total amount that the agents pay is equal to the total electricity bill that the collective has to pay to the utility (i.e., budget-balance is achieved). The design of this provably optimal budget-balanced algorithm in the presence of limited information is the key contribution of this paper. We also performed simulations to show the scalability of our algorithm based on real world consumption data.

2 Related Work

Our work is related to two streams of research: demand side and storage management. There is extensive research about demand response programs for managing consumer side demand, either through direct load control (DLC) or indirect incentive based control, such as real time prices (RTP) (see [8] and references therein). We will restrict our discussion on how to design virtual price signals to incentivize consumer demand shifting. The design of price signal is challenged by the possibility of the *herding* phenomenon, whereby agents shift their consumptions towards the low price times simultaneously and thus cause a new spike in demand, thereby increasing the energy cost [9]. Previous work has proposed various heuristics. Reference [10] uses an adaptive approach by sending both price and control signals to agents to control the rate and frequency of agents reacting to real-time prices so as to avoid simultaneous shifting. Unlike these papers, we study the problem from the perspective of demand schedule planning. In addition, we maintain privacy of consumers without relying on consumers learning and we keep budget balance without charging any addition fees.

Most storage management problems can be classified into two categories: storage management at the demand-side or at the supply-side. The demand-side storage management is related to how to coordinate end users each of whom owns a storage facility and makes overall decisions in terms of individual energy demand scheduling and charging/discharging of individual owned batteries; e.g., [11–14]. In contrast, in our model, the storage facility is not operated at the individual level but at the aggregate level: the coordinator who operates the storage facility finds it difficult to optimize storage decisions for the whole group as he has no information of the individual consumption constraints and preferences.

Prior work on the supply-side storage management has focused on how to use storage to stabilize the output of renewable energy supply when joining the conventional electricity markets (e.g., [15,16]), or to directly satisfy consumers demand (e.g., [17,18]). A common setting in these papers is that the energy demand is exogenous and independent of storage operations. In contrast, in our problem, the demand pattern of consumers is endogenous and interdependent with the storage policy, i.e., the demand pattern is consumers' optimal response to the virtual price, whose design depends on the coordinator's storage decisions. A succinct version of this work without the proofs and details appeared as an extended abstract in [19].

3   Problem Formulation

We consider a consumer collective [6] with N members with the planning period divided into M discrete time slots. Similar to [6], a central coordinator purchases electricity on behalf of the consumer collective from the market. In addition to [6], the collective also has a community owned renewable energy generation facility (e.g., solar panels, wind mills) which can only supply a part of the energy required by the collective, and has a community owned storage facility [15]. The storage capacity is also less than the amount of energy required by the collective. The coordinator is in charge of the storage and the generation facility. We assume that the forecast for the amount of generation for the planning horizon is accurate. Furthermore, the electricity prices from the market are also known. These assumptions are reasonable for a 24 h planning period with the electricity being bought from a day-ahead market.

Let $\mathbf{g}$ be the forecast vector for the amount of electricity to be generated over the M time slots with g_j the energy generated in time slot j. There is an upper bound on the amount of electricity that the agents can draw from the market (determined by the physical constraints of the distribution infrastructure) over the M time slots denoted by $\mathbf{h}$ with h_j the upper bound in time slot j. Let $\mathbf{p}$ be the market price vector which can vary over the M time slots. The component for the jth time slots of $\mathbf{h}$ and $\mathbf{p}$ are denoted by h_j and p_j respectively. The market price and the market supply capacity is common information for the central coordinator and all the agents in the group. Let $\mathbf{R}$ be an $N \times M$ matrix where each row of the matrix, $\mathbf{r}_i$ is the electricity demand of the agent i, $i \in \{1, 2, \ldots, N\}$. We call $\mathbf{r}_i$ the *demand profile* of agent i. Each entry r_{ij} is the electricity demand of agent i for time slot j. The total aggregated demand in time slot j is $\rho_j = \sum_{i=1}^{N} r_{ij}$.

The demand profile $\mathbf{r}_i$ of each agent i must satisfy their individual constraints. In general, it is assumed that the overall demand comes from two types of loads, shiftable loads and non-shiftable loads. For the shiftable loads, some can be interrupted after started and shifted to any other time slots; while other loads, once started, cannot be shifted before they are completed. In general loads may be dependent. We will assume that the constraints on the demands are given by a constraint set $\mathcal{X}_i$ which is private information of the agent i. **An agent does not share this constraint set $\mathcal{X}_i$, neither with other firms nor with the coordinator.** Unless otherwise specified we will assume $\mathcal{X}_i$ to be a convex polytope, which is a fairly general model for energy consumption constraints in this setting [6,12].

The central coordinator needs to make an energy storage decision. Specifically, in each time slot j it needs to decide whether and how much to charge or discharge. These (dis)charging decisions are represented by the (dis)charging amount at each time slot, denoted by $y_j \in [-d, d]$ for $j \in \{1, 2, \ldots, M\}$, where d denotes the maximum (dis)charging amount during each time slot, and if $y_j > 0$, at time slot j, the facility is charged with an amount of y_j; if $y_j \leq 0$, the facility is discharged by an amount of $-y_j$. The storage decision variable over the time horizon is the vector $\mathbf{y}$. Given the charging and discharging amount at each time slot, $\{y_j\}$, the storage level at the end of the time slot is $\sum_{k=1}^{j} y_k$. The storage level at the end of each time slot should be non-negative and also less than the capacity of the storage facility, denoted by $\mathbf{u}$. Therefore, the charging and discharging decision is also constrained by $\sum_{k=1}^{j} y_k \in [0, u], \forall j \in \{1, 2, \ldots, N\}$. The constraints for storage decisions is a system of linear inequalities: $0 \leq \mathbf{Ay} \leq \mathbf{u}$, $-\mathbf{d} \leq \mathbf{y} \leq \mathbf{d}$, where $\mathbf{A}$ is a lower triangular matrix with elements 1, and 0, $\mathbf{y} = [y_1, \ldots y_j, \ldots y_M]^T$, $\mathbf{u} = [u, \ldots u, \ldots u]^T$ and $\mathbf{d} = [d, \ldots d, \ldots d]^T$.

The amount of energy drawn from the market in time slot j is $\rho_j + y_j - g_j$. Thus the energy cost is $p_j \cdot (\rho_j + y_j - g_j)$. Since the objective is to minimize the sum of all agents costs, the central demand scheduling problem can be written as:

$$\min_{\mathbf{R}, \mathbf{y}} C^s(\mathbf{R}, \mathbf{y}) := \sum_{i=1}^{N} p_j \cdot (\rho_j + y_j - g_j)$$
$$\text{s.t. } \mathbf{r}_i \in \mathcal{X}_i, r_{ij} \geq 0$$
$$0 \leq \mathbf{Ay} \leq \mathbf{u}, -\mathbf{d} \leq \mathbf{y} \leq \mathbf{d} \tag{1}$$
$$0 \leq \rho + \mathbf{y} - \mathbf{g} \leq \mathbf{h}$$

The operating cost associated with the renewable energy is assumed to be constant, i.e., independent of the amount of energy produced, and is thus not a part of the objective function. Note that Problem (1) is defined on a convex set with a linear objective function, and is thus a solvable convex minimization problem.

4   Solution Approach

While Problem (1) is solvable, the central coordinator could not directly determine the optimal demand profile and storage solution because he has no information of consumers' consumption constraints $\mathcal{X}_i, \forall i \in \{1, 2, \ldots N\}$. We assume that the coordinator affects individual agents' energy consumption plans via *virtual price signals* and the individual agents *honestly* report their optimal demand based on the virtual price signals.

A simple virtual signal based on market price alone is ineffective to optimize coordination because in our problem the aggregate supply also depends on the coordinator's storage choices. The optimal storage solution and the optimal demand profile are coupled in the constraints and thus depend on each other. To address this issue, one approach is to decompose the problem into two subproblems: (1) **optimizing the storage solution** given the demand profile (**OSS**); (2) **optimizing the demand profile** given the storage solution (**ODP**). Correspondingly, we give two definitions:

Definition 1. $\sigma : \mathbf{R} \rightarrow \mathbf{y}$: *the function maps demand profile to optimal storage solution.*

Definition 2. $\delta : \mathbf{y} \rightarrow \mathbf{R}$: *the function maps storage policy to optimal demand profile.*

We design the coordination algorithm as (see Fig. 1):

1. Without storage, the central coordinator finds the demand of individual agents, $\mathbf{R}^*$, that minimizes the energy cost.
2. **OSS**: Given the demand profile $\mathbf{R}^*$, the central coordinator solves the optimal storage solution, $\mathbf{y}^* = \sigma(\mathbf{R}^*)$, to minimize the energy cost.
3. **ODP**: Given the storage solution $\mathbf{y}^*$, the central coordinator coordinates the demand of individual agents, $\mathbf{R}^\sharp = \delta(\mathbf{y}^*)$, to minimize the energy cost.

Stopping Criterion. If $\mathbf{R}^\sharp = \mathbf{R}^*$, stop. Otherwise, $\mathbf{R}^* = \mathbf{R}^\sharp$, and go back to step 2.

Note that step 1 is a special case of step 3 **ODP** with storage solution $\mathbf{y}^* = \mathbf{0}$. In step 2 **OSS**, only central coordinator makes decision on the storage solution. In step 3 **ODP**, the central coordinator determines virtual price signal and the agents calculate demand according to the virtual price signal.

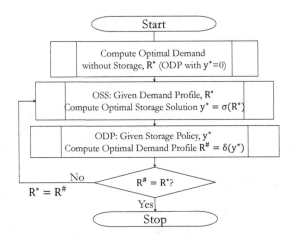

Fig. 1. Algorithm overview

The basic intuition of the overall coordination algorithm is that after either step 2 or step 3, the aggregate energy cost is reduced. For step 2, given the new demand profile, $\mathbf{R}^*$, the new storage solution $\mathbf{y}^* = \sigma(\mathbf{R}^*)$ should give a lower aggregate cost compared with the previous storage solution, because $\mathbf{y}^*$ is the storage solution minimizing the aggregate cost when the demand profile is $\mathbf{R}^*$. By similar argument, after step 3, the aggregate cost decreases too. Thus, the energy cost continues to decrease during the execution of the overall coordination algorithm. We will prove the basic algorithm keeps iterating and converges to the optimal solution to problem (1) under realistic conditions.

5   Overall Coordination Algorithm

In this section, we describe step 2 and 3 our algorithm works and prove its convergence.

5.1   Optimal Storage Solution (OSS)

Given the demand profile, $\mathbf{R}^*$, the central coordinator could obtain the optimal storage solution $\mathbf{y}^* = \sigma\left(\mathbf{R}^*\right)$ by solving the following problem 2. Problem 2 is a linear programming problem and thus solvable.

$$\min_{\mathbf{y}} C^s\left(\mathbf{R}^*, \mathbf{y}\right) = \sum_{i=1}^{N} p_j \cdot \left(\rho_j^* + y_j - g_j\right)$$
$$\text{s.t. } \mathbf{0} \leq \mathbf{Ay} \leq \mathbf{u}, -\mathbf{d} \leq \mathbf{y} \leq \mathbf{d} \tag{2}$$
$$\mathbf{0} \leq \rho^* + \mathbf{y} - \mathbf{g} \leq \mathbf{h}$$

5.2   Optimal Demand Profile (ODP)

In this section we discuss how to induce the optimal demand profile given a storage solution $\mathbf{y}^*$. The corresponding problem is expressed as follows:

$$\min_{\mathbf{R}} C^s\left(\mathbf{R}, \mathbf{y}^*\right) = \sum_{i=1}^{N} p_j \cdot \left(\rho_j + y_j - g_j\right)$$
$$\text{s.t. } \mathbf{r}_i \in \mathcal{X}_i, r_{ij} \geq 0 \tag{3}$$
$$\rho + \mathbf{y}^* - \mathbf{g} \leq \mathbf{h}$$

The iterative algorithm to solve Problem 3 is:

1. The central coordinator sends initial virtual price signal, denoted by s_{ij}, to the agents.
2. After receiving the virtual price signals, each agent individually calculates its optimal demand profile $\mathbf{r}_i$ by minimizing its energy cost computed based on the virtual prices and reports its profiles back to the coordinator.
3. Based on the reported demand profile $\mathbf{R}$, the central coordinator updates the virtual price signal and sends the new signal to each agent.
4. Given the new price signal, each agent chooses its new demand profile $\mathbf{r}_i'$.

Stopping Criterion. If $\mathbf{R}' = \mathbf{R}$, stop. Otherwise, set $\mathbf{R} = \mathbf{R}'$ and go back to step 3.

In order to minimize the total electricity cost, the virtual price signal has to induce consumers to shift their demands from time slots with high marginal cost to those with low marginal cost and keep the aggregate demands within the capacity limits. Therefore, first of all, we need to determine the marginal cost and the capacity limit of the electricity in each time slot. In our model there are three possible electricity sources: on-site generations, storage and the market. While the marginal cost of electricity from generations and the market is well defined, we need to find a method to calculate the marginal cost of electricity from storage.

First Charging First Discharging (FCFD). We use first charging first discharging (FCFD) method to *calculate the marginal unit cost* of the energy discharged from storage. The FCFD policy in our setting means that the electricity charged earliest is recorded as discharged first. Below we use an example to show how the FCFD policy works. Assume there are 5 time slots, and the prices **p** are $\{10, 20, 40, 10, 30\}$. The storage solution **y** is $\{3, 8, -10, 4, -5\}$, which implies the electricity is charged in time slot $1, 2, 4$ and discharged in time slot $3, 5$. Under the FCFD policy, the electricity charged earliest is recorded as discharged first. Therefore, the 10 units of electricity discharged in time slot 3 should consist two parts, i.e., 3 units charged in time slot 1 at price $p_1 = 10$ and 7 unit charged in time slot 2 at price $p_2 = 20$. The corresponding average price for the electricity discharged in time slot 3 should be $(10 \times 3 + 20 \times 7)/10 = 17$. Similarly, the 5 units of electricity discharged in time slot 5 should consist of two parts, i.e., 1 unit charged in time slot 2 at price $p_2 = 20$ and 4 units charged in time slot 4 at price $p_4 = 10$. The corresponding average price should be $(20 \times 1 + 10 \times 4)/5 = 12$.

Thus, we can calculate the average price of electricity discharged in each time slot. The unit cost of the energy discharged in each time slot should be the average cost of the corresponding energy still in the inventory and charged earliest. We use $p_j^{FCFD}(\mathbf{R}^*, \mathbf{y}^*)$ to denote the unit cost or the price of the discharging energy in time slot j with demand profile $\mathbf{R}^*$ and storage solution $\mathbf{y}^*$.

Cost Structure. Next we define the cost structure of the electricity in each time slot j, $\hat{p}_j(r)$, in an increasing order. By the following Proposition 1, in each time slot, the unit cost of electricity from storage, $p_j^{FCFD}(\mathbf{R}^*, \mathbf{y}^*)$, is smaller than the unit cost of electricity from market p_j.

Proposition 1. *If the storage solution, $\mathbf{y}^*$, is the optimal storage solution given demand profile, $\mathbf{R}^*$, then $p_j^{FCFD}(\mathbf{R}^*, \mathbf{y}^*) < p_j$.*

Thus, the cost structure of the energy under the demand profile $\mathbf{R}^*$ and the storage solution $\mathbf{y}^*$ should be

$$
\hat{p}_j(r) = \begin{cases} \infty & z_{3j} < r \leq z_{4j} \\ p_j & z_{2j} < r \leq z_{3j} \\ p_j^{FCFD}(\mathbf{R}^*, \mathbf{y}^*) & z_{1j} < r \leq z_{2j} \\ 0 & z_{0j} < r \leq z_{1j} \end{cases}
$$

where $z_{0j} = 0, z_{1j} = g_j, z_{2j} = \max\{g_j, g_j - y_j\}, z_{3j} = h_j + g_j - y_j, z_{4j} = \infty$.

Initial Virtual Price Signal. After we derive the formula of marginal cost, a straightforward approach is to set a quota for different sources in each time slot j for each agent i. Thus, the price signal is

$$
s_{ij}(r) = \begin{cases} \infty & z_{3ij} < r \leq z_{4ij} \\ p_j & z_{2ij} < r \leq z_{3ij} \\ p_j^{FCFD}(\mathbf{R}^*, \mathbf{y}^*) & z_{1ij} < r \leq z_{2ij} \\ 0 & z_{0ij} < r \leq z_{1ij} \end{cases}
$$

The initial quota of electricity from each resource in time slot j for agent i is proportional to agent i's demand in time slot j, r_{ij} (using aggregate demand r_i

for first iteration without previous $\mathbf{R}^*$). Therefore, the corresponding thresholds in the initial virtual price signal are:

$$z_{kij} = \frac{r_{ij}^*}{\sum_{i=1}^{N} r_{ij}^*} z_{kj}, k = 0, 1, 2, 3, 4$$

Individual Agent's Problem. After receiving the virtual price signal, each agent i computes $\mathbf{r}_i$ from

$$\min \sum_{j=1}^{M} s_{ij}(r_{ij}) r_{ij}, \quad \text{s.t. } \mathbf{r}_i \in \mathcal{X}_i, \quad r_{ij} \geq 0. \tag{4}$$

The above problem is a convex optimization problem and thus solvable. However, due to individual constraints, some agents might not use all the quota assigned to them, while others may need more quota to further reduce their energy costs. Thus we need to update the virtual price signals accordingly.

Update Rule of Virtual Price Signal. If the agent i does not fully use the quota assigned to her at time slot j, it may be due to either (a) constraints on electricity consumption or (b) lower price in other time slots. In either case, as long as the quotas on other time slots don't become smaller, which enforces agent i to shift the energy consumption from other time slots to time slot j, agent i prefers to keep the current consumption level to minimize the energy cost. Moreover, if the quotas on other time slots become larger, agent i may prefer to shift demand from time slot j to other time slot to reduce the energy cost. It is true if the prices on the time slots with larger quotas are lower than current marginal energy cost in time slot j. Therefore, to reduce the individual agent's energy cost and thus reduce the aggregate energy cost, the central coordinator needs to construct a new price signal by adjusting the quotas so as to share the excess quota among agents who use all of the quota in a time slot j. The new quotas for a demand profile $\mathbf{R}$ are determined by: let for $k = 1, 2, 3$, $\Phi_{kj} = \{i | r_{ij} < z_{kij}\}$,

$$z'_{kij} = \begin{cases} r_{ij} & i \in \Phi_{kj} \\ \dfrac{r_{ij}\left(z_{kj} - \sum_{i \in \Phi_j} r_{ij}\right)}{\sum_{i \in \Phi_j \cup \Psi_j} r_{ij}} & i \notin \Phi_{kj} \end{cases}$$

Under the new virtual price signal, for either electricity from generation or market supply, the agents give up their excess quota to the agents who use all of the quota with a sharing rule based on the current consumption level. Then the agents can continue optimizing their consumption schedule to reduce energy cost as they may get more quota in the time slots when they use all quota assigned to them previously.

5.3 Convergence Analysis

First, we prove the convergence of the ODP algorithm.

Theorem 1. *The ODP algorithm always converges.*

Theorem 1 is an immediate result of the two lemmas below:

Lemma 1. *The energy cost of individual agent based on the virtual price signal,* $\sum_{j=1}^{M} s_{ij}(r_{ij}) r_{ij}$, *is non-increasing during the execution of the iterative algorithm, so is the aggregate cost based on the virtual price signal,* $\sum_{i=1}^{N} \sum_{j=1}^{M} s_{ij}(r_{ij}) r_{ij}$.

Lemma 2. *If* $\forall i, j, r_{ij} > z_{2ij}$, *i.e., every agent use all quota of the electricity supply from energy generation and storage, then the aggregate cost based on the virtual price signal,* $\sum_{i=1}^{N} \sum_{j=1}^{M} s_{ij}(r_{ij}) r_{ij}$, *is equal to the aggregate cost based on the market price,* $C^s(\mathbf{R}, \mathbf{y}^*) = \sum_{i=1}^{N} p_j \cdot (\rho_j + y_j - g_j)$.

We now prove the convergence of the overall algorithm.

Theorem 2. *The overall algorithm always converges.*

Proof. From Definition 1, we have $C^s(\mathbf{R}, \sigma(\mathbf{R})) < C^s(\mathbf{R}, \mathbf{y})$, if $\mathbf{y} \neq \sigma(\mathbf{R})$. Also, from Theorem 1, we have $C^s(\mathbf{R}^\sharp, \mathbf{y}) < C^s(\mathbf{R}^*, \mathbf{y})$, if $\mathbf{R}^\sharp \neq \mathbf{R}^*$. Thus the energy cost $C^s(\mathbf{R}, \mathbf{y})$ is strictly decreasing as long as the optimal storage solution or optimal demand profile changes after each iteration. Thus, this algorithm converges.

Theorem 3. $\mathbf{R}^*$ *is optimal demand profile with storage, if* $\mathbf{y}^* = \sigma(\mathbf{R}^*)$ *and* $0 < \varrho^* + \mathbf{y}^* - g < h$.

Proof. Since $0 < \varrho^* + \mathbf{y}^* + g < h$, $0 \leq \varrho^* + y + g \leq h$ is not binding in the problem of optimal storage solution, which means $\mathbf{y}^*$ is also the solution to the following problem:

$$\min_{\mathbf{y}} C^s(\mathbf{R}^*, \mathbf{y}) - C^s(\mathbf{R}^*, \mathbf{0}) = \sum_{i=1}^{N} p_j \cdot y_j \tag{5}$$
$$\text{s.t. } 0 \leq AY \leq u, -d \leq y \leq d$$

Similarly, $\mathbf{R}^*$ is also the solution to the following problem:

$$\min_{\mathbf{R}} C^s(\mathbf{R}, \mathbf{y}^*) - C^s(\mathbf{0}, \mathbf{y}^*) = \sum_{i=1}^{N} p_j \cdot \rho_j \tag{6}$$
$$\text{s.t. } r_i \in \mathcal{X}_i, r_{ij} \geq 0$$

Assume C^* is the minimum cost of Problem (1) and $C^\#$ is the minimum cost of the following problem:

$$\min_{\mathbf{R}, \mathbf{y}} C^s(\mathbf{R}, \mathbf{y}) := \sum_{i=1}^{N} p_j \cdot (\rho_j + y_j - g_j)$$
$$\text{s.t. } r_i \in \mathcal{X}_i, r_{ij} \geq 0 \tag{7}$$
$$0 \leq AY \leq u, -d \leq y \leq d$$

Since problem (1) has more constraints than problem (7), we have $C^* \geq C^\#$. Moreover, by looking at problem (7), it actually can be decomposed to problem (6) and problem (5). Therefore, as we already show that $\mathbf{y}^*$ is also solution to problem (5) when $0 < \varrho^* + \mathbf{y}^* < h$, $(\mathbf{R}^*, \mathbf{y}^*)$ is the optimal solution to problem (7), i.e., $C^s(\mathbf{R}^*, \mathbf{y}^*) = C^\#$. As $C^* \geq C^\#$, we have $C^s(\mathbf{R}^*, \mathbf{y}^*) = C^\# \leq C^*$. Moreover, $(\mathbf{R}^*, \mathbf{y}^*)$ also satisfies all of the constraints in problem (1), which implies $C^s(\mathbf{R}^*, \mathbf{y}^*) \geq C^*$. Thus, $C^s(\mathbf{R}^*, \mathbf{y}^*) = C^*$, i.e., $(\mathbf{R}^*, \mathbf{y}^*)$ is the optimal solution to problem (1).

6   Simulation

In this section, we perform simulations to show the performance of our algorithm. We first state our assumptions on consumers' demand features, central coordinator's energy generation facility and storage facility, and the structure of market electricity price.

To reflect this diversity of the consumers' demand within the simulations, we used the Irish Commission for Energy Regulation (CER) electricity consumption data set to identify two important classes of consumers with shared characteristics. Here, Class 1 represents consumers that consume most of their electricity during the day and have a low load at night, whereas Class 2 represents consumers that have a stable consumption during the day, but have a higher consumption at night. We assume there are N agents, among them half are of Class 1 and the other half are of Class 2. Moreover, each agent is defined by its total electricity requirement over the whole planning horizon, r_i, and its constraints on electricity consumption over the planning horizon. We assume that the constraints are given by upper and lower bounds at each time slot: $r_{ij} \in [\underline{r_{ij}}, \overline{r_{ij}}]$. The parameters are determined based on a group of samples, $\mathbf{X}$, drawn from the distributions for the demand profiles. Specifically, $r_i = \sum_{j=1}^{24} x_{ij}$ and $\underline{r_{ij}} = 0.8x_{ij}, \overline{r_{ij}} = 1.2x_{ij}$.

For the central coordinator, we assume the capacity of the energy generation facility is fixed and the amount of energy generated in each time slot is between 0 and the capacity. Similarly, we also assume the capacity of the storage facility is u and the maximum amount of energy charged or discharged in each time slot is d.

For the market price, we use the real market prices from the EEX_dat_set[1] which is the average hourly day-ahead spot market prices gathered from the European Energy Exchange as the price input, p. These prices are fixed across all the simulations. The capacity of the electricity contract on the other hand are not fixed. The capacity of market supply is determined by $h_j = 1.2 \sum_{i=1}^{N} x_{ij}$. The simulations were considered to converge when the cost reduction in one iteration got less than 0.001 %.

Table 1. Average performance of coordination algorithm

Number of agents	Number of rounds	Convergence accuracy (%)
20	11.30	99.74
40	11.42	99.67
60	12.18	99.63
80	12.58	99.69
100	15.02	99.62

---

[1] The data is available at http://www.eex.com/en/Market%20Data.

Table 1 shows how the number of iterations and the convergence accuracy changes with the number of agents. The convergence accuracy is defined as the ratio of cost reduction achieved using our algorithm to the maximum cost reduction with all private demand profiles made available to the coordinator. The results indicate the scalability of our algorithm with respect to the number of agents.

7   Conclusion and Future Work

In this paper we proposed a novel multiagent coordination algorithm to shift the energy consumption of a consumer collective in the presence of energy generation and storage. In the collective, a central coordinator buys the electricity and decides the storage level for the whole group, and consumers make their own consumption decisions based on their private consumption constraints and preferences. To coordinate individual consumers under incomplete information and optimize the storage decision, we decompose the problem to two sub-problems: (a) optimizing demand profile of consumers given storage policy and (b) optimizing storage solution given demand profile of consumers. We proposed an iterative algorithm in which the two sub-problems above are solved alternately, and proved the convergence of our algorithm.

One possible direction of future work is to consider scenarios where storage and generation is under the control of the individual agents. In such cases does an approximation guarantee on the individual problem result in an approximation guarantee of the overall problem?

References

1. Jorgensen, J., Sorensen, S., Behnke, K., Eriksen, P.: Ecogrid eua prototype for european smart grids. In: 2011 IEEE Power and Energy Society General Meeting, pp. 1–7. IEEE (2011)
2. Chen, S., Weiss, G.: An efficient and adaptive approach to negotiation in complex environments. In: ECAI, pp. 228–233 (2012)
3. Zheng, R., Chakraborty, N., Dai, T., Sycara, K., Lewis, M.: Automated bilateral multi-issue negotiation with no information about opponent. In: Proceedings of Hawaii International Conference on Systems Sciences, January 2013
4. Klein, M., Faratin, P., Sayama, H., Bar-Yam, Y.: Negotiating complex contracts. Group Decis. Negot. **12**, 111–125 (2003)
5. Lai, G., Sycara, K.: A generic framework for automated multi-attribute negotiation. Group Decis. Negot. **18**, 169–187 (2009)
6. Veit, A., Xu, Y., Zheng, R., Chakraborty, N., Sycara, K.P.: Multiagent coordination for energy consumption scheduling in consumer cooperatives. In: AAAI (2013)
7. Veit, A., Xu, Y., Zheng, R., Chakraborty, N., Sycara, K.: Demand side energy management via multiagent coordination in consumer cooperatives. J. Artif. Intell. Res. **50**, 885–922 (2014)
8. Medina, J., Muller, N., Roytelman, I.: Demand response and distribution grid operations: opportunities and challenges. IEEE Trans. Smart Grid **1**(2), 193–198 (2010)

9. Ramchurn, S., Vytelingum, P., Rogers, A., Jennings, N.: Putting the 'smarts' into the smart grid: a grand challenge for artificial intelligence. Commun. ACM **55**(4), 86–97 (2012)
10. Ramchurn, S.D., Vytelingum, P., Rogers, A., Jennings, N.: Agent-based control for decentralised demand side management in the smart grid. In: The 10th International Conference on Autonomous Agents and Multiagent Systems, vol. 1, pp. 5–12 (2011)
11. Vytelingum, P., Voice, T., Ramchurn, S., Rogers, A., Jennings, N.: Theoretical and practical foundations of large-scale agent-based micro-storage in the smart grid. J. Artif. Intell. Res. **42**(1), 765–813 (2011)
12. Wu, C., Mohsenian-Rad, H., Huang, J.: Wind power integration via aggregator-consumer coordination: a game theoretic approach. In: 2012 IEEE PES Innovative Smart Grid Technologies (ISGT), pp. 1–6. IEEE (2012)
13. Atzeni, I., Ordóñez, L.G., Scutari, G., Palomar, D.P., Fonollosa, J.R.: Demand-side management via distributed energy generation and storage optimization. IEEE Trans. Smart Grid **4**(2), 866–876 (2013)
14. Atzeni, I., Ordóñez, L.G., Scutari, G., Palomar, D.P., Fonollosa, J.R.: Noncooperative and cooperative optimization of distributed energy generation and storage in the demand-side of the smart grid. IEEE Trans. Signal Process. **61**(10), 2454–2472 (2013)
15. Kim, J.H., Powell, W.B.: Optimal energy commitments with storage and intermittent supply. Oper. Res. **59**(6), 1347–1360 (2011)
16. Brunetto, C., Tina, G.: Optimal hydrogen storage sizing for wind power plants in day ahead electricity market. IET Renew. Power Gener. **1**(4), 220–226 (2007)
17. Su, H.I., El Gamal, A.: Modeling and analysis of the role of fast-response energy storage in the smart grid. In: 2011 49th Annual Allerton Conference on Communication, Control, and Computing (Allerton), pp. 719–726. IEEE (2011)
18. Harsha, P., Dahleh, M.: Optimal management and sizing of energy storage under dynamic pricing for the efficient integration of renewable energy (2012)
19. Zheng, R., Xu, Y., Chakraborty, N., Sycara, K.: Multiagent coordination for demand management with energy generation and storage. In: Proceedings of the 2014 International Conference on Autonomous Agents and Multi-agent Systems, International Foundation for Autonomous Agents and Multiagent Systems, pp. 1587–1588 (2014)

Market Mechanisms and Their Users

Warsaw School of Economics

Back-End Bidding for Front-End Negotiation: A Model

Réal A. Carbonneau and Rustam Vahidov[✉]

InterNeg Research Centre, Concordia University, Montreal, Canada
rustam.vahidov@jmsb.concordia.ca

Abstract. Negotiations allow parties to exchange offers in search for mutually agreeable solutions. The exchange process is usually flexible and ill-structured and it may involve a set of multiple issues, which may change in the course of negotiation. Auctions, on the other hand feature strict rules regarding bid submission and evaluation. Most of the existing auctions allow for single attribute bids. This paper proposes an approach by which a software agent solution could emulate a multi-attribute negotiation front-end while bidding in single-attribute auction marketplaces. The bidding model is based upon concession-making curve introduced in prior work on electronic negotiations. Using data collected from eBay the paper shows that bidding across several attributes would result in higher utility outcome, and faster results than bidding within a single attribute set.

Keywords: Electronic negotiations · Software agents · Auctions · eBay · Concession-making

1 Introduction

Negotiations allow parties to exchange offers in search for mutually agreeable solutions. Auctions feature well defined set of rules designed to automate the process of winning bid determination. Because of this well-structuredness they are amenable to automation, and, thus are well-represented in practice. Negotiations allow parties more flexibility in exchanging offers and making an agreement. Integrative negotiations are characterized by joint solution process aimed at incorporating the interests of all negotiating parties [1]. This is possible due to the involvement of multiple issues in a given negotiation instance. In fact, issues could be added and removed in the process of negotiation to facilitate the search for mutually acceptable agreements.

However, the flexibility of negotiations also implies that they are less amenable to automation as compared to auctions, and they require more effort on the negotiator's behalf. In single-sided single-attribute auctions, for example, a number of bidders have to decide on their reservation values and actively bid on the market (or delegate bidding to software), while the other party is not involved after the auction started, since the auction mechanism takes care of managing the process. In negotiations, on the other hand, both sides need to be involved, make offers on multiple issues, and consider adding issues in the process.

This is why electronic auction sites, such as eBay enjoy much better represented than e-negotiation sites. The latter often are limited to naming one's price on the

© Springer International Publishing Switzerland 2015
B. Kamiński et al. (Eds.): GDN 2015, LNBIP 218, pp. 385–393, 2015.
DOI: 10.1007/978-3-319-19515-5_30

available product, service, or bundle with little attempt to take full advantage of negotiation mechanism capabilities, such as multi-issue offers or preference modeling. Research on electronic negotiation systems, on the other hand, has been extensive [2], focusing on design and assessment of means of communication, presentation and analytical support for negotiators.

The aim of the current work is to propose an approach that borrows ideas from the area of e-negotiations and negotiation software agents while relying on widely available auction models. In the presence of multiple auctions for similar products, they can be seen as potential offers in the market. Furthermore, since these products may have varying attribute values (e.g., size, color, capacity, etc.), the process of deciding how to bid on these auctions can be viewed as multi-issue negotiations. In this sense the approach allows for the development of a solution concept that imitates negotiation front-end with auctions back-end.

Currently, the largest auction marketplace at the moment is eBay and it and many others are single-attribute marketplace. However, these markets often fill the need for multi-attribute auctions simply by the large quantity and diversity of auctions occurring. A single set of attributes can even be formally defined in these auctions and it may also be informally specified in the auction title. This fact opens the door to develop a mechanism for emulating multi-issue negotiation strategy for a set of single attribute auctions.

2 Background

Proliferation of computing and networking technologies had thus quite naturally led to the appearance and subsequent popularity of electronic auctions, the best known example of which is eBay. Bidders in different locations and time zones can participate in the same online auctions. Prompted by these developments, research on online auctions has been extensive. Since online auctions typically allow for some extended period to collect bids, it has been noted that late bidding, or "sniping" would be expected to be widespread and eBay would effectively turn into sealed-bid second-price auction [3, 4]. At eBay the auction closing time is set, which encourages the bidders to delay their. The evidence for the late bidding behavior was found in experimental lab settings [5].

In [6] data from a major Chinese auction site Taobao was analyzed for the analysis of bidding strategies. The results suggested that the bidders used different strategies, including early bidding, late-bidding, and "agent-supported ratchet" strategies. The need for mundane tasks of monitoring auction developments and timely bidding has led to work on designing bidding software agents. In [7] the authors have proposed a method for buying on eBay. When there are multiple auctions for similar products, an agent can choose which one to bid on. Auction clustering has also been proposed in [8]. Here the authors use clustering based on such attributes as initial price, average bid rate, and others for prediction of final price. Four strategies for bidding are proposed based on decision functions defined in [9].

An algorithm for bidding on multiple sealed-bid auctions featuring similar products has been presented in [10]. For fully or partially matching attributes, an agent sets the desired maximum (reservation) value, which could be expressed as a time-dependent function and selects the ones with relevant times and expected prices. Based on the user's risk attitude one could choose the minimal probability of winning or to maximize the expected utility. However, the assumption is made that the probability distribution functions of winning for each auction is known.

In [11] a concept for an agent that can monitor multiple auctions, and bid on them has been proposed that uses decision functions described in [9] and tactics, such as remaining time tactic, remaining auctions tactic, and others. A mix of tactics is used to calculate maximum current bid using the agent's strategy. An agent then chooses target auction that maximizes expected utility, and bids using maximum bid as a reservation value. The authors used simulations and Genetic Algorithms to evolve strategies. Another simulation study has been reported in [12] The performance of intelligent agents was compared with that of "human" bidders, which were simulated using different risk profiles. Multiple English auctions were simulated featuring similar units. Agents with greedy, sniping and heuristic behaviors were involved. Overall, it was reported that agents outperformed "standard" bidders.

The decision functions mentioned above have been introduced to devise negotiation agent tactics. Families of tactics that could be flexibly defined for software agents have been introduced [9]. According to the authors, the tactics are used to decide on what offer to make at a given point in the negotiation process. These tactics were divided into three categories: behavior-dependent, time-dependent, and resource-dependent. The first family bases its choice of offer on the moves made by the parties, e.g. tit-for-tat tactics. Resource-dependent tactics aimed at adjusting concession levels based on a given resource scarcity. Time-dependent tactics dictate concession-making as a function of time between the beginning of negotiation and the estimated ending. Functions that dictated small concessions in the beginning (negative second derivative over time) corresponded to tougher competitive behavior, and were named (perhaps, somewhat controversially) boulware tactics. Those that implied early large concessions were named conceder tactics. Time-dependent tactics can be visualized as curves showing concession-making pattern through time. In [9] the curve is used to show the acceptable level for an issue (e.g.) at a given point in time. In [13] a model for concession-making has been introduced, which features a symmetrical curve. The curve's shape can be set by defining the value for the center of a curve relative to its starting and ending points. This center-point represents the level of competitiveness of the tactic. The present work is employing this model to represent concession-making by a bidding agent.

3 Model for Bidding

There are several reasons to justify expanding the bidders search from a single attribute set to a multi-attribute set. The most trivial one being that there will be more auctions to bid on, thus more opportunities to find a good deal. More formally, since the final prices for auctions has previously been shown to follow a normal distribution using data from eBay and Yahoo auctions {Dumas, 2002 #30}, and the same picture has been observed in our data Błąd! Nie można odnaleźć źródła odwołania., expanding to multiple-attributes will expand the set of auctions that can be won with lower bids than the average.

In order to quantitatively compare auction listings with different attribute sets and to decide on bidding prices, it is necessary that the buyer defines her utility function (preferences). It is also necessary to specify a time frame for searching and bidding on auctions. We assume that the user has defined a quantitative utility function with which we can precisely calculate the utility of a given set of attributes (including price) of an auction. This utility is assumed to be static and assumed to not change over time in the case of this experiment.

Once these are defined, one could adjust utility cut-off for bidding over time until an auction is won. To introduce the bidding model, we start with utility (u) and current time (t) which is a percentage starting at the beginning of the search period ($t = 0$) progressing through towards the stopping time ($s = 100$). We use a simple measure, the concession curve center utility (w) [13] to specify how aggressive or collaborative we want to be with our bids over time. The following Eq. (1) permits the calculation of the c parameter for the utility concession curve formula (2). As one approaches the deadline, the curve will move towards the minimum utility the buyer is willing to accept. At any given point in time, the minimum utility accepted can be calculated using formula (2). The c parameter can be computed (1) once at the beginning as soon as the target utility concession curve center is known, whereas the threshold utility (2) needs to be recomputed anytime that the time (t) changes, thus for each auction listing evaluated.

$$c = \frac{sw^2}{200(w - 50)} \tag{1}$$

$$u = \frac{100c(t - s)}{s(t - c)} \qquad c > 0 \lor c > s \tag{2}$$

Thus, at a given time (t) for a given negotiation style (defined by concession curve center w), we can calculate the minimum utility (u) the buyer is willing to accept. With this minimum utility (u), a maximum bid for a set of attributes specified in a single-attribute auction can be calculated. As demonstrated with a real example in **Błąd! Nie można odnaleźć źródła odwołania.**, auction listings that feature better current utility than the utility concession curve threshold at that time will be considered for bidding up to a price that is at the minimum utility threshold. An auction that ends with a utility above the threshold could be won by the bidding model by sniping.

Ebay has two types of listings, "Buy It Now" and "Auction", our simulation includes both. Based on real Ebay data, we know that we can win the "Buy It Now" by offering the requested price before the item is sold. Thus in respects to "Buy It Now", our simulation timing is exact and representative of reality. For auctions, we propose only sniping in the very last seconds of the auctions. Since we know the exact ending time of the auction and the final wining price, it is quite probable that sniping at a higher price in the last seconds will win the auction. There is a positive relationship between the number of auction listings and the optimal utility concession curve center which maximizes expected utility. As buyers have more auctions listings to evaluate and bid on, their utility concession curve center (w) should be higher; in other words, they should be more competitive.

Considering the volume of listings in modern online auction markets, an autonomous agent is required to effectively implement multi-attribute bidding in a single attribute market. Although it would be possible for a user to use our proposed multi-attribute bidding model as a decision support tool, it would be too demanding and would probably not be worth the effort for most common uses other than high value and critical type purchases Fig. 1.

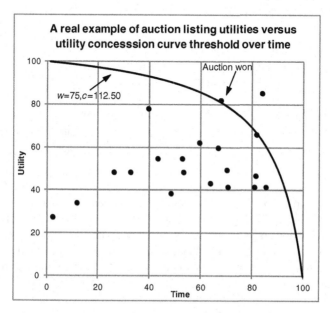

Fig. 1. A real example of auction listings utilities versus the utility concession curve threshold over time.

4 Findings

We used the data from eBay marketplace, which is essentially a Vickrey auction with two exceptions: the winning bid is the second-highest plus one bid increment (usually relatively small) and the highest bidder's bid is sealed but the current winning one is displayed to allow price discovery (second highest bid plus one bid increment). The scenario involves a purchase of an intelligent phone, specifically the Apple iPhone. Each auction listing has a set of attributes that are directly related to the utility that our simulated buyer will derive from the purchase. The iPhone Price, Model, Memory, Carrier and Color are the attributes included. The options for each attribute are as follows, although not all combinations are feasible (The data was collected from publicly posted auction listings on eBay from March 13th 2014 to April 21nd 2014. To remove the outliers, the data was cleaned by iteratively removing all listings with a price that deviated by more than 3 standard deviations from the average of the set of attributes. The final dataset contains 8157 listings. For the purpose of simulations, we assume that the auction could be won at the price of the winning bid. Although this is not a perfect representation of reality since our bid would increase competition in the marketplace and thus potentially increase the price, this impact would be minimal because the bid would be placed at the very last second (sniping); thus it would often be the winning bid by an insignificant increment and there would not be enough time to start a bidding war **Table** 1).

Table 1. List of options and attributes (not all combinations are possible)

Model	Memory	Carrier	Color
iPhone 4	16 GB	AT&T	Black
iPhone 4s	32 GB	Sprint	Blue
iPhone 5	64 GB	T-Mobile	Gold
iPhone 5c	8 GB	Unlocked	Green
iPhone 5s		Verizon	Pink
			White
			Yellow

The data was collected from publicly posted auction listings on eBay from March 13th 2014 to April 21nd 2014. To remove the outliers, the data was cleaned by iteratively removing all listings with a price that deviated by more than 3 standard deviations from the average of the set of attributes. The final dataset contains 8157 listings. For the purpose of simulations, we assume that the auction could be won at the price of the winning bid. Although this is not a perfect representation of reality since our bid would increase competition in the marketplace and thus potentially increase the price, this impact would be minimal because the bid would be placed at the very last second (sniping); thus it would often be the winning bid by an insignificant increment and there would not be enough time to start a bidding war.

Normally, a buyer would build her utility function in respect to her own needs and desires. However, for current research, we use the average market price to determine the utility (3) of each attribute set. Essentially, instead of just making up any utility function, or making random utility functions, we built utility function that represents the average that the market has determined. Thus the average price for a set of attributes is considered to have a utility of 50, a price of 3 standard deviations lower than the average is considered to have a utility of 100 (a great deal), while a price of 3 standard deviations above the average is considered to have a utility of 0 (a bad deal). Thus the utility (u) of a set of combinations is based on the market price (μ), its standard deviation (σ) and the auction price (p).

$$u = 50 + \frac{50(\mu - p)}{3\sigma} \tag{3}$$

For the purpose of the simulations in this article, we always consider only 1 day of auctions listings and repeat the simulation for each day of complete data. In many cases there are not many listings on a specific day and a specific set of attributes, however, to be conservative in our results, any day for which no auction was won, either because there were no listings or because no listings were over the utility threshold at the end time of the auction, we ignore this case instead of recording an entry for zero utility. We also set the utility concession curve center w to 75 to provide competitive concession profile.

First we have compared bidding on all the auctions within a single set of attributes vs. bidding on only one auction, as we expected the former case to lead to better outcome. The findings suggest that considering all of the auctions in an attribute set had an average utility of 68.86 and choosing a random listing has an average utility of 50.34, thus representing an improvement of 36.78 % in expected utility and demonstrates that using the utility concession curve bidding model to consider all of the auctions for a single set attributes is beneficial to the buyer.

Next, we have compared winning utility using the utility concession curve to evaluate the listings within a single set of attributes versus expanding across all sets of attributes. We have found an average utility of 68.86 for the bidding within a set of attributes vs. an average utility of 93.86 for bidding across all of the sets of attributes. This represents an improvement of 36.31 % in expected utility and demonstrates that using the utility concession curve bidding model to consider all of the auctions across all set of attributes is beneficial to the buyer.

Thirdly, we have compared the time necessary to reach an acceptable utility outcome using bidding across multiple sets of attributes vs. bidding within a single set. We have found the average time of 84.30 for the bidding within a set of attributes and an average time of 48.25 for bidding across all of the sets of attributes. Thus, when considering only one set of attributes at one time, it takes 74.73 % longer to get a winning bid, which demonstrates that using the utility concession curve bidding model to consider all of the auctions across all set of attributes is beneficial to the buyer from the efficiency perspective.

Since bidding across multiple sets of attributes implies higher number of auctions than bidding on a single set, we next considered limiting number of auctions in the former case to that in the single-attribute case. We randomly chose the same number of listings across all attribute sets as those found in the current single attribute set. The findings show an average utility of 68.86 for the bidding within a set of attributes and an average utility of 72.75 for bidding across all of the sets of attributes while limiting the number of auction listings considered. This represents an improvement of 5.66 % in expected utility, and demonstrates that even when limiting the number of auctions listings, using the utility concession curve bidding model to consider all of the auctions across all set of attributes is beneficial to the buyer because of the increase in price variances.

Finally, just like in the previous case we have limited the number of auctions in the multiple sets of attributes to the same as the single set of attributes to see if the acceptable bid will arrive earlier. We obtained an average time of 84.30 for the bidding within a set of attributes and an average time of 81.29 for bidding across all of the sets of attributes while limiting the number of auction listings considered. Thus, when considering only one set of attributes at one time, it takes 3.70 % longer to get a winning bid, and demonstrates even when limiting the number of auctions listings, using the utility concession curve bidding model to consider all of the auctions across all set of attributes is beneficial to the buyer because of the increase in price variances.

5 Conclusions

The aim of this paper was to introduce a model for facilitating back-end bidding by software agents while emulating front-end negotiation-like setting. It was motivated by the possibility of taking advantage of the existence multiple auction listings for similar products with different attribute sets. The ability to monitor and bid using sniping technique on multiple auctions essentially turns the problem of single-attribute bidding into a multi-attribute space. This makes possible to perform multi-attribute bidding in single-attribute auctions marketplace. The paper employs concession-making model from electronic negotiations research to guide the bidding process. The results tentatively suggest the advantage of expanding to a multi-attribute bidding model in a single-attribute market place since it greatly expands the number of auction listings that can be considered. Combining this with the proposed utility concession curve based bidding model, a buyer can easily take advantage of the large number of auction listings provided under a multi-attribute search model to win an auction with a high utility for the buyer. Even when we control for the number of auction listings, it is still beneficial to expand to multiple attribute sets because of the increase in price variance. Future research could be directed to implementation of an agent-based solution that would help the bidders to manage their bidding processes, and the subsequent experimental assessment of such solution.

References

1. Kersten, G.E., Noronha, S.J., Teich, J.: Are all e-commerce negotiations auctions? In: Designing Cooperative Systems: Proceedings of the 5th International Conference on the Design of Cooperative Systems (COOP 2000). IOS Press (2000)
2. Kersten, G.E., Lai, H.: Negotiation support and E-negotiation systems: an overview. Group Decis. Negot. 16(6), 553–586 (2007)
3. Bapna, R.: When snipers become predators: can mechanism design save online auctions? Commun. ACM 46(12), 152–158 (2003)
4. Roth, A.E., Ockenfels, A.: Last minute bidding and the rules for ending second-price auctions: evidence from eBay and Amazon auctions on the internet. Am. Econ. Rev. 92, 1093–1103 (2002)
5. Ariely, D., Ockenfels, A., Roth, A.E.: An experimental analysis of late-bidding in internet auctions. RAND J. Econ. 36, 890–907 (2005)
6. Cui, X., Lai, V.S.: Bidding strategies in online single-unit auctions: their impact and satisfaction. Inf. Manag. 50(6), 314–321 (2013)
7. Kehagias, D.D., Mitkas, P.A.: Efficient e-commerce agent design based on clustering eBay data. In: Proceedings of the 2007 IEEE/WIC/ACM International Conferences on Web Intelligence and Intelligent Agent Technology. IEEE Computer Society (2007)
8. Kaur, P., Goyal, M., Lu, J.: A proficient and dynamic bidding agent for online auctions. In: Cao, L., Zeng, Y., Symeonidis, A.L., Gorodetsky, V.I., Yu, P.S., Singh, M.P. (eds.) ADMI. LNCS, vol. 7607, pp. 178–190. Springer, Heidelberg (2013)

9. Faratin, P., Sierra, C., Jennings, N.R.: Negotiation decision functions for autonomous agents. Robot. Auton. Syst. **24**(3–4), 159–182 (1998)
10. Shehory, O.: Optimal bidding in multiple concurrent auctions. Int. J. Coop. Inf. Syst. **11**(3/4), 315–327 (2002)
11. Anthony, P., Jennings, N.R.: Developing a bidding agent for multiple heterogeneous auctions. ACM Trans. Internet Technol. **3**(3), 185–217 (2003)
12. Sow, J., Anthony, P., Ho, C.M.: Analyzing the performance of multiple agents with varying bidding behaviors and standard bidders in online auctions. Web Intell. Agent Syst. **11**(2), 185–199 (2013)
13. Carbonneau, R.A., Vahidov, R.M.: A utility concession curve data fitting model for quantitative analysis of negotiation styles. Expert Syst. Apps **41**(9), 4035–4042 (2014)

Lot-Rolling – Supply Chain Negotiation in a Two-Stage Multi-echelon System

Michael Filzmoser[(✉)]

Institute of Management Science, Vienna University of Technology,
Vienna, Austria
michael.filzmoser@tuwien.ac.at

Abstract. Interdependencies between procurement and production processes between buyers and sellers concerning order and production lots require coordination to minimize the costs in a supply chain. This paper compares distributed and central decision-making in lot determination to different negotiation mechanisms – with the aim to overcome shortcomings of the two former approaches – in a two-stage multi-echelon supply chain.

Keywords: Distributed decision making · Central decision making · Negotiation · Supply chain management · Simulation

1 Introduction

Interdependences exist between the procurement and production processes of buyers and sellers in a supply chain. The decisions when to order and produce which amount of a product – i.e. the determination of order and production lots – cause costs, such as set-up or order costs as well as inventory costs, at both sides. To achieve optimal results, i.e. minimum costs in the supply chain, coordination of order and production lots is necessary.

Distributed decision-making – i.e. the buyer and the seller determining order and production lots autonomously – can lead to results optimal for the partial systems but suboptimal for the entire supply chain system. On the other hand, central decision-making – i.e. a central instance collecting all demand and cost information and determining order and production lots based on full information – can be expensive and time consuming. Moreover, the benefits of central coordination could be distributed unevenly which can cause resistance to its implementation.

This paper suggest negotiation of order and production lots between the buyer and the seller in a supply chain as a viable alternative to lot determination by means of distributed or central decision making. The results of distributed decision-making can act as the best alternative to a negotiated agreement (BATNA) and the results of central decision-making as a performance benchmark for the negotiation process and outcome.

Computational experimentation and simulation of the three approaches in a variety of problem situations – determined by buyer demands and buyer and seller order, set-up and inventory costs – enables the evaluation of the performance of negotiation in supply chain lot determination in comparison to distributed and central decision-making and to analyze different (concession- or veto-based) negotiation procedures.

© Springer International Publishing Switzerland 2015
B. Kamiński et al. (Eds.): GDN 2015, LNBIP 218, pp. 395–401, 2015.
DOI: 10.1007/978-3-319-19515-5_31

The remainder of this paper is structured as follows: Sect. 2 introduces the specific supply chain system considered in this paper – a simple two-stage multi-echelon supply chain – and describes the distributed decision making, central decision-making and the negotiation approach to this problem. Section 3 provides a numerical example and Sect. 4 provides an outlook of the simulation planned to evaluate the three approaches. Section 5 concludes the paper with preliminary findings.

2 Lot Determination in a Multi-echelon Two-Stage Supply Chain

We use and extend the multi-echelon model, which was introduced by Wagner and within [1] together with an approach to determine the optimal dynamic lot size.[1] In this model a buyer faces – due to orders of customers, existing stocks, capacities, etc. – varying demands d_i over the T periods of a finite planning horizon. Buying (or production) costs are assumed to be constant during the planning horizon so that the problem reduces to a minimization of order and inventory costs. Placing an order causes fixed costs f_b for administration of the order, transportation etc. Order costs can be avoided by aggregating current and future demands, however, then future demands have to be stored which causes inventory costs v_b.

In a two-level supply chain context the identical situation is observed by the supplier, which also faces demands – i.e. the orders o_i of the buyer within the planning horizon – which she can fulfill by producing each order the time it is placed, causing set-up costs f_s only, or production for stock to be able to fulfill later orders which again involves inventory costs v_s.

2.1 Distributed Decision Making

In lot determination by distributed decision-making the buyer has to determine when to order the demanded quantities and thereby trades-off order costs and inventory costs. The buyer orders the demand of the next period if storing the products for one period is cheaper then placing a new order next period. In determining lots we do not have to consider all possible combinations of demands and periods. Clearly a demand of a period has to be ordered in the focal period at the latest. Furthermore, according to Theorem 1 in [1] it will never be optimal to shift (and store) a demand of the next period to the previous period, if the previous periods demand is not ordered in that period, otherwise the additional inventory costs could be saved by ordering in the next period with the exactly the same order costs. Also it never makes sense to split a demand over several periods according to Theorem 2 in [1]. Table 1 presents the possible order lots and the resulting costs for a planning period with four echelons – e.g. quarters of a year.

---

[1] The multi-echelon model with finite planning horizon is a classic inventory optimization problem, although stochastic systems are more appropriate in many cases, it still has relevance in domains that show complex non-stationary structures [2].

Table 1. Possible order lots of the buyer and the resulting costs.

No.	o_1	o_2	o_3	o_4	Costs buyer
1	d_1	d_2	d_3	d_4	$4*f_b$
2	d_1	d_2	d_3,d_4	–	$3*f_b + d_4*v_b$
3	d_1	d_2,d_3	–	d_4	$3*f_b + d_3*v_b$
4	d_1	d_2,d_3,d_4	–	–	$2*f_b + d_3*v_b + 2*d_4*v_b$
5	d_1,d_2	–	d_3	d_4	$3*f_b + d_2*v_b$
6	d_1,d_2	–	d_3,d_4	–	$2*f_b + d_2*v_b + d_4*v_b$
7	d_1,d_2,d_3	–	–	d_4	$2*f_b + d_2*v_b + 2*d_3*v_b$
8	d_1,d_2,d_3,d_4	–	–	–	$1*f_b + d_2*v_b + 2*d_3*v_b + 3*d_4*v_b$

In Table 1 d_1,d_2 in column o_1 indicates that the demand of periods 1 and 2 are covered in the order of period 1 – i.e. for these two periods order costs occur only once but the demand for period two has to be stored which leads to inventory costs of d_2*v_b –, a '–'means that no order is place in this period so not fixed costs occur.

Based on the orders in the different periods of the buyer the supplier has then to decide production lots to fulfill the orders in choosing the lowest cost variant among the remaining alternatives as depicted in Table 2.

2.2   Central Decision Making

As can be seen from Table 2 in distributed lot size determination the decision alternatives for the supplier for most of the order constellations of the buyer are restricted – except for the case that an order is placed each period. It is, therefore, likely that the lowest cost alternatives according to the buyer's and the supplier's distributed decisions are not optimal for the entire two-stage supply chain.

Differences in inventory and setup-up or ordering costs between the buyer and the supplier generate integrative potential. A central decision maker with full information about demands d_i, as well as fixed costs – i.e. the buyer's order costs f_c and the supplier's set-up costs f_s – and their inventory costs v_c and v_s, could exploit this integrative potential by determining the overall minimum cost order and production lots.

2.3   Negotiation

However, such a central decision might not always be realizable. The buyer and the supplier might be unwilling to disclose their cost structures. Furthermore, incentives to report wrong fixed and variable costs exist, as thereby the central decision can be influenced in the own favor. Moreover, central coordination could involve additional costs for information exchange or be time consuming. We therefore suggest (concession or veto-based) negotiation between the participants of the supply chain as a viable alternative.

A further argument for the use of negotiations is the possibility to transfer utility – measured in terms of order or set-up costs and inventory costs – by means of side

Table 2. Possible production lots of the supplier and the resulting costs.

No.	p_1	p_2	p_3	p_4	Costs supplier
1	o_1	o_2	o_3	o_4	$4*f_s$
	o_1	o_2	o_3,o_4	–	$3*f_s + o_4*v_s$
	o_1	o_2,o_3	–	o_4	$3*f_s + o_3*v_s$
	o_1	o_2,o_3,o_4	–	–	$2*f_s + o_3*v_s + 2*o_4*v_s$
	o_1,o_2	–	o_3	o_4	$3*f_s + o_2*v_s$
	o_1,o_2	–	o_3,o_4	–	$2*f_s + o_2*v_s + o_4*v_s$
	o_1,o_2,o_3	–	–	o_4	$2*f_s + o_2*v_s + 2*o_3*v_s$
	o_1,o_2,o_3,o_4	–	–	–	$1*f_s + o_2*v_s + 2*o_3*v_s + 3*o_4*v_s$
2	o_1	o_2	o_3	–	$3*f_s$
	o_1	o_2,o_3	–	–	$2*f_s + o_3*v_s$
	o_1,o_2	–	o_3	–	$2*f_s + o_2*v_s$
	o_1,o_2,o_3	–	–	–	$1*f_s + o_2*v_s + 2*o_3*v_s$
3	o_1	o_2	–	o_4	$3*f_s$
	o_1,o_2	–	–	o_4	$2*f_s + o_2*v_s$
	o_1	o_2,o_4	–	–	$2*f_s*2*o_4*v_s$
	o_1,o_2,o_4	–	–	–	$1*f_s + o_2*v_s + 3*o_4*v_s$
4	o_1	o_2	–	–	$2*f_s$
	o_1,o_2	–	–	–	$1*f_s + o_2*v_s$
5	o_1	–	o_3	o_4	$3*f_s$
	o_1,o_3	–	–	o_4	$2*f_s + 2*o_3*v_s$
	o_1	–	o_3,o_4	–	$2*f_s + o_4*v_s$
	o_1,o_3,o_4	–	–	–	$1*f_s + 2*o_3*v_s + 3*o_4*v_s$
6	o_1	–	o_3	–	$2*f_s$
	o_1,o_3	–	–	–	$1*f_s + 2*d_3*v_s$
7	o_1	–	–	o_4	$2*f_s$
	o_1,o_4	–	–	–	$1*f_s + 3*d_4*v_s$
8	o_1	–	–	–	$1*f_s$

payments. Instead of a sequential – in case of distributed decision-making – or central – in case of central decision-making – determination of order and production lots the buyer and the seller exchange proposals to overcome inefficiencies of distributed lot determination and to converge to or even achieve the optimal solution of central lot determination. The result of the distributed lot size determination can act as a starting point and BATNA for the negotiation process, the optimal result of central lot determination can be used as a benchmark to evaluate negotiated agreements.

3   Numerical Example

For a comparison of the three approaches introduced in Sect. 2 consider the following numerical example of a two-stage supply chain between a buyer and a supplier in a four period planning horizon: Demands for the four periods are $d_1 = 30$, $d_2 = 50$, $d_3 = 80$,

$d_4 = 60$, the costs of the buyer are her order costs $f_b = 125$ and her inventory costs $v_b = 3$ for storing one item one period, similarly the seller has set-up costs $f_s = 300$ and inventory costs $v_s = 1$. Table 3 presents the costs for the eight possible order lot constellations – according to the cost formulae provided in Table 1 – and the best responses in terms of production lot constellations – according to the cost formulae provided in Table 2.

Table 3. Order lot constellations, best production lot constellations and resulting costs for buyer and supplier.

| No. | Order lots o_i | | | | | Production lots p_i | | | | | |
	o_1	o_2	o_3	o_4	buyer costs	p_1	p_2	p_3	p_4	supplier costs	total costs
1	d_1	d_2	d_3	d_4	500	o_1,o_2,o_3,o_4	–	–	–	690	1,190
2	d_1	d_2	d_3,d_4	–	555	o_1,o_2,o_3	–	–	–	630	1,185
3	d_1	d_2,d_3	–	d_4	615	o_1,o_2,o_4	–	–	–	610	1,225
4	d_1	d_2,d_3,d_4	–	–	850	o_1,o_2	–	–	–	490	1,340
5	d_1,d_2	–	d_3	d_4	525	o_1,o_3,o_4	–	–	–	640	1,165
6	d_1,d_2	–	d_3,d_4	–	580	o_1,o_3	–	–	–	580	1,160
7	d_1,d_2,d_3	–	–	d_4	880	o_1,o_4	–	–	–	480	1,360
8	d_1,d_2,d_3,d_4	–	–	–	1,295	o_1	–	–	–	300	1,595

Under distributed lot determination, for the cost structure of the numerical example, the buyer will implement alternative 1, which leads to the lowest costs of 500 and order the demand for each period in exactly that period. The best response, given the cost structure of the numerical example, for the supplier is to produce all orders in the first period which results in the lowest costs of 690 for this production lots. The total costs under distributed decision-making are, therefore, 1,190, which is above the optimal solution of alternative 6 with total costs of 1,160. These minimal costs could be achieved by means of central lot determination for the entire supply chain system, which involves the communication of demands and costs of all parties to a central coordinator.

An alternative would be to engage in supply chain negotiations. A negotiation procedure would not require enclosing full cost information of both sides but just the communication of the actual demands d_i from the buyer to the supplier. Negotiations then start with the initial proposal by the buyer as if order lots would be determined by distributed decision making. This proposal serves as the BATNA in the negotiation process. The supplier subsequently can propose different order lot constellations, which reduce her costs, plus a side payment – as utility is transferable in this problem by means of money – as an incentive for the buyer to accept this proposal.

Assume a benevolent negotiation orientation of the parties, so that all cost savings beyond the BATNA are proposed by the supplier or only additional costs beyond the BATNA are demanded by the buyer as side payments when making proposals. In a concession-based alternating negotiation procedure the buyer's initial proposal of alternative 1 would be followed by the supplier's proposal of her favorite alternative 8 plus a side payment to the buyer of his savings of 390. This proposal and side payment would reduce the buyer's costs to 905 which is still above the BATNA. A counter-proposal of the buyer of her next best alternative 5 with a demanded side payment of 25 results in costs of 665 for the supplier which is already better than the BATNA for both parties and therefore an acceptable solution that is superior to the result of distributed decision making.

Alternatively, an alternating veto procedure could be applied for negotiations [3]. Starting randomly with the buyer or the seller the parties alternate in deleting alternatives until only one alternative remains which then is compared to the BATNA and implemented if it is superior for both parties. Starting with the buyer – and assuming a rational strategy to always eliminate the alternative with the highest costs from the set of remaining alternatives – the alternatives 8, 1, 7, 5, 4, 2 and finally 3 would be deleted to achieve the optimal alternative 6, which is better for both parties if the seller proposes a side payment of at least 80. If the seller starts the veto procedure the result would be the same with a slightly different elimination order.

Of course other negotiation strategies than the benevolent versions discussed above are also possible, which risk the opportunity to create value by claiming too much of this value. This trade-off results from the inherent mixed motive nature of negotiations in which competitive individualism and cooperative collectivism [4] are combined.

4   Simulation

Note that the values in the numerical example in Sect. 3 are chosen to demonstrate the three approaches presented in Sect. 2. There exists some integrative potential as the order costs of the buyer are less than the set-up costs of the supplier while the inventory costs of the supplier are less than the inventory costs of the buyer. Furthermore, note that the results strongly depend on the demand structure of the buyer. Therefore, computational experimentation and simulations, with varying demand structures as well as different ratios of fixed (order and set-up costs) and variable (inventory) costs of the buyer and supplier, are mandatory to evaluate the performance of negotiation compared to distributed and central order and production lot determination.

Furthermore a variety of negotiation approaches exist as already briefly mentioned in Sect. 3 besides concession based approaches – with an orientation that can range from competition to cooperation – also veto approaches are possible and have to be evaluated in a simulation study – preliminary results of the simulations will be presented at the GDN 2015.

5   Conclusion

This paper introduced a two-stage multi-echelon supply chain model and compared different order and production lot determination approaches theoretically and by means of a numerical example. Decentralized decision-making restricts the alternatives available to the supplier and thereby can cause suboptimal results for the entire system.

Centralized lot determination assures the optimal – i.e. minimum cost – solution but might be hard or expensive to implement.

Supply chain negotiation of order and production lots is proposed as an alternative that in some cases might outperform distributed decision-making and come close to or even implement the optimal solution without extensive information exchange or central coordination. Automated negotiation could enable to realize the benefits of negotiation over rigid coordination mechanisms – like in our case distributed lot determination – without causing the cost of human involvement, which is normally necessary in negotiations and could mitigate or outweigh potential benefits [5].

References

1. Wagner, H.M., Within, T.M.: Dynamic version of the economic lot size model. Manage. Sci. **5**(1), 89–96 (1958)
2. Li, X., Wang, Q.: Coordination mechanisms of supply chain systems. Eur. J. Oper. Res. **179**, 1–16 (2007)
3. Filzmoser, M., Gettinger, J.: Negotiation by veto. In: Proceedings of the International Conference on Group Decision and Negotiation 2013, Stockholm, pp. 348–350 (2013)
4. Bartos, O.J.: Simple model of negotiation. J. Conflict Resolut. **21**(4), 565–579 (1977)
5. Filzmoser, M.: Simulation of Automated Negotiation. Springer, Vienna (2010)

Procurement Auctions: Improving Efficient Winning Bids Through Multi-bilateral Negotiations

Gregory E. Kersten[(⊠)]

InterNeg Research Centre and J. Molson School of Business,
Concordia University, Montreal, PQ, Canada
gregory.kersten@jconcordia.ca

Abstract. Auctions have been used in the procurement of heterogeneous products, produced and delivered after the auctions conclude, as well as services. In these situations the quasi-linearity assumption of the buyer and the sellers is violated and the price and other attributes are interrelated. The relationship between price and other attributes is illustrated here with two exchanges in which the market participants are characterized by Cobb-Douglass production functions. It shown that even in the simplest case, when the contract curve is linear, the price and other attributes are interrelated. This relationship becomes more complex for non-linear contract curves. The paper shows that in these cases the auction does not maximize social welfare, i.e., it is an inefficient mechanism. Furthermore, even if the winning bid is an efficient solution, a win-win solution which dominates this bid may be possible. The buyer needs to engage in multi-bilateral negotiations in order to seek joint-improvements. The purpose of the negotiation is to search for side-payments.

Keywords: Reverse auctions · Negotiations · Procurement · Supply chain · Contract curve · Efficient frontier

1 Introduction

Automation of production and real-time communication among the supply chain participants is necessary for lean manufacturing and just-in-time-production. Internet of things (or of everything) provides the necessary communication platform for integrating components, services, processes, people and organizations. In the past it was common to first produce goods and then deliver and sell them on the markets. Today, markets are often virtual places or ad hoc venues set-up by one organization for a specific purchase and closed after the purchase is completed [1, 2].

Goods and services are exchanged between the supply chain members through one or more market mechanisms. Three types of mechanisms have been used in the marketplace: (1) posted price; (2) negotiations; and (3) auctions. What mechanism is deemed appropriate depends on the good complexity and importance, type of exchange, and the relationship between the participants [3]. Traditionally, simple and low importance products and services (called goods here) were exchanged through auctions and posted price while negotiations were used for complex and strategically

© Springer International Publishing Switzerland 2015
B. Kamiński et al. (Eds.): GDN 2015, LNBIP 218, pp. 403–416, 2015.
DOI: 10.1007/978-3-319-19515-5_32

important goods. The continuously increasing pace of the movement of goods and services through supply chain and the search for efficiencies has resulted in the increased use of auctions. The use of auctions in the procurement of complex goods has been enabled with the increased sophistication of models that include multi-attribute and heterogeneous auctions, combinatorial auctions, and a variety of hybrid auction-negotiation models [eg., 4–7].

Auctions offer advantages. They are cost-effective and efficient exchange mechanisms and although they may be difficult to set up, but they can be easily run leveraging competition. While in practice auction results may not be theoretically efficient solutions because of the participants' limited rationality, winner curse, and bias [8, 9], these limitations may be alleviated with training, support tools and incentives [2, 10, 11]. Auction theory proves that, under certain conditions, the outcomes of reverse auctions have three characteristics: (1) they are Pareto-efficient solutions, which (2) maximize buyers' surplus, and which (3) maximize social welfare. Taken together, these three characteristics mean that auctions are efficient mechanisms. No other mechanism can have all three characteristics.

The "under certain conditions" requirement is, however, of key importance. If these conditions are not met, then auctions are inefficient. Moreover, they can produce suboptimal results for both buyers and sellers, i.e., it is possible to determine outcomes that are better from both sides of the exchange [12].

The purpose of this paper is to show that in an economy in which goods are produced to order, i.e., the members of the supply chain are just-in-time producers, then the three key characteristics of auctions cannot be met. The means that auctions' efficiency may be severely impaired; they neither maximize the buyers' surplus nor social welfare. Furthermore, joint improvements are possible, if the exchange problem can be augmented. Such auctions are inefficient from the social (supply-chain) perspective contributing to a loss of social welfare. They also may be inefficient from the buyers' perspective causing loss of potential gains.

This paper focuses on exchanges between supply chain members through reverse auctions. The focus is narrowed to the transactions involving products that are produced after the auction's winner has been selected. This type of transactions includes services (e.g., web services, computing services, and spectrum trading), which have been often auctioned [13, 14]. The underlying assumption is that the goods are sufficiently complex so that different configurations are possible. This means that, for example, nails that can be made from only one type of material and of one size are not considered, but fastening goods that can be either nails or screws made either from steel, bronze, or aluminum, and with different heads and of different sizes are included. The implication of the assumption is two-fold: (1) the costs of production and delivery depends on the good's configuration; and (2) the configuration of the good used in production and its delivery influences the processing costs and the good's value. To illustrate, the production costs of aluminum nails production is lower than the production costs of bronze screws but the boxes in which nails are used as fasteners have lower value (can be sold for less) than boxes fastened with bronze screws.

The paper proceeds in four sections. Input configurability and the reliance on multiple attributes (not solely on price) in procurement, requires the consideration of multi-attribute exchange mechanisms. Section 2 compares single- and multi-attribute auctions and different utility functions. Two types of exchanges characterized by Cobb-Douglass utilities are presented in Sect. 3. Section 4 discusses the auction inefficiency and shows the ways in which the winning bids can be improved. Conclusions recommendations for future research are given in Sect. 5.

2   Price-Only and Multi-attribute Auctions

The assumption that buyers consider different configurations of the good they need to purchase is based on studies showing that most of the purchasing managers base their decisions on both price and non-price attributes, e.g., durability, service, lead-time, and transportation [15, 16]. This suggests that price-only auctions are inappropriate to many sourcing decisions and they should be replaced with multi-attribute auctions.

Auctions may be price-only and yet be multi-attribute. This is the case when the buyer requests the sellers to bid on several different price (cost) items, e.g., price of the good, price of the service, replacement price, and delivery. Risk-attitude, position on the market, credit rating and other elements may contribute to the situation that the different price items have different monetary value (e.g., net-present value) for the buyer and for the sellers.

2.1   Multi-attribute Auctions

Multi-attribute auctions have been experimentally shown to produce better results than single-attribute auctions. Chen-Ritzo [17] experimentally compared three-attribute with single-attribute auctions and showed that the buyer's utility increased in the former type of auctions in comparison to the latter. They also made an interesting observation about the supplier's gains: the multi-attribute auctions have the potential to increase the contractors' revenue or profit. The flexibility achieved due to the consideration of multiple attributes (which can be of different importance for the buyers and the sellers) creates an opportunity for joint gains.

Arguably, the most compelling argument for the use of multi-attribute auctions rather than single-attribute auctions can be derived from the Lewis and Bajari's [18] study. They analysed over 1300 contracts awarded by the California Department of Transportation (Caltrans) between 2003 and 2008. The mechanisms used were single- and two-attribute auctions. Their analysis showed that two-attribute auctions resulted in savings of about 40 % as compared to the Caltrans's estimate. These savings amount to $6.1 million per contract (op. cit., p. 1175). The average price in the A + B winning contracts is $1.5 million higher than the price in the price-only contracts. Hence, A + B auctions increase social welfare by about $4.6 million on average.

Lewis and Bajari [18] also conducted counterfactual analysis. The results show that if Caltrans had employed A + B rather than price-only auctions, it would have saved them $1.03 billion, that is, 20 % of the total value of these contracts and the contractors' costs would have, not increased. Clearly two-attribute auctions are better for social welfare than single-attribute ones.

2.2   Linear and Quasi-linear Utilities

The limitation of the single- and multi-attribute studies is the reliance on quasi-linear utility functions. In the field studies the buyer was assumed to be a risk-neutral government agency concerned with price as well as the project completion time [18]. This allowed representing the buyer's utility with a linear function. Utility of the sellers (contractors) was unknown but it was assumed to be quasi-linear.

For the buyer (b), quasi-linear utility is of the form:

$$u_b(x, y) = r_b(x) - y, \tag{1}$$

where: x ($x \in X$) is the good, $r_b(x)$ is the valuation function of the good, which is strictly concave (twice differentiable with $v_b' > 0$; $v_b'' < 0$, and bounded from above), and x_1 is the price.

Correspondingly, the seller's (s) utility function is:

$$u_s(x, y) = y - c_i(x), \tag{2}$$

where: $c_s(x)$ is the seller's valuation of good $x \in X$, the cost function is assumed convex (twice differentiable with $v_i' > 0$; $v_i'' \geq 0$,).

The assumption that the market participants have quasi-linear utilities has significant implications for both the market efficiency and the participants' behavior.

Monetary valuations: The buyer's and the sellers' valuations are expressed in the same monetary terms as price. This requirement follows directly from (1) and (2) in which price is subtracted from (in (1)) or added to (in (2)) the valuation. For the sellers this can be interpreted as $c_s(x)$ being the cost required to produce and deliver the good $x \in X$. For the buyer this can be interpreted as $r_b(x)$ being the revenue achieved from using good $x \in X$.

Unique efficient configuration: Given quasi-linear buyer's and sellers' utilities for each buyer-seller pair there exist only one efficient configuration [12].

Mechanism's efficiency: If a mechanism produces efficient winning bids, then the mechanism is efficient, i.e., it maximizes social welfare, providing that its measure is the sum of the buyer and the winning seller utility. This result follows from unique efficient configuration and the social welfare price-independence. Social welfare defined as a difference between the utility values depends only on the difference between the buyer's revenue and the seller's costs:

$$u_b(x, y) + u_i(x, y) = v_b(x) - y + y - v_i(x) = v_b(x) - v_i(x). \tag{3}$$

2.3   Sellers' Costs and Buyer's Revenue

Quasi-linear utilities describe the situation in which the producers' costs do not influence the price and the price has no effect on the buyer's valuation of the good. This may happen but only in specific situations.

For the producer, price is independent of costs when the product was made prior to the exchange. Alternatively, price may not influence the good subject to exchange, if the good is a commodity and the production process cannot be modified in a short time, e.g., prior its delivery.

For the buyer, price may not affect the good's valuation if these two components of (1) are in a non-Archimedean order, which may be the case with the good being, for example, a musical composition. The other possibility is when the buyer has unlimited budget and is price-insensitive; as long as the price of a good does not exceed its valuation the buyer is willing to purchase the good.

Increasingly, goods composed of different components can be configured after the transaction takes place. This leads to the price affecting the configuration and vice versa. The producers' ability to propose different configurations of the good during the exchange process has two implications:

1. The price cannot be the sole attribute because the sellers may compete on price as well as such attributes as quality, composition, warranty, and service.
2. The price and other attributes, are inter-related; the sellers may trade-off the price they ask for, for attribute levels they can provide.

Flexibility and agility of production, which is due to such developments as internet-of-things, just-in-time production, bundling of services and products requires that markets move beyond dynamic pricing and include dynamic specification of all attributes, including various price-types. Optimization of design, development and service delivery is predicated on optimization of all variables, i.e., attributes.

2.4 Inefficiency of Auctions

Given that multiple attributes need to be considered in procurement and that the values of these attributes need to be determined during the exchange process, the two typical mechanisms are auctions and negotiations. A number of multi-attribute reverse auctions models have been proposed [e.g., 19–22]. Many e-procurement systems provide multi-attribute bidding functionality (e.g., Ariba, Trade Extensions, Epicor Software and Perfect Commerce).

The assumption on which the theoretical models and software are based is the quasi-linearity of the market participants' utilities [12]. Either both sides (i.e., the buyers and the sellers) have quasi-linear utilities or one side (e.g., the buyers) has linear utilities and the other side (e.g., sellers) has quasi-linear utilities.

Let's distinguish between configurable and non-configurable goods (both products and services. Note that configurability is not limited to the set of attributes that describe the goods' features and qualities (e.g., color, weight, fitness) but also attributes associated with delivery terms, warranties, and other post-sale services. Non-configurable goods are produced prior the buyers and the sellers engaging in the exchange. The delivery terms and post-sale services of these goods are either non-negotiable or irrelevant to the buyer.

While both configurable and non-configurable goods may be described by multiple attributes, the sellers cannot modify attribute values of non-configurable goods with the exception of their price. They can offer different (heterogeneous) goods, but they seller can neither modify them nor change the sale and post-sale conditions. In contrast, configurable goods have attributes whose values are determined during the exchange process. There is no difference in price and non-price attributes in that their values are determined during the process.

The argument here is that *auctions are inefficient mechanisms for the procurement of configurable goods*. They should not be used in exchanges in which attribute values are determined because:

1. They cannot maximize social welfare,[1] which means that it is possible to achieve better results at least for one side (i.e., the buyer or the winning seller); and
2. If it is possible to extend the exchange beyond the good and/or consider its enhancement, then the efficient result may be further improved.

The theoretical underpinnings behind these two claims are given in [12] and they rely on the assumption that the buyer-seller pairs of utilities produce a concave efficient frontier in the decision space. In this paper we show that these claims are valid for other shapes of the efficient frontier. Specifically, we show that when at least one side of the exchange has Cobb-Douglas production function, then both of the above claims apply.

3   Two Exchanges

An analysis of two exchanges in which the buyers' and the sellers' production functions are Cobb-Douglass functions shows that price and other attributes are interrelated.

3.1   Cobb-Douglass Production Functions

Cobb-Douglass functions have often been used in economics to model production at the macro- and micro-economic levels and also in decision science to represent multi-attribute utility. The function is:

$$u = kx^\alpha y^\beta, \quad \alpha + \beta \gtreqless 1 \tag{4}$$

where, u is the value of the total production, x is labor, and y is the value of capital. The coefficients α and β are elasticity of the inputs x and y, respectively. We use here production functions that has a constant return to scale ($\alpha + \beta = 1$).

The inputs may be aggregated into two components as in (4) or there may be multiple different inputs. To simplify the discussion and provide graphical illustrations two components are used here.

---

[1] Auction theory employs social welfare which is an additive function. When the goods are non-configurable, then auctions maximize additive social welfare. They do not, however, maximize multiplicative social welfare. When the goods are configurable then auctions maximize neither additive nor multiplicative social welfare.

Observe that in exchanges in which the good is produced and delivered after the transaction the agreed price may be considered as capital. The buyer either pays the price up-front as is the case with some services or the seller uses its own or the bank's capital on the assumption that it will be replenished.

3.2   Equal Elasticity

Let's assume that the auction is over a simple good, which is described by only one attribute, for example the quality of the good or the time to deliver a service. Thus, the sellers bid on price x and good y. We consider here a simple case in which the buyer's and the sellers' production functions are given by (4) with $k = 1$ and $\alpha = \beta = 0.5$. Buyer b has a limited amount of money m_b and seller s can at most allocate time t_s. Their production functions are given by, respectively,

$$u_b = x_b^{0.5} y_b^{0.5} \text{ and } u_s = x_s^{0.5} y_s^{0.5}, \tag{5}$$

where: $x_b + x_s = m_b$ and $y_b + y_s = t_s$.

We interpret x_b as the amount of money that buyer b keeps after paying x_s to seller s and y_s as the time that seller s keeps after allocating y_b to buyer b.

The formula for the contract curve for the production functions (5) is given by [23]:

$$(1-\alpha)\,\beta(m_b-x_s)\,y_b \ - \ \alpha(1-\beta)x_s(t_s - y_b) =$$
$$= \ 0.25(m_b-x_s)\,y_b \ - \ 0.25\,x_s(t_s-y_b) \ = \ 0 \tag{6}$$

Equation 6 can be simplified to the following equation of a straight line:

$$\frac{m_b}{t_s} y_b - x_s \ = \ 0. \tag{7}$$

3.3   Different Elasticity

We consider now a case in which the buyer's and the sellers' production functions are given by (4) with $k = 1$ and $\alpha = 0.7$ and $\beta = 0.4$. As before buyer b has a limited amount of money m_b and seller s can at most allocate time t_s. Their production functions are given by, respectively,

$$u_b = x_b^{0.3} y_b^{0.7} \text{ and } u_s = x_s^{0.8} y_s^{0.2}, \tag{8}$$

where: $x_b + x_s = m_b$ and $y_b + y_s = t_s$.

The formula for the contract curve for the production functions (9) is given by

$$00.6\,(m_b-x_s)\,y_b \ - \ 0.56\,x_s(t_s-y_b) \ = \ \frac{0.06\,m_b y_b}{0.56\,t_s + 0.5\,y_b} - x_s \ = \ 0. \tag{9}$$

Contract curve (9) is illustrated in Fig. 2. The relationship between price and time is non-linear; for every quantity y_b there is a corresponding price x_s. The initial small change in price x first requires a significant time increase which levels off. An increasingly greater price is subsequently required to obtain an additional service time.

In this exchange it is no longer solely the money and time available to the buyer and the seller, respectively, that determines the winning bid. In addition to these resources it is also the sellers elasticity (i.e., β_s, $s \in S$) which, for a given elasticity α of the buyer that is used to determine the efficient winning contract. A trade-off between the price and time is, for a given resource allocation and the buyer's elasticity efficiency, determined by seller's s elasticity β_s.

3.4   Relationship Between Price and Time

The two exchanges between participants characterized by the Cobb-Douglass utility functions demonstrates that price and time are interdependent and cannot be treated separately in bidding. An increase in price allows the seller to increase service time and vice versa.

4   Efficient & Inefficient Mechanisms

We assume that the incentives are such that every exchange leads to an efficient (i.e., Pareto-optimal) contract. Furthermore, we measure social welfare as the sum of the utilities rather than their product. This latter simplification is often used in auction theory. The reason is, as we show, that the use of a product would result in no auction mechanism maximizing social welfare.

4.1   Efficient Frontier

Contract curves comprise efficient solutions. These curves illustrated in Figs. 1 and 2 were shown in the bidding (attribute) space. We can also show the corresponding efficient frontier (utility possibility frontier) in the utility space. Three different efficient frontiers are shown in Fig. 3.

When the pairs buyer-seller utilities are quasi-linear or when one is quasi-linear and the other utility is linear, then the efficient frontier is linear and has -1 slope [24]. Because it is linear and its slope is -1 every contract yields the same social welfare, i.e., the sum of the utilities is constant. Note that this takes place when social welfare is the sum; if it is a product (as postulated by e.g., Nash), then social welfare is different for different contracts.

When the utility functions are concave or quasi-concave, e.g., they are Cobb-Douglass functions, then the efficient frontier is (quasi)concave [25].

Other pairs of utilities result in linear or some combinations of convex, linear, and concave functions. For example, when two utilities are linear and the attributes are equally important for the buyer and the seller, or the same attributes are more important, then the efficient frontier is linear as well as for the quasi-linear utilities (Fig. 3). However, when the pairs of utilities are linear but attributes have different

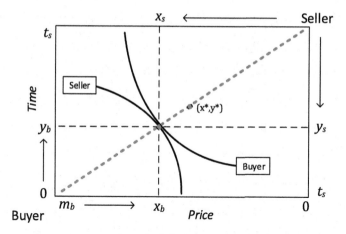

Fig. 1. Edgeworth box: contract curve for identical production functions; $\alpha = \beta = 0.5$.

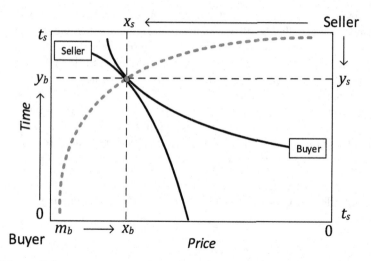

Fig. 2. Edgeworth box: contract curve for identical production functions and $\alpha = 0.3$; $\beta = 0.8$.

importance for the seller than the buyer (e.g., time is more important for the buyer than money and money are more important for the seller than time; Sect. 3.3), then the efficient frontier is piece-wise linear (P-linear in Fig. 3).

When two utilities are convex, then the efficient frontier is piece-wise convex (P-convex in Fig. 3). Other possibilities include combinations of convex and concave functions (e.g., when one utility is concave and the other utility is convex).

4.2   Winning Bids

In auctions the bidders have to compete in order to win a contract. This competition takes the form of the bidders making bids which increase the buyer's utility and

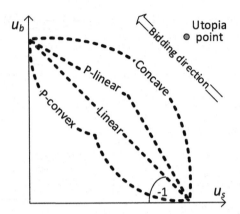

Fig. 3. Three efficient frontiers.

typically decrease the bidders' utility. In auction, we can thus say that the bidding process moves along the North-West direction (Fig. 3). This shows that with the exception of the quasi-linear utilities, the bidders move away from the contracts that maximize social utility; in order to move towards these contracts they would have to move along the North-East direction, towards the Utopia point.

The fact that auctions are likely to produce inferior social welfare is important from the macro-economic, social, and market-makers' perspective. For the buyer who sets up an auction this may not be important as long as the buyer is assured that the contract she awards maximizes her utility. This, however, may not be the case.

We use two examples to illustrate the possible loss of utility by the buyer.

Example 1. Assume that the auction winner is seller s and that the efficient frontier for buyer b and seller s is concave. Point a shown in Fig. 4 is the winning contract and it yields $u_b(a)$.

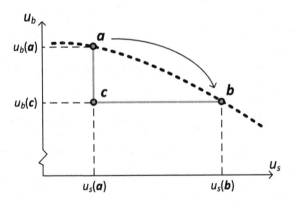

Fig. 4. Concave efficient frontier and winning bid **a** and win-win contract **b**.

Normally, the auction terminates and s gets the contract. An analysis of the winning bid and alternative contracts shows that it may be possible to suggest a contract that may make both the buyer and the seller better off.

Compare contracts a and b. The buyer's utility of contract b is $u_b(b) = u_b(a) - \Delta_1$. The seller's utility is $u_s(b) = u_s(a) + \Delta_2$, and $\Delta_2 = 2.3\Delta_1$. If both utilities are expressed in money, then the buyer may suggest a change of the contract. The buyer may propose contract b providing that she is compensated in the amount of $0.65\Delta_1$, that is she accepts a loss of $\Delta 1$ providing that the seller splits his gain with her. If the seller accepts this solution, then the result of this post-auction negotiation is contract b with a side payment, and this revised contract dominates a, i.e.,

$$u_b(b) > u_b(a) + 1.65\Delta_1 \text{ and } u_b(b) > u_b(a) - 1.65\Delta_1.$$

Utility may not be expressed in money. In this case the analysis is somewhat more difficult and the participants may have to consider attributes that were not included in bidding but that may be offered by the seller and are valuable to the buyer.

The inefficiency of auction and the possibility to seek win-win agreements through post-auction negotiations is not limited to the case where the efficient frontier is concave. When the frontier is quasi-linear and the set of alternatives is convex, then such a possibility also exists. The case of piece-wise linear frontier (P = linear) shown in Fig. 3 presents possibility of joint improvements.

Example 2. Joint improvements may also be possible when the participants' utilities are convex. Two winning bids d and e are shown in Fig. 5. If the winning bid is d, then it is not possible find an alternative contract in the neighbourhood that may lead to joint improvements. For example, contract e increases seller's s utility but the loss of buyer's b utility is much greater. If however, the winning bid is e, then it may be possible to move to win-win solution f. The process would be the same as the one illustrated in Fig. 4.

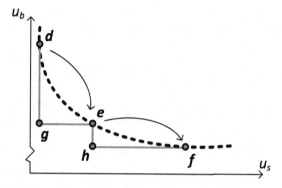

Fig. 5. Convex efficient frontier, winning bid **d** and win-lose contract **e**, and winning bid **e** and win-win contract **f**.

4.3   Buyer-Determined Auctions

The complexities involved with multi-attribute auctions led to the popularity of buyer-determined auctions, which are price-only auctions followed by negotiations with the selected sellers over non-price attributes [26]. Although this format of auctions has been frequently used, multi-attribute auctions have been shown to outperform price-only auctions [18]. Santamaría [27] compares buyer-determined and multi-attribute auctions and shows that multi-attribute auctions yield lower price on average.

The discussion in this section gives grounds for the claim that even when multi-attribute auctions are employed, the winner should not be automatically selected by the mechanism. When the auction is followed by negotiations and either a monetary side-payment or additional attributes may be introduced, then a win-win contract that dominates the winning bid may be obtained.

The post-auction negotiation should not be limited to the winning seller. Instead, the buyer should create a short-list of sellers and initiate multi-bilateral negotiations with them. This is because there may be a seller who did not win but who may offer a contract with side-payment that dominates the winning bid. This situation is illustrated in Fig. 6. Contract *a* is the winning bid as in the situation depicted in Fig. 4. This contract is offered by seller *s1*. If only this seller is invited to the negotiation then the win-win contract may be *b*. If, however, seller *s2* is also invited, then this seller may offer contract *b'*. Contract *b'* allows offering greater side-payment than contract *b*, which is consequently better for both buyer *b* and seller *s1*.

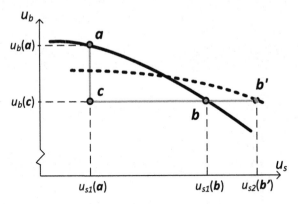

Fig. 6.   Convex efficient frontiers for two sellers: s1 and s2, winning bid **a** and win-win contract **b'**.

5   Conclusions

Auction mechanisms have been used for selling objects of art, flowers, fish and other goods that were available on the market at the time of the transaction. Increasingly, however, reverse auctions have been used in the procurement of services, commodities, and often complex products that would be produced and delivered after the completion of the transaction. If these goods are uniform and no have attributes, other than price, that are relevant to the buyer and the sellers, then the underlying assumption that the utilities are quasi-linear may be accepted. If, however, there are non-price attributes which describe the good's characteristics, terms of payments, delivery conditions, and warranties, and if these attributes are important for the buyer and the sellers, then the quasi-linearity assumption must be rejected. The key reason for the rejection is that price and attributes are inter-related.

This implication for the rejection of the quasi-linearity assumption includes the mechanism's inefficiency—auction mechanism can no longer maximize social welfare. This implication should be of interest to public organizations which frequently use auction to allocate contracts for the construction and renovation of infrastructure, maintenance and other services, etc. While private organizations may be less concerned with social welfare, another implication is that the winning bid efficiency does no longer mean that the buyer and the winning seller obtain the best possible deal. Moreover, there may be another seller who did not win the auction but who may offer the buyer a better deal. In effect the use of auctions to allocate contracts leads to underutilization of resources and forgone deals. The scope and degree of losses made due to the use of inefficient auctions in procurement needs to be studied. This requires an assessment of the market participants' utilities followed by the counterfactual analysis of the deals.

In order to address these shortcomings the buyer needs to engage in a post-auction negotiation. This negotiation should include the winning seller as well as sellers who did not win the auction.

The post-auction negotiation process resembles the buyer-determined auctions that have often been used in practice. There is an important difference between these two processes. While buyer-determined auctions have been employed in order to address the difficulty in multi-attribute bidding, the auction is price-only and the negotiation concern the attribute values as well as the price. The post-auction negotiation proposed here is due to the inherent limitations of the auction mechanisms and follows a multi-attribute auction. The multi-bilateral negotiation process requires the parties to seek side-payments so that they can achieve a deal that dominates the winning bid.

Acknowledgments. This work has been supported by the grants from the Natural Sciences and Engineering Research Council of Canada (NSERC), Concordia University (Canada). I am grateful to the reviewer for their insightful comments.

References

1. Christopher, M.: The agile supply chain: competing in volatile markets. Ind. Mark. Manage. **29**(1), 37–44 (2000)
2. Jain, V., Panchal, G.B., Kumar, S.: Universal supplier selection via multi-dimensional auction mechanisms for two-way competition. Omega **47**, 127–137 (2014)
3. Kraljic, P.: Purchasing must become supply management. Harvard Bus. Rev. **83**(5), 109–117 (1983)
4. Bellantuono, N., et al.: Multi-attribute auction and negotiation for e-procurement of logistics. Group Decis. Negot. **23**(3), 421–441 (2014)

5. Li, Z., Ryan, J.K., Sun, D.: Multi-attribute procurement contracts. Int. J. Prod. Econ. **159**, 137–146 (2014)
6. Huang, H., et al.: Hybrid mechanism for heterogeneous e-procurement involving a combinatorial auction and bargaining. Electron. Commun. Res. Apps **12**(3), 181–194 (2013)
7. Rao, C., Zhao, Y., Ma, S.: Procurement decision making mechanism of divisible goods based on multi-attribute auction. Electron. Commun. Res. Apps **11**(4), 397–406 (2012)
8. Romero-Morales, D., Steinberg, R.: Revenue deficiency under second-price auctions in a supply-chain setting. Eur. J. Oper. Res. **233**(1), 131–144 (2014)
9. Holt, C.A., Sherman, R.: Risk aversion and the winner's curse. South. Econ. J. **81**, 7–22 (2014)
10. Hass, C., Bichler, M., Guler, K.: Optimization-based decision support for scenario analysis in electronic sourcing markets. Electron. Commun. Res. Apps **12**(3), 152–165 (2013)
11. Yang, N., Liao, X., Huang, W.W.: Decision support for preference elicitation in multi-attribute electronic procurement auctions through an agent-based intermediary. Decis. Support Syst. **57**, 127–138 (2014)
12. Kersten, G.E.: Multi-attribute procurement auctions: efficiency and social welfare in theory and practice. Decis. Anal. **11**(4), 215–232 (2014)
13. Smart, A., Harrison, A.: Reverse auctions as a support mechanism in flexible supply chains. Int. J. of Logist. **5**(3), 275–284 (2002)
14. Duenyas, I., Hu, B., Beil, D.R.: Simple auctions for supply contracts. Manage. Sci. **59**(10), 2332–2342 (2013)
15. Ferrin, B.G., Plank, R.E.: Total cost of ownership models: an exploratory study. J. Supply Chain Manage. **38**(3), 18–29 (2002)
16. Johnson, M.D., Sawaya, W.J., Natarajarathinam, M.: A methodology for modelling comprehensive international procurement costs. Int. J. of Prod. Res. **51**(18), 5549–5564 (2013)
17. Chen-Ritzo, C.H., et al.: Better, faster, cheaper: an experimental analysis of a multiattribute reverse auction mechanism with restricted information feedback. Manage. Sci. **51**(12), 1753–1762 (2005)
18. Lewis, G., Bajari, P.: Procurement contracting with time incentives: theory and evidence. Q. J. Econ. **126**(3), 1173–1211 (2011)
19. Bichler, M., Kalagnanam, J.: Configurable offers and winner determination in multi-attribute auctions. Eur. J. Oper. Res. **160**(2), 380–394 (2005)
20. David, E., Azoulay-Schwartz, R., Kraus, S.: An english auction protocol for multi-attribute items. In: Padget, J., Shehory, O., Parkes, D.C., Sadeh, N.M., Walsh, W.E. (eds.) AMEC 2002. LNCS (LNAI), vol. 2531, pp. 52–68. Springer, Heidelberg (2002)
21. Kameshwaran, S., et al.: Multiattribute electronic procurement using goal programming. Eur. J. Oper. Res. **179**(2), 518–536 (2007)
22. Kersten, G.E., Pontrandolfo, P., and Wu, S.: A multiattribute auction procedure and its implementation. In: HICSS 45, Hawaii. IEEE (2012)
23. Besanko, D., Braeutigam, R.: Microeconomics. Wiley, NewYork (2010)
24. Kersten, G.E.: Are procurement auctions good for society and for buyers? In: Hernández, J. E., Kersten, G.E., Zaraté, P. (eds.) GDN 2014. LNBIP, vol. 180, pp. 30–40. Springer, Heidelberg (2014)
25. Mumpower, J.L.: The judgement policies of negotiators and the structure of negotiation problems. Manage. Sci. **37**(10), 1304–1324 (1991)
26. Jap, S.D.: Online reverse auctions: issues, themes, and prospects for the future. J. Acad. Mark. Sci. **30**(4), 506–525 (2002)
27. Santamaría, N.: An Analysis of Scoring and Buyer-Determined Procurement Auctions. Production and Operations Management, (2015, in print)

Author Index